10 0
D0265015
HAM
FROM THE LIBRARY

PROGRESS IN BRAIN RESEARCH

VOLUME 118

NITRIC OXIDE IN BRAIN DEVELOPMENT, PLASTICITY, AND DISEASE

Other volumes in PROGRESS IN BRAIN RESEARCH

Volume 95: The Visually Responsive Neuron: From Basic Neurophysiology to Behavior, by T.P. Hicks, S. Molotchnikoff and T. Ono (Eds.) – 1993, ISBN 0-444-89492-6.

Volume 96: Neurobiology of Ischemic Brain Damage, by K. Kogure, K.-A. Hossmann and B.K. Siesjö (Eds.) – 1993, ISBN 0-444-89603-1.

Volume 97: Natural and Artificial Control of Hearing and Balance, by J.H.J. Allum, D.J. Allum-Mecklenburg, F.P. Harris and R. Probst (Eds.) – 1993, ISBN 0-444-81252-0.

Volume 98: Cholinergic Function and Dysfunction, by A.C. Cuello (Ed.) – 1993, ISBN 0-444-89717-8.

Volume 99: Chemical Signalling in the Basal Ganglia, by G.W. Arbuthnott and P.C. Emson (Eds.) – 1993, ISBN 0-444-81562-7.

Volume 100: Neuroscience: From the Molecular to the Cognitive, by F.E. Bloom (Ed.) – 1994, ISBN 0-444-81678-X.

Volume 101: Biological Function of Gangliosides, by L. Svennerholm et al. (Eds.) – 1994, ISBN 0-444-81658-5.

Volume 102: The Self-Organizing Brain: From Growth Cones to Functional Networks, by J. van Pelt, M.A. Corner, H.B.M. Uylings and F.H. Lopes da Silva (Eds.) – 1994, ISBN 0-444-81819-7.

Volume 103: Neural Regeneration, by F.J. Seil (Ed.) – 1994, ISBN 0-444-81727-1.

Volume 104: Neuropeptides in the Spinal Cord, by F. Nyberg, H.S. Sharma and Z. Wiesenfeld-Hallin (Eds.) – 1995, ISBN 0-444-81719-0.

Volume 105: Gene Expression in the Central Nervous System, by A.C.H. Yu et al. (Eds.) – 1995, ISBN 0-444-81852-9.

Volume 106: Current Neurochemical and Pharmacological Aspects of Biogenic Amines, by P.M. Yu, K.F. Tipton and A.A. Boulton (Eds.) – 1995, ISBN 0-444-81938-X.

Volume 107: The Emotional Motor System, by G. Holstege, R. Bandler and C.B. Saper (Eds.) – 1996, ISBN 0-444-81962-2.

Volume 108: Neural Development and Plasticity, by R.R. Mize and R.S. Erzurumlu (Eds.) – 1996, ISBN 0-444-82433-2.

Volume 109: Cholinergic Mechanisms: From Molecular Biology to Clinical Significance, by J. Klein and K. Löffelholz (Eds.) – 1996, ISBN 0-444-82166-X.

Volume 110: Towards the Neurobiology of Chronic Pain, by G. Carli and M. Zimmermann (Eds.) – 1996, ISBN 0-444-82149-X.

Volume 111: Hypothalamic Integration of Circadian Rhythms, by R.M. Buijs, A. Kalsbeek, H.J. Romijn, C.M.A. Pennartz and M. Mirmiran (Eds.) – 1996, ISBN 0-444-82443-X.

Volume 112: Extrageniculostriate Mechanisms Underlying Visually-Guided Orientation Behavior, by M. Norita, T. Bando and B.E. Stein (Eds.) – 1996, ISBN 0-444-82347-6.

Volume 113: The Polymodal Receptor: A Gateway to Pathological Pain, by T. Kumazawa, L. Kruger and K. Mizumura (Eds.) – 1996, ISBN 0-444-82473-1.

Volume 114: The Cerebellum: From Structure to Control, by C.I. de Zeeuw, P. Strata and J. Voogd (Eds.) – 1997, ISBN 0-444-82313-1.

Volume 115: Brain Function in Hot Environment, by H.S. Sharma and J. Westman (Eds.) – 1998, ISBN 0-444-82377-8.

Volume 116: The Glutamate Synapse as a Therapeutical Target: Molecular Organization and Pathology of the Glutamate Synapse, by O.P. Ottersen, I.A. Langmoen and L. Gjerstad (Eds.) – 1998, ISBN 0-444-82754-4.

Volume 117: Neuronal Degeneration and Regeneration: From Basic Mechanisms to Prospects for Therapy, by F.W. van Leeuwen, A. Salehi, R.J. Giger, A.J.G.D. Holtmaat and J. Verhaagen (Eds.) – 1998, ISBN 0-444-82817-6.

PROGRESS IN BRAIN RESEARCH

VOLUME 118

NITRIC OXIDE IN BRAIN DEVELOPMENT, PLASTICITY, AND DISEASE

With a digital atlas accessible on the World Wide Web

EDITED BY

R. RANNEY MIZE

Department of Cell Biology and Anatomy and the Neuroscience Center, Louisiana State University Medical Center, New Orleans, LA, USA

TED M. DAWSON
VALINA L. DAWSON

Department of Neurology, Johns Hopkins University School of Medicine, Baltimore, MD, USA

MICHAEL J. FRIEDLANDER

Department of Neurobiology, University of Alabama at Birmingham, Birmingham, AL, USA

ELSEVIER
AMSTERDAM – LAUSANNE – NEW YORK – OXFORD – SHANNON – SINGAPORE – TOKYO
1998

ISBN 0-444-82885-0 (volume)
ISBN 0-444-80104-9 (series)

Published by:
Elsevier Science B.V.
P.O. Box 211
1000 AE Amsterdam
The Netherlands

Library of Congress Cataloging in Publication Data
A catalog record from the Library of Congress has been applied for.

The paper used in this publication meets the requirements of ANSI/NISO Z39.48-1992 (Permanence of Paper).

Printed in The Netherlands.

List of Contributors

E. Aizenman, Department of Neurobiology, University of Pittsburgh School of Medicine, E1456-BST, 3500 Terrace St., Pittsburgh, PA 15261, USA
C. Aoki, Center for Neural Science, New York University, 4 Washington Place, Room 809, New York, NY 10003, USA
O. Arancio, Center for Neurobiology and Behavior, College of Physicians and Surgeons, Columbia University, New York, NY 10032, USA
L. Barbeito, Seccion Neurociencias and Division Neurobiologia Celular y Molecular, Instituto Clemente Estable, Montevideo, Uruguay
N.G. Bazan, Louisiana State University Medical Center, School of Medicine, Neuroscience Center of Excellence, 2020 Gravier St., Suite D, New Orleans, LA 70112, USA
J.S. Beckman, Department of Anesthesiology, Department of Biochemistry and Molecular Genetics, Department of Neurobiology, and the UAB Center for Free Radical Biology, Research Division, The University of Alabama at Birmingham, 1900 University Blvd., THT 958, Birmingham, AL 35233, USA
D.S. Bredt, Department of Physiology, University of California, 513 Parnassus Ave., San Francisco, CA 94143, USA
D. Bridges, Department of Cell Biology and Anatomy, Louisiana State University Medical Center, 1901 Perdido Street, New Orleans, LA 70112, USA
J.C. Brimecombe, Department of Neurobiology, University of Pittsburgh School of Medicine, E1405 BST, 3500 Terrace St., Pittsburgh, PA 15261, USA
H.S.-V. Chen, The CNS Research Institute, Brigham and Women's Hospital and Program in Neuroscience, Harvard Medical School, Boston, MA 02115, USA
Y.-B. Choi, The CNS Research Institute, Brigham and Women's Hospital and Program in Neuroscience, Harvard Medical School, Boston, MA 02115, USA
R.J. Cork, Department of Cell Biology and Anatomy, Medical Center, 1901 Perdido St., New Orleans, LA 70112, USA
K.S. Cramer, Virginia Merrill Bloedel Hearing Research Center, University of Washington, Box 357923, Seattle, WA 98195, USA
T. Dalkara, Department of Neurology, Hacettepe University Hospitals, Ankara, Turkey
T.M. Dawson, Departments of Neurology and Neuroscience, Johns Hopkins University School of Medicine, 600 N. Wolfe St., Pathology 2-210, Baltimore, MD 21287, USA
V.L. Dawson, Departments of Neurology, Neuroscience and Physiology, Johns Hopkins University School of Medicine, 600 N. Wolfe St., Pathology 2-210, Baltimore, MD 21287, USA
D.M. Egelman, Center for Theoretical Neuroscience, Division of Neuroscience, Baylor College of Medicine, One Baylor Plaza, Houston, TX 77030, USA
A.E-D. El-Husseini, The Graduate Program in Neuroscience, Department of Psychiatry, The University of British Columbia, Vancouver, BC V5K 1Z3, Canada
M. Endres, Stroke and Neurovascular Regulation Laboratory, Massachusetts General Hospital, Harvard Medical School, 149 13th St., Room 6403, Charlestown, MA. 02129 USA

A.F. Ernst, Department of Cell Biology and Neuroanatomy, University of Minnesota, 4-144 Jackson Hall, 321 Church St. SE, Minneapolis, MN 55455, USA
A.G. Estévez, Department of Anaesthesiology, The University of Alabama at Birmingham, Center For Free Radical Biology, Birmingham, AL 35233, USA and Sección Neurociencias, Facultad de Ciencias, Montevideo, Uruguay
S. Fenstemaker, Center for Neural Science, New York University, 4 Washington Place, Rm. 809, New York, NY 10003, USA
M.J. Friedlander, Department of Neurobiology, The University of Alabama, Birmingham, AL 35233, USA
M. Gonzalez-Zulueta, Department of Neurology, Johns Hopkins University School of Medicine, 600 N. Wolfe St., Path 2-210, Baltimore, MD 21287, USA
R.D. Hawkins, Center for Neurobiology and Behavior, College of Physicians and Surgeons, Columbia University, and New York State Psychiatric Institute, New York, NY 10032, USA
P.L. Huang, Cardiovascular Research Center, Department of Medicine, Massachusetts General Hospital-East, Harvard Medical School, Charlestown, MA 02129-2060, USA
Y. Izumi, Department of Psychiatry and Neurobiology, Washington University School of Medicine, 4940 Children's Place, St. Louis, MO 63110, USA
W.M. Jurney, Department of Cell Biology and Neuroanatomy, University of Minnesota, 4-144 Jackson Hall, 321 Church St. SE, Minneapolis, MN 55455, USA
P. Kara, Department of Neurobiology and Department of Physiology and Biophysics, University of Alabama at Birmingham, 1719 6th Ave. S, Birmingham, AL 35294, USA
R.D. King, Center for Theoretical Neuroscience, Division of Neuroscience, Baylor College of Medicine, One Baylor Plaza, Houston, TX 77030, USA
C.A. Leamey, Department of Brain and Cognitive Sciences, Massachusetts Institute of Technology, Cambridge, MA 02139, USA
S.A. Lipton, The CNS Research Institute, Brigham and Women's Hospital and Program in Neuroscience, Harvard Medical School, 221 Longwood Ave. – LMRC First Floor, Boston, MA 02115, USA
E.H. Lo, Neuroprotection Research Lab, Departments of Neurology and Radiology, Massachusetts General Hospital, Harvard Medical School, Charlestown, MA 02129-2000, USA
A.K. Loihl, Department of Pharmacology and Neuroscience Program, University of Iowa College of Medicine, Iowa, City, IA 52242, USA
M. Lubin, Center for Neural Science, New York University, 4 Washington Pl. Rm. 809, New York, NY 10003, USA
S.M. Manuel, School of Pharmacy, Texas Tech University Health Sciences Center, Amarillo, TX 7920-1797, USA
S.C. McLoon, Department of Cell Biology and Neuroanatomy, University of Minnesota, 4-144 Jackson Hall, 321 Church St. SE, Minneapolis, MN 55455, USA
R.R. Mize, Department of Cell Biology and Anatomy and the Neuroscience Center, Louisiana State University Medical Center, 1901 Perdido St., New Orleans, LA 70112, USA
P.R. Montague, Center for Theoretical Neuroscience, Division of Neuroscience, Baylor College of Medicine, One Baylor Plaza, Houston, TX 77030, USA
M.A. Moskowitz, Stroke and Neurovascular Regulation Laboratory, Massachusetts General Hospital, 149 13th St., Rm. 6403, Charlestown, MA 02129, USA
S. Murphy, Department of Pharmacology and Neuroscience Program, University of Iowa College of Medicine, Iowa City, IA 52242, USA

M.L. Perrone, Department of Cell Biology and Anatomy, Louisiana State University Medical Center, 1901 Perdido Street, New Orleans, LA 70112, USA
W.K. Potthoff, Department of Neurobiology, University of Pittsburgh School of Medicine, E1456-BST, 3500 Terrace St., Pittsburgh, PA 15261, USA
R. Radi, Departmento de Bioquímica, Facultad de Medicina, Universidad de la Republica, Montevideo, Uruguay
P.V. Rayudu, The CNS Research Institute, Brigham and Women's Hospital, and Program in Neuroscience, Harvard Medical School, Boston, MA 02115, USA
P.B. Reiner, The Graduate Program in Neuroscience, Department of Psychiatry, The University of British Columbia, Vancouver, BC V5K 1Z3, Canada
P.A. Rosenberg, Department of Neurology and Program in Neuroscience, Children's Hospital and Harvard Medical School, Boston, MA 02115, USA
M. Sasaki, Department of Neurology, Johns Hopkins University School of Medicine, 600 N Wolfe St., Pathology 2-210, Baltimore, MD 21287, USA
C.A. Scheiner, Department of Cell Biology and Anatomy, Louisiana State University Medical Center, 1901 Perdido St., New Orleans, LA 70112-1393, USA
H. Son, Department of Biochemistry, College of Medicine, Hanyang University, 17 Haengdang-dong, Seong-dong-Gu, Seoul 133-791, Korea
H. Sontheimer, Department of Neurobiology, The University of Alabama at Birmingham, Birmingham, AL 35233, USA
N. Spear, Department of Anesthesiology, The UAB Center for Free Radical Biology, The University of Alabama at Birmingham, 1900 University Blvd., THT 958, Birmingham, AL 35233, USA
N.J. Sucher, Department of Biology, Hong Kong University of Science and Technology, Hong Kong
M. Sur, Department of Brain and Cognitive Sciences, Massachusetts Institute of Technology, Cambridge, MA 02139, USA
S.R. Vincent, The Graduate Program in Neuroscience, Department of Psychiatry, University of British Columbia, Vancouver, BC V5T 1Z3, Canada
J. Wandell, Department of Cell Biology and Anatomy, Louisiana State University Medical Center, 1901 Perdido St., New Orleans, LA 70112, USA
J.A. Williams, The Graduate Program in Neuroscience, Department of Psychiatry, The University of British Columbia, Vancouver, BC V5K 1Z3, Canada
H.H. Wu, Department of Cell Biology and Anatomy and the Neuroscience Center, Louisiana State University Medical Center, 1901 Perdido St., New Orleans, LA 70112, USA
Zu-Cheng Ye, Department of Neurobiology, The University of Alabama at Birmingham, Birmingham, AL 35233, USA
C.F. Zorumski, Departments of Psychiatry and Neurobiology, Washington University School of Medicine, 4940 Children's Place, St. Louis, MO 63110, USA

Preface

Nitric oxide has been of intense interest to neuroscientists since its discovery as a messenger molecule in the brain during the last decade. NO, chosen by Science Magazine as the "Molecule of the Year" in 1995, continues to be one of the most widely studied molecules in brain research. This free radical gas has been implicated in a variety of brain processes, including the regulation of glutamate and its receptors; in synaptic plasticity that occurs during long-term potentiation and depression; as a retrograde messenger to stabilize or refine synapses during development; as a regulatory second messenger involved in the control of brain blood flow; and as an agent that contributes to both brain degeneration and neuroprotection.

We have divided the volume into five sections. The first, entitled "Nitric oxide: brain distribution, production, and signaling", includes chapters that discuss the mechanisms underlying the release and action of NO in the CNS and others that describe the distribution of NO in brain using various techniques. The second section, "Nitric oxide and NMDA receptor interaction", contains chapters that describe how nitric oxide interacts with and regulates the *N*-methyl-D-aspartate (NMDA) receptor. The third section, "Nitric oxide in brain development", includes chapters that review the role of NO in pathway refinement and synapse stabilization in the developing CNS. Section four, "Nitric oxide in synaptic plasticity", contains chapters that describe the role of NO in adult brain plasticity, including long term potentiation and depression. The final section of the book, "Nitric oxide and other signals in neurodegeneration and neuroprotection" includes chapters on the role of nitric oxide in brain injury such as occurs during ischemia and other events produced by glutamate neurotoxicity, and in neuroprotective mechanisms that promote neuron survival. The final chapter focuses upon another agent, platelet activating factor, that functions in both degeneration and plasticity.

This Progress in Brain Research volume is an important milestone in the evolution of the Elsevier PBR series. "Nitric oxide in brain development, plasticity, and disease" is the first PBR volume to include a digital atlas that can be accessed on the World Wide Web (http://nadph.anatomy.lsumc.edu). The brain atlas illustrates the distribution of nitric oxide-containing neurons in the central nervous system of the mouse. The atlas is a powerful addition to this volume in that it is possible for readers to view the brain in a three-dimensional full-color rendition. The atlas contains digital images of histological sections stained by nicotinamide adenine dinucleotide phosphate diaphorase (NADPHd), a marker of nitric oxide; digital drawings of section outlines upon which the distribution of NADPHd labeled neurons have been plotted; and 3-D representations of the cell distributions contained within the volume of brain represented by translucent shading of its surface (Cork et al., Ch. 4).

"Nitric oxide in brain development, plasticity, and disease" is an outgrowth of a satellite symposium that was held in conjunction with the twenty-seventh annual meeting of the Society for Neuroscience, held in New Orleans, LA on October 24–25, 1997. The symposium received the generous support of individuals, universities, and corporations. The Corporate Guarantor was Olympus America, Inc., University Sponsors were the Louisiana State University Medical Center Foundation, the LSUMC Neuroscience Center, the LSUMC Department of Cell Biology and Anatomy, and the University of Alabama at Birmingham Department of Neurobiology. Corporate Sponsors were Alexis Biochemicals Corporation, Calbiochem-Novabiochem International, Inc., Chemicon International, Inc., Nikon, Inc., Noran Instruments, Inc., Sievers Instruments, Inc., Transduction Labs, and Carl Zeiss. Corporate supporters were Cayman Chemical Company, Inc., Harlan Sprague Dawley, Inc., Transmolecular, Inc., and Vector Laboratories, Inc.

We are also grateful for the administrative and organizational support of Anne Marie Hesson, Business Manager, and Claudia Hendricks and Richard Brown, Assistant Business Managers, of the Department of Cell Biology and Anatomy, LSU Medical Center, New Orleans. Neither the symposium nor this PBR volume would have been completed without their skillful organization and effort.

R. Ranney Mize
Ted M. Dawson
Valina M. Dawson
Michael J. Friedlander
April, 1998

Contents

III. Nitric oxide in brain development

IV. Nitric oxide in synaptic plasticity

V. Nitric oxide and other signals in neurodegeneration and neuroprotection

SECTION I

Nitric oxide: Brain distribution, production and signaling

R.R. Mize, T.M. Dawson, V.L. Dawson and M.J. Friedlander (Eds.)
Progress in Brain Research, Vol 118

CHAPTER 1

Regulation of neuronal nitric oxide synthase and identification of novel nitric oxide signaling pathways

Ted M. Dawson[1,2,*], Masayuki Sasaki[1], Mirella Gonzalez-Zulueta[1] and Valina L. Dawson[1,2,3]

[1]*Department of Neurology*, [2]*Department of Neuroscience*, [3]*Department of Physiology, Johns Hopkins University School of Medicine, 600 N. Wolfe Street, Carnegie 2-214, Baltimore, MD 21287, USA*

Abstract

Neuronal nitric oxide synthase (nNOS) participate in a variety of physiologic and pathologic processes in the nervous system. nNOS was originally felt to be a constitutively expressed enzyme, but recent observations suggest that its levels are dynamically controlled in response to neuronal development, plasticity and injury. nNOS expression is regulated through alternative promoter usage through alternative mRNA splicing and it is likely that this plays an important role in the inducibility of gene expression in response to extracellular stimuli. Emerging data also suggests that NO may be the key mediator linking activity to gene expression and long-lasting neuronal responses through NO activating $p21^{Ras}$ through redox-sensitive modulation.

Introduction

Neuronal nitric oxide (NO) synthase participates in diverse biologic processes including synaptic plasticity and remodeling, neurotransmission, neuronal development, behavior, and neuroendocrine regulation, among others. Nitric oxide is produced by a family of NO synthases (NOS) in which there are three distinct isoforms: neuronal NOS (nNOS, Type 1), inducible NOS (iNOS, Type 2), and endothelial NOS (eNOS, Type 3). The three enzymes are expressed in different cells and tissues and have distinct physiologic functions (Garthwaite and Boulton, 1995; Yun et al., 1996). nNOS and eNOS are calcium/calmodulin-dependent enzymes that are predominantly expressed in neurons and endothelial cells, respectively. Both nNOS and eNOS are sometimes referred to as constitutive NOS to distinguish them from iNOS which is an inducible enzyme in response to immunologic activation and subsequent transcription (Nathan and Xie, 1994; Bredt and Snyder, 1994a,b).

nNOS is dynamically regulated

Although nNOS is thought to be a constitutively expressed enzyme, recent data suggest that its expression is dynamically regulated (Dawson and Snyder, 1994; Forstermann et al., 1995). For instance, during olfactory neuronal development nNOS is expressed in most olfactory neurons, but nNOS expression is absent in adult olfactory

*Corresponding author. Tel.: +1 410 614 3359; fax: +1 410 614 9568; e-mail: ted.dawson@qmail.bs.jhu.edu

neurons (Roskams et al., 1994). nNOS protein is also highly expressed in embryonic cortical plate neurons from E15 to E19, as well as, in dorsal root ganglion cells during development (Bredt and Snyder, 1994a,b). As the animal matures, the expression switches from being in essentially all cortical plate neurons to an adult phenotype in which the neurons are expressed in 1–2% of cortical neurons (Dawson et al., 1991). nNOS is also dynamically regulated in response to a variety of neuronal injury paradigms including olfactory neurons in response to bulbectomy (Roskams et al., 1994), ibogaine-mediated Purkinje cell toxicity (O'Hearn et al., 1995), axotomy of motor neurons in the brainstem (Herdegen et al., 1993; Rossiter et al., 1996; Yu, 1994), ventral root avulsion (Wu, 1993; Wu et al., 1994), axotomy of sympathetic nerves (Verge et al., 1995) and neuronal damage in the cerebral cortex induced by the middle cerebral artery occlusion (Zhang et al., 1994).

Neurotrophins regulate the expression of nNOS

The molecular mechanisms controlling the dynamic regulation of nNOS levels are poorly understood, but recent data suggest that growth factors and neurotrophins may dynamically control the expression of nNOS. The association of the regulation of nNOS and neurotrophins may have profound implications for neuronal survival. Neurotrophins consist of a family of growth factors including nerve growth factor (NGF), brain derived neurotrophic factor (BDNF), neurotrophin-3 (NT-3), and neurotrophin-4/5 (NT-4/5) which act on high affinity receptor tyrosine kinases, TrkA, TrkB and TrkC, to promote neurite extension, neuronal survival and differentiation (Barbacid, 1995; Barde, 1994; Thoenen, 1995). Besides having a role in neuronal differentiation and survival, neurotrophins can attenuate neuronal cell death caused by excitotoxins, glucose deprivation, and ischemia (Anderson et al., 1996; Burke et al., 1994; Cheng and Mattson, 1994; Davies and Beardsall, 1992; Frim et al., 1993; Lindholm, 1994; Mattson et al., 1995; Nakao et al., 1995; Shigeno et al., 1991). Despite the abundance of in vitro and in vivo data that neurotrophins enhance the survival of neurons following neuronal injury, Koh et al. (1995) recently provided provocative data that neurotrophins under certain conditions enhance excitotoxic insults.

Glutamate neurotoxicity acting via NMDA receptors mediates cell death in focal cerebral ischemia (Choi, 1988) and may also play a part in neurodegenerative diseases such as Huntington's disease and Alzheimer's disease (Meldrum and Garthwaite, 1990). "Delayed or rapidly triggered neurotoxicity" in which irreversible processes are set in motion by a short (5 min) application of NMDA in primary cortical neuronal cultures as assessed 24 h later (Choi and Rothman, 1990) is exquisitely dependent upon calcium. NO plays a major role in this form of toxicity as NOS inhibitors, removal of L-arginine or reduced hemoglobin which scavenges NO attenuates NMDA neurotoxicity (Samdani et al., 1997a). Compelling evidence of the role of NO in NMDA neurotoxicity was our observation that primary neuronal cultures from nNOS null mice are relatively resistant to NMDA neurotoxicity as well as combined oxygen and glucose (Dawson et al., 1996). Our recent studies also indicate that the conditions under which neurons are cultured can profoundly influence the expression of nNOS and subsequent involvement of NO in excitotoxicity (Samdani et al., 1997b). In primary cultures the expression of nNOS is dependent on both the method of culture and the age of the cultures. If cultures are grown on glial feeder layers or grown in serum-free conditions, nNOS will not be expressed at sufficient levels to mediate neurotoxicity (Samdani et al., 1997b). The expression of nNOS in primary cultures at levels equivalent to in vivo expression (1–2% of the neuronal population) is critical for the ability to observe NO mediated neurotoxicity. Neurotrophins, despite their general role in attenuating excitotoxic neuronal injury, were recently shown by us to increase nNOS neurons, nNOS protein and NOS catalytic activity in cortical cultures grown on glial feeder layers and render neurons more sensitive to NMDA (Fig. 1). This finding is consistent with the previous report by Koh et al. (1995) on the enhancement of excitotoxic neuronal injury by neurotrophins under certain conditions. However, when neurons are grown on a poly-ornithine matrix, neurotrophins fail to enhance the expression of nNOS and are neuroprotective, consistent with the neuroprotective role of neurotrophins in neurotoxicity studies in intact animals (Lindvall et al., 1994). Confirming the role of neurotrophin induced nNOS expression in mediating the enhanced excitotoxic response to NMDA in cultures grown on glial feeder layers was our observation that cultures from nNOS null mice failed to show any alteration in NMDA-induced toxicity following neurotrophin administration (Fig. 2). Thus, nNOS expression and NMDA, NO mediated, neurotoxicity are dependent, in part, upon the culture paradigm; and neurotrophins regulate the susceptibility to NMDA neurotoxicity through modulation of nNOS. Furthermore, NMDA-neurotoxicity in culture is critically dependent on the developmental state of the neurons being assessed and when cortical neurons are cultured on a glial feeder layer, they do not reach nearly as mature a phenotype as when grown on a Poly-O-matrix (Samdani et al., 1997b).

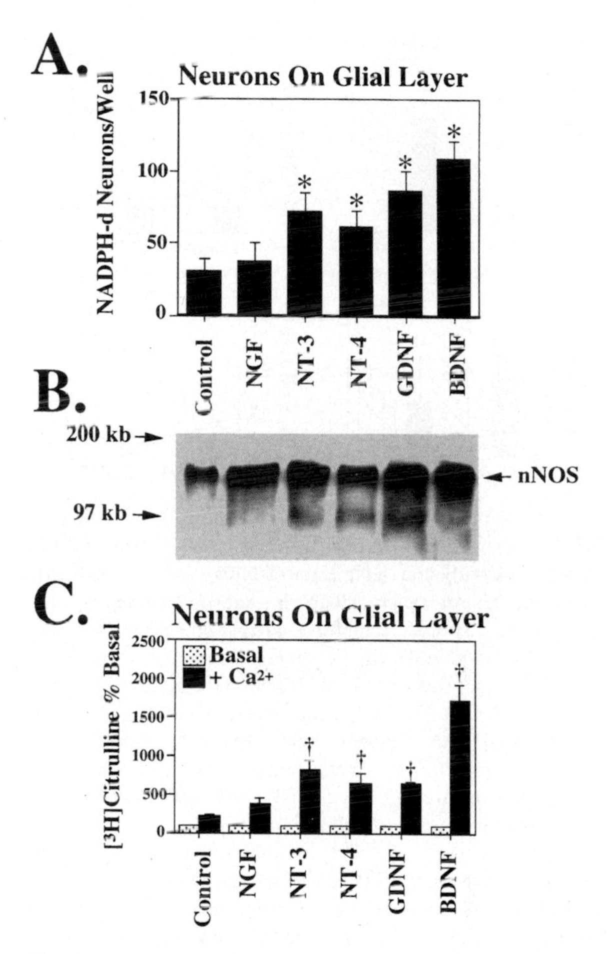

Fig. 1. Neurotrophins enhance nNOS expression in cortical neurons: 24 h pre-treatment with neurotrophins (100 ng/ml) markedly enhances the number of nNOS neurons as indicated by NADPH-diaphorase staining of cortical neurons. Each data point represents the mean ± SEM (n = 8–12) of at least two separate experiments. $^{*}P < 0.0001$ for NT-3, NT-4, GDNF and BDNF compared to control cultures. NGF is not significantly different from control. Neurotrophins increase the amount of nNOS protein as measured by Western blot analysis. A parallel increase in NOS catalytic activity was observed. $^{\dagger}P < 0.02$ for NT-3, NT-4, GDNF and BDNF compared to control cultures. Each data point represents the mean ± SEM of at least two separate experiments performed in duplicate. Reprinted with permission (Samdani et al., 1997b).

Characterization of the molecular mechanisms controlling nNOS expression

The molecular mechanisms controlling the expression of nNOS are beginning to be elucidated. Human nNOS is localized to chromosome 12 q24.2 and is a 160 kb genome comprising 29 exons and 28 introns (Hall et al., 1994). Several alternative transcripts are derived from the nNOS locus yielding diverse mRNA species that are controlled by numerous promoter domains. The greatest diversity of nNOS transcripts may occur through the 5′ untranslated region. Human nNOS has nine alternative first exons which may lead to differential expression of nNOS in neuronal and non-neuronal tissues as well as expression due to toxic or stressful insults and developmental regulation (Wang and Marsden, 1995). The fine control and complex patterns of expression of nNOS probably occurs through the alternative splicing and the diversity of promoters (Hall et al., 1994; Xie et al., 1995).

Rodents also have 5′ terminal region diversity (Brenman et al., 1997; Sasaki et al., 1998). Wild

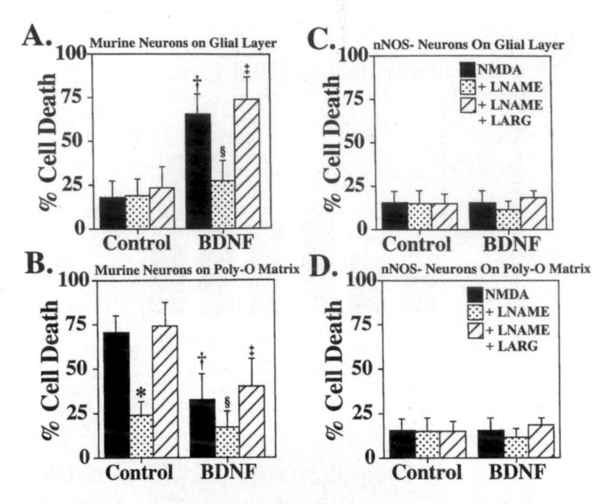

Fig. 2. Increases in nNOS levels mediate potentiation of NMDA neurotoxicity by neurotrophins: (A) Murine cortical neurons grown on a glial layer exhibit NMDA (500 μM) mediated neurotoxicity which is not influenced by the competitive nNOS inhibitor nitroarginine-methyl ester (LNAME) (500 μM) or by excess L-arginine (LARG) (5 mM). 24 h pre-treatment with BDNF (100 ng/ml) markedly enhances NMDA (500 μM) neurotoxicity ($^{\dagger}P < 0.0001$) which is prevented with LNAME (500 μM) ($^{\S}P < 0.0001$), and LARG (5 mM) ($^{\ddagger}P < 0.0001$) reverses this neuroprotection. (B) Cortical neurons grown on a Poly-O matrix demonstrate an equivalent amount of NMDA (500 μM), NO dependent, neurotoxicity ($^{*}P < 0.0001$) as that observed in cortical neurons grown on a glial layer pretreated with BDNF (100 ng/ml). BDNF treatment of neurons grown on a Poly-O matrix attenuates the neurotoxic response to NMDA (500 μM) by approximately 50% ($^{\dagger}P < 0.0001$), LNAME (500 μM) ($^{\S}P < 0.01$) provides further protection and LARG (5 mM) ($^{\ddagger}P < 0.0001$) reverses this protection. (C) nNOS^{-} cultures are markedly resistant to NMDA neurotoxicity and BDNF (100 ng/ml) fails to enhance NMDA neurotoxicity in cortical cultures grown on a glial layer. (D) BDNF fails to provide neuroprotection to cortical neurons grown on a Poly-O matrix. Each data point represents the mean ± SEM (n = 6–12) of at least two separate experiments and a minimum of 4,000 to 10,000 neurons counted. Reprinted with permission (Samdani et al., 1997b).

type nNOS has amazing structural diversity in the brain. We and others have identified multiple alternative 5′ transcripts, the majority of which are spliced into a common exon 2 (Brenman et al., 1997; Sasaki et al., 1998). 5′ RACE, RNase protection of mRNA and Northern blot analysis indicates that some of these 5′ species are expressed in a tissue and developmental specific pattern. Despite the unusual 5′ structural diversity all the transcripts seem to code for a 155 kD protein. Thus, it is most likely that the 5′ structural diversity controls the developmental and tissue-specific expression of nNOS as well as the expression of nNOS following injury, stress, environmental changes and toxic insults. Consensus binding sites for a number of transcription factors were identified in the human nNOS exon 1a 5′-regulatory sequence including AP-2, TEF-1/MCBF, CREB/ATF/c-Fos, NRF-1, Ets, NF-1 and NF-κB sequences (Hall et al., 1994).

To gain a better understanding of the molecular mechanisms that control nNOS transcription and subsequent protein levels, we cloned and characterized the 5′ flanking region of mouse nNOS (Sasaki et al., 1998). We identified multiple promoters which control nNOS expression in a temporal and tissue/region-specific pattern in mouse brain. Numerous binding sites for ubiquitous and brain specific transcription factors were identified within exon 1a and exon 1b. Comparison of mouse nNOS exon 1 with human nNOS exon 1 identified three areas of significant homology. Due to the high conservation of these regions between mouse and human genomic DNA, these regions likely represent important regulatory regions that control the expression of nNOS perhaps in a temporal and tissue-specific manner. Future analyses with DNA footprinting and electrophoretic mobility shift assays will be necessary to confirm the potential importance of these putative 5′-regulatory regions. Several binding sites for transcription factors were also identified within exon 2 including GATA, Sp1, NF-1, AP-1, AP-2, ETS, p53, and CRE. Comparison of human and mouse exon 2 revealed two regions of high homology. One of the regions contains two CRE sites which are immediately upstream of a transcription initiation start site identified by 5′-RACE and the other region contains an Sp1, 2 ETS and a CRE site which is immediately upstream of the initiation start site identified by 5′-RACE in adult and cerebellar cortex. Due to the high sequence homology it is likely that these regions are also important in controlling nNOS transcription.

nNOS promoter analysis through construction of promoter–reporter constructs confirmed that the multiple 5′-flanking regions of mouse nNOS contains multiple promoters that direct tissue and temporal-specific expression of the nNOS gene (Sasaki et al., 1998). nNOS is structurally related to the classical P450 gene and it is interesting to note that some gene members of the P450 superfamily also employ alternative promoters for their tissue or temporal specific expression patterns. For instance, the *CYP*19 gene encoding aromatase P450 has been found to have six different primary transcripts and expression is driven by tissue specific promoters (Simpson et al., 1997). These transcripts differ in only the first exon and it is these multiple first exons that are spliced to a common exon 2 containing the translation initiation codon so as to generate the same polypeptide from each transcript. This pattern of untranslated first exon usage appears to be utilized by the nNOS gene, although the regulation of nNOS gene transcription seems more complex due to the presence of minor variants that are translated from other exons (Ayoubi and Van De Ven, 1996). The physiological significance of these unusual cases of alternative promoter usage largely remains unclear (Brenman and Bredt, 1997). However, it is interesting to note that exon 2 codes for a PDZ domain which enables nNOS to be targeted to the post-synaptic density, thus playing a pivotal role in NMDA receptor mediated synaptic plasticity (Brenman and Bredt, 1997) and neurotoxicity (Dawson et al., 1996) while other nNOS isoforms lack the PDZ domain contained in exon 2. These observations reveal the importance of alternative promoter and exon usage in the molecular control of nNOS gene expression and regulation. Alternative promoter usage is also likely to play an important role in the inducibility of gene expression in response to extracellular stimuli that is characteristic of the P450 gene family and the nNOS gene. Fruitful areas for future study would be the characterization of the transcriptional regulatory regions in in vitro and in vivo promoter–reporter analyses.

Nitric oxide activates $p21^{Ras}$: Implications for neuronal development and cell survival

NO may play a key role in nervous system morphogenesis and developmental synaptic plasticity. Cerebral cortical subplate neurons transiently express nNOS from embryonic day E15–E19 of rats (Bredt and Snyder, 1994). These cells extend their processes to the corpus striatum and thalamus. nNOS positive immunostaining of cortical subplate neurons and their processes decreases rapidly and vanishes by the 15th postnatal day. Embryonic sensory ganglia are nNOS positive and this decreases to less than 1% of the cells at birth. nNOS is also expressed during embryonic development in neurons of the developing olfactory epithelium during migration and the establishment of primary synapses in the olfactory bulb (Roskams et al., 1994). Olfactory nNOS expression rapidly declines after birth and is undetectable by post-natal day 7. Regenerating olfactory receptor neurons express nNOS after bulbectomy and nNOS is particularly enriched in their outgrowing axons. Thus, NO may play a role in activity dependent establishment of connections in both developing and regenerating olfactory neurons. Development of proper patterns of connections in the retinotectal system may involve NO as nNOS expression peaks at the time when refinement of the initial pattern of connections is occurring (Williams et al., 1994). Adult spinal cord motor neurons molecular maturation may also involve NO (Kalb and Agostini, 1993). Growth arrest during differentiation of neuronal cells may get triggered by NO; and thus NOS may serve as a growth arrest gene initiating the switch to cytostasis during differentiation (Peunova and Enikolopov, 1995). NOS is highly expressed in developing imaginal discs of drosophila and inhibition of NOS in larvae causes organ hypertrophy and ectopic NOS expression in larvae has the opposite effects, implicating NO as an antiproliferative signal during development (Kuzin et al., 1996).

The molecular mechanism for NO's role in neural development is not known. A key as to how NO regulates neuronal growth, differentiation, survival and death may come from recent observations that NO activates p21^{Ras} (Ras) (Fig. 3) (Lander et al., 1995; Yun et al., 1998). Our recent studies indicate that stimulation of NMDA receptor in cultured cortical neurons activates Ras–ERK pathway via calcium-dependent activation of nNOS and NO generation through non-cGMP dependent mechanisms (Fig. 4) (Yun et al., 1998). Activation of Ras/ERK pathway by NO may be mediated by direct activation of Ras GTPase activity presumably by nitrosylation of cysteine through a redox-sensitive interaction (Lander et al., 1996a). NO may be the key mediator linking activity to gene expression and long-term plasticity as calcium-dependent activation of Ras–ERK pathway is thought to be a major pathway of neural activity-dependent long-term changes in the nervous system (Finkbeiner and Greenberg, 1996). NO dependent activation of Ras may also mediate activity-dependent survival of immature cortical neurons (Gonzalez-Zulueta et al., 1997). NO mediated survival may

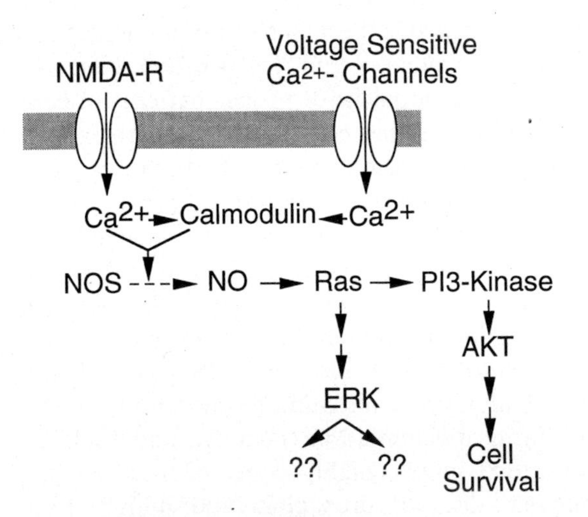

Fig. 3. A schematic diagram showing the role of NO in coupling NMDA receptor-mediated and voltage sensitive calcium influx to Ras and downstream effectors. *Abbreviations*: AKT, serine/threonine protein kinase; PI3-Kinase, phosphoinositide 3-kinase; ERK, extracellular signal-regulated kinase.

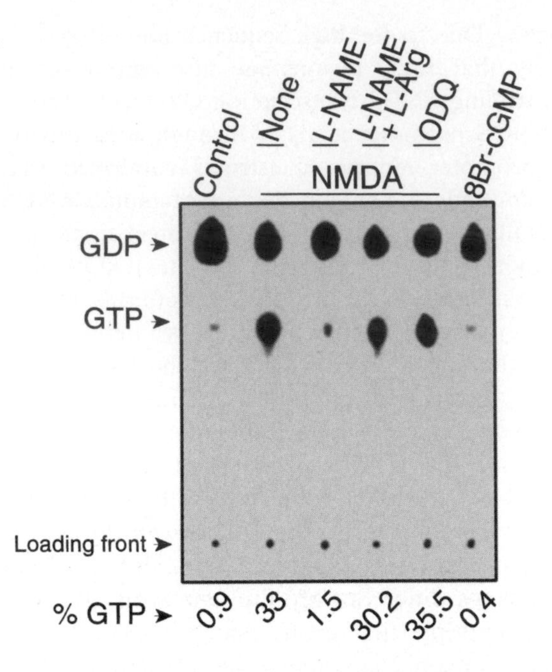

Fig. 4. NMDA induces Ras activation through an NO-dependent, cGMP-independent mechanism. Activated Ras is indicated by the detection of GTP on TLC (Thin Layer Chromatography) plates. Cells were exposed to control saline solution (Control), or treated with 50 μM NMDA for 5 min with no previous pretreatment (None), or pretreated for 10 min with 500 μM L-NAME, 500 μM L-NAME plus 5 mM L-Arg, or 10 μM ODQ before exposure to 50 μM NMDA for 5 min. Cells were also treated with 5 mM 8Br-cGMP for 5 min. Cells were lysed and collected for in situ Ras activation assay 0.5 min following each of the treatments. The NOS inhibitor, L-NAME, blocks NMDA induced activation of Ras, and this blockade is reversed by the NOS substrate, L-Arg. Inhibition of guanylyl cyclase by ODQ has no effect on NMDA induced activation of Ras, and 8Br-cGMP does not activate Ras. Reproduced with permission (Yun et al., 1998).

be through activation of PI3 kinase as Ras directly activates PI3 kinase to promote cellular survival through activation of AKT (Kaufmann-Zeh et al., 1997). NO donors also activate all three types of MAPK (ERK, JNK/ SAPK, p38 MAPK) to different extents and time courses in a T-cell leukemic cell line (Lander et al., 1996b) and PC12 cells (Yun et al., 1998). NO's role as a promoter of cellular survival and cell death may occur through NO's activation of the Ras–PI3 kinase–AKT pathway and Ras–ERK pathway, respectively. Thus, NO–Ras signaling may underline NO's role in neuronal survival, differentiation and apoptotic cell death during development. These processes may occur through redox-sensitive modulation of Ras and suggest that Ras is a potential endogenous NO-redox-sensitive effector molecule mediating the intercellular actions of NO in the CNS.

Conclusion

NO has revolutionized our concepts about neuronal signaling. Recent advances indicate that NO's actions are more diverse than originally appreciated. Uncovering the targets of NO and the mechanisms that regulate the expression of nNOS will contribute to a greater understanding of CNS physiology and pathological neuronal functions.

Acknowledgements

We thank Ann Schmidt for typing assistance. VLD is supported by US PHS Grant NS 33142, the American Heart Association. TMD is an Established Investigator of the American Heart Association and is supported by US PHS Grants, NS 37090, and the Paul Beeson Faculty Scholar Award in Aging Research. Under an agreement between the Johns Hopkins University and Guilford Pharmaceuticals, TMD and VLD are entitled to a share of sales royalty received by the University from Guilford. TMD and the University also own Guilford stock, and the University stock is subject to certain restrictions under University policy. The terms of this arrangement are being managed by the University in accordance with its conflict of interest policies.

References

Anderson, K.D., Panayotatos, N., Corcoran, T.L., Lindsay, R.M. and Wiegand, S.J. (1996) Ciliary neurotrophic factor protects striatal output neurons in an animal model of Huntington disease. *Proc. Natl. Acad. Sci, USA*, 93: 7346–7351.

Ayoubi, T.A. and Van De Ven, W.J. (1996) Regulation of gene expression by alternative promoters. *FASEB J.*, 10: 453–460.

Barbacid, M. (1995) Neurotrophic factors and their receptors. *Curr. Opin. Cell. Biol.*, 7: 148–155.

Barde, Y.A. (1994) Neurotrophins: A family of proteins supporting the survival of neurons. *Prog. Clin. Biol. Res.*, 390: 45–56.

Bredt, D.S. and Snyder, S.H. (1994a) Nitric oxide: A physiologic messenger molecule. *Ann. Rev. Biochem.*, 63: 175–195.

Bredt, D.S. and Snyder, S.H. (1994b) Transient nitric oxide synthase neurons in embryonic cerebral cortical plate, sensory ganglia and olfactory epithelium. *Neuron*, 13: 301–313.

Brenman, J.E. and Bredt, D.S. (1997) Synaptic signaling by nitric oxide. *Curr. Opin. Neurobiol.*, 7: 374–378.

Brenman, J.E., Xia, H., Chao, D.S., Black, S.M. and Bredt, D. S. (1997) Regulation of neuronal nitric oxide synthase through alternative transcripts. *Dev. Neurosci.*, 19: 224–231.

Burke, M.A., Mobley, W.C., Cho, J., Wiegand, S.J., Lindsay, R.M., Mufson, E.J. and Kordower, J.H. (1994) Loss of developing cholinergic basal forebrain neurons following excitotoxic lesions of the hippocampus: Rescue by neurotrophins. *Exp. Neurol.*, 130: 178–195.

Cheng, B. and Mattson, M.P. (1994) NT-3 and BDNF protect CNS neurons against metabolic/excitotoxic insults. *Brain Res.*, 640: 56–67.

Choi, D.W. (1988) Glutamate neurotoxicity and diseases of the nervous system. *Neuron*, 1: 623–634.

Choi, D.W. and Rothman, S.M. (1990) The role of glutamate neurotoxicity in hypoxic-ischemic neuronal death. *Ann. Rev. Neurosci.*, 13: 171–182.

Davies, S.W. and Beardsall, K. (1992) Nerve growth factor selectively prevents excitotoxin induced degeneration of striatal cholinergic neurones. *Neurosci. Lett.*, 140: 161–164.

Dawson, T.M., Bredt, D.S., Fotuhi, M., Hwang, P.M. and Snyder, S.H. (1991) Nitric oxide synthase and neuronal NADPH diaphorase are identical in brain and peripheral tissues. *Proc. Natl. Acad. Sci. USA*, 88: 7797–7801.

Dawson, V.L., Kizushi, V.K., Huang, P.L., Snyder, S.H. and Dawson, T.M. (1996) Resistance to neurotoxicity in cortical neuronal cultures from neuronal nitric oxide synthase deficient mice. *J. Neurosci.*, 16: 2479–2487.

Dawson, T.M. and Snyder, S.H. (1994) Gases as biological messengers: Nitric oxide and carbon monoxide in the brain. *J. Neurosci.*, 14: 5147–5159.

Finkbeiner, S. and Greenberg, M.E. (1996) Ca^{2+}-dependent route to Ras: Mechanisms for neuronal survival, differentiation and plasticity. *Neuron*, 16: 233–236.

Forstermann, U., Gath, I., Schwarz, P., Closs, E.I. and Kleinert, H. (1995) Isoforms of nitric oxide synthase. Properties, cellular distribution and expressional control. *Biochem. Pharmacol.*, 50: 1321–1332.

Frim, D.M., Yee, W.M. and Isacson, O. (1993) NGF reduces striatal excitotoxic neuronal loss without affecting concurrent neuronal stress. *Neuroreport*, 4: 655–658.

Garthwaite, J. and Boulton, C.L. (1995) Nitric oxide signaling in the central nervous system. *Annu. Rev. Physiol.*, 57: 683–706.

Gonzalez-Zulueta, M., Yun, H.-Y., Dawson, V.L. and Dawson, T.M. (1997) Nitric oxide mediates activity-dependent neuronal survival. *Soc. Neurosci. Abst.*, 23: 630.

Hall, A.V., Antoniou, H., Wang, Y., Cheung, A.H., Arbus, A.M., Olson, S.L., Lu, W.C., Kau, C.L. and Marsden, P.A. (1994) Structural organization of the human neuronal nitric oxide synthase gene (NOS1). *J. Biol. Chem.*, 269: 33082–33090.

Herdegen, T., Brecht, S., Mayer, B., Leah, J., Kummer, W., Bravo, R. and Zimmermann, M. (1993) *J. Neurosci.*, 13: 4130–4145.

Kalb, R.G. and Agostini, J. (1993) Molecular evidence for nitric oxide-mediated motor neuron development. *Neuroscience*, 57: 1–8.

Kauffmann-Zeh, A., Rodriguez-Viciana, P., Ulrich, E., Gilbert, C., Coffer, P., Downward, J. and Evan, G. (1997) Suppression of c-Myc-induced apoptosis by Ras signalling through PI(3)K and PKB. *Nature*, 385: 544–548.

Koh, J.Y., Gwag, B.J., Lobner, D. and Choi, D.W. (1995) Potentiated necrosis of cultured cortical neurons by neurotrophins. *Science*, 268: 573–575.

Kuzin, B., Roberts, I., Peunova, N. and Enikolopov, G. (1996) Nitric oxide regulates cell proliferation during Drosophila development. *Cell*, 87: 639–649.

Lander, H.M., Jacovina, A.T., Davis, R.J. and Tauras, J.M. (1996a) Differential activation of mitogen-activated protein kinase by nitric oxide-related species. *J. Biol. Chem.*, 271: 19705–19709.

Lander, H.M., Milbank, A.J., Taurus, J.M., Hajjar, D.P., Hemstead, B.L., Schwartz, G.D., Kraemer, R.T., Mirza, U.A., Chait, B.T., Burk, S.C. and Quilliam, L.A. (1996b) Redox regulation of cell signalling. *Nature*, 381: 380–381.

Lander, H.M., Ogiste, J.S., Pearce, S.F.A., Levi, R. and Novogrodsky, A. (1995) Nitric oxide-stimulated guanine nucleotide exchange on $p21^{ras}$. *J. Biol. Chem.*, 270: 7017–7020.

Lindholm, D. (1994) Role of neurotrophins in preventing glutamate induced neuronal cell death. *J. Neurol.*, 242: S16–S18.

Lindvall, O., Kokaia, Z., Bengzon, J, Elmer, E. and Kokaia, M. (1994) Neurotrophins and brain insults. *Trends Neurosci.*, 17: 490–496.

Mattson, M.P., Lovell, M.A., Furukawa, K. and Markesbery, W.R. (1995) Neurotrophic factors attenuate glutamate-induced accumulation of peroxides, elevation of intracellular Ca^{2+} concentration and neurotoxicity and increase antioxidant enzyme activities in hippocampal neurons. *J. Neurochem.*, 65: 1740–1751.

Meldrum, B. and Garthwaite, J. (1990) Excitatory amino acid neurotoxicity and neurodegenerative disease. *Trends Pharmacol.*, 11: 379–387.

Nakao, N., Kokaia, Z., Odin, P. and Lindvall, O. (1995) Protective effects of BDNF and NT-3 but not PDGF against hypoglycemic injury to cultured striatal neurons. *Exp. Neurol.*, 131: 1–10.

Nathan, C. and Xie, Q.W. (1994) Nitric oxide synthases: Roles, tolls and controls. *Cell*, 78: 915–918.

O'Hearn, E., Zhang, P. and Molliver, M.E. (1995) Excitotoxic insult due to ibogaine leads to delayed induction of neuronal NOS in Purkinje cells. *Neuroreport*, 6: 1611–1616.

Peunova, N. and Enikolopov, G. (1995) Nitric oxide triggers a switch to growth arrest during differentiation of neuronal cells. *Nature*, 375: 68–73.

Roskams, A.J., Bredt D.S., Dawson, T.M. and Ronnett, G.V. (1994) Nitric oxide mediates the formation of synaptic connections in developing and regenerating olfactory receptor neurons. *Neuron*, 13: 289–299.

Rossiter, J.P., Riopelle, R.J. and Bisby, M.A. (1996) Axotomy-induced apoptotic cell death of neonatal rar facial motoneurons: Time course analysis and relation to NADPH diaphorase activity. *Exp. Neurol.*, 138: 33–44.

Samdani, A., Dawson, T.M. and Dawson, V.L. (1997a) Nitric oxide synthase in models of focal ischemia. *Stroke*, 28: 1283–1288.

Samdani, A., Newcamp, C., Resink, A., Facchinetti, F., Hoffman, B.E., Dawson, V.L. and Dawson, T.M. (1997b) Differential susceptibility to neurotoxicity mediated by neurotrophins and neuronal nitric oxide synthase. *J. Neurosci.*, 17: 4633–4641.

Sasaki, M., Huang, H., Wei, X., Dillman, J.F., Dawson, V.L. and Dawson, T.M. (1998) Transcriptional regulation of nNOS during neuronal development and plasticity is controlled by multiple promoters. *Soc. Neurosci. Abst.*, 23: 225.

Shigeno, T., Mima, T., Takakura, K., Graham, D.I., Kato, G., Hashimoto, Y. and Furukawa, S. (1991) Amelioration of delayed neuronal death in the hippocampus by nerve growth factor. *J. Neurosci.*, 11: 2914–2919.

Simpson, E.R., Michael, M.D., Agarwal, V.R., Hinshelwood, M.M., Bulun, S.E. and Zhao, Y. (1997) Cytochrome p450 11: Expression of the CYP19 (aromatase) gene: An unusual case of alternative promoter usage. *FASEB J.*, 11: 29–36.

Thoenen, H. (1995) Neurotrophins and neuronal plasticity. *Science*, 270: 593–598.

Verge, V.M.K., Richardson, P.M., Wisenfeld-Hallin, Z. and Hokfelt, T. (1995) Differential influence of nerve growth factor on neuropeptide expression in vivo: A novel role in peptide suppression in adult sensory neurons. *J. Neurosci.*, 15: 2081–2096.

Wang, Y. and Marsden, P.A. (1995) Nitric oxide synthases: Gene structure and regulation. *Adv. Pharmacol.*, 34: 71–90.

Williams, C.V., Nordquist, D. and McLoon, S.C. (1994) Correlation of nitric oxide synthase expression with changing patterns of axonal projections in the developing visual system. *J. Neurosci.*, 14: 1746–1755.

Wu, W. (1993) Expression of nitric-oxide synthase (NOS) in injured CNS neurons as shown by NADPH diaphorse histochemistry. *Exp. Neurol.*, 120: 153–159.

Wu, W., Liuzzi, F.J., Schinco, F.P., Depto, A., Li, Y., Mong, J.A., Dawson, T.M. and Snyder, S.H. (1994) Induction of neuronal nitric oxide synthase in spinal neurons by traumatic injury. *Neuroscience*, 61: 719–726.

Xie, J., Roddy, P., Rife, T.K., Murad, F. and Young, A.P. (1995) Two closely linked but separable promoters for human neuronal nitric oxide synthase gene transcription. *Proc. Natl. Acad. Sci. USA*, 92: 1242–1246.

Yu, W.H. (1994) Nitric oxide synthase in motor neurons after axotomy. *J. Histochem. Cytochem.*, 42: 451–457.

Yun, H.-Y., Dawson, V.L. and Dawson, T.M. (1996) Neurobiology of nitric oxide. *Crit. Rev. in Neurobiol.*, 10: 291–316.

Yun, H.-Y., Gonzalez-Zulueta, M., Dawson, V.L. and Dawson, T.M (1998) Nitric oxide mediates NMDA receptor induced activation of p21^{ras}. *Proc. Natl. Acad. Sci. USA*, 95: 5773–5778.

Zhang, Z.G., Chopp, M., Gautam, S., Zaloga, C., Zhang, R.L., Schmidt, H.H.H.W., Pollock, J.S. and Forstermann, U. (1994) Upregulation of neuronal nitric oxide synthase and mRNA and selective sparing of nitric oxide synthase-containing neurons after focal cerebral ischemia in rat. *Brain Res.*, 654: 85–95.

R.R. Mize, T.M. Dawson, V.L. Dawson and M.J. Friedlander (Eds.)
Progress in Brain Research, Vol 118

CHAPTER 2

Genetic analysis of NOS isoforms using nNOS and eNOS knockout animals

Paul L. Huang* and Eng H. Lo

[1]*Cardiovascular Research Center and Cardiology Division, Massachusetts General Hospital, Boston, MA 02114, USA*

Abstract

All three major isoforms of nitric oxide synthase (NOS) are expressed in the brain. Because of complex and overlapping expression patterns (Marletta, 1994; Nathan and Xie, 1994), the particular NOS isoform involved in many processes is not clear. In fact, NO generated by separate isoforms may have different roles and potentially opposing effects (Iadecola et al., 1994). We have taken a genetic approach, to disrupt or knockout the genes for NOS isoforms to circumvent some of the limitations of pharmacologic agents. This approach allows the study of each individual NOS isoform in physiologic processes in the context of intact animals. It gives insights into possible developmental roles for NO and parallel processes that may compensate for the absence of each NOS isoform.

We have made nNOS and eNOS knockout mice, as well as double knockout mice that lack both nNOS and eNOS isoforms (Huang et al., 1993; Huang et al., 1995; Son et al., 1996). In this chapter, we review some of the physiologic roles for NO that have been elucidated making use of these mice, including regulation of cerebral blood flow, response to cerebral ischemia, regulation of neurotransmitter release in the brain, and development of synaptic plasticity. Other chapters will discuss results using NOS knockout animals in studies of long term potentiation (see Hawkins, this volume), neuronal development (see Mize, this volume), and potential mechanisms for protection in nNOS knockout mice (Moskowitz, M.A.; Dawson, V.L, this volume).

* Corresponding author.

History of nitric oxide

The fascinating history of nitric oxide as a biological messenger molecule spans several distinct fields, including vascular biology, nutrition science, immunology, and neurobiology. Alfred Nobel first synthesized nitroglycerin as an explosive from glycerol and nitric acid in the 1860s. Nitroglycerin soon became known for its medicinal properties and even within Nobel's lifetime, it was established as a first-line medication to treat angina pectoris. Despite their widespread use, the mechanism by which nitrovasodilators relax blood vessels has not been well understood. Furchgott and Zawadzki (1980) made a major advance with the first description of endothelium derived relaxing factor (EDRF), later demonstrated to be NO (Furchgott, 1988; Ignarro et al., 1988).

A second independent line of research involved the study of dietary nitrite and nitrate as precursors for carcinogenic nitrosamines. Tannenbaum and coworkers found that rats and humans fed a diet low in nitrates still excrete significant amounts of nitrate (Green et al., 1981a,b). Furthermore, the excreted nitrate levels of one human subject rose markedly when he fell ill with an infection, establishing a link between infection with nitrate production. Subsequent work established that macrophages utilize L-arginine to generate nitric oxide, which results in nitrite and nitrate excretion (Stuehr and Marletta, 1985). We now know that inducible nitric oxide synthase plays an important role in cellular immunity (Nathan and Xie, 1994).

Parallel work in neurobiology showed that NMDA receptor activation can stimulate neuronal production of nitric oxide, which among other functions, activates soluble guanylate cyclase and increases cGMP levels (Garthwaite, 1995; Garthwaite and Boulton, 1995). Bredt and Snyder purified neuronal NOS, cloned its gene, and localized its expression to discrete populations of neurons using immunohistochemistry (Bredt and Snyder, 1994). Neuronal NOS expression colocalizes with the histochemical activity NADPH diaphorase, and purified neuronal NOS has NADPH diaphorase activity. This is of interest because NADPH diaphorase positive neurons are resistant to ischemia, toxins, and neurodegenerative diseases (Vincent and Hope, 1992). Some have speculated that NO produced by these neurons may mediate damage to neighboring neurons.

NO: Multiple biological roles and tissue origins

NO is synthesized from L-arginine by the NOS enzymes. The names of these isoforms – neuronal NOS (nNOS), endothelial NOS (eNOS) and macrophage NOS (macNOS) or inducible NOS (iNOS) – reflect the original tissues in which they were first described, although we now know that the expression patterns of these isoforms overlap greatly. For example, certain hippocampal neurons express eNOS (Dinerman et al., 1994; O'Dell et al., 1994), whereas non-neuronal cells may express the nNOS isoform (Schmidt et al., 1992; Kobzik et al., 1994). Within the brain, parenchymal neurons, perivascular nerves, vascular endothelium, microglia, astrocytes, and oligodendroglia are all potential sources of NO (Murphy et al., 1993; Iadecola et al., 1994). In many cases, nNOS and eNOS isoforms are constitutively expressed whereas the iNOS isoform is inducible. However, whether the expression of a given isoform is inducible or constitutive varies in different tissues.

Each NOS isoform has many postulated roles (Dawson et al., 1992; Bredt and Snyder, 1994). The neuronal NOS isoform is thought to play a role in neurotransmission in autonomic non-adrenergic, non-cholinergic nerves in the gastrointestinal system, genitourinary system, and respiratory tract. In the central nervous system, nNOS is thought to regulate cerebral blood flow and mediate responses to excitatory neurotransmission. In blood vessels, eNOS is responsible for EDRF activity, and regulates vascular tone and blood pressure. In addition, vascular NO inhibits smooth muscle proliferation, inhibits platelet aggregation, and blocks leukocyte adhesion and activation. The iNOS isoform is thought to play a role in cellular immunity and defense mechanisms against tumor cells and pathogens. iNOS may be responsible for the hypotension seen in sepsis.

Rationale and limitations of genetic approach

Transgenic and knockout rodents have proven to be useful tools to study mediators of brain function (Chan, 1996). In a series of elegant studies, Chan and coworkers have defined the importance of Cu/Zn superoxide dismutase (SOD) and mitochondrial Mg–SOD using transgenic and knockout mice (Kinouchi et al., 1991; Chan et al., 1993; Yang et al., 1994). Transgenic mice expressing basic FGF (Macmillan et al., 1993), bcl-2 (Martinou et al., 1994; Farlie et al., 1995), and TGF-β1 (Hennrich-Noack et al., 1996), as well as knockout mice for p53 (Crumrine et al., 1994) have also been applied to the study of cerebral ischemia.

We have taken the approach of generating neuronal NOS knockout mice and endothelial NOS knockout mice to examine the effects of single gene deletions in intact animals (Huang et al., 1993, 1995). Others have generated inducible NOS knockout mice (Laubach et al., 1995; MacMicking et al., 1995; Wei et al., 1995). The gene knockout approach has several advantages. First, this approach is not subject to the lack of specificity of NOS inhibitors and the potential effects of such agents on more than one NOS isoform. Second, the genetic approach pinpoints the roles of individual NOS genes, since many tissues, including the brain, contain all three of the major NOS isoforms.

There are certain limitations to using knockout animals. First, there may be developmental abnormalities due to the gene knockout. If one of the NOS isoforms plays a critical role in embryonic development, its absence may lead to other secondary abnormalities that are difficult to predict. Although we have not found any structural abnormalities in the brain of these animals, there may be subtle, as yet undetected, developmental changes. Second, there may be physiologic compensation for the absence of individual NOS genes. Compensation may be due to pre-existing redundant pathways, or altered expression of genes in the setting of the gene deletion. For example, the cerebrovascular responses to hypercapnia and whisker stimulation (Irikura et al., 1995; Ma et al., 1996) are normally mediated by nNOS and blocked by NOS inhibitors such as L-NA. In the nNOS knockout mice, these processes are normal but they are no longer sensitive to L-NA, suggesting that they are now subserved by parallel, redundant pathways. Another form of compensation is the use of one NOS isoform to substitute for another. Evidence suggests that nNOS may substitute for eNOS in the pial vessel response to acetylcholine (Meng et al., 1996), and that nNOS and eNOS may substitute for one another in long-term potentiation (Son et al., 1996). Finally, genetic background may confound the observed phenotype, particularly when the parental background strains of the knockout mice behave differently from one another. These considerations must be borne in mind when interpreting results from gene knockout mice. We are addressing some of these limitations directly, for example, by making gene knockouts with a uniform genetic background, or by creating conditional and tissue-specific gene knockout mice using Cre recombinase and loxP methodology.

Neuronal NOS knockout mice

We generated nNOS mutant mice mice by deleting the exon that contains the translation initiation codon ATG of nNOS. These mice lack nNOS as assessed by several important criteria. Catalytic NOS activity in the brain is reduced to less than 5% of wild-type levels, and residual activity has a different distribution than native nNOS (Huang et al., 1993). This residual activity is in part due to eNOS in the brain, both in neurons and blood vessels (Dinerman et al., 1994; O'Dell et al., 1994), and in part, to splice variants that do not contain the deleted exon (Brenman et al., 1996). These variants, termed nNOS-β and nNOS-γ are present in wild-type animals at the same low levels, and are not upregulated in the nNOS knockout animals. Although they have catalytic activity in vitro, they lack the critical PDZ domain that is required for nNOS to associate with cell membranes, so their functional significance is not known (Brenman et al., 1996). NADPH diaphorase staining is entirely absent in the brain and in peripheral tissues. A more sensitive electron micrograph diaphorase technique using formazan precipitation also fails to detect any nNOS expression (Darius et al., 1995).

Using electron paramagnetic resonance spin trapping methods with diethyldithiocarbamate to detect levels of NO in the brain, no basal production is seen in the nNOS knockout mice and no elevation is seen following ischemia. Similarly, basal cGMP production is lower in nNOS knockout mice than in wild-type mice, and no elevation is seen after ischemia (Huang et al., 1994). Taken together, these results indicate that disruption of the nNOS gene results in loss of over 95% of NOS activity in the brain, and absence of functional nNOS activity by immunohistochemistry, diaphorase staining, spin trapping, or assay for the secondary messenger. The role of the remaining splice variants in the nNOS knockout mice, and whether they are involved in development or survival are not known.

There are no neuroanatomic differences between the nNOS mutant mice and wild-type animals using cresyl violet staining and light microscopy (Huang et al., 1993). nNOS colocalizes with somatostatin and NPY in the striatum, and with choline acetyltransferase in the pedunculopontine nucleus. Immunoreactivity for these markers is preserved despite loss of nNOS immunoreactivity, showing that neurons that normally express nNOS are preserved in the nNOS mutant mice. Furthermore, there are no differences in the cerebrovascular anatomy of wild-type and nNOS mutant mice in whom the circle of Willis and major tributaries were filled with carbon black particles through cardiac injection (Huang et al., 1994).

Endothelial NOS knockout mice

In making the eNOS mutant mice, the NADPH ribose and adenine binding sites, which are essential to the function of the protein, were deleted. Lack of eNOS was documented by Northern blot, Western blot, immunohistochemistry, and NOS catalytic assay (Huang et al., 1995). No truncated message or alternative splice variants are seen. Isolated aortic rings from eNOS mutant mice do not relax in response to acetylcholine, although the vessels do relax to sodium nitroprusside and papaverine. This provides genetic evidence that the eNOS gene is required for EDRF activity.

Endothelial NOS mutant mice have mean arterial blood pressures that are about 30% higher than wild-type littermates, consistent with a role for basal eNOS activity in blood pressure regulation and vascular tone. It is unclear why other mechanisms involved in blood pressure control – other vasodilators or vasoconstrictors – do not compensate fully for the absence of eNOS to return the blood pressure to normal. One possibility is that the renin–angiotensin system and the autonomic nervous system evolved primarily as a defense against hypotension, and diminution of their activity is a poor buffer against hypertension. Another is that eNOS may be involved in establishing the baroreceptor setpoint.

Treatment of eNOS mutant mice with L-NA results in a paradoxic drop in blood pressure, an effect that is blocked by L-arginine and not seen with D-NA. If this effect is due to NOS inhibition, it implicates an isoform other than eNOS, which is absent in the eNOS mutant mice. Neuronal NOS mutant mice have normal blood pressure, but they have a tendency toward hypotension under anesthesia (Irikura et al., 1995). A role for nNOS in maintenance of blood pressure would explain both the hypotensive tendency in nNOS mutant mice and the abnormal response to L-NA in the eNOS mutant mice.

NO and cerebral blood flow

Roy and Sherrington first proposed that local neuronal activity in the brain is coupled to cerebral blood flow (CBF) (Roy and Sherrington, 1890). Multiple mediators may be involved, including products of neuronal metabolism (H^+ and adenosine), ions released following neuronal activation, or neurotransmitters released adjacent to blood vessels. NO is an attractive candidate for coupling cerebral blood flow to brain metabolism because it is a vasodilator that diffuses freely across cell membranes and has a short biological half-life (Edelman and Gally, 1992; Dirnagl et al., 1993; Northington et al., 1995).

Studies using nNOS mutant mice confirm the role of the nNOS isoform in the hyperemic response to hypercapnia (Irikura et al., 1995) and whisker stimulation (Ayata et al., 1996; Ma et al., 1996). They also underscore the importance of parallel mechanisms of vasodilation that compensate for the absence of nNOS in the mutant mice. In wild-type mice, the response of cerebral blood flow to moderate hypercapnia (5% CO_2) is mediated by NO, since inhibitors of NOS such as L-NA and L-NAME attenuate the rCBF increase (Iadecola et al., 1994). However, NO is not the only vasoactive agent involved, since NOS inhibitors can only block the response within a range of CO_2 levels, and even in the most susceptible range, there is residual vasodilation.

nNOS mutant mice have equivalent responses to 5% CO_2 as wild-type mice. However, while topical L-NA blocks the response of wild-type mice, it has no effect on nNOS mutant mice (Irikura et al., 1995). Thus, the response of nNOS mutant mice is not due to eNOS or another NOS isoform. Mediators other than NO, such as hydrogen ions and arachidonic acid metabolites, must therefore mediate the response in nNOS mutant mice. The same mediators may participate in wild-type animals as well, since NOS inhibition blocks the rCBF response only over a very narrow range of $PaCO_2$. Thus, nNOS plays a role in hypercapnic hyperemia in the wild-type animal, but in the chronic absence of nNOS in the nNOS mutant mice, other mechanisms compensate to maintain a quantitatively normal response.

Another model of cerebral blood flow coupling to metabolism is the cortical barrel blood flow response to whisker stimulation. In wild-type animals, whisker stimulation at 2–3 Hz for 60 s results in a predictable increase in rCBF, which can be attenuated by topical superfusion with L-NA. In nNOS mutant animals, the response to whisker stimulation is very similar to wild-type animals, but topical L-NA has no effect (Ma et al., 1996). Endothelium-dependent relaxation, as assessed by pial dilation to acetylcholine, is the same in wild-type and nNOS mutant mice. These results indicate that endothelial NOS does not mediate the whisker response in nNOS mutant mice, and that the effect of L-NA in wild-type mice is due to nNOS inhibition. In addition, they suggest that NO-independent mechanisms couple rCBF with metabolism in nNOS mutant mice. In separate studies, eNOS mutant mice have a response indistinguishable from that in wild-type animals, including sensitivity to L-NA, confirming that nNOS mediates blood flow responses in the cortical barrel fields (Ayata et al., 1996).

Taken together, these studies demonstrate that the nNOS isoform, but not the eNOS isoform, mediates the cerebrovascular response to hypercapnia and whisker stimulation. Furthermore, nNOS-derived NO acts in parallel with other vasodilators, since NOS inhibition attenuates but does not obliterate each response. In the chronic absence of nNOS expression, redundant vasodilatory mechanisms compensate, so the response is preserved and no longer L-NA sensitive. Other examples of physiologic compensation have been noted in the anesthesia minimum alveolar concentration and nociception of the nNOS mutant mice (Crosby et al., 1995; Ichinose et al., 1995).

Another form of compensation is seen in the acetylcholine dependent dilation of pial arterioles (Meng et al., 1996). Acetylcholine 10 μM superfusion causes atropine-sensitive dose-dependent arteriolar dilation in wild-type mice. Superfusion with L-NA inhibits cortical NOS activity by over 70% in wild-type mice and abrogates the vasodilation response, consistent with the role for NO in mediating the response, and in agreement with results found in other species. The expected source for NO in this process is the eNOS isoform expressed in vascular endothelial cells. However, results in the eNOS mutant mice produced unexpected findings. The eNOS mutant mice still demonstrate atropine-sensitive dilation in response to acetylcholine. Moreover, L-NA diminished the response by about 50%, suggesting that one component of the response in eNOS mutant mice is NO-dependent, while another is NO-independent. Tetrodotoxin (TTX) blocks the response in eNOS mutant mice, while it has no effect in wild-type or nNOS mutant mice. These results suggest that nNOS may compensate for the absence of eNOS. This type of alternative isoform compensation differs from compensation for nNOS activity by non-NO vasodilatory mediators seen in hypercapnia or whisker response. Another example NOS isoforms compensating for one another is seen in long-term potentiation in the nNOS knockout, eNOS knockout, and nNOS/eNOS double knockout mice (Son et al., 1996), as detailed elsewhere in this volume (Chapter 11).

NO and cerebral ischemia

Malinski et al. (1993) found that NO levels increase by several orders of magnitude following middle cerebral artery occlusion (MCAO). Significant evidence points to the nNOS isoform as the major source of this increase in the first 24 h after ischemia. Early studies on the effect of NOS inhibition on outcome of ischemia led to variable results, in part because pharmacologic agents have effects on multiple NOS isoforms, some of which may be protective and others detrimental.

nNOS mutant mice demonstrate significantly reduced infarct size following MCAO, with a 38% reduction over SV129 and C57 BL/6 wild-type mice (Huang et al., 1994). The functional outcome was also significantly improved. There were no neuroanatomic or vascular differences between the nNOS mutant mice and wild-type mice. Measurements of rCBF by laser Doppler demonstrate that the filament causes the same reduction in rCBF in both the core ischemic area as well as the penumbra in both wild-type and nNOS mutant mice. Thus, differences in blood flow do not account for the reduction in infarct size. These results support the notion that nNOS contributes to neuronal damage after focal ischemia. Similar protection is seen in the nNOS mutant mice using a model of transient focal ischemia followed by reperfusion (Hara et al., 1996) and a model of global ischemia (Panahian et al., 1996). Treatment of nNOS mutant mice with L-NA, which inhibits remaining isoforms of NOS including eNOS, caused larger strokes, consistent with the protective role of eNOS.

In contrast, the eNOS mutant mice develop larger infarcts following permanent MCAO than either C57BL/6 or SV129 wild-type controls, even when their blood pressure was reduced to normotensive levels for the duration of the experiment (Huang et al., 1996). These results confirm that eNOS plays a protective role following cerebral ischemia, and its absence is detrimental. Measurements of rCBF by laser Doppler flowmetry show that the hemodynamic effect of MCAO is greater in the eNOS mutant mice than in wild-type mice. Blood flow kinetics can also be mapped by functional CT scanning using a contrast agent (Lo et al., 1996). Each pixel in a coronal view was mapped as normal flow (normal kinetics), core infarct (absent flow), or penumbra (flow present, but kinetics abnormal). The total area of abnormality (infarct plus penumbra) was the same in wild-type mice as in eNOS mutant mice, but the size of the core infarct area was greater in the eNOS mutant mice and the ischemic penumbra was correspondingly smaller. These results suggest that the larger infarcts in the eNOS mutant mice are due to absence of the normal protective effects of eNOS which serve to preserve cerebral blood flow.

Although calcium fluxes may activate pre-existing NOS isoforms, the expression of each of the major NOS isoforms is affected by cerebral ischemia. The level of nNOS mRNA increases as early as 15 min after MCAO and peaks at 1 h, while the number of nNOS-containing neurons peak in number at 4 h after MCAO (Zhang et al., 1995a,b). ^{3}H-L-NA binding to brain sections increases by up to 250% after MCAO in wild-type and eNOS mutant mice, but is very low in nNOS mutant mice (Hara et al., 1997). The level of nNOS mRNA in wild-type brain also increases following MCAO. These results suggest that ischemic insult upregulates the expression of the nNOS isoform and this upregulation may be important to pathogenesis. The expression of the eNOS isoform, as detected by immunohistochemistry, increases in cerebral microvessels after MCAO as well (Zhang et al., 1993). Thus, nNOS and eNOS, which have been thought of as constitutive NOS isoforms, are induced in the brain following cerebral ischemia.

Rapid induction of the iNOS isoform after MCAO occurs in rats (Iadecola et al., 1995a,b, 1996), coincident with invasion of ischemic areas by leukocytes and activation of microglial cells and astrocytes. Aminoguanidine inhibits iNOS with some selectivity and reduces infarct size in several models (Iadecola et al., 1996; Zhang et al., 1996), suggesting that iNOS mediates cytotoxicity following focal ischemia. Another study also showed a benefit of aminoguanidine, but the effects were too rapid to be consistent with iNOS inhibition (Cockroft et al., 1996), raising the possibility of an independent mechanism of protection. In support of a role for iNOS in cytotoxicity, induction of iNOS potentiates NMDA receptor mediated neuronal cell death following hypoxic–ischemic insult (Hewett et al., 1996).

Recently, Iadecola and coworkers reported that iNOS knockout mice show reduced infarct size following MCAO in a focal ischemia model, and this is not due to vascular or blood flow differences. Thus, the emerging picture is that nNOS may mediate early injury following cerebral ischemia, and iNOS may mediate later injury, corresponding to influx of inflammatory cells into the ischemic or infarcted area. In contrast, the eNOS isoform appears to be protective.

Mechanisms of toxicity and protection

The data support potentially toxic roles for nNOS (early) and iNOS (later), and protective roles for eNOS. These differences may be due in part to the redox state of NO produced (free radical NO· versus nitrosonium ion NO^+) (Lipton et al., 1993, 1994). The precise mechanisms by which NO· mediates toxicity remains to be explored, but there is evidence for several possibilities (Pelligrino, 1993; Iadecola et al., 1994; Chan, 1996). NO reacts with superoxide to form peroxynitrite anion, which may participate in lipid oxidation, DNA damage, and other toxic sequelae of reactive oxygen species. Peroxynitrite nitrosylates tyrosine residues in proteins, and these changes may affect protein function, for example as a substrate for tyrosine kinases. NO activates poly-ADP ribose polymerase (PARP), leading to the depletion of cellular energy stores. Reactive oxygen species, including NO and its derivatives, may also activate the cellular pathways of apoptosis and may participate in the mediation of apoptosis.

There are several potential mechanisms for the protective effects of NO. Studies in eNOS mutant mice suggest that vasodilation and maintenance of rCBF in the face of ischemia is important (Huang et al., 1996). NO inhibits platelet aggregation and adhesion, which may prevent microvascular occlusion by platelet aggregates (Radomski et al., 1993). NO blocks leukocyte adhesion, which might reduce the recruitment of inflammatory cells to the region. The expression of the adhesion molecule ICAM-1 (Zhang et al., 1995b,c), and the chemokines MIP-1a and MCP-1 (Kim et al., 1995) are upregulated following focal ischemia, and may play a role in recruitment of leukocytes and post-ischemic inflammatory processes. NO may also be protective by inhibiting calcium fluxes through NMDA receptors (Lipton et al., 1994).

NO and neurotransmitter release in the brain

NO may serve as an inter-neuronal messenger that modulates neurotransmitter release in mammalian brain. Stimulation of post-synaptic NMDA receptors leads to elevations in intracellular calcium that in turn activates NOS enzymes (Moncada et al., Dawson et al., 1992). NO then diffuses in a retrograde manner to the pre-synaptic neuron where it is able to modify neurotransmitter release (Garthwaite and Boulton, 1995). Some studies suggest that NO can diffuse away from one original synapse and influence neurotransmitter release in adjacent synapses as well (Schuman and Madison, 1994).

In synaptosomes prepared from guinea pig cortex, NMDA-stimulated release of glutamate and norepinephrine was significantly attenuated by inhibition of NOS with L-NOARG and L-NMMA (Montague et al., 1994). Extracellular diffusion of NO was implicated since hemoglobin, a potent NO chelator restricted to extracellular space, was also able to reduce glutamate and norepinephrine release after NMDA stimulation. In hippocampal slices, NO donors and NOS inhibitors increased and decreased NMDA-stimulated release of aspartate and norepinephrine respectively (Jones et al., 1995). Similar results have been obtained in cerebellar slices where NO potentiated NMDA-stimulated release of aspartate (Dickie et al., 1992).

The contribution of NO to neurotransmitter release can be studied in vivo using microdialysis. Perfusion of NO donors or NO gas through microdialysis probes amplifies release of aspartate, glutamate, and GABA in rat striatum (Guevara-Guzman et al., 1994). Similar results have been obtained in other brain regions; microdialysis perfusion of NO donors increased the release of aspartate and glutamate in rat brainstem (Lawrence and Jarrott, 1993), and induced release of acetylcholine in the rat basal forebrain (Prast and Phillipu, 1992).

We have used perfusion of NMDA via microdialysis probes to activate NOS in vivo in nNOS and eNOS mutant mice to assess the contributions of the two isoforms towards neurotransmitter release. nNOS mutants, eNOS mutants, and wild type SV129 and C57BL/6 mice were anesthetized with halothane and microdialysis probes were used to examine the cerebral cortex and hippocampus. Concentrations of the excitatory neurotransmitter glutamate were measured in the microdialysis samples using fluorescent HPLC. No differences in basal concentrations were found between the various strains. To induce glutamate release in vivo, depolarizing stimuli (100 mM K^+ or 1 mM NMDA) were administered locally through the probes. High K^+ stimulation induced 4–6 fold elevations in glutamate concentrations and no significant differences were seen in the various wild type and mutant strains (Fig. A). However, release in the cortex appeared to be slightly reduced in the nNOS mutants. High K^+ stimulation via the microdialysis probes results in widespread and clamped depolarizations that induce massive neurotransmitter release via both pre-synaptic vesicular efflux as well as reversal of reuptake carriers (Nicholls, 1989; Nicholls and Attwell, 1990). In such situations, NO would not be expected to make significant contributions towards neurotransmitter release. On the other hand, direct stimulation of the NMDA receptor should reveal the contribution of NO as the retrograde amplifying signal since NOS is primarily activated via the NMDA type glutamate receptor (Garthwaite and Boulton, 1995). In this study, NMDA-stimulated glutamate release in the cortex was significantly attenuated in the nNOS mutants compared to wild type and eNOS mutant mice (Fig. B). However, no significant effects were observed in the hippocampus. These results indicate that in the cortex, NO may serve as an amplifier of glutamate release following NMDA receptor stimulation. In the hippocampus, both nNOS and eNOS isoforms may contribute and within the limits of the microdialysis assay, loss of a single isoform did not lead to any detectable effects on glutamate release. These results are consistent with the study showing that long-term potentiation in the hippocampus was altered only in mutant mice where both nNOS and eNOS genes were disrupted (Son et al., 1996).

The precise mechanism by which NO modulates neurotransmitter release remains to be elucidated. A study using synaptosomal preparations derived from rat hippocampus has suggested that the mechanism may involve calcium-independent pathways (Meffert et al., 1994). The state of neuronal activity may also be important. In hippocampal slices, NO amplifies synaptic transmission only when paired with weak tetanic stimulation (Zhuo et al., 1993). Furthermore, not all studies show that NO is a signal that amplifies neurotransmitter release. Under certain conditions, NO donors induce synaptic depression in the hippocampus (Boulton et al., 1994). Therefore, it is possible that NO can modulate neurotransmitter release in a positive or negative direction depending on the levels of synaptic activity present at the time. Finally, in addition to effects on release, NO may also influence neurotransmitter dynamics by modifying reuptake, as NO donors reversibly inhibit uptake of glutamate, dopamine and serotonin in rat striatal synaptosomes (Pogun et al., 1994).

NO and synaptic plasticity

Ever since it was found in the brain, NO has been proposed as the elusive "retrograde messenger" for long-term depression and long-term potentiation, models of how learning and memory may work. The retrograde messenger is an essential component of a feedback loop in models that invoke a pre-synaptic component to synaptic plasticity. The proposed mechanism is that post-synaptic neurons, upon stimulation, would send the retrograde messenger back to the pre-synaptic neurons, to indicate that the signal had been received and to strengthen or weaken the synapse in question. NO is a natural candidate for the retrograde messenger because of the very properties that make it an unusual signaling molecule: it diffuses freely across cell membranes, and has a short half-life, limiting its sphere of influence to a precise volume in space. Moreover, pre-existing NOS in the pre-synaptic neuron could easily be activated by receptor-mediated calcium influx that occurs on stimulation.

Inhibitors of NOS and scavengers of NO block LTP, and application of exogenous NO produces activity-dependent enhancement of synaptic transmission (Bohme et al., 1991; O'Dell et al., 1991; Schuman and Madison, 1991; Haley et al., 1992). Since nNOS accounts for over 95% of NOS catalytic activity in the brain, it would be logical to assume that the neuronal isoform would be the one involved in LTP. However, one critical controversy earlier on was that the nNOS isoform could not be detected in the appropriate cells, the hippocampal pyramidal neurons. The nNOS mutant mice show preserved LTP, suggesting that the nNOS isoform may not be only one responsible for LTP (O'Dell et al., 1994). Immunohistochemical studies localized eNOS to pyramidal cells (Dinerman et al., 1994; O'Dell et al., 1994), giving rise to the possibility that eNOS might generate NO as the retrograde messenger. Still other studies demonstrated that using gentler methods of fixation, the nNOS isoform could indeed be detected in pyramidal cells (Wendland et al., 1994), so both nNOS and eNOS could be involved in LTP. Double mutant mice that lack both nNOS and eNOS mice showed reduced LTP in the stratum radiatum, whereas mice lacking nNOS or eNOS alone did not, showing that both isoforms are involved in LTP and that each can compensate for absence of the other (Son et al., 1996).

Tsien and coworkers used whole cell patch clamp in cerebellar slices to examine long-term depression (LTD) in nNOS knockout mice (Lev-Ram et al., 1997). LTD could not be induced in animals lacking nNOS, confirming that nNOS is required for LTD. However, caged compounds designed to deliver NO and cGMP inside Purkinje cells could not rescue the nNOS knockout mice, even though these compounds can circumvent pharmacologic inhibition of NO synthesis or guanylate cyclase in wild-type mice. Thus, chronic absence of nNOS in the mutant mice may have allowed "atrophy" of the signaling pathways downstream of cGMP.

Conclusion

NO has many cellular sources and complex roles in the brain. Each of the three known NOS isoforms which can be expressed in the brain, serve different biological functions. Some of their roles have been elucidated using animals that lack an NOS isoform as an specific molecular genetic tool.

The nNOS isoform mediates the responses of the NMDA glutamate receptors, by becoming activated during transient conditions of calcium influx, and generating NO which activates soluble guanylate cyclase, either in the same cell or in neighboring cells. nNOS is at least one of the NOS isoforms that participates as a retrograde messenger for synaptic plasticity. nNOS also mediates coupling of cerebral blood flow with cerebral metabolism, as shown by the dependence of hypercapnic vasodilation and whisker barrel blood flow responses on nNOS. However, excessive nNOS activity, as may occur with excitotoxicity or cerebral ischemia, has severe deleterious consequences. The nNOS mutant mice demonstrate examples of both non-NO, non-cGMP dependent compensation (blood flow responses to hypercapnia and whisker stimulation), as well as alternative NOS isoform use as compensation (LTP).

The eNOS isoform mediates blood flow through pial vessels in response to acetylcholine, bradykinin, and other mediators. Its continued expression is protective in models of cerebral ischemia, as shown by exacerbation of infarct size and degree of neuronal damage in the eNOS knockout mice. The eNOS isoform also plays an important role as one source of NO as a retrograde messenger for LTP. The iNOS isoform is not expressed at baseline in the brain, but is induced following ischemia as inflammatory cells enter the damaged area. iNOS contributes to later damage following cerebral ischemia and infarction.

The use of NOS knockout animals has also demonstrated the importance of documenting compensatory mechanisms in the mutant mice in the chronic absence of the NOS genes, as well as the possible comfounding factors of developmental abnormalities and genetic background effects. These issues need to be considered in any studies that use knockout mice, and will hopefully be addressed by the next generation of knockout mice that bear conditional or tissue-specific gene deletions.

Acknowledgements

This work was supported by the Interdepartmental Stroke Program Project P01 NS-10828 (PLH), R01 NS33335 (PLH), R01 HL57818 (PLH), R29 NS32806 (EHL) and R01 NS37074 (EHL). PLH is an Established Investigator of the American Heart Association.

References

Ayata, C., Ma, J., Meng, W., Huang, P. and Moskowitz, M.A. (1996) L-NA sensitive rCBF augmentation during vibrissal stimulation in type III nitric oxide synthase mutant mice. *J Cereb. Blood Flow Metab.*, 16: 539–541.

Bohme, G.A., Bon, C., Stutzman, J.M., Doble, A. and Blanchard, J.C. (1991) Possible involvement of nitric oxide in long-term potentiation. *Eur. J. Pharmacol.*, 199: 379–381.

Boulton, C.L., Irving, A.J., Southam, E., Pottier, B., Garthwaithe, J. and Collingridge, G. (1994) The nitric oxide-cyclic GMP pathway and synaptic depression in rat hippocampal slices. *Eur. J. Neurosci.*, 6: 1528–1535.

Bredt, D.S. and Snyder, S.H. (1994) Nitric oxide: A physiologic messenger molecule. *Ann. Rev. Biochem*, 63: 175–195.

Brenman, J.E., Chao, D.S., Gee, S.H., McGee, A.W., Craven, S.E., Santillano, D.R., Wu, Z., Huang, F., Xia, H., Peters, M.F., Froehner, S.C. and Bredt, D.S. (1996) Interaction of nitric oxide synthase with the post synaptic density protein PSD-95 and α1-syntrophin mediated by PDZ domain. *Cell*, 84: 757–767.

Chan, P.H. (1996) Role of oxidants in ischemic brain damage. *Stroke*, 27 (6): 1124–1129.

Chan, P.H., Kamii, H., Yang, G., Gafni, J., Epstein, C.J., Carlson, E. and Reola, L. (1993) Brain infarction is not reduced in SOD-1 transgenic mice after a permanent focal cerebral ischemia. *Neuroreport*, 5 (3): 293–296.

Cockroft, K.M., Meistrel, M., Zimmerman, G.A., Risucci, D., Bloom, O., Cerami, A. and Tracey, K.J. (1996) Cerebroprotective effects of aminoguanidine in a rodent model of stroke. *Stroke*, 27: 1393–1398.

Crosby, G., Marota, J.J.A. and Huang, P.L. (1995) Intact nociception-induced neuroplasticity in transgenic mice deficient in neuronal nitric oxide synthase. *Neuroscience*, 69: 1013–1017.

Crumrine, R.C., Thomas, A.L. and Morgan, P.F. (1994) Attenuation of p53 expression protects against focal ischemic damage in transgenic mice. *J. Cereb. Blood Flow Metab.*, 14: 887–891.

Darius, S., Wolf, G., Huang, P.L. and Fishman, M.C. (1995) Localization of NADPH-diaphorase/nitric oxide synthase in the rat retina: An electron microscopic study. *Brain Res.*, 690: 231–235.

Dawson, T.M., Dawson, V.L. and Snyder, S.H. (1992) A novel neuronal messenger molecule in brain: The free radical, nitric oxide. *Ann. Neurol.*, 32 (3): 297–311.

Dickie, B.G.M., Lewis, M.J. and Davies, J.A. (1992) NMDA-induced release of nitric oxide potentiates aspartate overflow from cerebellar slices. *Neurosci. Lett.*, 138: 145–148.

Dinerman, J.L., Dawson, T.M., Schell, M.J., Snowman, A. and Snyder, S.H. (1994) Endothelial nitric oxide synthase localized to hippocampal pyramidal cells: Implications for synaptic plasticity. *Proc. Natl. Acad. Sci. USA*, 91 (10): 4214–4218.

Dirnagl, U., Lindauer, U. and Villringer, A. (1993) Role of nitric oxide in the coupling of cerebral blood flow to neuronal activation in rats. *Neurosci. Lett*, 149 (1): 43–6.

Edelman, G.M. and Gally, J.A. (1992) Nitric oxide: Linking space and time in the brain. *Proc. Natl. Acad. Sci. USA*, 89 (24): 11651–11652.

Farlie, P.G., Dringen, R., Rees, S.M., Kannourakis, G. and Bernard, O. (1995) bcl-2 transgene expression can protect neurons from developmental and induced cell death. *Proc. Natl. Acad. Sci. USA*, 92: 4397–4401.

Furchgott, R.F. (1988) Studies on the relaxation of the rabbit aorta by sodium nitrate: Basis for the proposal that the acid-activatable component of the inhibitory factor from retractor penis is inorganic nitrate and the endothelium-derived relaxing factor is nitric oxide. In: P.M. Vanhoutte (Ed.), *Mechanisms of Vasodilatation.* Raven: New York, pp. 401–414.

Furchgott, R.F. and Zawadzki, J.V. (1980) The obligatory role of endothelial cells in the relaxation of arterial smooth muscle by acetylcholine. *Nature*, 288: 373–376.

Garthwaite, J. (1995) Neural nitric oxide signalling. *Trends in Neurosciences*, 18 (2): 51–52.

Garthwaite, J. and Boulton, C.L. (1995) Nitric oxide signaling in the central nervous system. *Ann. Rev. Physiol.*, 57: 683–706.

Green, L.C. Ruiz de Luzuriaga, K., Wagner, D.A., Rand, W., Istfan, N., Young, V.R. and Tannenbaum, S.R. (1981a) Nitrate biosynthesis in man. *Proc. Natl. Acad. Sci. USA*, 78: 7764–7768.

Green, L., Tannenbaum, S. and Goldman, P. (1981b) Nitrate synthesis in the germfree and conventional rat. *Science*, 212: 56–58.

Guevara-Guzman, R., Emson, P.C. and Kendrick, K.M. (1994) Modulation of in vivo striatal transmitter release by NO and cyclic GMP. *J. Neurochem.*, 62: 807–810.

Haley, J.E., Wilcox, G.L. and Chapman, P.F. (1992) The role of nitric oxide in hippocampal long-term potentiation. *Neuron*, 8: 211–216.

Hara, H., Ayata, C., Huang, P.L., Waeber, C., Ayata, G., Fujii, M. and Moskowitz, M.A. (1997) [^{3}H]L-NG-nitroarginine binding after transient focal ischemia and NMDA-induced excitotoxicity in type I and type III nitric oxide synthase null mice. *J. Cereb. Blood Flow Metab.*, 17: 515–526.

Hara, H., Huang, P.L., Panahian, N., Fishman, M.C. and Moskowitz, M.A. (1996) Reduced brain edema and infarction volume in mice lacking the neuronal isoform of nitric oxide synthase after transient MCA occlusion. *J. Cereb. Blood Flow Metab.*, 16: 605–611.

Henrich-Noack, P., Prehn, J.H.M. and Krieglstein, J. (1996) TGF-B1 protects hippocampal neurons against deterioration caused by transient glocal ischemia. *Stroke*, 27: 1609–1615.

Hewett, S.J., Muir, J.K., Lobner, D., Symons, A. and Choi, D.W. (1996) Potentiation of oxygen-glucose deprivation-induced neuronal death after induction of iNOS. *Stroke*, 27: 1586–1591.

Huang, P.L., Dawson, T.M., Bredt, D.S., Snyder, S.H. and Fishman, M.C. (1993) Targeted disruption of the neuronal nitric oxide synthase gene. *Cell*, 75 (7): 1273–1286.

Huang, P.L., Huang, Z., Mashimo, H., Bloch, K.D., Moskowitz, M.A., Bevan, J.A. and Fishman, M.C. (1995) Hypertension in mice lacking the gene for endothelial nitric oxide synthase. *Nature*, 377 (6546): 239–242.

Huang, Z., Huang, P.L., Ma, J., Meng, W., Ayata, C., Fishman, M.C. and Moskowitz, M.A. (1996) Enlarged infarcts in endothelial nitric oxide synthase knockout mice are attenuated by nitro-L-arginine. *J. Cereb. Blood Flow Metab.*, 16: 981–987.

Huang, Z., Huang, P.L., Panahian, N., Dalkara, T., Fishman, M.C. and Moskowitz, M.A. (1994) Effects of cerebral ischemia in mice deficient in neuronal nitric oxide synthase. *Science*, 265 (5180): 1883–1885.

Iadecola, C., Pelligrino, D.A., Moskowitz, M.A. and Lassen, N.A. (1994) Nitric oxide synthase inhibition and cerebrovascular regulation. *J. Cereb. Blood Flow Metab.*, 14: 175–192.

Iadecola, C., Zhang, F. and Xu, X. (1995a) Inhibition of inducible nitric oxide synthase ameliorates cerebral ischemic damage. *Am. J. Physiol.*, 268 (1 Pt 2): R286–R292.

Iadecola, C., Zhang, F., Xu, S., Casey, R. and Ross, M.E. (1995b) Inducible nitric oxide synthase gene expression in brain following cerebral ischemia. *J. Cereb. Blood Flow Metab.*, 15 (3): 378–384.

Iadecola, C., Zhang, F., Casey, R., Clark, H.B. and Ross, M.E. (1996) Inducible nitric oxide synthase gene expression in vascular cells after transient focal cerebral ischemia. *Stroke*, 27 (8): 1373–1380.

Ichinose, F., Huang, P.L. and Zapol, W.M. (1995) Effects of targeted neuronal nitric oxide synthase gene disruption and nitroG-L-arginine methylester on the threshold for isoflurane anesthesia. *Anesthesiology*, 83 (1): 101–108.

Ignarro, L.J., Bugas, G.M., Wood, K.S., Byrns, R.E. and Chaudhuri, G. (1988) Endothelium-derived relaxing factor produced by and released from artery and vein is nitric oxide. *Proc. Natl. Acad. Sci. USA*, 84. 9265–9269.

Irikura, K., Huang, P.L., Ma, J., Lee, W.S., Dalkara, T., Fishman, M.C., Dawson, T.M., Snyder, S.H. and Moskowitz, M.A. (1995) Cerebrovascular alterations in mice lacking neuronal nitric oxide synthase gene expression. *Proc. Nat. Acad. Sci. USA*, 92, (15): 6823–6827.

Jones, N.M., Loiacono, R.E. and Beart, P.M. (1995) Roles for nitric oxide as an intra and interneuronal messenger at NMDA release-regulating receptors: Evidence from studies of NMDA evoked release of ^{3}H-noradrenaline and D-^{3}H-aspartate from rat hippocampal slices. *J. Neurochem.*, 64: 2054–2063.

Kim, J.S., Gautam, S.C., Chopp, M., Zaloga, C., Jones, M.L., Ward, P.A. and Welch, K.M. (1995) Expression of monocyte chemoattractant protein-1 and macrophage inflammatory protein-1 after focal cerebral ischemia in the rat. *J. Neuroimmunol.*, 56 (2): 127–134.

Kinouchi, H., Epstein, C.J., Mizui, T., Carlson, E.J., Chen, S.F. and Chan, P.H. (1991) Attenuation of focal cerebral ischemia in transgeneic mice overexpressing CuZn superoxide dismutase. *Proc. Natl. Acad. Sci. USA*, 88: 11158–11162.

Kobzik, L., Reid, M.B., Bredt, D.S. and Stamler, J.S. (1994) Nitric oxide in skeletal muscle. *Nature*, 372 (6506): 546–548.

Laubach, V.E., Shesely, E.G., Smithies, O. and Sherman, P.A. (1995) Mice lacking inducible nitric oxide synthase are not resistant to lipopolysaccaride-induced death. *Proc. Natl. Acad. Sci. USA*, 92: 10688–10692.

Lawrence, A.J. and Jarrott, B. (1993) Nitric oxide increases interstitial excitatory amino acid release in the rat dorsomedial medulla oblongata. *Neurosci. Lett.*, 151: 126–129.

Lev-Ram, V., Nebyelul, Z., Ellisman, M.H., Huang, P.L. and Tsien, R.Y. (1997) Absence of cerebellar long-term depression in mice lacking neuronal nitric oxide synthase. *Learning and Memory*, 3: 169–177.

Lipton, S.A., Choi, Y.B., Pan, Z.H., Lei, S.Z., Chen, H.S., Sucher, N.J., Loscalzo, J., Singel, D.J. and Stamler, J.S. (1993) A redox-based mechanism for the neuroprotective and neurodestructive effects of nitric oxide and related nitroso-compounds [see comments]. *Nature*, 364, (6438): 626–632.

Lipton, S.A., Singel, D.J. and Stamler, J.S. (1994) Neuroprotective and neurodestructive effects of nitric oxide and redox congeners. *Ann. N.Y. Acad. Sci.*, 738: 382–387.

Lo, E., Hara, H., Rogowska, J., Trocha, M., Pierce, A.R., Huang, P.L., Fishman, M.C., Wolf, G.L. and Moskowitz, M.A. (1996) Temporal correlation mapping analysis of the hemodynamic penumbra in mutant mice deficient in endothelial nitric oxide synthase gene expression. *Stroke*, 27: 1381–1385.

Ma, J., Ayata, C., Huang, P.L., Fishman, M.C. and Moskowitz, M.A. (1996) Regional cerebral blood flow response to vibrissal stimulation in mice lacking type I NOS gene expression. *Am. J. Physiol.*, 270: H1085–H1090.

MacMicking, J.D., Nathan, C., Hom, G., Chartrain, N., Fletcher, D.S., Trumbauer, M., Stevens, K., Xie, Q.W., Sokol, K., Hutchinson, N. et al. (1995) Altered responses to bacterial infection and endotoxic shock in mice lacking inducible nitric oxide synthase. *Cell*, 81 (4): 641–50.

MacMillan, V., Judge, D., Wiseman, A., Settle, D., Swain, J. and Davis, J. (1993) Mice expressing a bovine basic fibroblast growth factor transgene in the brain show increased resistance to hypoxemic-ischemic cerebral damage. *Stroke*, 24: 1735–1739.

Malinski, T., Bailey, F., Zhang, Z.G. and Chopp, M. (1993) Nitric oxide measured by a porphyrinic microsensor in rat brain after transient middle cerebral artery occlusion. *J. Cereb. Blood Flow Metab.*, 13 (3): 355–358.

Marletta, M.A. (1994) Nitric oxide synthase: Aspects concerning structure and catalysis. *Cell*, 78 (6): 927–930.

Martinou, J.C., Dubois-Dauphin, M., Staple, J.K., Rodriguez, I., Frankowski, H., Missotten, M., Albertini, P., Talabot, D., Catsicas, S., Pietra, C. and Huarte, J. (1994) Overexpression of BCL-2 in transgenic mice protects neurons from naturally occurring cell death and experimental ischemia. *Neuron*, 13: 1017–1030.

Meffert, M.K., Premarck, B. and Schulman, H. (1994) Nitric oxide stimulates Ca-independent synaptic vesicle release. *Neuron*, 12: 1235–1244.

Meng, W., Ma, J., Ayata, C., Hara, H., Huang, P.L., Fishman, M.C. and Moskowitz, M.A. (1996) Acetylcholine dilates pial arterioles in endothelial and neuronal nitric oxide synthase knockout mice by nitric oxide-dependent mechanisms. *Am. J. Physiol.*, 271: H1145–H1150.

Moncada, S., Palmer, R.M.J. and Higgs, E.A. (1991) Nitric oxide: physiology, pathophysiology, and pharmacology. *Pharmacol. Rev.*, 43: 109–142.

Montague, P.R., Gancayco, C.D., Winn, M.J., Marchase, R.B. and Freidlander, M.J. (1994) Role of NO production in NMDA receptor mediated neurotransmitter release in cerebral cortex. *Science*, 263: 973–977.

Murphy, S., Simmons, M.L., Agullo, L., Garcia, A., Feinstein, D.L., Galea, E., Reis, D.J., Minc-Golomb, D. and Schwartz, J.P. (1993) Synthesis of nitric oxide in CNS glial cells. *Trends Neurosci.*, 16 (8): 323–328.

Nathan, C. and Xie, Q.W. (1994) Nitric oxide synthases: Roles, tolls, and controls. *Cell*, 78 (6): 915–918.

Nicholls, D. and Attwell, D. (1990) The release and uptake of excitatory amino acids. *Trends Pharmacol Sci.*, 11: 462–478.

Nicholls, D.G. (1989) Release of glutamate, aspartate, and GABA from isolated nerve terminals. *J. Neurochem.*, 52: 331–337.

Northington, F.J., Tobin, J.R., Koehler, R.C. and Traystman, R.J. (1995) In vivo production of nitric oxide correlates with NMDA-induced cerebral hyperemia in newborn sheep. *Am. J. Physiol.*, 269 (1 Pt 2): H215–H221.

O'Dell, T.J., Hawkins, R.D., Kandel, E.R. and Arancio, O. (1991) Tests of the roles of two diffusable substances in long term potentiation: Evidence for nitric oxide as a possible early retrograde messenger. *Proc. Natl. Acad. Sci. USA*, 88: 11285–11289.

O'Dell, T.J., Huang, P.L., Dawson, T.M., Dinerman, J.L., Snyder, S.H., Kandel, E.R. and Fishman, M.C. (1994) Endothelial NOS and the blockade of LTP by NOS inhibitors in mice lacking neuronal NOS. *Science*, 265 (5171): 542–546.

Panahian, N., Yoshida, T., Huang, P.L., Hedley-Whyte, E.T., Fishman, M. and Moskowitz, M.A. (1996) Attenuated hippocampal damage after global cerebral ischemia in knock-out mice deficient in neuronal nitric oxide synthase. *Neuroscience*, 72: 343–354.

Pelligrino, D.A. (1993) Saying NO to cerebral ischemia [editorial]. *J. Neurosurg. Anesthesiol.*, 5 (4): 221–231.

Pogun, S.K., Baumann, M.H. and Kuhns, M.J. (1994) Nitric oxide inhibits 3H dopamine uptake. *Brain Res.*, 641: 83–91.

Prast, H. and Phillipu, A. (1992) Nitric oxide releases acetylcholine in the basal forebrain. *Eur. J. Pharmacol.*, 216: 139–140.

Radomski, M.W., Vallance, P., Whitley, G., Foxwell, N. and Moncada, S. (1993) Platelet adhesion to human vascular endothelium is modulated by constitutive and cytokine induced nitric oxide. *Cardiovasc. Res.*, 27, (7): 1380–1382.

Roy, C.W. and Sherrington, C.S. (1890). On the regulation of the blood supply of the brain. *J. Physiol.*, 11: 85–108.

Schmidt, H.H., Warner, T.D., Ishii, K., Sheng, H. and Murad, F. (1992) Insulin secretion from pancreatic B cells caused by L-arginine-derived nitrogen oxides. *Science*, 255: 721–723.

Schuman, E.C. and Madison, D.V. (1994) Locally distributed synaptic potentiation in the hippocampus. *Science*, 263: 532–536.

Schuman, E.M. and Madison, D.V. (1991) A requirement for the intercellular messenger nitric oxide in long-term potentiation. *Science*, 254: 1503–1506.

Son, H., Hawkins, R.D., Martin, K., Kiebler, M., Huang, P.L., Fishman, M.C. and Kandel, E.R. (1996) Long-term potentiation is reduced in mice that are doubly mutant in endothelial and neuronal nitric oxide synthase. *Cell*, 87: 1015–1023.

Stuehr, D. and Marletta, M. (1985) Mammalian nitrate biosynthesis: Mouse macrophages produce nitrite and nitrate in response to *Escherichia coli* lipopolysaccharide. *Proc. Natl. Acad. Sci. USA*, 82: 7738–7742.

Vincent, S. and Hope, B. (1992) Neurons that say NO. *TINS*, 15: 108–113.

Wei, X.Q., Charles, I.G., Smith, A., Ure, J., Feng, G.J., Huang, F.P., Xu, D., Muller, W., Moncada, S. and Liew, F.Y. (1995) Altered immune responses in mice lacking inducible nitric oxide synthase. *Nature*, 375 (6530): 408–411.

Wendland, B., Schweizer, F.E., Ryan, T.A., Nakane, M., Murad, F., Scheller, R.H. and Tsien, R.W. (1994) Existence of nitric oxide synthase in rat hippocampal pyramidal cells. *Proc. Natl. Acad. Sci. USA*, 91: 2151–2155.

Yang, G., Chan, P.H., Chen, J., Carlson, E., Chen, S.F., Weinstein, P., Epstein, C.J. and Kamii, H. (1994) Human copper-zinc superoxide dismutase transgenic mice are highly resistant to reperfusion injury after focal cerebral ischemia. *Stroke*, 25 (1): 165–170.

Zhang, F., Casey, R.M., Ross, M.E. and Iadecola, C. (1996) Aminoguanidine ameliorates and L-arginine worsens brain damage from intraluminal middle cerebral artery occlusion. *Stroke*, 27 (2): 317–323.

Zhang, F., Xu, S. and Iadecola, C. (1995a) Time dependence of effect of nitric oxide synthase inhibition on cerebral ischemic damage. *J. Cereb. Blood Flow Metab.*, 15 (4): 595–601.

Zhang, R.L., Chopp, M., Jiang, N., Tang, W.X., Prostak, J., Manning, A.M. and Anderson, D.C. (1995b) Anti-intercellular adhesion molecule-1 antibody reduces ischemic cell damage after transient but not permanent middle cerebral artery occlusion in the Wistar rat. *Stroke*, 26 (8): 1438–1442.

Zhang, R.L., Chopp, M., Zaloga, C., Zhang, Z.G., Jiang, N., Gautam, S.C., Tang, W.X., Tsang, W., Anderson, D.C. and Manning, A.M. (1995c) The temporal profiles of ICAM-1 protein and mRNA expression after transient MCA occlusion in the rat. *Brain Res.*, 682 (1-2): 182–188.

Zhang, Z.G., Chopp, M., Zaloga, C., Pollock, J.S. and Forstermann, U. (1993) Cerebral endothelial nitric oxide synthase expression after focal cerebral ischemia in rats. *Stroke*, 24 (12): 2016–2021.

Zhuo, M., Small, S.A., Kandel, E.R. and Hawkins, R.D. (1993) Nitric oxide and carbon monoxide produce activity-dependent long term synaptic enhancement in hippocampus. *Science*, 260: 1946–1950.

R.R. Mize, T.M. Dawson, V.L. Dawson and M.J. Friedlander (Eds.)
Progress in Brain Research, Vol 118

CHAPTER 3

Monitoring neuronal NO release in vivo in cerebellum, thalamus and hippocampus

Steven R. Vincent*, Julie A. Williams, Peter B. Reiner and Alaa El-Din El-Husseini

The Graduate Program in Neuroscience, Department of Psychiatry, The University of British Columbia, Vancouver, B.C. V5K 1Z3, Canada

Abstract

A variety of methods has been developed based on in vivo microdialysis which allow one to examine the NO/cGMP signal transduction system in action in behaving animals. The extracellular levels of cGMP, the NO oxidative products nitrate and nitrite, and NO itself can all be determined. Using these methods changes in NO and cGMP production in response to pharmacological manipulations can be examined in vivo. In addition, it has been discovered that the activity of this system varies with the behavioral state of the animal. NO and cGMP appear to act via distinct downstream effectors in different brain regions. This opens up the possibility of selectively manipulating NO and cGMP signaling in discrete neuronal populations.

Introduction

It has been known for more than 20 years that NO can stimulate guanylyl cyclase activity in the brain (Arnold et al., 1977; Miki et al., 1977). However, it was only with the discovery by Garthwaite et al. (1988) that excitatory amino acids could increase cGMP levels via the formation of a factor with properties similar to the endothelium-derived relaxing factor (EDRF) that interest in NO in the nervous system really began. At the same time, EDRF was shown to be NO, and its enzymatic synthesis from arginine was demonstrated (Palmer et al., 1988). Subsequent work identified the neurons throughout the brain which synthesize and release NO (Hope et al., 1991; Vincent and Kimura, 1992). It is now clear that NO is a unique form of cellular messenger. It is not stored or released in a quantal fashion. Instead, NO is produced by neuronal NO synthase in response to an increase in intracellular calcium, and then rapidly diffuses in all directions from its source. Furthermore, rather than acting on a cell surface receptor, NO passes readily through membranes to bind to and activate soluble guanylyl cyclase to increase cGMP levels in target cells. Just as NO can pass through cell membranes, it can readily enter a microdialysis probe in the brain, providing a mechanism for the in vivo assay of NO production.

Cyclic nucleotides are the classic intracellular second messengers. However, the efflux of cyclic nucleotides from stimulated cells has been noted many times. Davoren and Sutherland (1963) first reported the efflux of cAMP when erythrocytes were stimulated by noradrenaline. Efflux of cAMP

*Corresponding author. Tel.: +1 604 822 7038; fax: +1 604 822 7981; e-mail: svincent@unixg.ubc.ca.

from cerebellar slices was also detected (Kakiuchi and Rall, 1968). Likewise, the release of cGMP from a variety of tissues has been described, and efflux of cGMP from cerebellar slices was noted in response to an NO donor or stimulation with glutamate (Tjörnhammar et al., 1986). Based on these observations, the monitoring of cGMP using in vivo microdialysis has been introduced as an index of nitric oxide synthase activity in freely-moving animals (Vallebuona and Raiteri, 1993, Luo et al., 1994).

NO and cGMP signaling in the cerebellum

The cerebellar cortex has been an important system for the examination of NO action. Biochemical and histochemical studies have demonstrated that NO synthase is expressed in the granule cells and their processes, the parallel fibers, which provide a glutamate input to the Purkinje cells. In addition, the basket cell GABA interneurons in the molecular layer also contain NO synthase (Bredt et al., 1990; Vincent and Kimura, 1992). In situ hybridization indicates that soluble guanylyl cyclase is highly expressed in Purkinje cells, with moderate levels seen in stellate, basket and Golgi cells, and only low levels in granule cells (Furuyama et al., 1993; Matsuoka et al., 1992). This is consistent with results from immunohistochemical studies as well (Ariano et al., 1982; Nakane et al., 1983). Thus one would predict that NO produced in the parallel fibers and basket cells would act primarily within the Purkinje cells to generate cGMP.

Microdialysis studies indicate that in the cerebellar cortex there is a basal extracellular level of cGMP of about 6–9 nM (Vallebuona and Raiteri, 1993; Luo et al., 1994), corresponding to about 1–5% of the tissue concentration of cGMP in the adult rat cerebellar cortex (Lorden et al., 1985; De Vente and Steinbusch, 1992). This extracellular cGMP was largely dependent on tonic NO production since it could be decreased by NOS inhibitors. In contrast, local administration of NO donors produced a dose-dependent increase in extracellular cGMP (Vallebuona and Raiteri, 1993, Luo et al., 1994; Laitinen et al., 1994). Direct activation of NMDA, AMPA or metabotropic receptors in the cerebellar cortex also dramatically increased the extracellular levels of cGMP. These increases could be blocked by receptor antagonists or by NOS inhibitors. An increase in cGMP could also be produced by the activation of the climbing fibers following administration of harmaline (Luo et al., 1994). This increase could be attenuated by blocking either NMDA or AMPA receptors, indicating that activation of glutamate receptors by the release of endogenous glutamate can lead to NO production in the cerebellum.

What is the role of extracellular cGMP? Experiments using probenecid and bromosulfophthalein indicate that an energy-dependent anion carrier mediates the removal of cGMP from the extracellular space (Luo et al., 1994). This may represent uptake into glial cells, and could account for the strong staining of these cells for cGMP (De Vente and Steinbusch, 1992; Southern et al., 1992). Thus, the efflux of cGMP could simply be a mechanism for cells to rapidly regulate intracellular levels of the cyclic nucleotide. This could be of particular relevance in the cerebellum, where levels of cGMP phosphodiesterase are very low (Greenberg et al., 1978).

Another method that has been used as an index of NOS activity is to monitor the extracellular levels of the oxidative products of NO, nitrite and nitrate (NO_x^-; Luo et al., 1993; Ohta et al., 1994; Shintani et al., 1994; Yamada and Nabeshima, 1997). These are stable anions, and can be quantified by the azo dye method, using an automated procedure (Green et al., 1982). This method is based on the Griess reaction in which nitrite reacts with a primary aromatic amine to yield a diazonium salt (Griess, 1864), which can then react stoichiometrically with a coupling agent to generate a colored azo dye, which can be monitored spectrophotometrically. Nitrate must be reduced, i.e. with a cadmium reduction column, to nitrite before it can be detected. Using this method, we have found that basal extracellular levels of NO_x^- are about 3.5 μM; half in the form of nitrite and half as nitrate (Luo et al., 1993). This basal level can be reduced by NOS inhibitors or blockade of

NMDA receptors, and elevated by local application of NMDA (Luo et al., 1993; Shintani et al., 1994).

The results of these microdialysis experiments, together with other anatomical, biochemical, pharmacological and electrophysiological studies have led to the development of a model of where and when NO and cGMP are produced in the cerebellum (Vincent, 1996). NO synthase is found in the granule cells and basket cells of the cerebellar cortex (Bredt et al., 1990; Vincent and Kimura, 1992), while soluble guanylyl cyclase is present mainly in Purkinje cells (Ariano et al., 1982; Furuyama et al., 1993; Matsuoka et al., 1992; Nakane et al., 1983). Furthermore, the Purkinje cells are unique in the brain in expressing very high levels of type I cGMP-dependent protein kinase (Lohmann et al., 1981). Together, these observations indicate that NO is produced by the calcium-dependent activation of NO synthase in granule cells and basket cells, and that its principal target is the soluble guanylyl cyclase in Purkinje cells, activation of which leads to the activation of the type I cGMP-dependent protein kinase (Fig. 1). The key question then is, what are the downstream targets of this kinase?

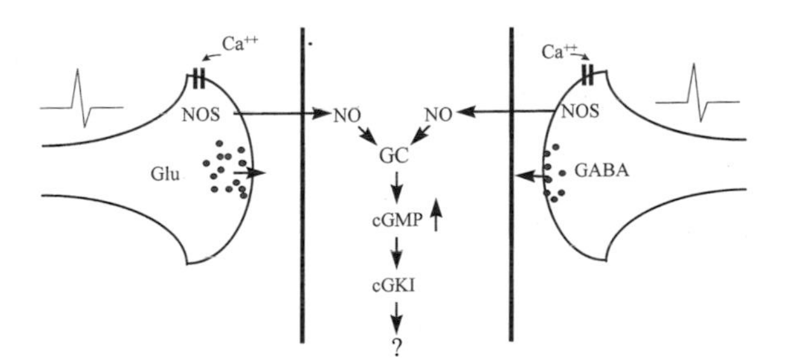

Fig. 1. Diagram of the NO/cGMP system in the cerebellar cortex. NO synthase is present in the glutamatergic parallel fibers which arise from the granule cells, and in the GABAergic basket cell interneurons in the molecular layer. NO produced in these compartments can act within Purkinje cell dendrites to stimulate the formation of cGMP. This in turn activates the type I cGMP-dependent protein kinase, which is concentrated in the Purkinje cells. The downstream targets of this kinase are largely unknown.

NO and cGMP signaling in the thalamus

The thalamus contains very few NOS positive neurons. These are primarily found in midline nuclei and the lateral geniculate nucleus (Bertini and Bentigovlio, 1997; Vincent and Kimura, 1992). However, thalamus does receive input from a number of regions known to contain NOS positive neurons, including the hypothalamus, tectum, dorsal raphe and the laterodorsal and pedunculopontine tegmental nuclei. These latter brainstem cell groups express among the highest levels of NO synthase found in any neuronal population, and appear to provide the majority of NO synthase inputs to the thalamus (Vincent and Kimura, 1992; Vincent et al., 1983, 1986).

The neurons of the laterodorsal and pedunculopontine tegmental nuclei are well known to be cholinergic, and to play an important role in regulating thalamic activity across behavioral states. Initially Moruzzi and Magoun (1949) demonstrated that stimulation in this region elicited a desynchronized EEG. Shute and Lewis (1967) found that these cells stained for acetylcholinesterase, and suggested that this was a cholinergic cell group which formed part of an ascending reticular activating system. Subsequent studies have confirmed that these are indeed cholinergic cells, and that they project heavily to the thalamus (Hallanger and Wainer, 1988; Satoh and Fibiger, 1986; Sofroniew et al., 1985). Electrophysiological studies have shown that cells in these regions fire action potentials during wake and REM sleep (El Mansari et al., 1989; Kayama et al., 1992; Steriade et al., 1990). It is now apparent that acetylcholine release from these neurons in the thalamus is high during periods of wake and REM sleep and is significantly lower during slow-wave sleep (Williams et al., 1994).

We used the hemoglobin trapping technique, combined with in vivo microdialysis to examine NO production in the thalamus. NO is able to stochiometrically oxidize the ferrous form of hemoglobin (oxyhemoglobin, HbO_2) to the ferric form, metHb, yielding nitrate. Indeed, this is probably the source of the nitrate seen in the microdialysis studies mentioned above. There is a linear relationship between NO concentration and the conversion of HbO_2 to MetHb (Murphy and Noack, 1994; Noack et al., 1992). The oxidation of HbO_2 can be assayed using absorbance spectrophotometry, by following the color change from red to brown associated with oxidation. We monitored the absorbance at 577 nm, an absorbance maximum for HbO_2, and at 591, an isobestic point at which the absorbances of HbO_2 and MetHb are equal, thereby providing an internal control. Thus, a decrease in absorbance at 577 nm relative to 591 nm should be proportional to the amount of NO present. Details regarding the preparation of HbO_2, calibration of the assay using NO standards, and the preparation of the microdialysis probes are presented elsewhere (Williams et al., 1997b).

The extracellular levels of NO in the thalamus were reduced by systemic treatment with a NO synthase inhibitor, or by removing extracellular calcium by including BAPTA in the microdialysate (Williams et al., 1997a). In contrast, thalamic NO levels could be elevated by local depolarization with potassium, or by electrical stimulation of the NO synthase containing afferents from the laterodorsal tegmental nucleus (Miyazaki et al., 1996; Williams et al., 1997a). When thalamic NO levels were examined in freely moving animals, they were found to vary significantly with behavioral state. NO levels were high during wake and REM sleep, but decreased during slow-wave sleep (Williams et al., 1997a).

What does NO do in the thalamus? NO production is enhanced in the brainstem afferents to the thalamus during wake and REM sleep, periods when these mesopontine tegmental neurons are firing action potentials. At these times, one would also predict an increase in the release of atriopeptides which are also highly expressed in these cells (Saper et al. 1989; Standaert et al., 1986). These peptides can act on particulate guanylyl cyclases to increase cGMP. NO-releasing compounds and cGMP agonists depolarize thalamic relay cells by shifting the activation of the hyperpolarization-activated cation current (I_h) to more positive potentials (Pape and Mager, 1992). This may allow NO to mediate the switch in firing from bursting to a tonic mode which occurs in thalamic neurons during wake and REM sleep (Steriade et al., 1990). This suggests that NO would play an excitatory role in arousal. Indeed, there is pharmacological evidence that this is so (Bagetta et al., 1993; Dzolijic and De Vries, 1994; Dzolijic et al., 1996; Kapas et al., 1994; Kapas and Kruger, 1996; Nistico et al., 1994). Furthermore, NO has been shown to facilitate the responses of thalamic neurons to excitatory amino acids (Shaw and Salt, 1997), and to visual (Cudeiro et al., 1996) and tactile (Do et al., 1994) stimuli. The decrease in NO release during slow-wave sleep may also contribute to the dramatic decrease in regional cerebral blood flow which occurs in the pontomesencephalic tegmentum and thalamus during this state (Koyama et al., 1994; Hofle et al., 1997). Soluble guanylyl cyclase mRNA has been detected in thalamic neurons (Matsuoka et al., 1992; Smigrodzki and Levitt, 1996) and the actions of NO and cGMP in thalamic neurons would appear to be mediated by the type II cGMP-dependent protein kinase, which is very highly expressed in these cells (El-Husseini et al., 1995) (Fig. 2). The substrates on which this kinase acts remain to be determined (Wang and Robinson, 1997).

NO and cGMP signaling in the hippocampus

Within the hippocampus and dentate gyrus, NO synthase is present in interneurons scattered in stratum oriens, the pyramidal cell layer and stratum radiatum (Bredt et al., 1990; Vincent and Kimura, 1992). In addition, the hippocampal formation receives a major input from NO synthase-positive neurons in the medial septum and the nucleus of the diagonal band (Brauer et al., 1991; Kinjo et al., 1989; Kitchener and Diamond, 1993; Pasqualotto and Vincent, 1991; Schöber et al., 1989). In situ hybridization indicates the presence of soluble guanylyl cyclase in the pyramidal and granule cells (Matsuoka et al., 1992)

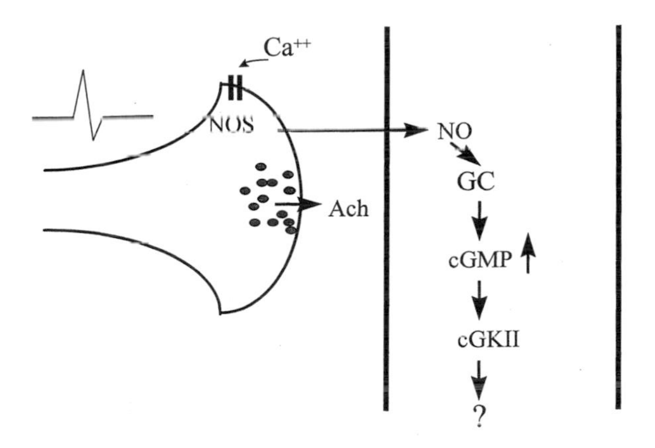

Fig. 2. Diagram of the NO/cGMP system in the thalamus. NO synthase is present in the afferents arising in the pedunculopontine and laterodorsal tegmental nucleus, and its activity is increased during periods when these cells fire action potentials, i.e. during wake and REM sleep. At this time, NO can act within the thalamic neurons to activate the type II cGMP-dependent protein kinase, which is concentrated there. The targets of this kinase are unknown.

Microdialysis for extracellular cGMP has been applied to the hippocampus and other cortical areas as an index of NO production (Laitinen et al., 1994; Vallebuona and Raiteri, 1994). The results were similar to those seen in the cerebellum, with NO donors elevating and NO synthase inhibitors decreasing extracellular cGMP, although the basal extracellular levels of cGMP were about 18-fold lower in the hippocampus than in the cerebellum (Vallebuona and Raiteri, 1993; 1994). This reflects the differences in tissue levels of this cyclic nucleotide between these regions, and is likely due to differences in phosphodiesterase activity (Greenberg et al., 1978). Local application of NMDA in the hippocampus also produced a NO synthase-dependent increase in extracellular cGMP (Vallebuona and Raiteri, 1994). The cGMP production in response to NMDA decreases dramatically with aging, and is associated with a decrease in hippocampal NO synthase activity (Vallebuona and Raiteri, 1995).

As in the cerebellum, the changes in extracellular cGMP in the hippocampus are paralleled by changes in NO oxidation products (NO_x^-; Luo and Vincent, 1994). In contrast to cGMP, the extracellular levels of nitrite and nitrate are similar in the hippocampus and cerebellum. Local administration of NMDA produces a NO synthase-dependent increase in NO_x^-, which could be prevented by blockade of the NMDA receptor.

The hemoglobin-trapping assay has also been used to monitor changes in NO production in the hippocampus. The NO concentration in the hippocampus was increased dramatically by systemic treatment with kainate, and this could be prevented by pretreatment with a NO synthase inhibitor (Balcioglu and Maher, 1993; Balcioglu et al., 1998). We found that NO levels in the hippocampus varied with behavioral state, high during periods of REM sleep and wake, and low during slow-wave sleep (Williams et al., 1997a,b). The hippocampal EEG is dominated by theta activity during REM sleep and in the waking state during exploration and information gathering behaviors (Vanderwolf, 1969). Theta activity is also well known to occur during urethane anesthesia (Kramis et al., 1975), and hippocampal NO levels are high in conscious animals and in animals anesthetized with urethane (Balcioglu et al., 1998). There is evidence that NO and cGMP may themselves modulate hippocampal theta activity (Bawin et al., 1994). The increased NO production in the hippocampus during periods of theta activity may be associated with the heightened state of synaptic plasticity associated with these oscillations (Huerta and Lisman, 1993). Indeed, there is evidence that hippocampal theta activity promotes both the induction and reversal of LTP (Larson et al., 1986; 1993).

It is clear from numerous immunohistochemical studies that the type 1 cGMP-dependent protein kinase is predominantly expressed in the Purkinje cells of the cerebellar cortex (Lohmann et al., 1981). Hippocampal cells immunoreactive for this kinase have not been described. The type 2 cGMP-dependent protein kinase is widely expressed in the brain, being particularly highly expressed in the thalamus (El-Husseini et al., 1995). However, the hippocampus shows very little, if any expression of this kinase. These observations suggest that cGMP may not act via activation of a protein kinase in hippocampal neurons. Indeed, although some studies have suggested a role for cGMP-dependent protein kinase in long-term potentiation (Zhuo et al., 1994), others have found that cGMP-dependent protein kinase does not affect synaptic transmission or long-term potentiation in the hippocampus (Shuman et al., 1994).

A large and diverse family of phosphodiesterases has been identified, which catalyze the hydrolysis of cAMP and cGMP to their corresponding 5′-nucleotide monophosphates. The different isoforms are differentially regulated, derive from distinct genes, and have distinct pharmacologies. Phosphodiesterases stimulated by cGMP have been purified from heart, adrenal gland, liver, platelets and brain. A cGMP-stimulated phosphodiesterase was initially cloned from a bovine adrenal cortex library (Sonnenburg et al., 1991). Northern blot analysis indicated that this phosphodiesterase is highly expressed in hippocampus, cortex and basal ganglia, but is undetectable in cerebellum (Sonnenburg et al., 1991). The corresponding rat gene was subsequently obtained and shown to have a similar distribution in the rat brain (Repaske et al., 1993). The brain isoform appears to differ from that in other organs in being generated via alternative mRNA splicing, such that a hydrophobic N-terminal domain is present which may target the neuronal form to the membrane (Yang et al., 1994). In situ hybridization studies examining the expression of this phosphodiesterase showed strong labeling in both the granule cells of the dentate gyrus, and in the hippocampal pyramidal cells (Repaske et al., 1993). Indeed, the distribution appears complementary to that seen for the cGMP-dependent protein kinases.

The cGMP-stimulated phosphodiesterase allows relatively low intracellular concentrations of cGMP (1–5 μM) to stimulate cAMP hydrolysis. Thus the presence of abundant cGMP-stimulated phosphodiesterase, and the relative absence of cGMP-dependent protein kinases in the hippocampal pyramidal neurons suggests that perhaps for these neurons, the antagonistic, or "yin–yang" hypothesis of cyclic nucleotide action should be reconsidered (Goldberg et al., 1975; Stone et al., 1975) (Fig. 3). Indeed, increases in intracellular cGMP are thought to depress a high voltage-activated calcium current in hippocampal neurons through this mechanism (Doerner and Alger, 1988), as previously demonstrated in frog cardiac cells (Hartzell and Fischmeister, 1986) and *Aplysia* neurons (Levitan and Levitan, 1988). The recent development of an inhibitor of the cGMP-stimulated phosphodiesterase (Mery et al., 1995; Podzuweit et al., 1995) should allow further analysis of the interactions between cyclic nucleotides in the regulation of hippocampal function.

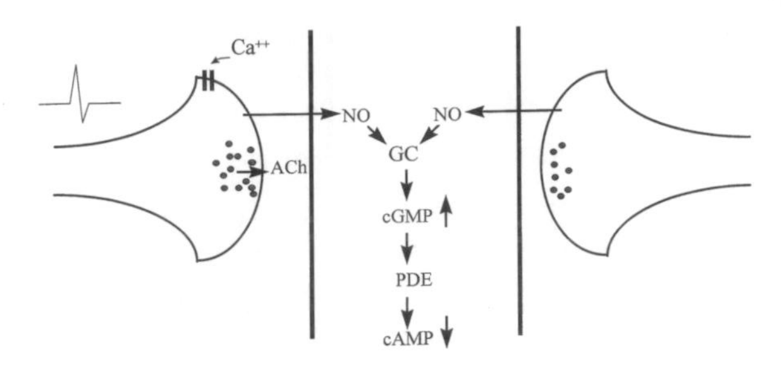

Fig. 3. Diagram of the NO/cGMP system in the hippocampus. NO synthase is present in hippocampal interneurons and in the septal cholinergic neurons which innervate the hippocampus. NO released from the septohippocampal system during periods of theta activity such as REM sleep, and exploratory behavior, would be able to act within the hippocampal pyramidal cells to increase cGMP. This would lead to activation of the cGMP-stimulated phosphodiesterase, which is concentrated in these cells, thereby decreasing intracellular cAMP.

Acknowledgements

Supported by grants from the Medical Research Council of Canada. PBR is an MRC Scientist and SRV an MRC Senior Scientist.

References

Ariano, M.A., Lewicki, J.A., Brandwein, H.J. and Murad, F. (1982) Immunohistochemical localization of guanlyate cyclase within neurons of rat brain. *Proc. Natl. Acad. Sci. USA*, 79: 1316–1320.

Arnold, W.P., Mittal, C.K., Katsuki, S. and Murad, F. (1977) Nitric oxide activates guanlyate cyclase and increases guanosine 3′-5′-cyclic monophosphate levels in various tissue preparations. *Proc. Natl. Acad. Sci. USA*, 74: 3203–3207.

Bagetta, G., Iannonc, M., Del Duca, C. and Nistico, G. (1993) Inhibition by N^{ω}-nitro-L-arginine methyl ester of the electrocortical arousal response in rats. *Br. J. Pharmacol.*, 108: 858–860.

Balcioglu, A. and Maher, T.J. (1993) Determination of kainic acid-induced release of nitric oxide using a novel hemoglobin trapping technique with microdialysis. *J. Neurochem.*, 61: 2311–2313.

Balcioglu, A., Watkins, C.J. and Maher, T.J. (1998) Use of a hemoglobin-trapping approach in the determination of nitric oxide in in vitro and in vivo systems. *Neurochem. Res.* (in press).

Bawin, S.M., Satmary, W.M. and Adey, W.R. (1994) Nitric oxide modulates rhythmic slow activity in rat hippocampal slices. *NeuroReport*, 5: 1869–1872.

Bertini, G. and Bentivoglio, M. (1997) Nitric oxide synthase in the adult and developing thalamus – histochemical and immunohistochemical study in the rat. *J. Comp. Neurol.*, 388: 89–105.

Brauer, K., Schöber, A., Wolff, J.R., Winkelmann, E., Luppa, H., Lüth, J.-J. and Böttcher, H. (1991) Morphology of neurons in the rat basal forebrain nuclei: Comparison between NADPH-diaphorase histochemistry and immunohistochemistry of glutamic acid decarboxylase, choline acetyltransferase, somatostatin and parvalbumin. *J. Hirnforsch.*, 32: 1–18.

Bredt, D.S., Hwang, P.M. and Snyder, S.H. (1990) Localization of nitric oxide synthase indicating a neural role for nitric oxide. *Nature*, 347: 768–770.

Cudeiro, J., Rivadulla, C., Rodriguez, R., Martinez-Conde, S., Grieve, K.L. and Acuña, C. (1996) Further observations on the role of nitric oxide in the feline lateral geniculate nucleus. *Eur. J. Neurosci.*, 8: 144–152.

Davoren, P.R. and Sutherland, E.W. (1963) The effect of *L*-epinephrine and other agents on the synthesis and release of adenosine 3′ -5′-phosphate by whole pigeon erythrocytes. *J. Biol. Chem.*, 238: 3009–3015.

De Vente, J. and Steinbusch, H.W.M. (1992) On the stimulation of soluble and particulate guanylate cyclase in the rat brain and the involvement of nitric oxide as studied by cGMP immunocytochemistry. *Acta Histochem.*, 92: 13–38.

Do, K.-Q., Binns, K.E. and Salt, T.E. (1994) Release of the nitric oxide precursor, arginine, from the thalamus upon sensory afferent stimulation and its effect on thalamic neurons in vivo. *Neurosci.*, 60: 581–586.

Doerner, D. and Alger, B.E. (1988) Cyclic GMP depresses hippocampal calcium current through a mechanism independent of cGMP-dependent protein kinase. *Neuron*, 1: 693–699.

Dzolijic, M.R. and De Vries, R. (1994) Nitric oxide synthase inhibition reduces wakefulness. *Neuropharmacol.*, 33: 1505–1509.

Dzoljic, M.R., De Vries, R. and van Leeuwen, R. (1996) Sleep and nitric oxide: Effects of 7-nitro indazole, inhibitor of brain nitric oxide synthase. *Brain Res.*, 718: 145–150.

El-Husseini, A.E.-D., Bladen, C. and Vincent, S.R. (1995) Molecular characterization of a type II cGMP-dependent protein kinase expressed in the rat brain. *J. Neurochem.*, 64: 2814–2817.

El Mansari, M., Sakai, K. and Jouvet, M. (1989) Unitary characteristics of presumptive cholinergic tegmental neurons during the sleep-waking cycle in freely moving cats. *Exp. Brain Res.*, 76: 519–529.

Furuyama, T., Inagaki, S. and Takagi, H. (1993) Localizations of α_1 and β_1 subunits of soluble guanylate cyclase in the rat brain. *Molec. Brain Res.*, 20: 335–344.

Garthwaite, J., Charles, S.L. and Chess-Williams, R. (1988) Endothelium-derived relaxing factor release on activation of NMDA receptors suggests a role as intercellular messenger in the brain. *Nature*, 336: 385–388.

Goldberg, N.D., Haddox, M.K., Nicol, S.E., Glass, D.B., Sanford, C.H., Kuehl, F.A. Jr. and Estensen, R. (1975) Biologic regulation through opposing influences of cyclic GMP and cyclic AMP: The Yin Yang hypothesis. In: P. Greengard and G.A. Robison (Eds.), *Advances in Cyclic Nucleotide Research*, Vol. 5, Raven, New York, pp. 307–330.

Green, L.C., Wangner, D.A., Glogowksi, J., Skipper, P.L., Wishniok, J.S. and Tannenbaum, S.R. (1982) Analysis of nitrate, nitrite, and [^{15}N]nitrate in biological fluids. *Analyt. Biochem.*, 126: 131–138.

Greenberg, L.H., Troyer, E., Ferrendelli, J.A. and Weiss, B. (1978) Enzymatic regulation of the concentration of cyclic GMP in mouse brain. *Neuropharmacol.*, 17: 737–745.

Griess, J.P. (1864) On a new series of bodies in which nitrogen is substituted for hydrogen. *Phil. Trans. R. Soc. Lond.*, 154: 667–731.

Hallanger, A. and Wainer, B.H. (1988) Ascending projections from the pedunculopontine tegmental nucleus and the adjacent mesopontine tegmentum in the rat. *J. Comp. Neurol.*, 274: 483–515.

Hartzell, H.C. and Fischmeister, R. (1986) Opposite effects of cyclic GMP and cyclic AMP on calcium current in single heart cells. *Nature*, 3223: 273–275.

Hofle, N., Paus, T., Reutens, D., Fiset, P., Gotman, J., Evans, A.C. and Jones, B.E. (1997) Regional cerebral blood flow changes as a function of delta and spindle activity during slow wave sleep in humans. *J. Neurosci.*, 17: 4800–4808.

Hope, B.T., Michaels, G.J., Knigge, K.M. and Vincent, S.R. (1991) Neuronal NADPH diaphorase is a nitric oxide synthase. *Proc. Natl. Acad. Sci. USA*, 88: 2811–2814.

Huerta, P.T. and Lisman, J.E. (1993) Heightened synaptic plasticity of hippocampal CA1 neurons during a cholinergically induced rhythmic state. *Naturė*, 364: 723–725.

Kakiuchi, S. and Rall, T.W. (1968) The influence of chemical agents on the accumulation of adenosine 3′,5′-phosphate in slices of rabbit cerebellum. *Mol. Pharmacol.*, 4: 367–378.

Kapas, L., Fang, J. and Krueger, J.M. (1994) Inhibition of nitric oxide synthesis inhibits rat sleep. *Brain Res.*, 664: 189–196.

Kapas, L. and Kruger, J.M. (1996) Nitric oxide donors Sin-1 and Snap promote non-rapid-eye-movement sleep in rats. *Brain Res. Bull.*, 41: 293–298.

Kayama, Y., Ohta, M. and Jodo, E. (1992) Firing of "possibly" cholinergic neurons in the rat laterodorsal tegmental nucleus during sleep and wakefulness. *Brain Res.*, 569: 210–220.

Kinjo, N., Sakinner, R.D. and Powell, E.W. (1989) A study of NADPH-diaphorase positive septohippocampal neurons in rat. *Neurosci. Res.*, 7: 154–158.

Kitchener, P.D. and Diamond, J. (1993) Distribution and colocalization of choline acetyltransferase immunoreactivity and NADPH diaphorase reactivity in neurons within the medial septum and diagonal band of Broca in the rat basal forebrain. *J. Comp. Neurol.*, 335: 1–15.

Koyama, Y., Toga, T., Kayama, Y. and Sato, A. (1994) Regulation of regional blood flow in the laterodorsal thalamus by ascending cholinergic nerve fibers from the laterodorsal tegmental nucleus. *Neurosci. Res.*, 20: 79–84.

Kramis, R., Vanderwolf, C.H. and Bland, B.H. (1975) Two types of hippocampal rhythmical slow activity in both the rabbit and the rat: Relations to behavior and effects of atropine, diethyl ether, urethane, and petnobarbital. *Exp. Neurol.*, 49: 58–85.

Laitinen, J.T., Laitinen, K.S.M., Tuomisto, L. and Airaksinen, M.M. (1994) Differential regulation of cyclic GMP levels in the frontal cortex and the cerebellum of anesthetized rats by nitic oxide – An in vivo microdialysis study. *Brain Res.*, 668: 117–121.

Larson, J., Wong, D. and Lynch, G. (1986) Patterned stimulation at the theta frequency is optimal for the induction of hippocampal long-term potentiation. *Brain Res.*, 368: 347–350.

Larson, J., Xiao, P. and Lynch, G. (1993) Reversal of LTP by theta frequency stimulation. *Brain Res.*, 600: 97–102.

Levitan, E.S. and Levitan I.B. (1988) A cyclic GMP analog decreases the currents underlying bursting activity in the Aplysia neuron R15. *J. Neurosci.*, 8: 1162–1171.

Lohmann, S.M., Walter, U., Miller, P.E., Greengard, P. and De Camilli, P. (1981) Immunohistochemical localization of cyclic GMP-dependent protein kinase in mammalian brain. *Proc. Natl. Acad. Sci. USA*, 78: 653–657.

Lorden, J.F., Oltmans, G.A., McKwon, T.W., Lutes, J., Beales, M. (1985) Decreased cerebellar 3′5′-cyclic guanosine monophosphate levels and insensitivity to harmaline in the genetically dystonic rat (dt). *J. Neurosci.*, 5: 2628–2625.

Luo, D., Knezevich, S. and Vincent, S.R. (1993) *N*-methyl-D-aspartate-induced nitric oxide release: An in vivo microdialysis study. *Neurosci.*, 57: 897–900.

Luo, D., Leung, E. and Vincent, S.R. (1994) Nitric oxide-dependent efflux of cGMP in rat cerebellar cortex: An in vivo microdialysis study. *J. Neurosci.*, 14: 263–271.

Luo, D. and Vincent, S.R. (1994) NMDA-dependent nitric oxide release in the hippocampus in vivo: Interactions with noradrenaline. *Neuropharmacol.*, 33: 1345–1350.

Matsuoka, I., Giuili, G., Poyard, M., Stengel, D., Parma, J., Guellaen, G. and Hanoune, J. (1992) Localization of adenylyl and guanlylyl cyclase in rat brain by in situ hybridization: Comparison with calmodulin mRNA distribution. *J. Neurosci.*, 12: 3350–3360.

Mery, P.F., Pavoine, C., Pecker, F. and Fischmeister, R. (1995) Erythro-9-(2-hydroxy-3-nonyl)adenine inhibits cyclic GMP-stimulated phosphodiesterase in isolated cardiac myocytes. *Mol. Pharmacol.*, 48: 121–130.

Miki, N., Kawabe, Y. and Kuriyama, K. (1977) Activation of cerebral gunalylate cyclase by nitric oxide. *Biochem. Biophys. Res. Commun.*, 75: 851–855.

Miyazaki, M., Kayama, Y., Kihara, T., Kawasaki, K., Yamaguchi, E., Wada, Y. and Ikeda, M. (1996) Possible release of nitric oxide from cholinergic axons in the thalamus by stimulation of the rat laterodorsal tegmental nucleus as measured with voltametry. *J. Chem. Neuroanat.*, 10: 203–207.

Moruzzi, G. and Magoun, H.W. (1949) Brainstem reticular formation and activation of the EEG. *Electroencephalogr. Clin. Neurophysiol.*, 1: 455–473.

Murphy, M.E. and Noack, E. (1994) Nitric oxide assay using hemoglobin method. *Meth. Enzymol.*, 233: 240–250.

Nakane, M., Ichikawa, M. and Deguchi, T. (1983) Light and electron microscopic demonstration of guanylate cyclase in rat brain. *Brain Res.*, 273: 9–15.

Nistico, G., Bagetta, G., Iannone, M. and Del Duca, C. (1994) Evidence that nitric oxide is involved in the control of electrocortical arousal. *Ann. N.Y. Acad. Sci.*, 738: 191–200.

Noack, E., Kubitzek, D. and Kojda, G. (1992) Spectrophotometric determination of nitric oxide using hemoglobin. *NeuroProtocols*, 1: 133–139.

Ohta, K., Araki, N., Shibata, M., Hamada, J., Komatsumoto, S., Shimazu, K. and Fukuchi, Y. (1994) A novel in vivo assay system for consecutive measurement of brain nitric oxide production combined with the microdialysis technique. *Neurosci. Lett.*, 176: 165–168.

Palmer, R.M.J., Ashton, D.S. and Moncada, S. (1988) Vascular endothelial cells synthesize nitric oxide from L-arginine. *Nature*, 333: 664–666.

Pape, H.C. and Mager, R. (1992) Nitric oxide controls oscillatory activity in thalamocortical neurons. *Neuron*, 9: 441–448.

Pasqualotto, B.A. and Vincent, S.R. (1991) Galanin and NADPH-diaphorase coexistence in cholinergic neurons of the rat basal forebrain. *Brain Res.*, 551: 78–86.

Podzuweit, T., Nennstiel, P. and Muller, A. (1995) Isozyme selective inhibition of cGMP-stimulated cyclic nucleotide phosphodiesterases by erythro-9-(2-hydroxy-3-nonyl) adenine. *Cell. Signal.*, 7: 733–738.

Repaske, D.R., Corbin, J.G., Conti, M. and Goy, M.F. (1993) A cyclic GMP-stimulated cyclic nucleotide phosphodiesterase gene is highly expressed in the limbic system of the rat brain. *Neurosci.*, 56: 673–686.

Saper, C.B., Hurley, K.M., Moga, M.M., Holmes, H.R., Adams, S.A., Leahy, K.M. and Needleman, P. (1989) Brain natriuretic peptides: Differential localization of a new family of neuropeptides. *Neurosci. Lett.*, 96: 29–34.

Satoh, K. and Fibiger, H.C. (1986) Cholinergic neurons of the laterodorsal tegmental nucleus: Efferent and afferent connections. *J. Comp. Neurol.*, 253: 277–302.

Schöber, A., Brauer, K. and Luppa, H. (1989) Alternate coexistence of NADPH-diaphorase with choline acetyltransferase or somatostatin in the rat neostriatum and basal forebrain. *Acta Histochem. Cytochem.*, 22: 669–674.

Schuman, E.M., Mefert, M.K., Shulman, H. and Madison, D.V. (1994) An ADP-ribosyltransferase as a potential target for nitric oxide action in hippocampal long-term potentiation. *Proc. Natl. Acad. Sci. USA*, 91: 11958–11962.

Shaw, P.J. and Salt, T.E. (1997) Modulation of sensory and excitatory amino acid responses by nitric oxide donors and glutathione in the ventrobasal thalamus of the rat. *Eur. J. Neurosci.*, 9: 1507–1513.

Shintani, F., Kanba, S., Nakaki, T., Sato, K., Yagi, G., Kato, R. and Asai, M. (1994) Measurement by in vivo brain microdialysis of nitric oxide release in the rat cerebellum. *J. Psychiatr. Neurosci.*, 19: 217–221.

Shute, C.C.C. and Lewis, P.R. (1967) The ascending cholinergic reticular system: Neocortical, olfactory, and subcortical projections. *Brain*, 90: 497–520.

Smigrodzki, R. and Levitt, P. (1996) The α1 subunit of soluble guanylyl cyclase is expressed prenatally in the rat brain. *Develop. Brain Res.*, 97: 226–234.

Sofroniew, M.V., Priestly, J.V., Consaloazione, A., Eckenstein, F. and Cuello, A.C. (1985) Cholinergic projections from the midbrain and pons to the thalamus in the rat, identified by combined retrograde tracing and choline acetyltransferase immunohistochemistry. *Brain Res.*, 329: 213–223.

Sonnenburg, W.K., Mullaney, P.J. and Beavo, J.A. (1991) Molecular cloning of a cyclic GMP-stimulated cyclic nucleotide phosphodiesterase cDNA. *J. Biol. Chem.*, 266: 17655–17661.

Southern, E., Morris, R. and Garthwaite, J. (1992) Sources and targets of nitric oxide in the cerebellum. *Neurosci. Lett.*, 137: 241–244.

Standaert, D.G., Needleman, P and Saper, C.B. (1986) Organization of atriopeptin-like immunoreactive neurons in the central nervous system of the rat. *J. Comp. Neurol.*, 253: 315–341.

Steriade, M., Datta, S., Pare, D., Oakson, G. and Curro Dossi, R. (1990) Neuronal activites in brainstem cholinergic nuclei related to tonic activation processes in thalamocortial systems. *J. Neurosci.*, 10: 2541–2559.

Stone, T.W., Taylor, D.A. and Bloom, F.E. (1975) Cyclic AMP and cyclic GMP may mediate opposite neuronal responses in the rat cerebral cortex. *Science*, 187: 845–847.

Tjörnhammar, M.-L., Lazardis, G. and Bartfai, T. (1986) Efflux of cyclic guanosine 3′,5′-monophosphate from cerebellar slices stimulated by L-glutamate or high K$^+$ or *N*-methyl-*N*′-nitro-*N*-nitrosoguanidine. *Neurosci. Lett.*, 68: 95–99.

Vallebuona, F. and Raiteri, M. (1993) Monitoring of cyclic GMP during cerebellar microdialysis in freely-moving rats as an index of nitric oxide synthase activity. *Neurosci.*, 57: 577–585.

Vallebuona, F. and Raiteri, M. (1994) Extracellular cGMP in the hippocampus of freely moving rats as an index of nitric oxide (NO) synthase activity. *J. Neurosci.*, 14: 134–138.

Vallebuona, F. and Raiteri, M. (1995) Age-related changes in the NMDA receptor/nitric oxide/cGMP pathway in the hippocampus and cerebellum of freely moving rats subjected to transcerebral microdialysis. *Eur. J. Neurosci.*, 7: 694–701.

Vanderwolf, C.H. (1969) Hippocampal electrical activity and voluntary movement in the rat. *Electroencephalogr. Clin. Neurophysiol.*, 407–418.

Vincent, S.R. (1996) Nitric oxide and synaptic plasticity: NO news from the cerebellum. *Behav. Brain Sci.*, 19: 362–367.

Vincent, S.R. and Kimura, H. (1992) Histochemical mapping of nitric oxide synthase in the rat brain. *Neurosci.*, 46: 755–784.

Vincent, S.R., Satoh, K., Armstrong, D.M. and Fibiger, H.C. (1983) NADPH-diaphorase: A selective histochemical marker for the cholinergic neurons of the pontine reticular formation. *Neurosci. Lett.*, 43: 31–36.

Vincent, S.R., Satoh, K., Armstrong, D.M., Panula, P., Vale, W. and Fibiger, H.C. (1986) Neuropeptides and NADPH-diaphorase activity in the ascending cholinergic reticular system of the rat. *Neurosci.*, 17: 167–182.

Wang, X. and Robinson, P.J. (1997) Cyclic GMP-dependent protein kinase and cellular signaling in the nervous system. *J. Neurochem.*, 68: 443–456.

Williams, J.A., Comisarow, J., Day, J., Fibiger, H.C. and Reiner, P.B. (1994) State-dependent acetylcholine release in rat thalamus as measured by in vivo microdialysis. *J. Neurosci.*, 14: 5236–5242.

Williams, J.A., Vincent, S.R. and Reiner, P.B. (1997a) Nitric oxide production in rat thalamus changes with behavioral state, local depolarization, and brainstem stimulation. *J. Neurosci.*, 17: 420–427.

Williams, J.A., Vincent, S.R. and Reiner, P.B. (1997b) Measurement of nitric oxide in brain using the hemoglobin trapping technique coupled with in vivo microdialysis. In: R. Lydic (Ed.), *Molecular Regulation of Arousal States*, CRC Press, Boca Raton, pp. 201–212.

Yamada, K. and Nabeshima, T. (1997) Simultaneous measurement of nitrite and nitrate levels as indices of nitric oxide release in the cerebellum of conscious rats. *J. Neurochem.*, 68: 1234–1243.

Yang, Q., Paskind, M., Bolger, G., Thompson, W.J., Repaske, D.R., Cutler, L.S. and Epstein, P.M. (1994) A novel cyclic GMP stimulated phosphodiesterase from rat brain. *Biochem. Biophys. Res. Commun.*, 205: 1850–1858.

Zhuo, M., Hu, Y., Schultz, C., Kandel, E.R. and Hawkins, R.D. (1994) Role of guanylyl cyclase and cGMP-dependent protein kinase in long-term potentiation. *Nature*, 368: 635–639.

R.R. Mize, T.M. Dawson, V.L. Dawson and M.J. Friedlander (Eds.)
Progress in Brain Research, Vol 118

CHAPTER 4

A web-accessible digital atlas of the distribution of nitric oxide synthase in the mouse brain[1]

R.J. Cork*, M.L. Perrone, D. Bridges, J. Wandell, C.A. Scheiner and R.R. Mize

Department of Cell Biology and Anatomy, and the Neuroscience Center, Louisiana State University Medical Center, 1901, Perdido Street, New Orleans, Louisiana, LA 70112, USA

Abstract

We have produced a digital atlas of the distribution of nitric oxide synthase (NOS) in the mouse brain as a reference source for our studies on the roles of nitric oxide in brain development and plasticity. NOS was labeled using nicotinamide adenine dinucleotide phosphate diaphorase (NADPHd) histochemistry. In addition, choline acetyltransferase (ChAT) immunocytochemistry was used to identify cholinergic cells because many of the NADPHd positive cells were thought to colocalize acetylcholine. Some sections were also labeled with antibodies to either the neuronal (nNOS) or endothelial (eNOS) isoforms of NOS. Series of sections from 11 C57/BL6 mice were collected and labeled for NADPHd and/or ChAT. We collected two types of data from this material: color digital photographs illustrating the density of cell and fiber labeling, and computer/microscope plots of the locations of all the labeled cells in selected sections. The data can be viewed as either a series of single-section maps produced by combining the plots with the digital images, or as 3-D views derived from the cell plots. The atlas of labeled cell maps, together with selected color photographs and 3-D views, is available for viewing via the World Wide Web (http://nadph.anatomy.lsumc.edu).

Examination of the atlas data has revealed several points about the distribution of NOS throughout the mouse brain. Firstly, different populations of NADPHd-positive neurons can be distinguished by different patterns of staining. In some brain areas neurons are intensely stained by the NADPHd technique where label fills the cell bodies and much of the dendritic trees. In other brain regions labeling is much lighter, is principally confined to the cytoplasm of the cell soma, and extends only a short distance within proximal dendrites. Intense labeling is typical of neurons in the caudate/putamen and mesopontine tegmental nuclei. Most of the labeled neurons in the cortex also stain this way. Lighter, "granular" label is found in many other nuclei, including the medial septum, hippocampus, and cerebellum.

In addition to staining pattern, we have also noted that different subpopulations of NOS-neurons can be distinguished on the basis of colocalization with ChAT. Substantial overlap of the distributions of these two substances was observed although very little colocalization was found in most cholinergic cell groups except the mesopontine tegmental nuclei.

Other points of interest arising from this project include the apparent lack of NADPHd labeling in

*Corresponding author. Tel.: +1 504 568 7059; fax: +1 504 568 4392; e-mail: jcork@lsumc.edu

[1]Accessible on the Web at: http://nadph.anatomy.lsumc.edu

the CA1 pyramidal cells of the hippocampus or the Purkinje neurons in the cerebellum. This observation is especially relevant given that synaptic plasticity in these regions is reported to be nitric-oxide dependent.

Introduction

Nitric oxide (NO) is a short-lived, rapidly diffusable, free radical that has been implicated as an important signal messenger having roles during neuronal development and synaptic plasticity (Gally et al., 1990; Garthwaite, 1991). The synthetic enzyme for NO is nitric oxide synthase (NOS), and cells containing NOS are intensely labeled using nicotinamide adenine dinucleotide phosphate diaphorase (NADPHd) histochemistry (Thomas and Pearse, 1961; Bredt et al., 1991; Schmidt et al., 1992) . Much of the research on NO has been done using the rat brain as a model system, and the distribution of NADPHd labeling in the rat brain has been described by a number of investigators (e.g. Vincent and Kimura, 1992). In recent years, the advent of molecular genetic techniques has led to the availability of several strains of mice in which one or more of the genes encoding for NOS have been deleted (e.g. Huang et al., 1993, see Huang et al., Chapter 2). The use of such NOS-knockout mice has been of great benefit to investigations into the roles of NO. Our laboratory has begun to use several types of NOS knockout mice for various physiological, developmental and anatomical studies (see Mize et al., Chapter 10). Surprisingly, we noted some quite distinct differences between the distributions of NADPHd labeling in the mouse and the rat. We therefore set out to produce an atlas of NADPHd labeling in the mouse brain as a reference for our studies on nitric oxide. Because of the popularity of the NOS knockout mouse as a model in a variety of experimental studies, this atlas should prove to be of value to many investigators.

The digital atlas we have produced by combining computer/microscope plotting techniques with digital imaging, shows the distributions of NADPHd labeled cells throughout the mouse brain (excluding the cerebellum and spinal cord). The atlas comprises a series of maps showing representative sections with the locations of every NADPHd labeled neuron marked on an image of the section. The atlas is available on the World Wide Web at http://nadph.anatomy.lsumc.edu. It is interactive in that any section can be selected by clicking on an image of the whole brain to bring up the required image. Clicking on parts of the maps themselves will bring up higher magnification maps of specific areas or digital images of the original sections. In addition to the maps, the data collected from the whole series of sections was used to construct a virtual 3-dimensional model of the brain and NADPHd distribution. This atlas is an on-going project and more section-maps and 3-D views will be added in the future; in this chapter we briefly describe the methods used to construct the atlas, and then illustrate some of its uses with some interesting observations based upon the maps. Some of our observations from the atlas are relevant to the use of NADPHd labeling as a specific marker for NOS, and to the specific distributions of the different NOS isoforms, so we also briefly mention some recent experiments we have done to address these issues.

Histochemistry and immunocytochemistry

Eleven normal adult C57-BL6 mice were anesthetized with ketamine hydrochloride/xylazine and perfused transcardially with 4.0% paraformaldehyde/0.2% glutaraldehyde in 0.1 M phosphate buffer. The brains were removed, post-fixed, and stored in phosphate buffer with 8.0% dextrose prior to sectioning. All brains were blocked and cut into 50 μm sections with a Vibratome.

Six brains were used for nicotinamide adenine dinucleotide phosphate diaphorase NADPH-d histochemistry to label cells and fibers that contain NOS. Sections were rinsed in Tris buffer (pH 7.1) for 2×5 min, and then incubated in Tris buffer (with 0.3% Triton X-100) containing nitroblue tetrazolium and NADPH for 30 min to 2 hr, then rinsed again in buffer, mounted on slides, dehydrated, and coverslipped.

Two brains were used for choline acetyltransferase (ChAT) labeling. Sections were blocked in 4.0% normal rabbit serum in PBS containing 0.25% Triton X-100, incubated in primary antibody (Chemicon goat anti-ChAT, 1 : 500), rinsed 3 × 10 min in PBS, and incubated in biotinylated rabbit anti-goat (Vector; 1 : 100). Sections were stained with Vector ABC Elite kit and nickel cobalt intensified DAB.

Sections from a further three brains were labeled with ChAT, NADPHd, or both ChAT and NADPHd; two sets of serial sections were sequentially labeled with ChAT, both ChAT/NADPHd, and NADPHd to more accurately compare the differences in labeling between these three markers. Double-labeled sections were processed as described for ChAT immunocytochemistry and then after the final DAB staining they were processed for NADPH histochemistry as described above.

Computer plotting, imaging and atlas construction

Plots of the labeled neuron distributions were made using a Eutectic Neuron Tracing System (NTS) [Eutectic Electronics, Raleigh, NC]. In this system, a computer-generated cursor on a video monitor is superimposed on the microscope field-of-view by way of a drawing tube attached to the microscope (a Nikon Optiphot). Objects of interest in the sections are traced by moving the cursor and/or the microscope stage such that the cursor moves like a pen around edges. Outlines were traced from every other section from the original series (i.e. a section every 100 μm), and included the outer border of the whole section, the borders of the lateral and third ventricles, and other clearly defined landmarks of the brain. The distributions of individual labeled cells were also plotted from alternate outlined sections (i.e. every 200 μm). The cursor was positioned over each cell and the location (x, y coordinates) recorded with a keystroke. The final NTS data files for each section included the coordinates for the outlines, the coordinates of each labeled cell, and a code tag specifying the intensity and pattern of labeling in the cell.

A complete series of low magnification images was obtained by combining sections from several animals. The sections were viewed on a Zeiss microscope with a 1 × objective and photographed using a Kodak digital camera (DCS420, with 24-bit 1.5 K resolution) mounted on the microscope. Images were processed using Adobe Photoshop. All images were adjusted to match color balance, contrast and brightness. Extraneous background was removed, and a sharpening filter was applied to accentuate edges. The final maps showing the cell distributions were also produced using Adobe Photoshop. The NTS plots were scanned, color coded, and overlaid onto the digital photographs of the sections.

For the 3-D reconstruction, the NTS data files were converted into a format suitable for importing into our 3-D reconstruction package. The files were then exported to a Silicon Graphics IRIS workstation running the Skandha 3-D reconstruction software package (Washington University, Seattle, WA). Sections were individually rotated and aligned to bring them into register. Outlines could be tiled and displayed as surfaces with varying degrees of opacity. Cell distributions could be color-coded and displayed as dots in their correct 3-dimensional positions. The reconstructed model could be rotated for viewing from different angles and different surfaces could be removed to show internal cell distributions.

Overview of the atlas – intense and granular labeling

A representative section map from the complete atlas is illustrated in Fig. 1A. These maps show the locations of all NADPHd labeled neurons at selected levels throughout the complete mouse brain. During plotting, labeled neurons were categorized based on the intensity and pattern of labeling. The most obviously labeled cells had dark Golgi-like staining throughout the cell soma, dendrites and axons. Although in some cases they were occluded by labeled cytoplasm, nuclei were

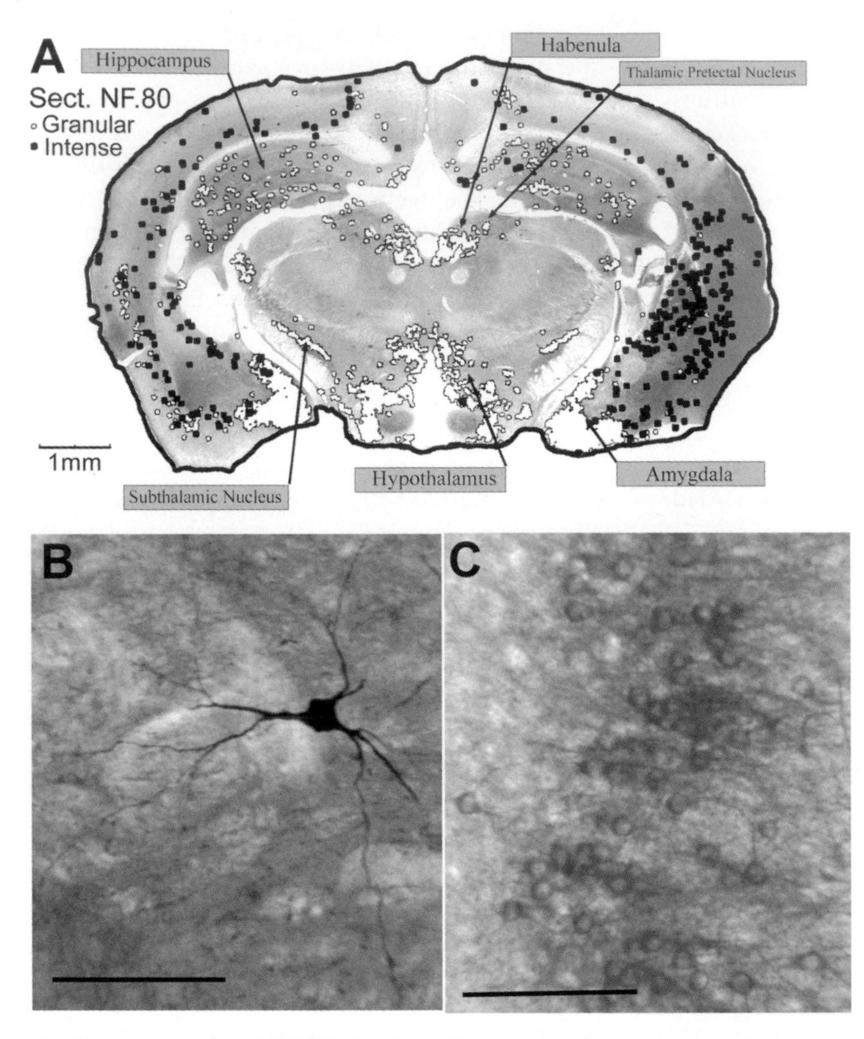

Fig. 1. Atlas maps show NADPHd-labeled cells categorized by staining pattern: **A** A representative map from the atlas that has the locations of all the labeled cells superimposed on an image of the section. Intensely labeled cells are represented by black dots and granular labeled cells are shown as white dots. **B** A typical intensely labeled cell illustrating the dense staining of the soma and much of the dendritic tree. The particular cell shown is from the mesopontine tegmental nuclei, but most labeled cells in the cortex have a similar appearance. **C** Granular labeling of pyramidal cells in the hippocampus CA3 region. Scale bars in B and C are 100 μm.

not usually stained. Cells with this pattern of labeling were categorized as "intense" (Fig. 1B). They are marked on the maps as black dots.

At the other end of the spectrum of labeling were those cells we categorized as "granular". Labeling in these cells was pale, and in general confined to the cell soma (Fig. 1C). This pattern of labeling was particularly noticeable in areas where neurons were clustered together; the light cytoplasmic labeling made the cell bodies appear as clear beads outlined by a ring of staining, and the clustering of many similarly labeled cells together gave those regions a granular look. (Fig. 1C). Cells with this type of labeling are shown on the maps as white dots.

For the purposes of plotting we categorized all cells as "intense" or "granular", based on both the intensity and pattern of labeling. Although staining intensity varied greatly, the type of labeling could usually be determined by also taking the pattern into account. Thus, although some "intensely" labeled neurons were paler than others, they always had significant dendritic labeling. Granular labeling could be fairly dark but such cells always appeared as rings of staining with little or no dendritic labeling. As a broad generalization most "intensely" labeled neurons were isolated cells scattered throughout the neuropil while most "granular" neurons were clustered into densely packed groups.

A survey of the maps reveals quite specific distributions of the two types of NADPHd labeling in different parts of the brain. Intensely labeled neurons are found scattered throughout the cortex (Figs. 1A, 3A, 4 and 5) and striatum (Figs. 3A,D and 5), and clustered in several groups in the dorsal lateral periaqueductal gray and the mesopontine tegmental nuclei (Figs. 2A,C). Paler intensely labeled neurons were found in many nuclei of the basal forebrain, the hypothalamus, and the colliculi. Granular labeling was particularly apparent in hippocampal pyramidal cells (Fig. 1C), the amygdala (Fig. 1A), and the molecular and granular layers of the cerebellum (Fig. 8).

Co-localization with ChAT

Preliminary observations made on the mouse material suggested that, as in the rat brain, there is considerable overlap of NADPHd labeling in brain regions containing cholinergic neurons. To study this phenomenon more closely we produced some maps of cells labeled by ChAT immunore-

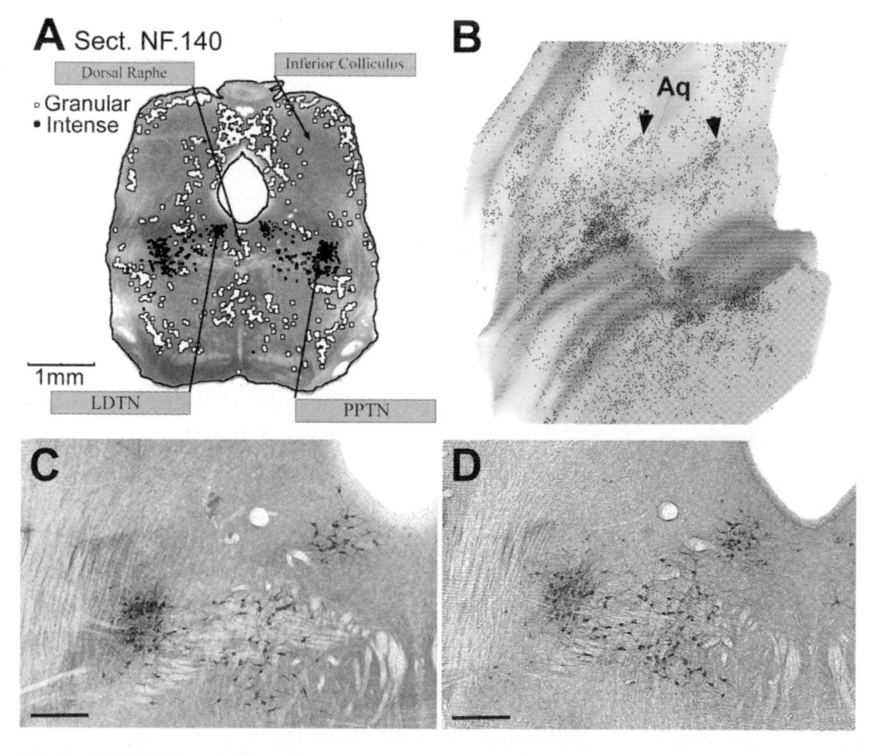

Fig. 2. NADPHd and ChAT labeling in the mesopontine tegmental nuclei. **A** Representative atlas map of NADPHd labeling at the level of the caudal inferior colliculus. **B** View of the 3-D brain reconstruction showing the inferior colliculus and the brainstem with the cerebellum removed. The transparency of the model surface has been increased to show the locations of the labeled cells. Extensive clusters of cells, representative of the pedunculopontine tegmental nucleus and the lateral dorsal tegmental nucleus, are visible (arrow heads). Aq = aqueduct. **CD** Labeling of adjacent sections with either NADPHd histochemistry (C) or ChAT immunocytochemistry (D). Note the almost identical distributions of labeled cells in the two sections. Scale bars are 250 μm.

activity in the normal mouse brain. ChAT is the synthetic enzyme for acetylcholine, and immunocytochemical labeling for ChAT has been shown to be a specific marker for cholinergic neurons (Armstrong et al., 1983). In addition to staining some complete sets of serial sections with ChAT immunocytochemistry, we also did some experiments where adjacent sections were labeled with either NADPHd histochemistry or ChAT immunocytochemistry, and every third section was double-labeled for both neurochemicals.

The overall distribution of ChAT labeling throughout the rostrocaudal extent of the mouse brain was similar to that described for the rat brain (Armstrong et al., 1983). Most ChAT-immunoreactive neurons were located in one of four cell groups: the caudate/putamen and associated nuclei, the basal forebrain nuclei comprising the magnocellular basal nucleus, the pontine tegmental nuclei, and the cranial nerve motor nuclei.

Mesopontine tegmental nuclei

The most intensely NADPHd-labeled cells in the mouse brain were found clustered in some of the mesopontine tegmental nuclei (Fig. 2A). Large, intensely labeled, multipolar neurons were found throughout the pendunculopontine tegmental nucleus (PPTN) from the level of the caudal superior colliculus for about 600 μm to where the PPTN merged with the lateral parabrachial nucleus (Fig. 2B). The LPB also contained NADPH-labeled cells, but these were much smaller than those in the PPTN, and were only lightly labeled. Both dorsal and ventral parts of the lateral dorsal tegmental nucleus (LDTNd and LDTNv) were also filled with similar intensely labeled neurons (Fig. 2C). The anterior, ventral, and dorsal tegmental nuclei were generally unlabeled, but NADPHd-labeled fibers were seen running from the PPTN through the microcellular tegmental nucleus and the parabigeminal nucleus.

In the ChAT labeled material there were also densely labeled cells throughout the PPTN and LDTN; in fact, these were some of the most intensely labeled cells seen in the mouse brain. The ChAT-labeled cells and NADPHd-labeled cells were similar in shape and size and had very similar distributions throughout the tegmental nuclei (Figs. 2C,D). A comparison of the numbers of labeled cells in adjacent sections suggested that most, if not all, of the labeled neurons in the PPTN and LDTN contained both substances (data not shown). Close examination of some of the double labeled sections revealed a few NADPHd labeled neurons at the most caudal end of the PPTN that did not contain ChAT. There were also a few NADPHd-only neurons clustered in caudal regions of the LDTN. Occasional ChAT-only neurons were also seen, but the vast majority of the neurons in the PPTN/LDTN complex appeared to be double-labeled. This degree of colocalization has also been reported in cat where direct counts in double-labeled sections revealed that 94.6% of labeled cells in PPTN/LDTN contained both NADPHd activity and ChAT (Butler et al., 1996).

Caudate putamen

Besides the PPTN/LDTN complex another region where intense NADPHd-labeled neurons were concentrated was in the striatum. Medium-sized multipolar neurons were found throughout the caudate/putamen (CP), globus pallidus and accumbens nucleus (Fig. 3A and 5). The neuropil of the CP was darkly stained by the NADPHd reaction. While the intense-labeled neurons in the PPTN/LDTN were clustered together, very often overlapping each other, the NADPHd-positive neurons in the CP were more evenly distributed throughout the region.

Comparison of the NADPHd-labeled map with a map of ChAT labeling (Fig. 3B), and examination of different labeled sections (Figs. 3C,D), revealed very similar distributions of labeled cells scattered throughout the CP. Despite the similarity of the distributions, examination of double-labeled sections revealed that there were two populations of cells in this region. There was a population of blue colored neurons that was clearly labeled only with NADPHd. These NADPHd cells were interspersed amongst a second population of neurons

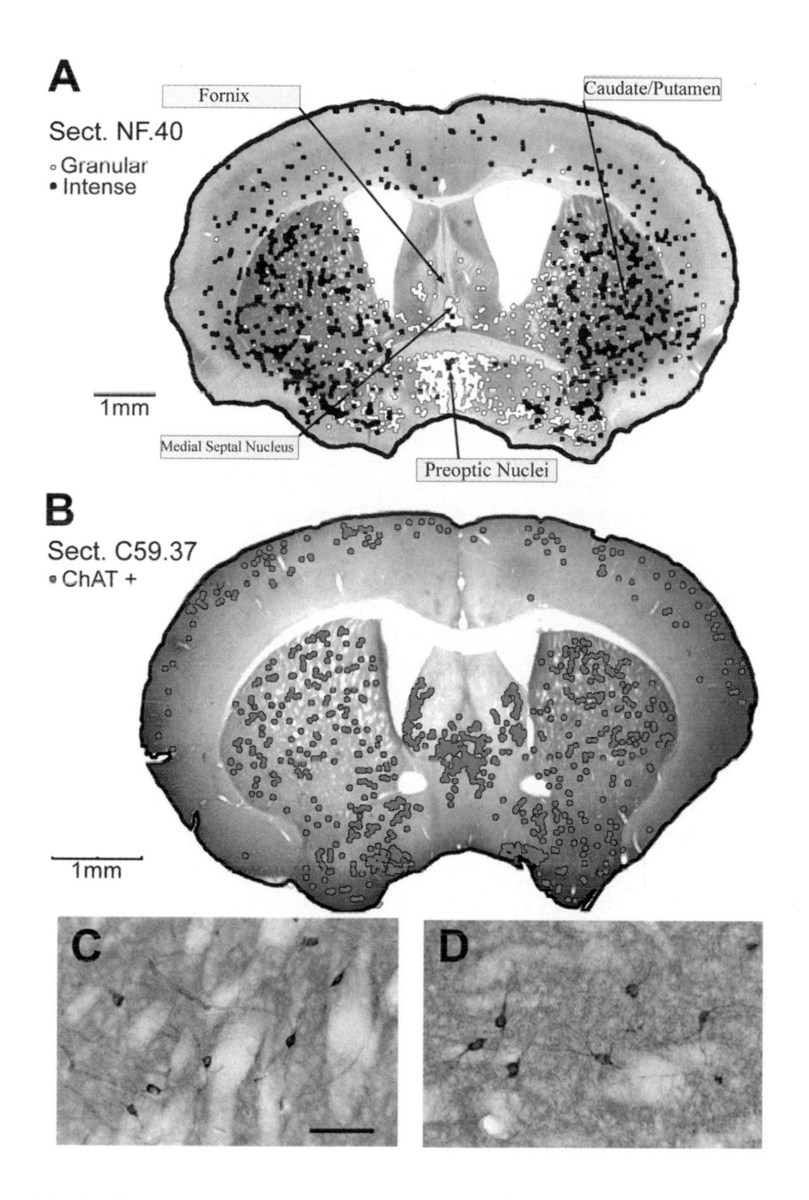

Fig. 3. There are separate populations of ChAT- and NADPHd-positive neurons in the caudate putamen. **A** NADPHd atlas map of a section containing the caudate putamen. **B** ChAT atlas map at a similar rostro-caudal level to A. **CD** Micrographs of the NADPHd-labeled (C), and ChAT-labeled (D) neurons scattered through the caudate putamen. Scale bars are 100 μm.

that appeared to be slightly larger and were stained dark blue brown or black. Because NADPHd gave the CP neuropil a dense blue background staining, it was not certain if the blue–brown population of neurons were double-labeled or only contained ChAT. Examination of single-labeled ChAT sections showed that most ACh-containing neurons were similar in size and morphology to this second population of cells To further examine this, we counted the numbers of labeled cells in the CP from several sets of adjacent sections labeled for either ChAT or NADPHd, or double-labeled for both. The data in Table 1 show that there were approximately twice as many labeled cells in the double-labeled sections as there were in either the single-labeled ChAT or NADPHd sections. Thus

TABLE 1

Numbers of labeled cells in the caudate putamen. Counts of cells were made from 5 sets of 3 sequential sections labeled for NADPHd, NADPHd and chAT, and chAT alone.

NADPHd	ChAT	Double labeled Sections		
		Total	Blue	Black
175	104	218	139	79
156	173	214	75	139
134	172	259	68	191
105	106	227	136	91
107	107	220	132	88
134 ± 30	132 ± 37	228 ± 18	110 ± 35	118 ± 47

it appears that the two populations of labeled neurons are separate and do not contain both NOS and ACh. Measurements of the areas of cell bodies in the single-labeled sections further confirmed that there were, in fact, two populations of neurons. The ChAT labeled cells were larger with a mean area of 230.8 ± 35.6 μm^2 (mean ± s.d.) while NADPHd labeled cell bodies had an average area of 120.1 ± 26.5 μm^2. Sample measurements taken from the double-labeled sections showed that the blue cells belonged to the population of NADPHd-labeled cells and the blue–brown cells belonged to the ChAT-labeled population.

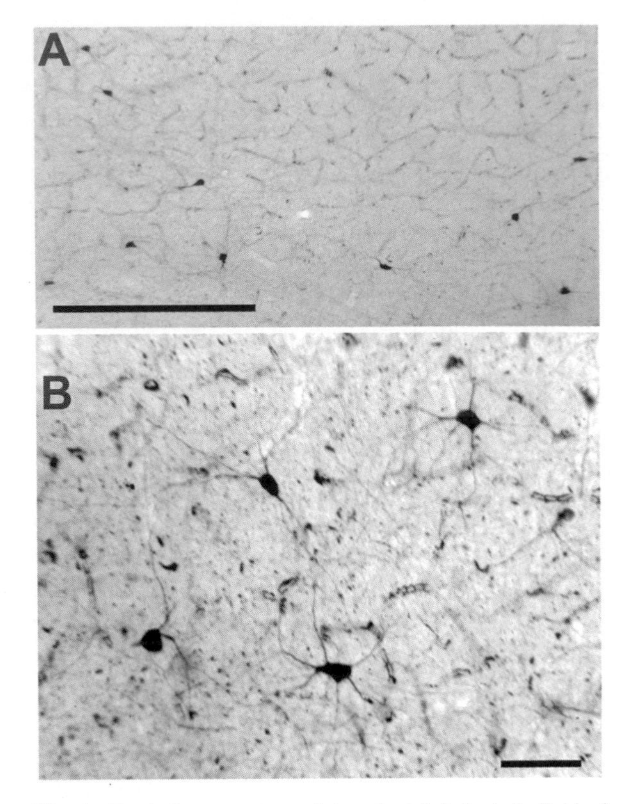

Fig. 4. Cortical neurons are intensely labeled. **A** Individual NADPHd-labeled cells are scattered throughout the cortex, mostly in the deeper cortical layers. Scale bar is 500 μm. **B** Cortical neurons are typically large multipolar neurons with intensely labeled somas and extensive dendritic labeling. Scale bar is 50 μm. NB both of these figures are from brains fixed with 2.5% gluteraldehyde/0.2% paraformaldehyde. This fixative reduces the background NADPHd labeling but accentuates blood vessel labeling.

Cerebral cortex

Intense NADPHd labeled neurons were found widely distributed throughout the cerebral cortex (Figs. 1A, 2A and 4). Typically, they were the "isolated" neurons first described by Thomas and Pearse (1961) as solitary active cells. Most had medium to large multipolar cell bodies with extensive dendritic trees (Fig. 4B). Diffuse neuropil labeling was light to moderate, but intensely labeled fibers were found throughout all regions of the cortex (Fig. 4B). There were also isolated ChAT-labeled neurons scattered throughout the cortex (e.g. Fig. 3B) but practically no double-labeled cells were observed. In fact, there appeared to be somewhat complementary patterns of NADPHd and ChAT expression in the cortex. In regions where labeled neurons were concentrated in particular cortical layers, NADPHd-positive neurons were concentrated in the inner layers (4–6) while most ChAT labeled cells were found in the outer cortical layers (1–3) (for example compare Figs. 3A, 3B).

The distribution of NADPHd-positive neurons in cortex should be of particular relevance to studies investigating possible roles for NO in neuroprotection, excitotoxicity and cortical plasticity. Although NO has been implicated as being involved in synaptic plasticity in the cerebral cortex (e.g. Harsanyi and Friedlander, 1997), the sparse distribution of NADPHd-labeled neurons would suggest that the plasticity is not mediated by high concentrations of NO in large numbers of post-synaptic neurons. Indeed, the low density of NOS-positive neurons in cortex was cited as one possible explanation for the failure to find any involvement of NO in the rearrangement of ocular dominance columns following monocular deprivation (Ruthazer et al., 1996).

Basal forebrain nuclei

A further concentration of NADPHd-positive neurons was found in the basal forebrain nuclei and parts of the hypothalamus (Figs. 3A, 5 and 6A). The intensity of labeling varied greatly in these nuclei, there were many single intense-labeled cells, groups of moderately labeled cells and some clusters of granular labeling. Intense cells were found in the lateral septal nucleus, the dorsal endopiriform nucleus, and in both horizontal and vertical limbs of the diagonal band. Mixed granular and intense labeling was found in the magnocellular pre-optic nucleus, olfactory tubercle, and the basomedial amygdaloid nucleus. Light granular labeling was seen throughout the lambdoid septal zone and some portions of the medial septum; the amygdaloid nucleus, the lateral olfactory tract, the ventral endopiriform nucleus, and the anterior hypothalamic area also contained some granular labeling. Several pre-optic nuclei, including the medial preoptic area, the median pre-optic nucleus, and the ventromedial preoptic nucleus contained clusters of granular labeling.

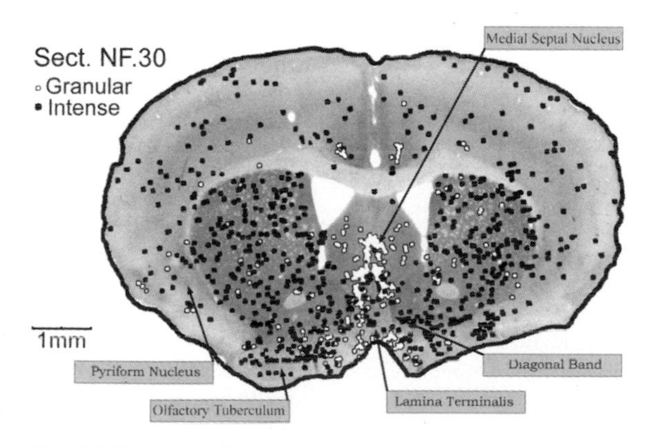

Fig. 5. NADPHd atlas map showing the distribution of labeled neurons in the basal forebrain.

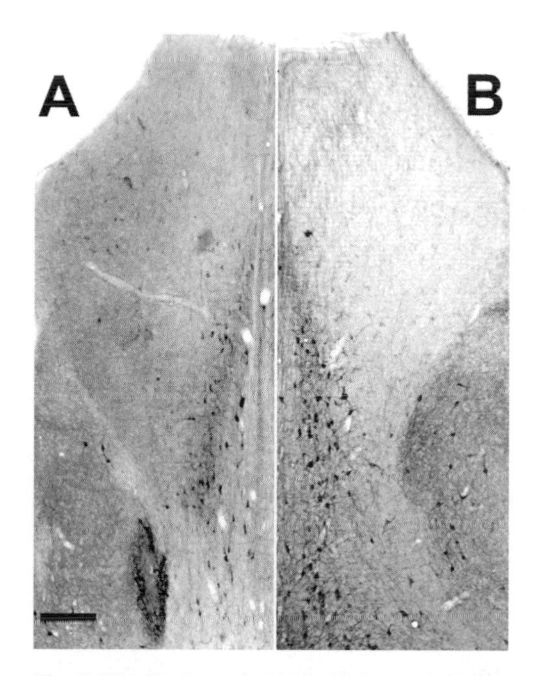

Fig. 6. Distributions of NADPHd- (A) and ChAT-labeled (B) in the medial septum and vertical limb of the diagonal band. Scale bar is 100 μm.

ChAT labeling (e.g. Fig. 6B) was found in the medial septum and the dorsal tenia tecta. ChAT labeled cells were also located in the ventral pallidum, both limbs of the diagonal band, the olfactory tubercle, and the lateral hypothalamic area. Several of these nuclei, including the diagonal band, the magnocellular preoptic nucleus, and the medial septum, comprise the magnocellular basal nucleus, one of the four major groupings of cholinergic neurons in the rat brain (Armstrong et al., 1983). While there was considerable overlap of the ChAT and NADPHd distributions, we found very little evidence for colocalization of these two neurochemicals in this part of the brain (e.g. Fig. 6). Although we have not studied this question in great detail, it seems clear that there are two distinct populations of neurons expressing ACh or NOS in close proximity to each other.

Hippocampus

The hippocampus had a consistent pattern of neuropil labeling (Fig. 7A). In CA1 the lacunosum molecular layer (LMOL) was the most darkly stained and the pyramidal cell layer was outlined by a thin band of neuropil, darker than the stratum oriens (SO) and stratum radiatum (SR). In CA3 the SO and SR was often more strongly stained. A comparable pattern of neuropil labeling was seen in the dentate gyrus. The molecular layer was the darkest, and the granular layer was edged by two thin bands of darker neuropil. The neuropil of the polymorphic layer of the dentate gyrus was usually quite lightly stained.

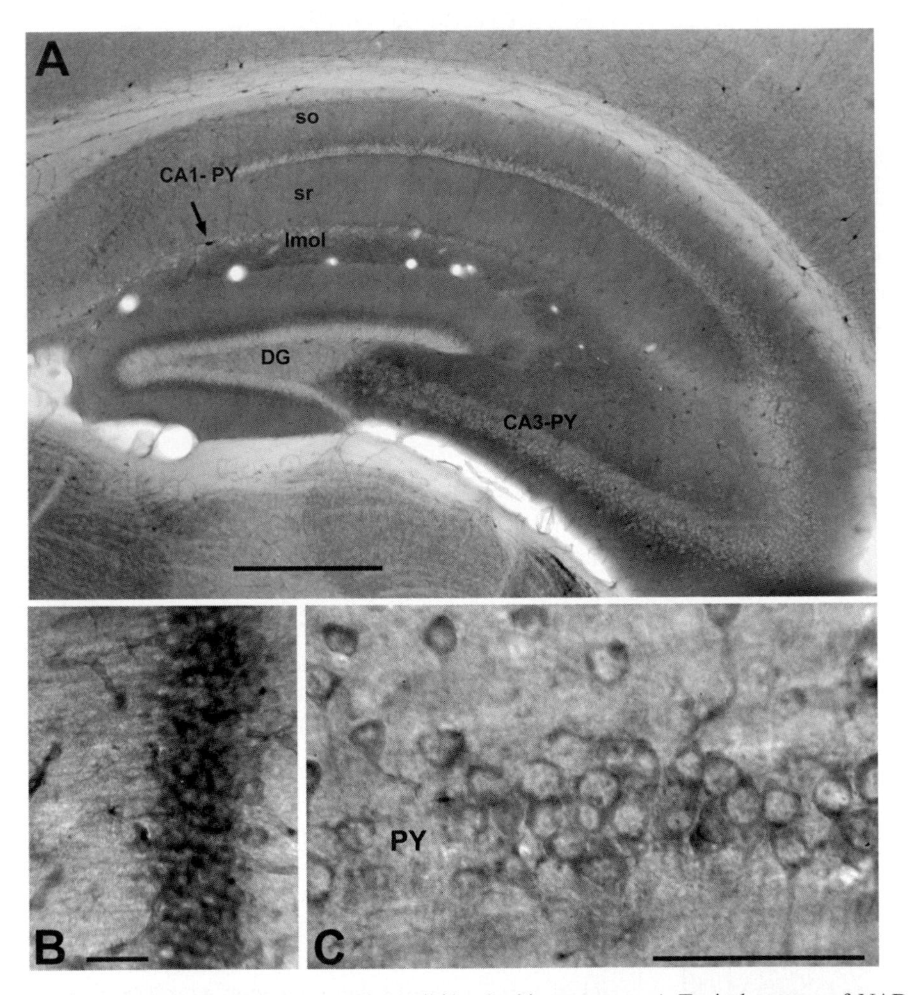

Fig. 7. Distribution of NOS-containing cells in the hippocampus. **A** Typical pattern of NADPHd labeling in the normal mouse hippocampus. Note the very pale granular labeling of the CA1 pyramidal cell (CA1-PY) layer that is outlined by darker neuropil. There is stronger staining in the CA3 pyramidal cell layer (CA3-PY). There is very little clear cellular labeling; moderately labeled neurons are widely spread throughout the stratum oriens (so) and stratum radiatum (sr) with more seen in the CA3 sr. Occasional intensely labeled cells are seen in the medial part of the CA1 (arrow). lmol = lacunosum molecular layer, DG = dentate gyrus. Scale bar is 500 μm. **B** Maximal NADPHd-staining (2.5% glutaraldehyde fixation, 1.5 h at 37°C NADPHd reaction) reveals dark granular labeling of the CA1 pyramidal cells, however, note that this is a knockout mouse in which one of the genes for neuronal NOS has been deleted. Scale bar = 25 μm. **C** Light labeling of the cytoplasm and proximal dendrites of cells in the CA1 pyramidal cell layer (PY) with an antibody to endothelial NOS (Transduction Laboratories). Scale bar = 50 μm.

The pattern of cellular labeling was also quite consistent. Most striking was the apparent lack of labeling in the pyramidal cells, especially in CA1. Occasional light to moderate labeled cells were seen in the pyramidal cell layer and scattered throughout the stratum radiatum. Such cells were more common in CA2/CA3; there were also more similar cells in the polymorphic layer of the dentate gyrus. There were very few intensely labeled cells visible in the hippocampus; a few were widely scattered in the CA1 SR or along the dorsal edge of the CA1 SO. Some of the latter cells could have been in the corpus callosum or alveus rather than the hippocampus proper. There were more of these large, intensely labeled neurons in the most medial portion of the CA1 SO, or the most rostral part of the subiculum. A fairly extensive network of intensely labeled fibers, many of which appeared to belong to the intensely labeled cells, was widely distributed throughout the SO and SR of CA1. There was also a finer network of labeled fibers throughout the LMOL.

Given that NO has been shown to participate in hippocampal LTP it is surprising that there are relatively few NADPHd-positive neurons in the mouse hippocampus. It is even more surprising that the neurons most often used as a model of LTP in the hippocampus, the CA1 pyramidal cells (CA1P), are distinguished by an almost complete lack of NADPHd labeling. Several other studies have used NADPHd histochemistry or NOS immunocytochemistry to localize NOS-positive neurons in the hippocampus. There is general agreement that there is less than expected NADPHd labeling in the hippocampus, and that most of the strongly stained neurons are the scattered non-pyramidal interneurons (Vaid et al., 1996). There has, however, been no consensus about the staining of the pyramidal cells, particularly in CA1 (see Dinerman et al., 1994). It has been suggested that some variability might be due to differing NADPHd activities depending upon fixation and processing of the tissue (Valtschanoff et al., 1992; Gonzalez-Hernandez et al., 1996; Vaid et al., 1996; Young et al., 1997). It has also been suggested that the CA1P contain eNOS rather than nNOS (Dinerman et al., 1994), and that under typical fixation protocols NADPHd does not reliably label eNOS activity. Nevertheless, even with optimal fixation (Vaid et al., 1996) the CA1P were not intensely labeled and still had granular type label (Fig. 7B; see also Fig. 1C in Vaid et al., 1996). We are currently investigating this situation using NADPHd histochemistry, nNOS and eNOS immunohistochemistry, and several strains of NOS-knockout mice. Preliminary immunocytochemistry experiments using eNOS and nNOS antibodies have not yet resolved this matter, but we do have some evidence for eNOS immunoreactivity in the CA1 pyramidal cells (Fig. 7C).

Cerebellum

The pattern of NADPHd labeling we observed in the cerebellum was generally similar to that reported previously for rat (Southam et al., 1992; Vincent and Kimura, 1992) and mouse (Bruning, 1993; Hawkes and Turner, 1994). Purkinje cell bodies were conspicuously unlabeled (Fig. 8A,D), and there was no evidence for labeling of Purkinje cell dendrites. The neuropil of the molecular layer was darkly stained and contained an array of what appeared to be labeled fibers extending orthogonally from the Purkinje cell layer to the surface of the folium (Fig. 8D). The identity of these fibers is unknown, though they might be the vertically oriented portions of parallel fibers from granule cells. Some individual cell bodies, with granular labeling, were seen in the molecular layer, particularly in the inner half adjacent to the Purkinje cell layer (Fig. 8D). The labeling of these cells was often obscured against the background of the molecular layer neuropil staining, but there was little or no dendritic labeling in these cells. We have not been able to positively identify these cells but from their distribution in the molecular layer they are most likely basket cells. The granular cell layer was labeled very strongly, and filled with small dark patches of label (Fig. 8A,D), however, discrete cell bodies could not be clearly distinguished in this region. In spite of the interest in the role of NOS in the cerebellum, the fact that very

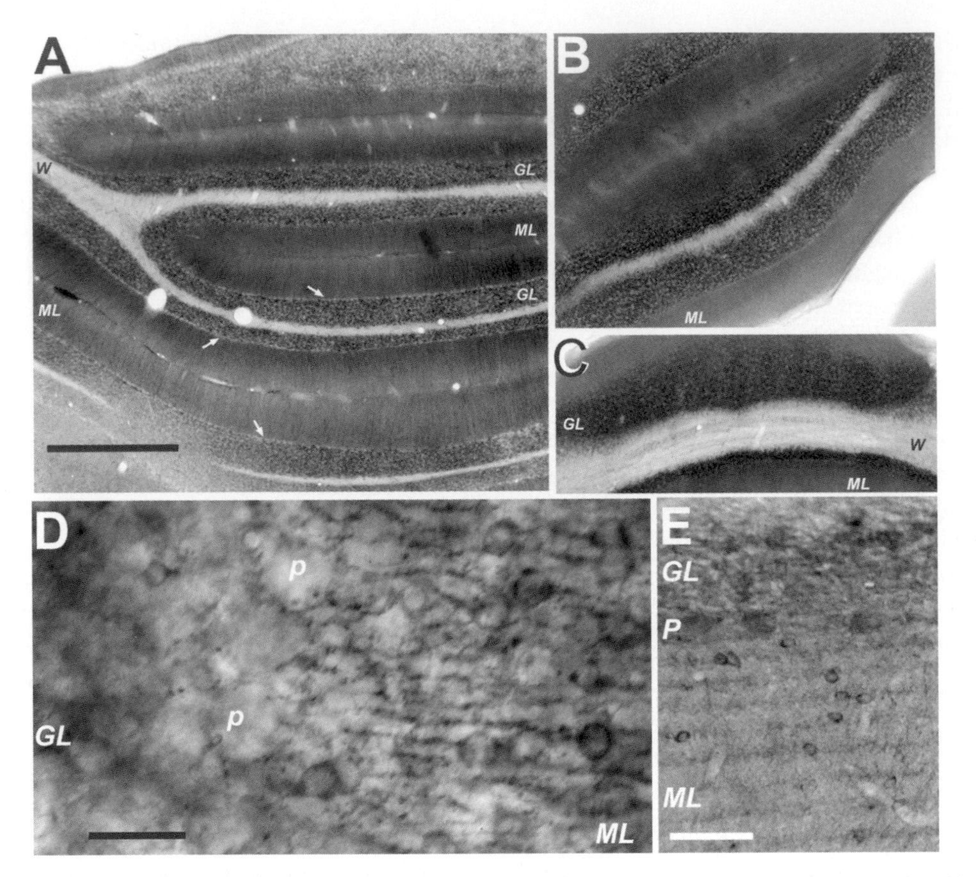

Fig. 8. Cerebellar labeling. **A** Typical appearance of NADPHd labeling in the cerebellar cortex. The white matter (W) is unlabeled, as is the Purkinje cell layer (arrows), the granular layer (GL) has a dark granular appearance, and the molecular layers (MLs) have a more uniform background staining with a regular pattern of fine parallel lines running across them. **B** and **C** Very similar patterns of labeling are seen in the cerebellar cortices of single nNOS knockout mice (**B**) and double nNOS/eNOS knockout mice (**C**). **D** NADPHd labeling of brains fixed with high glutaraldehyde (2.5%) reveals a clearer pattern of cellular labeling in the cerebellum. Purkinje cells (p) are noticeable as they lack any label, a few cells with light granular labeling are seen in the GL, parallel rows of label are seen running across the ML, and darker granular labeling is seen in cells scattered throughout the inner half of the ML. **E** nNOS immunohistochemistry reveals cellular labeling without the typical NADPHd background. Scattered cells are seen in the inner half of the ML, and some clusters of small labeled cells are seen in the GL. The Purkinje cells (P) are slightly darker than the neuropil but do not appear to have any specific nNOS labeling. Scale bars: A, B and C (shown in A) is 500 μm; D is 25 μm; E is 50 μm.

few individual cells could be resolved, particularly in the granular layer, made it impractical to plot cell locations for our maps. We therefore did not include the cerebellum in the atlas.

Sections from NOS-knockout mice revealed that not all of the typical NADPHd staining in the cerebellum represents NOS. Cerebellar NADPHd-staining patterns seen in sections cut from either nNOS single-knockout or eNOS/nNOS double-knockout mice were quite similar to those of wild-type mice (cf. Figs. 8A–C). The dark neuropil labeling that so dominates the pattern of NADPHd labeling in both the molecular and granular layers, appears not to be due to NOS at all, but probably represents some other diaphorase activity. Nevertheless, the granular NADPHd-labeling of cell bodies in the molecular and granular cell layers was not seen in sections from either knockout animal, suggesting that it does represent NOS in normal animals. Indeed, when background staining was reduced to a minimum, the labeling of the granular layer neuropil was seen to be somewhat patchy with relatively few labeled granule cells and the labeled cells in the molecular layer were even more clearly seen (Fig. 8D).

Preliminary experiments using an nNOS antibody from Research and Diagnostic Antibodies (Berkeley, CA) revealed a limited pattern of labeling (Fig. 8E) that corresponded to the best NADPHd-labeling. There were occasional clusters of labeled granule cells in the granule cell layer but the most strongly labeled cells were scattered throughout the innermost portion of the molecular cell layer; apparently corresponding to the NADPHd-labeled cells we thought were probably basket cells.

Although the cerebellum is the region of the brain expressing the highest levels of nNOS (Bredt, et al., 1990), and NO has been strongly implicated in cerebellar long term depression (LTD, e.g. Shibuki and Okada, 1991), there is, as yet, no consensus of opinion as to the distribution of NOS in the cerebellum, or the mechanisms whereby NO-release affects LTD. The distribution of NADPHd-labeling in the cerebellum has been of little assistance in elucidating any such role of NO. Given that NO is thought to lead to LTD of synapses between granule cell parallel fibers and Purkinje cell dendrites in response to climbing fiber firing, the expected site of NOS would be the Purkinje cells. An alternative site for NO release could be the climbing fibers themselves, however we have no evidence for NADPHd-labeling in either of these cell types. While the evidence for NO's involvement in LTD is equivocal there is growing evidence that NO has a role in the synaptic development of the cerebellum. There are several reports that reveal developmental changes in NADPHd-labeling suggesting NO affects the topographic refinement that occurs during cerebellar development (Bruning, 1993; Hawkes and Turner, 1994)

Future development of the atlas

The current atlas, consists of a limited number of section maps. More sections will be added to the complete atlas, especially in regions containing discrete concentrations of NOS-containing neurons. The present 3-D model is low resolution in the *z*-axis, as cells were plotted on sections that were spaced about 200 μm apart. For better resolution of 3-D distributions, specific areas will be plotted at 50 μm intervals. As the current 3-D model is not interactive, we are exploring the possibility of using the virtual reality modeling language (VRML) to construct a completely interactive model that the user can browse through on a CD-ROM or via the world wide web.

Summary and conclusions

We have constructed a digital atlas of NADPHd labeling throughout the mouse brain. This atlas consists of a series of hyperlinked section maps with high resolution digital micrographs of selected brain regions. The atlas is available via the world-wide-web (http://nadph.anatomy.lsumc.edu). Our laboratory is currently using the data from the atlas in several projects concerning the role of NO in brain development; it should also provide a useful reference for other investigators interested in the distribution of NOS in the brain. Some of the data from the atlas have led us to re-investigate the specificity of NADPHd labeling for NOS and to look at the possibility that other isoforms of NOS may be involved in neuronal plasticity and synaptic development.

We have also noted considerable overlap of the distributions of NADPHd labeling and cholinergic neurons, although in all areas except the mesopontine tegmental nuclei there seems to be little actual colocalization of ChAT and NADPHd activity in the same cells. The patterns of NADPHd labeling in several regions where NO is thought to have an active role in synaptic plasticity are generally unsupportive of such a role. In some regions there is growing evidence that NO may have a major role during synaptic development and topographic refinement, and we will be extending our NADPHd labeling studies to look at the ontogeny of NOS activity.

Acknowledgements

The authors are indebted to the following persons for invaluable technical assistance in constructing this digital atlas: Brett Wilson, Sam Buhler, Cheryl Dietze, Melvin Suggs, and Ebony Price. The nNOS antibody was kindly provided by Doug Webber. This work was supported by USPHS grants, EY-02973 & NS36000 from the National Institutes of Health, and agreement DAMD 17-93-V-3013 from the Department of Defense.

References

Armstrong, D.M., Saper, C.B., Levey, A.I., Wainer, B.H. and Terry, R.D. (1983) Distribution of cholinergic neurons in rat brain: Demonstrated by the immunocytochemical localization of choline acetyltransferase. *J. Comp. Neurol.*, 216: 53–68.

Bredt, D.S., Hwang, P.M. and Snyder, S.H. (1990) Localization of nitric oxide synthase indicating a neural role for nitric oxide. *Nature*, 347: 768–770.

Bredt, D.S., Glatt, C.E., Hwang, P.M., Fotuhi, M., Dawson, T.M. and Snyder, S.H. (1991) Nitric oxide synthase protein and mRNA are discretely localized in neuronal populations of the mammalian CNS together with NADPH diaphorase. *Neuron*, 7: 615–624.

Bruning, G. (1993) NADPH-diaphorase histochemistry in the postnatal mouse cerebellum suggests specific developmental functions for nitric oxide. *J. Neurosci. Res.*, 36: 580–587.

Butler, G.D., Scheiner, C.A., Whitworth, R.H. and Mize, R.R. (1996) Nitric oxide and choline acetyltransferase are colocalized in neurons projecting to the patch-cluster system of the cat superior colliculus. *Soc. Neurosci. Abstr.*, 22(1): 635.

Dinerman, J.L., Dawson, T.M., Schell, M.J., Snowman, A and Snyder, S.H. (1994) Endothelial nitric oxide synthase localized to hippocampal pyramidal cells: Implications for synaptic plasticity. *Proc. Natl. Acad. Sci. USA*, 91: 4214–4218.

Gally, J.A., Montague, P.R., Reeke, G.N. Jnr. and Edelman, G.M. (1990) The NO hypothesis: Possible effects of a short-lived, rapidly diffusable signal in development and function of the central nervous system. *Proc. Natl. Acad. Sci. USA*, 87: 3547–3551.

Garthwaite, J. (1991) Glutamate, nitric oxide and cell–cell signalling in the nervous system. *Trends Neurosci.*, 14: 60–67.

Gonzalez-Hernandez, T., Perez de la Cruz, M.A. and Mantolan-Sarmiento, B. (1996) Histochemical and immunohistochemical detection of neurons that produce nitric oxide: Effect of different fixative parameters and immunoreactivity against non-neuronal antisera. *J. Histochem. Cytochem.*, 44: 1399–1413.

Harsanyi, K. and Friedlander M.J. (1997) Transient synaptic potentiation in the visual cortex I. Cellular mechanisms. *J. Neurophysiol.*, 77: 1269–1283.

Hawkes, R. and Turner, R.W. (1994) Comparmentation of NADPH-diaphorase activity in the mouse cerebellar cortex. *J. Comp. Neurol.*, 343: 499–516.

Huang, P.L., Dawson, T.M., Bredt, D.S., Snyder, S.H. and Fishman, M.C. (1993) Targeted disruption of the neuronal nitric oxide synthase gene. *Cell*, 75: 1273–1286.

Ruthazer, E.S., Gillespie, D.C., Dawson, T.M., Snyder, S.H., and Stryker, M.P. (1996) Inhibition of nitric oxide synthase does not prevent ocular dominance plasticity in kitten visual cortex. *J. Physiol.*, 494: 519–527.

Schmidt, H.H.H.W., Gagne, G.D., Nakane, M., Pollock, J.S., Miller, M.F. and Murad, F. (1992) Mapping of neural nitric oxide synthase in rats suggests frequent co-localization with NADPH diaphorase but not soluble guanylyl cyclase, and novel paraneural functions of nitrinergic signal transduction. *J. Histochem. Cytochem.*, 40: 1439–1456.

Shibuki, K. and Okada, D. (1991) Endogenous nitric oxide release required for long-term depression in the cerebellum. *Nature*, 349: 326–328.

Southam, E., Morris, R. and Garthwaite, J. (1992) Sources and targets of nitric oxide in rat cerebellum. *Neurosci. Lett.*, 137: 241–244.

Thomas, E. and Pearse, A.G.E. (1961) The fine localization of dehydrogenases in the nervous system. *Histochemie.*, 2: 266–282.

Vaid, R.R., Yee, B.K., Rawlins, J.N.P. and Totterdell, S. (1996) NADPH-diaphorase reactive pyramidal neurons in Ammon's horn and the subiculum of the rat hippocampal formation. *Brain Res.*, 733: 31–40.

Valtschanoff, J.G., Weinberg, R.J. and Rustioni, A. (1992) NADPH diaphorase in the spinal cord of rats. *J. Comp. Neurol.*, 321: 209–222.

Valtschanoff, J.G., Weinberg, R.J., Kharazia, V.N., Nakane, M. and Schmidt, H.H.H.W. (1993) Neurons in the rat hippocampus that synthesize nitric oxide. *J. Comp. Neurol.*, 331: 111–121.

Vincent, S.R. and Kimura, H. (1992) Histochemical Mapping of nitric oxide synthase in the rat brain. *Neuroscience*, 46: 755–784.

SECTION II

Nitric oxide and NMDA receptor interaction

R.R. Mize, T.M. Dawson, V.L. Dawson and M.J. Friedlander (Eds.)
Progress in Brain Research, Vol 118

CHAPTER 5

Why is the role of nitric oxide in NMDA receptor function and dysfunction so controversial?

Elias Aizenman[1,*], Jessica C. Brimecombe[1], William K. Potthoff[1] and Paul A. Rosenberg[2]

[1]*Department of Neurobiology, University of Pittsburgh School of Medicine, E1456-BST, 3500 Terrace street, Pittsburgh, PA 15261, USA*
[2]*Department of Neurology and Program in Neuroscience, Children's Hospital, and Harvard Medical School, Boston, MA 02115, USA*

Introduction

This chapter will encompass two separate, albeit related, topics, both of which are highly controversial. The first one is the mechanism of inhibition of *N*-methyl-D-aspartate (NMDA) receptor function by nitric oxide, while the second one is the role of nitric oxide in NMDA receptor mediated neuronal injury. With respect to the first, some investigators have proposed that NO, or a related species, directly interacts with a redox modulatory site present on the NMDA receptor molecule. However, other studies have shown that when that site is chemically eliminated, the inhibitory actions of NO remain unaltered, suggesting an alternative mechanism of action. All the published evidence supporting the action of NO at this site will be re-evaluated. In addition, data from recent experiments using site-directed mutagenesis of recombinant NMDA receptors will be presented. With respect to the second topic, the extent of NO-dependent injury during NMDA receptor-mediated excitotoxicity has proven to be highly inconsistent, and in no case accounts for all of the observed injury. We review the possible sources of this controversy, which include the variable expression of NOS and NOS-containing neurons, neglect of alternate mechanisms, and also the fact that different subcellular targets of excitotoxicity may involve different mechanisms of cell death. A relevant question is which model systems best reflect the pathogenesis of neuronal death in specific diseases.

There is a vast literature that examines the structure, function and regulation of the NMDA receptor. This is due to the dazzling complexity of the protein and its critical involvement in a myriad of physiological and pathophysiological processes in the brain, including synaptic transmission, long-term potentiation, neuronal differentiation, and neuronal cell death (McBain and Mayer, 1994). This chapter will summarize our knowledge of the interrelationships between the diffusible messenger NO and the NMDA receptor gained from two different, but closely connected areas of study, each best epitomized by a question: (1) Does NO act as a modulator of NMDA receptor activity via direct interaction with the protein? and (2) Under what circumstances is NO in the causal chain leading from NMDA receptor activation to neurotoxicity?

*Corresponding author. Tel.: +1 412 648 9434; fax: +1 412 648 1441; e-mail: redox+@pitt.edu

Does NO directly interact with the NMDA receptor?

Since the so-called "redox modulatory site" (Aizenman et al., 1989; Tang and Aizenman, 1993a) has been suggested as a target for NO in modulating NMDA receptor function (Lei et al., 1992; Lipton et al., 1993; Stamler et al., 1997), we will first review our current understanding of the nature and function of this important site. Readers are also referred to a previous review for a more comprehensive historical perspective of redox modulation of NMDA receptor function (Aizenman, 1994).

Reduction, oxidation and alkylation of the NMDA receptor

Physiological responses mediated via activation of the NMDA receptor have been shown to be substantially altered by redox substances that can modify sulfhydryl groups of proteins. Disulfide-reducing agents, such as dithiothreitol (DTT; Tang and Aizenman, 1993a), dihydrolipoic acid (Tang and Aizenman, 1993b), and tri(carboxyethyl) phosphine (TCEP; Gozlan et al., 1994), among others, can dramatically potentiate NMDA-induced responses in neurons. Conversely, thiol oxidizing agents can rapidly reverse the effects of reductants, and, interestingly, sometimes depress native responses (Aizenman et al., 1989). This suggests that NMDA receptors in neurons may normally exist in alternate redox states (Aizenman et al., 1989; Sinor et al., 1997). Oxidizing agents shown to be effective in modifying NMDA receptor-mediated responses include 5,5′-dithio-bis-(2-nitrobenzoic acid) (DTNB; Aizenman et al., 1989), as well as several endogenous substances, such as oxygen-derived free radicals (Aizenman et al., 1990; Aizenman, 1995), the essential nutrient pyrroloquinoline quinone (PQQ; Aizenman et al., 1992, 1994; Jensen et al., 1994), oxidized glutathione (Gilbert et al., 1991; Sucher et al., 1991), and lipoic acid (Tang and Aizenman, 1993b). A critical feature of the action of oxidizing substances is that their effects are not reversible, at least not until a reducing agent is introduced into the preparation following the oxidation treatment (Aizenman et al., 1989; Gozlan et al., 1994). Thus, although NMDA receptors can spontaneously oxidize (Tang and Aizenman, 1993a), they have never been observed to spontaneously reduce following the addition of a thiol oxidant. This will become an important issue to consider when the effects of NO are discussed in a later section.

Following reduction with DTT, NMDA-evoked responses in neurons can, in most instances, be "permanently" potentiated after the addition and subsequent washout of an alkylating agent, such as *n*-ethylmaleimide (NEM; Majewska et al., 1990; Hoyt et al., 1992; Tang and Aizenman, 1993a; Fagni et al., 1995). In the alkylated state, NMDA receptors cannot be further modified by either reducing or oxidizing substances, and thus, in essence, become redox insensitive. In such a state, most, but not all of the other properties of the NMDA receptor, remain unaltered. Thus, although the potentiating actions of glycine and inhibitory effects of protons are unaffected by the chemical modification of the receptor, the blocking actions of Mg^{2+} and Zn^{2+} are slightly impaired in the alkylated state (Tang and Aizenman, 1993a). Alkylation has served as a useful tool for establishing redox modulation as the mechanism of action of substances like PQQ (Aizenman et al., 1994), and dihydrolipoic acid (Tang and Aizenman, 1993b) in altering NMDA receptor behavior.

A physiological role for the redox site?

The fact that native NMDA receptors can exist naturally in either reduced or oxidized configurations strongly suggests that endogenous factors can alter NMDA receptor function via the redox site. As mentioned earlier, several substances endogenous to the brain have been shown to interact with this site (Aizenman, 1994). In addition, a series of recent studies by Gozlan, Ben-Ari and colleagues (Gozlan and Ben-Ari, 1995) have demonstrated that modulation of the redox site may be responsible for certain forms of long-term potentiation (LTP) or depression (LTD) in the hippocampus. These authors first reported that a selective form of LTP, expressed primarily at the NMDA-component of the synaptic response in CA1 hippocampal neurons, could be induced by a brief anoxic–aglycemic period (Crepel et al., 1993). This group went on to show that this form of LTP could be reversed by the oxidizing agent DTNB, or inhibited by the reducing agent TCEP (Gozlan et al., 1994). More significantly, non-anoxic LTP or LTD of NMDA receptor-mediated synaptic responses in CA1 neurons after certain types of stimulation of Schaffer collaterals were also reported to be due to the alteration of the redox site (Gozlan et al., 1995; Bernard et al., 1997). The redox-active substances responsible for these effects in intact brain tissue remain to be identified.

Effect of redox agents on NMDA-evoked unitary currents

Single-channel recordings in outside-out neuronal patches revealed that thiol modification of native NMDA receptors produced mainly a large change in the frequency of opening, with little or no change in channel amplitude or open dwell-time (Nowak and Wright, 1992; Tang and Aizenman, 1993a; Aizenman et al., 1994). The one exception to this, at least in native receptors, was the observation that open time could be substantially prolonged in NMDA-activated channels exposed to sequential reducing and oxidizing treatments at potentials more positive than 0 mV (Tang and Aizenman, 1993c). For reasons not yet understood, this effect on open time, which appeared to be permanent or at least very long-lasting, could not only be induced just at positive potentials, but could only be observed at these membrane voltages. The nature, significance, and relationship to the more usual form of redox modulation of the modification of the NMDA receptor under such unusual circumstances have not yet been clarified.

Molecular localization of the redox site

Two studies have appeared in the literature (Köhr et al., 1994; Sullivan et al., 1994) which attempted to localize the redox-sensitive sites of recombinant NMDA receptors with the use of chimeric constructs and site-directed mutagenesis. Köhr et al. (1994), performed a detailed characterization of the effects of redox substances on recombinant NMDA receptors expressed in human embryonic kidney (HEK 293) cells. These investigators discovered pronounced subunit-dependent differences in the action of thiol substances on these receptors. The effects of DTT on NR1/NR2A receptors had two components: a rapidly developing and spontaneously reversible component, and a slowly developing and DTNB-reversible component. These authors went on to show that the reversible component was localized somewhere on the extracellular N-terminal of NR2A. Interestingly, NR1/NR2A receptors could not be alkylated to any degree with NEM following DTT treatment. In contrast, responses mediated by NR1/NR2B, NR1/NR2C and NR1/NR2D subunit configurations presented only the slowly developing potentiation during reduction, which was readily reversible by DTNB. Additionally, receptors composed of these three latter subunit combinations could be fully alkylated by NEM, thus becoming completely insensitive to further redox treatments.

Utilizing site-directed mutagenesis and the oocyte expression system, Sullivan et al. (1994) were able to demonstrate that a pair of cysteines (744 and 798) on NR1, altered independently or in tandem, constituted the redox-sensitive site of NR1/NR2B, NR1/NR2C and NR1/NR2D receptors. These cysteines are situated in the extracellular region between transmembrane domains II and III. As expected, mutated NR1(C744A,-C798A)/NR2A receptors remained redox sensitive as they still possessed the second putative site on NR2A. Interestingly, mutation of cys744 and cys798 of NR1 also decreased spermine-induced potentiation and decreased pH sensitivity of NR1/NR2B receptors, possibly suggesting that several other facets of NMDA receptor function may be allosterically related to these sites. The changes in pH sensitivity in the mutant are somewhat surprising in light of the alkylation studies in native receptors which showed no alteration in proton inhibition following the chemical modification (Tang and Aizenman, 1993a). Although similar alkylation studies have not been performed in recombinant receptors, these results do suggest that mutating the receptor's cysteine residues is not entirely akin to chemically modifying them.

A very recent study (Paoletti et al., 1997) has suggested that the rapidly reversible DTT effect on NR1/NR2A receptors, as reported by Köhr et al. (1994), actually represents the removal and re-introduction of a constitutive block by background levels of contaminating Zn^{2+} (or another metal) on a high affinity binding site on NR2A. As DTT can indeed effectively chelate this metal, the existence of a separate redox site on NR2A was questioned in that study (Paoletti et al., 1997). Nonetheless, since NR1/NR2A receptors still contain the NR1 site (cys744 and cys798), new experiments must be designed to address the question of why no component of the response mediated by this subunit configuration was sensitive to NEM treatment.

Redox sensitivity as a potential marker of NMDA receptor subunit composition

In a recent study (Brimecombe et al., 1997) we examined the effects of DTT and DTNB on the single channel properties of recombinant NMDA receptors composed of various subunit combinations. We noted that both the open dwell time and open channel frequency of unitary events elicited by NMDA in patches excised from Chinese hamster ovary (CHO) cells transfected with NR1/NR2A subunits were affected by redox agents. There was a significant (2 ms) difference between the open-dwell time constants of the reduced and oxidized states. Furthermore, DTT produced nearly a 6-fold increase in the open channel frequency when compared to the DTNB-treated patches. In contrast, redox agents affected only the frequency of opening (2-fold), and not the open time, of NMDA-activated events in patches containing either NR1/NR2B or NR1/NR2C receptor configurations, similar to what is seen in native channels. In parallel studies, we evaluated the effects of a novel non-competitive NMDA receptor antagonist, CP101,606 (Chenard et al., 1995) on NMDA-elicited channels. This drug dramatically reduced both the open time and frequency of channel opening of NR1/NR2B receptors, but only modestly inhibited NR1/NR2A and NR1/NR2C frequencies. Patch records were then obtained from cells transfected with three subunits (NR1, NR2A and NR2B) and treated with DTT, DTNB and CP101,606. Some of these patches had channels resembling either the NR1/NR2A or NR1/NR2B configurations (Table 1), while a new group emerged which was NR1/NR2B-like in its redox sensitivity (i.e. no change in open time, only a change in frequency), but NR1/NRA-like in its relative insensitivity to CP101,606 block, suggesting NR1/NR2A/NR2B co-assembly.

As mentioned earlier, we have never observed a change in open time with redox agents in NMDA-activated channels obtained from cortical neuronal patches (Tang and Aizenman, 1993a). This suggests that pure NR1/NR2A receptors are not

TABLE 1

Functional properties of NMDA receptors

Proposed subunit composition	ΔOT^{a}_{RDX}	F^{b}_{RDX}	ΔOT^{c}_{CP}	F^{d}_{CP}
NR1/NR2A	+	+	–	+
NR1/NR2B	–	+	+	++
NR1/NR2C	–	+	–	+
NR1/NR2A/NR2B	–	+	–	–
NR1/NR2A/NR2B*	–	+	–	++

[a] Difference in the open dwell-time constants of 10 μM NMDA-activated events recorded under reduced (1.0 mM DTT) and oxidized (0.1 mM DTNB) conditions.
[b] Change in the frequency of NMDA channel opening between reduced and oxidized states.
[c] Difference in the open dwell-time constants of 10 μM NMDA-induced events recorded under control conditions and in the presence of 1 μM CP.
[d] Change in the frequency of channel opening in the absence and presence of CP. +: Significant difference between the two experimental conditions. –: No significant difference between the two experimental conditions. Adapted from Brimecombe et al. (1997).

found in our cultures, at least not at the cell bodies, from where patches are usually excised. This appears to be the case even in cultures where, developmentally, NR2A message is already abundant (Zhong et al., 1994). We therefore performed a series of studies evaluating the effects of CP101,606 on these same neurons in an effort to establish the putative subunit composition of the receptors they expressed. Some patches had clearly a single population of NR1/NR2B receptors, affected both in their open time and frequency of opening by CP101,606 (Fig. 1). Surprisingly, a second group of patches had a novel profile: a lack of sensitivity to CP101,606 as far as open time, but presenting a pronounced susceptibility to frequency of opening to the drug (Fig. 2A). We returned to data previously obtained during experiments conducted for our published study (Brimecombe et al., 1997) and noted that two patches obtained from NR1, NR2A, NR2B-transfected CHO cells presented with this same pattern (Fig. 2B). As expected, the channel open time in these patches was unaffected by redox agents. This leads us to propose an additional variant of NR1/NR2A/NR2B co-assembly (NR1/NR2A/NR2B* in Table 1). The significance of this finding remains to be established, although it may provide information regarding the subunit order of the assembled receptors, or perhaps clues about subunit stoichiometry

NO and the NMDA receptor

Three manuscripts published in 1992 described the effects of NO on NMDA-stimulated responses in neurons in culture (Hoyt et al., 1992; Manzoni et al., 1992; Lei et al., 1992). These studies reported that NO or NO donors could inhibit NMDA receptor-mediated intracellular calcium elevations and whole-cell currents. However, the proposed mechanism responsible for this inhibition was not the same in all these studies. Our group (Hoyt et al., 1992) utilized either authentic NO gas or the NO donors isosorbide dinitrate (ISDN) and sodium nitroprusside (SNP) in rat cortical neurons in culture. We concluded, as had others (Fujimori and Pan-Hou, 1991; Kiedrowski et al., 1991), that the ferrocyanate moiety of SNP directly interacted with the NMDA receptor by competitively binding to the glutamate binding site, and thus this donor was not useful for examining the effects of NO at the NMDA receptor. Utilizing either NO or ISDN we noted a readily reversible, dose-dependent inhibition of NMDA-stimulated calcium responses in these neurons. Furthermore, the effects of NO were just as pronounced following pre-oxidation

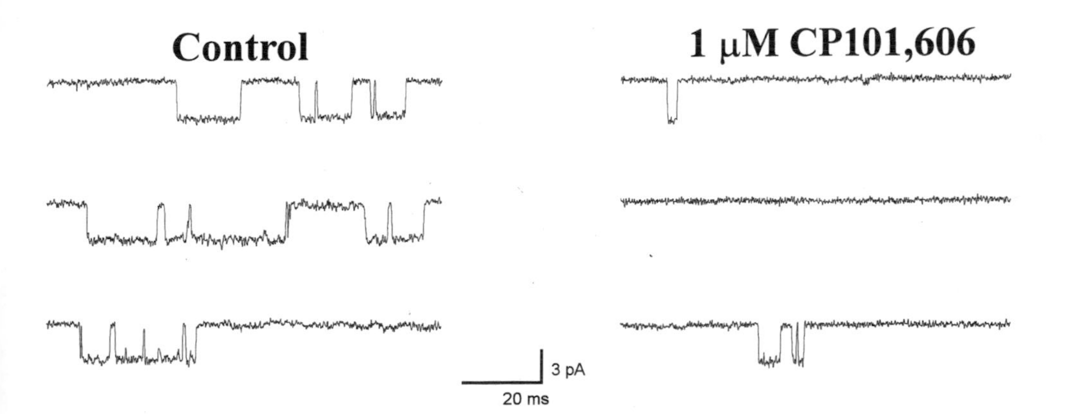

Fig. 1. NR1/NR2B NMDA receptors in neurons. NMDA (10 μM + 10 μM glycine)-activated unitary currents in an outside-out membrane patch excised from a cultured rat cortical neuron. Events at −60 mV were elicited under control conditions and in the presence of 1 μM CP101,606, an NR2B-selective non-competitive antagonist. Note that both the frequency of channel opening, and the mean channel open dwell-time are decreased by the antagonist. For this particular patch, the mean open time was decreased from 4.7 to 2.5 ms by CP101,606, while the open channel frequency was decreased by 92% by this drug.

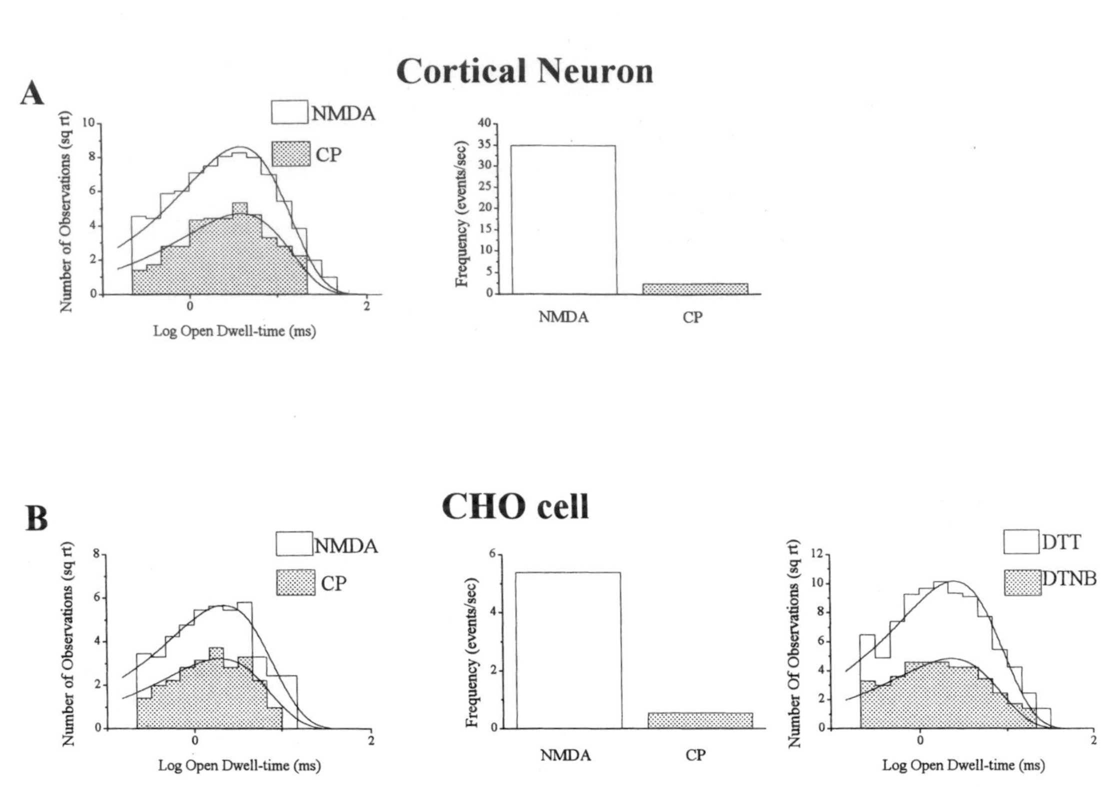

Fig. 2. NR1/NR2A/NR2B NMDA receptor co-assembly in neurons and in a transfected cell line. (A) **Left**: Open dwell-time histogram for events elicited by 10 μM NMDA in the absence or presence of 1 μM CP101,606 (CP) revealed that the open dwell-times of these channels were unaffected by the antagonist (−60 mV). **Right**: In contrast, CP101,606 produced a 93% decrease in the frequency of channel opening in the same patch. (B) Similar measurements were obtained from a patch excised from a CHO cells which had been transfected with NR1, NR2A and NR2B. No changes in open time (**left**), but a 90% decrease in open channel frequency produced by CP101,606 were observed. **Right**, This same patch was also subject to treatments with both 1 mM DTT and 0.1 mM DTNB; these redox agents produce no change in the open time of the channel.

of the receptor with DTNB or after the receptor had been fully alkylated with NEM. Most importantly, a concentration of ISDN which was observed to produce pronounced blockade of the calcium response was noted *not* to inhibit NMDA-induced whole-cell responses in the preparation. We concluded that the effects of NO on the NMDA response were due to a disruption of Ca^{2+} homeostatic mechanisms and not through a direct interaction with the receptor molecule itself. In support of this idea, several other studies have reported similar effects of NO on intracellular Ca^{2+} homeostasis in various cell types, including neurons (Furukawa et al., 1988; Morgan and Newby, 1989; Garg and Hassid, 1991; Clapp and Gurney, 1991; Pasqui et al., 1991; Stoyanovsky et al., 1997; Aghdasi et al., 1997; Brorson et al., 1997).

Manzoni and co-workers (1992) utilized the NO-producing agents 3-morpholino-sydnonimine (SIN-1) and 1-nitrosopyrrolidine (1-NP) to examine the inhibitory effects of NO on NMDA-stimulated responses in cultured mouse striatal or cerebellar granule neurons. These investigators noted that SIN-1 produced a rapidly reversible inhibition in intracellular Ca^{2+} responses in these neurons, as well as on NMDA receptor-mediated whole-cell currents. In addition, expired SIN-1, or SIN-1 in the presence of hemoglobin, failed to produce such inhibition. Similar results were obtained with 1-NP. SIN-1 was also noted to produce readily reversible blockade of NMDA-activated unitary current open probability in outside-out patches, suggesting that the actions of NO did not require any soluble intracellular messenger to produce inhibition of the receptor; the effects of NO on other single channel parameters were not analyzed in this report. These authors concluded that NO could directly interact with the NMDA receptor. However, Manzoni et al. (1992) could not determine the receptor site of action of NO during their investigations, but ruled out the redox site as a potential target, as their observed effects were readily reversible and not affected by DTT. In addition, the glutamate recognition site was also ruled out since SIN-1 did not affect the binding of the NMDA competitive antagonist [^{3}H]CGS-19755. In retrospect, we may have been somewhat premature in our conclusion as a later study by this same group (Fagni et al., 1995) showed that much higher concentrations of ISDN (10-fold higher than we had used) could in fact inhibit NMDA-evoked whole-cell responses. The results obtained in our investigation (Hoyt et al., 1992) together with this information suggest a dual action of NO on NMDA-induced physiological responses in neurons: one effect mediated by the direct interaction of the gas with the receptor molecule, and a second one which affects a downstream component of the response.

Lei et al. (1992) proposed that NO mediated its effects on NMDA receptor function via a direct interaction with the redox site. Similar to the other two reports (Hoyt et al., 1992; Manzoni et al., 1992), these investigators reported that the NO donors nitroglycerin (NTG) and SNP could produce a pronounced inhibition of NMDA-stimulated increases in intracellular Ca^{2+}. Lei and co-workers pointed out that in nearly 40% of their Ca^{2+} experiments, the effects of NTG and SNP spontaneously reversed (atypical for the effect of oxidants upon the redox site), which they attributed to a rapid decomposition of putative *S*-nitrosothiol groups formed by the reaction of an NO-related species with the sulfhydryl moieties of the receptor. In addition, this group noted that NTG and SNP could depress NMDA responses in DTNB pre-treated preparations, which is inconsistent with an effect at the redox site. Nonetheless, Lei and co-workers argued that NTG may have reacted with a separate set of thiols that are not recognized, or perhaps spared by DTNB. However, the strongest argument utilized by this group in support of a redox site of action by NO is based on a series of alkylation studies conducted in part with intracellular Ca^{2+} measurements and in part with whole-cell recordings. In their calcium studies, these investigators appeared to have produced partial alkylation of the receptor and, under these conditions, did not observe a pronounced effect of NTG, when compared with the action of this compound prior to alkylation. Similar experiments utilizing whole-cell currents are not as convincing, as they achieved only a very partial alkylation with NEM and the actions of NTG in depressing maximal (DTT-treated) responses are not very different between control (approximately 62% of maximum) and alkylated (approximately 74% of maximum) states. The results with the currents are of crucial importance since they directly measure receptor function independent of downstream events that may modulate calcium homeostasis. As mentioned earlier, interpretation of data obtained with calcium imaging is complicated by the potential effects of NO, and possibly NEM (which is membrane-permeable), on intracellular homeostatic mechanisms. Additional experiments by Lei et al. utilized the NO donor *S*-nitrosocysteine (SNOC), reproducing essentially what was observed with NTG, but only using the Ca^{2+} imaging paradigm.

Perhaps the most convincing evidence against the redox site being the site of action of NO or a related species comes from a study by Fagni et al. (1995). These investigators showed in a series of well-controlled and clearly described experiments the effects of various NO donors in modifying NMDA receptor-mediated whole-cell currents. This study is important as Lipton and co-workers had previously argued (Lipton et al., 1993) that the particular NO species generated in the various studies may account for the observed differences, with NO^+ being primarily responsible for down-regulating NMDA receptor activity via *S*-nitrosylation of the redox modulatory site. Fagni et al. showed that various NO donors (SIN-1, ISDN), including a substance thought to be an NO^+ donor, *S*-nitrosoacetylpenicillamine (SNAP; Stamler et al., 1992), produced a readily reversible inhibition of NMDA-activated currents. The effects of all these substances were similar, had voltage-dependent and voltage-independent components, and could be abolished by allowing the donors to exhaust their NO and by hemoglobin. At the single-channel level, SIN-1 was observed to produce both a voltage-independent decrease in the probability of channel opening, and a voltage-dependent decrease in channel amplitude. The latter effect is consistent with an unresolvable channel flickering block of NO or perhaps an NO-containing metal complex (Butler et al., 1995), as metal chelators also inhibit the effect of NO at the channel (Fagni et al., 1995). Neither the breakdown product of SIN-1, nor the degradation products of NO, nitrate and nitrite, mimicked the effects of SIN-1 on these currents. Furthermore, addition of superoxide dismutase and catalase failed to prevent the actions of the NO donors, suggesting that peroxynitrite was not the agent responsible for NMDA receptor inhibition. These authors also utilized SNOC, but due to its inherent instability, it could only be used in single channel recordings, where it mimicked the effects of SIN-1. Fagni and co-workers also performed alkylation experiments during their whole-cell measurements. These investigators continued to observe pronounced voltage-dependent and voltage-independent block by SIN-1 following the DTT and NEM treatments, which produced a complete alkylation of the receptor. In fact, there were no significant differences in the effects of SIN-1, at either negative or positive holding voltages, between the control and alkylated states. These studies thus argue strongly against an action of NO or a related species on the NMDA receptor redox site, at least the site which has been shown to be susceptible to alkylation (Tang and Aizenman, 1993a; Köhr et al., 1994). A recent review by Stamler et al. (1997) argues that cys744 and cys798 on NR1 are ideal targets for *S*-nitrosylation, based

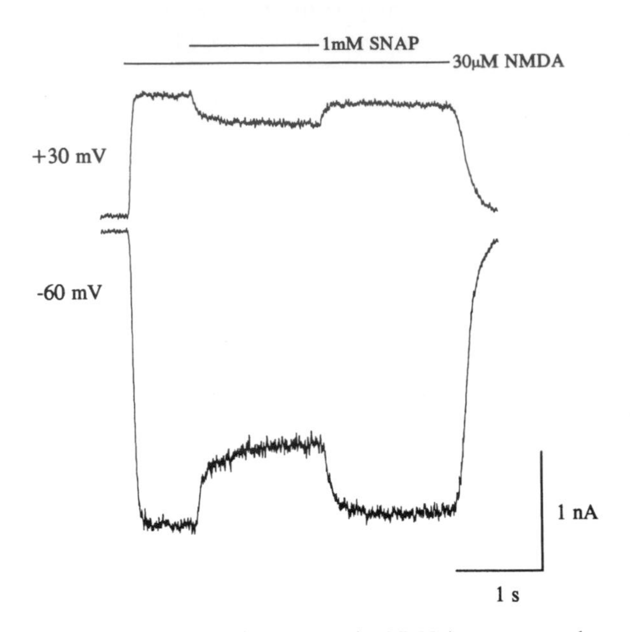

Fig. 3. NO does not interact at the NMDA receptor redox modulatory site. Whole-cell currents measured in a CHO cell transfected with a mutated NR1 subunit which lacks redox sensitivity, together with NR2B, in response to 30 μM NMDA alone or in combination with 1 mM SNAP. In this cell, SNAP produced a 27.1% block of the response at −60 mV, and 21.5% at +30 mV. A similar degree of block is observed with either this drug, or 1 mM SIN-1, at wild-type NR1/N2B receptors, and at the NR1/NR2A subunit combination.

on nearby sequence homology to similar proposed targets of NO at other proteins. As pointed out earlier in this chapter, mutation of either one or both of these cysteines is sufficient to remove the redox sensitivity of NR1/NR2B, NR1/NR2C and NR1/NR2D receptors (Sullivan et al., 1994), which are the subunit configurations that can be alkylated (Köhr et al., 1994). Hence, testing the effects of NO donors on mutated NR1 (cys744, cys798) with either NR2B, NR2C, or NR2D receptors should finally resolve this issue. We thus have recently evaluated the effects of SNAP and SIN-1 on NMDA-induced whole-cell currents in CHO cells transiently expressing either NR1/NR2A, NR1/NR2B or NR1(C744A,C798A)/NR2B (Fig. 3) subunit configurations. We observed that both 1 mM SNAP and 1 mM SIN-1 similarly produced a reversible 30.8 ± 1.8% and 24.4 ± 2.1% block of currents elicited by 30 μM NMDA at −60 and +30 mV, respectively, at all three subunit configurations tested.

From the above discussion, it appears that the hypothesis that NO blocks NMDA receptor function by *S*-nitrosylation of cys744 and cys798 of NR1 (Stamler et al., 1997) cannot be substantiated. These cysteine residues appear to constitute the only functional redox site to be localized to date on the NMDA receptor. The following points highlight the critical evidence in support of this conclusion:

1. *Oxidation*: Thiol oxidants, such as DTNB, PQQ, oxidized glutathione, or lipoic acid rapidly reverse the effects of reductants and can sometimes depress native (previously untreated) responses; their effects do not spontaneously reverse, that is, a re-application of a reducing agent is required for reversal from oxidation. The actions of NO and NO donors appear to be readily reversible under most circumstances.

2. *Alkylation*: At native (neuronal) receptors, complete alkylation has been repeatedly demonstrated. Following such treatment, neither reducing nor oxidizing agents can further alter NMDA-induced responses. In studies where full alkylation has been demonstrated, especially in experiments measuring whole-cell currents (which is the paradigm most important for testing for direct effects on NMDA receptor function), NO or NO donors still produce a reversible inhibition of the response.

3. *Physiological assay*: The particular physiological response assayed may complicate the interpretation of the actions of NO and NO donors of NMDA receptors. For example, a given concentration of an NO donor may have profound effects on NMDA-stimulated intracellular calcium increases but have no effect on NMDA receptor-mediated whole-cell currents. The fact that NO has been shown to directly alter intracellular calcium homeostatic mechanisms seriously confounds any conclusions drawn that are based on calcium measurements.

4. *Single channels*: Redox agents alter primarily the open channel frequency, not the amplitude or open time, of native NMDA channels. The published studies that examined the effects of NO on single NMDA-activated channels showed that this substance acted by influencing both frequency of opening and single channel conductance.

5. *Recombinant receptors*: Electrophysiological responses mediated by NR1/NR2B, NR1/NR2C and NR1/NR2D receptors are redox sensitive primarily via modification of cys744 and cys798 of NR1. In addition, these particular receptor configurations are fully susceptible to alkylation. However, currents mediated by mutated NR1(C744A,C798A)/NR2B receptors are blocked by NO donors to the same extent as wild-type NR1/NR2B subunits, suggesting these residues are not relevant to the actions of NO.

6. *NR1/NR2A receptors*: NR1/NR2A receptors cannot be alkylated to any extent, and appear to remain redox-sensitive even after mutating cys744 and cys798 of NR1. However, the previously proposed redox site on the amino terminal of NR2A may represent a high affinity zinc blocking site, seriously complicating any studies utilizing redox-active substances on receptors containing this subunit (although more work is needed to finally resolve the existence of a separate redox site on NR2A). More significantly however, the actions of NO donors on this receptor configuration are virtually indistinguishable from those observed at NR1/NR2B, making it very unlikely that NO works differently on the two subunit configurations.

Conclusion

We conclude then, in the first part of this chapter, that the redox modulatory site is not a target for NO. This gas, or a related species, does interact with the NMDA receptor, possibly at multiple sites, including the ion channel. However, functionally significant modification by NO of the NMDA receptor protein at any site has not been demonstrated. In addition, NO may affect NMDA receptor function by disrupting downstream Ca^{2+}

homeostatic mechanisms in neurons. NO-mediated inhibition of NMDA receptor may have physiological relevance (Izumi et al., 1992; Kato and Zorumski, 1993; Manzoni and Bockaert, 1993). Additional experiments are required to fully understand the complex effects of this novel neural messenger on glutamatergic function in the brain.

Role of NO as a mediator of NMDA receptor mediated neurotoxicity

The emergence of the hypothesis that excitotoxic injury is a major cause of neuronal death in the central nervous system in both acute and chronic neurologic disorders is one of the most important developments in neuroscience in the past 25 years (Choi, 1988; Meldrum and Garthwaite, 1990). Not surprisingly, this advance has been driven by an increase in our understanding of excitatory synaptic transmission in all its aspects, and a paradox that has come into focus with the development of this understanding is that excitatory neurotransmitters are also potent neurotoxins. In the last 10 years, the most significant development in the study of excitotoxicity has been the demonstration that nitric oxide mediates a component of NMDA receptor mediated neurotoxicity (Dawson et al., 1991), and this advance has paralleled a recognition and understanding of the important role played by nitric oxide in the central nervous system (Garthwaite et al., 1988; Garthwaite and Boulton, 1995). However, the role of nitric oxide in NMDA receptor mediated neurotoxicity has been a controversial subject, and the goal of this section is to examine the causes of the controversy as well as to put this discovery in perspective.

Defining the controversy

That a controversy exists is ascertained by the publication of papers failing to demonstrate a role for nitric oxide in NMDA receptor mediated neurotoxicity in cortical cultures (e.g. Demerle-Pallardy et al., 1991; Hewett et al., 1993), or in vivo after a stroke (Buchan et al., 1994), following the initial publication of results showing that nitric oxide does have an important role in this process. Additionally, the specificity of nitric oxide synthase inhibitors has been questioned (Peterson et al., 1992). Ultimately, the importance of nitric oxide production as a cause of cell death in stroke was demonstrated by using targeted gene disruption to produce mice that lacked neuronal nitric oxide synthase (Huang et al., 1994). These mice had smaller infarctions in a middle cerebral artery occlusion model of stroke compared to controls that expressed this enzyme. Importantly, these studies do not establish that nitric oxide is involved in NMDA receptor mediated injury, only that it is involved in the neuronal death that occurs in stroke, which is not the same thing. However, using dissociated cell cultures derived from the knockout animals, it was found that neurons in vitro were remarkably resistant to NMDA receptor mediated injury (Dawson et al., 1996). These latter studies, together with a variety of experiments demonstrating the effect of scavengers of NO, inhibition of NOS, and elimination of nNOS containing neurons demonstrated convincingly that nitric oxide has an important role in excitotoxicity (Dawson et al., 1991, 1993). In addition, this work has generated a great deal of interest in the possibility that nitric oxide is involved in the neurodegeneration that occurs in chronic diseases as well, and evidence to support this hypothesis has been forthcoming from a variety of approaches, including immunostaining using antibodies directed against nitrotyrosine as well as the biochemical demonstration of products of nitric oxide's reaction with cell constituents (Bruijn et al., 1997). Finally, a role in neurotoxicity for nitric oxide derived from non-neuronal sources has also been demonstrated (Hewett et al., 1994).

How to account for the controversy?

There are several factors that appear to contribute to the variable expression of NO mediated excitotoxicity:

1. *Variable expression of nitric oxide synthase in dissociated cell culture*. Recently Samdani et al.

(1997b) showed that the expression of nNOS is dependent upon the culture paradigm, and that plating of neurons on glial feeder layers, as is a common practice, is associated with an impoverishment of nNOS containing neurons and associated nNOS protein and catalytic activity. They also showed that exposure of neurons cultured in this manner to neurotrophins is associated with an increase in all of these parameters as well as an enhancement of NMDA neurotoxicity. These results could account for a previous observation that BDNF exposure enhances NMDA receptor mediated neurotoxicity, although mediation by nitric oxide in that report was not examined (Koh et al., 1995).

In order to make the case that nitric oxide mediated neurotoxicity is inhibited by culturing neurons on feeder layers, Samdani et al. compared vulnerability to NMDA toxicity between neurons plated directly on a poly-ornithine matrix, neurons plated on feeder layers, and neurons plated on feeder layers treated with BDNF. It is noteworthy that these authors only observed 20% neuronal death 24 h after a 5-min exposure of neurons on feeder layers to 500 μM NMDA. Previous work demonstrated 88% neuronal loss with a 5-min exposure to 1 mM NMDA (92% loss with a 24 h exposure to 300 μM) and EC_{50} values of 120 μM for a 5-min exposure and 16 μM for a 24 h exposure using similar cultures derived from mouse (Koh and Choi, 1988). Thus, the cultures used which did not show a nitric oxide mediated component of the toxicity were, surprisingly, relatively insensitive to short exposures of NMDA. The insensitivity of the cultures made a striking contrast with the cultures that were plated directly on poly-ornithine. An examination of the cytotoxicity paradigm used by these authors reveals that they used cultures at 14 days in vitro. However, sensitivity to NMDA receptor mediated neurotoxicity is developmentally regulated and may reach full expression in cultures only after weeks in vitro (Choi, 1985; Sinor et al., 1997). Therefore, it is likely that the resistance of neurons in the cultures on feeder layers is due to the developmental age of the cultures used.

The authors conclude that the resistance to NMDA toxicity of neurons plated on feeder layers is due to an immature phenotype with respect to NOS expression in that total NOS activity and number of NOS positive neurons is low. However, it has been shown that the developmental regulation of sensitivity to NMDA excitotoxicity (Choi, 1985) appears to be due to age-dependent changes in the redox state of the NMDA receptor (Sinor et al., 1997). Therefore an alternate explanation is that the resistance to NMDA toxicity in feeder layer cultures at 2 weeks in vitro can be explained by an *immature phenotype of NMDA receptor expression*, and if the cultures had been tested at 4 weeks in vitro instead of at 2 weeks in vitro, then they would have shown the sensitivity to NMDA demonstrated many times previously. Neurons in 2-week cultures, although relatively insensitive to NMDA (acting via nitric oxide independent mechanisms) appear to be quite sensitive to nitric oxide. Hence, if enough nitric oxide synthase containing neurons are present, then the vulnerability to NMDA stimulation will be increased. However the magnitude of the contribution of nitric oxide mediated death could be exaggerated by the use of immature cultures for these experiments. In order to sort out the relative contributions of nitric oxide mediated and nitric oxide independent mechanisms of toxicity, and to have some basis for comparison with work from other laboratories, what is needed is a careful examination of the concentration dependence of NMDA toxicity in the two culture systems performed with 5-min and 24 h exposures at different ages in vitro (2 and 4 weeks). Such studies would provide an understanding of the place of nitric oxide mediated neurotoxicity in the context of the changing sensitivity of neurons to NMDA receptor mediated neurotoxicity due to developmental regulation of NMDA receptor expression and perhaps intracellular death pathways as well. This understanding is currently lacking.

2. *Timing and intensity of insult is critical.* It follows from the previous discussion that the timing and intensity of exposure to NMDA is likely to be crucial to the detection of an NO-

dependent component of the toxicity. Recently it has been shown that this is clearly the case (Strijbos et al., 1996). Thus, in striatal cultures it was shown that the NO-mediated component of NMDA toxicity diminished as duration of exposure to the agonist was extended longer than 5 min. This study is noteworthy for carefully defining the parameters required to demonstrate NO mediation of cell death, but also in clearly demonstrating that an alternate nitric oxide independent mechanism of toxicity was present in the same cultures as well. Interesting also is the demonstration that the nitric oxide mediated component was due to the post-exposure elevation of extracellular glutamate levels in the extracellular space and thus was blocked, not by nitric oxide synthase antagonists *during* the NMDA exposure, but only during the first few hours *following* NMDA exposure. Thus, a careful characterization of the triggering insult is necessary to reveal a nitric oxide mediated component of excitotoxicity as well as nitric oxide-independent mechanisms.

3. *Cellular localization of the insult may be important*. Studies using a variety of approaches suggest that the most vulnerable site of excitotoxic injury is the dendrite, which is consistent with the fact that excitatory synapses are localized on this structure (Peters et al., 1991). For example, morphological studies have shown that the first signs of injury in response to exposure to excitotoxins is to be found on dendrites (Park et al., 1996; Faddis et al., 1997). In addition, other studies comparing the toxicity of transported and non-transported glutamate agonists (Rosenberg and Aizenman, 1989; Rosenberg et al., 1992) and the effects of antagonists (Speliotes et al., 1994), suggest that the most sensitive target for excitotoxic injury is the dendrite and not the cell body. In toxicity paradigms in which dendrites are protected by the presence of astrocytes, such as when mixed cultures of astrocytes and neurons are used, an agonist that is readily transported by glutamate transporters, such as glutamate itself, may target only the exposed cell body because it may never reach receptors on dendrites. In contrast, in a paradigm in which dendrites are exposed to the extracellular medium or a non-transported agonist such as NMDA is used, access to the dendrites is free, and then these will be the primary site of toxicity. This distinction may be important because the initial site at which agonists exert their toxicity may be critical to the type of mechanism that is actually activated. This hypothesis is supported by recent evidence showing that NOS is localized in dendrites in close association with NMDA receptors (Aoki et al., 1997). Sequestration of excitatory synapses may also be related to the observed developmental changes in sensitivity to NMDA toxicity.

Nitric oxide generation is not necessary to produce NMDA receptor mediated neurotoxicity

The available evidence supports the view that NMDA receptor mediated neurotoxicity can proceed without mediation by nitric oxide. The several papers that failed to confirm a role for nitric oxide all are relevant to this point. In addition, the fact that it has been clearly demonstrated that longer exposures to NMDA produce neurotoxicity independent of nitric oxide generation in the same culture system in which brief exposure produces NO dependent toxicity also is an evidence against a necessary role for nitric oxide. Therefore there is at least one NMDA receptor mediated mechanism of neurotoxicity that proceeds independent of nitric oxide production, and we can say with certainty that NO generation is not *necessary* to produce NMDA receptor mediated neurotoxicity.

A summary of conditions promoting NO-dependent and NO-independent NMDA receptor-mediated toxicity is presented in Table 2. What clearly is lacking at this point is a hypothesis concerning the downstream pathway(s) of nitric oxide independent NMDA toxicity that is as powerful a stimulant of new experiments, approaches, and ideas as the nitric oxide hypothesis has been. The existence of nitric oxide independent pathways is, however, indisputable and indisputably important.

TABLE 2

Conditions favoring expression of nitric oxide dependent or independent NMDA neurotoxicity

Condition	NO toxicity dominant	Non-NO toxicity dominant
Exposure time	≤ 5 min	> 5 min
Culture paradigm	No feeder layer	Feeder layer
NOS (+) neurons	Rare	Even rarer
Developmental status (assessed by efficacy of NMDA as a neurotoxin)	Immature (≤ 2weeks; relatively insensitive to NMDA in presence of NOS inhibitors)	Mature (> 2 weeks; relatively sensitive to NMDA in presence of NOS inhibitors)

Nitric oxide generation is not sufficient to cause toxicity

We know that exposure of neurons in vitro to nitric oxide donors results in neuronal death (Lipton et al., 1993). Therefore endogenous nitric oxide generation could be sufficient to account for a component of the neurotoxicity that follows NMDA receptor activation. However since superoxide dismutase blocks the toxicity of nitric oxide (Radi et al., 1991), it is presumed that reaction with superoxide is required for the formation of peroxynitrite. Thus generation of nitric oxide per se is not sufficient to account for the toxicity that results; rather, the combination with superoxide is required. This might account for the puzzling fact that although NOS containing neurons are vulnerable to rapidly activated calcium dependent toxicity due to the presence of calcium permeable AMPA/kainate receptors (Weiss et al., 1993), kainate activation is not sufficient to trigger nitric oxide mediated toxicity to surrounding (NOS-negative) neurons (Dawson et al., 1993). Kainate evoked calcium influx, after all, would be expected to activate the calcium dependent NOS in these neurons. The explanation may be that NMDA receptor mediated activation of superoxide production (Lafon-Cazal et al., 1993) is a necessary concomitant of nitric oxide toxicity. This idea has been put forward as a possible explanation of an interesting and pertinent result derived from studies of excitotoxicity in mixed cultures of cortical neurons from mice with a disrupted NR1 gene and from wild-type mice (Tokita et al., 1996). Pure cultures from wild-type mice showed nearly complete loss of neurons in response to NMDA treatment, whereas cultures from the knockout mice were completely resistant to NMDA toxicity, although vulnerable to kainate toxicity. Mixed cultures from tissue obtained from both types of mice showed 50% loss of neurons to NMDA treatment, and the neurons that survived were deficient in NR1 protein. If nitric oxide were mediating the death of wild-type neurons in these cultures, then one would also expect the NR1- neurons to be killed as well, but this was not the case. Thus, either insufficient nitric oxide was being produced, or it was not getting to neighboring neurons, or as the authors state, the "neurotoxic effect of nitric oxide may require the simultaneous activation of the NMDA receptor" (and, for example, superoxide production) in the target cell (Tokita et al., 1996).

Multiple mechanisms of neuronal death are activated in hypoxia/ischemia

1. *Involvement of nitric oxide in vivo is likely to be model-dependent.* One of the classically recognized features of injury in the nervous system is regional specificity. For example it has been shown that anoxia produces selective injury to specific layers of cortex, the CA1 region of hippocampus, the amygdala, and the striatum, and any hypothesis meant to explain the neuronal death that occurs must be able to account also for this selective vulnerability. A remarkable study has appeared that suggests that the selective vulnerability seen in

a global ischemia model of stroke is due to the release of zinc from pre-synaptic sites of accumulation and uptake by post-synaptic neurons (Koh et al., 1996). Entry of zinc into post-synaptic targets is presumably due to permeation through voltage dependent calcium channels as well as glutamate receptors, and previous studies demonstrated that zinc influx was associated with neurotoxicity (Weiss et al., 1993; Koh and Choi, 1994; Choi et al., 1988). The mechanism of zinc neurotoxicity is unclear, but there is little to suggest that it involves nitric oxide, although this has not been ruled out. Since NMDA antagonists are of limited effectiveness in global ischemia models (Buchan, 1992), and since nitric oxide mediates a form of NMDA receptor mediated injury, one would not expect that NMDA receptor dependent generation of nitric oxide has much of a role in the injury that occurs in global ischemia. Inducible NOS, however, contributes to the pathogenesis of injury in focal ischemia (Iadecola et al., 1997; Samdani et al., 1997a; Iadecola 1997), and may in global ischemia as well.

2. *Does nitric oxide dependent NMDA receptor mediated injury account for all the NMDA receptor mediated injury in focal ischemia models*? Multiple mechanisms, NO-dependent and NO-independent, appear to be able to cause neuronal death in response to NMDA receptor overstimulation. Under certain circumstances, as has been discussed, the NO-dependent form may predominate, for example in immature neurons that otherwise are relatively resistant to excitotoxicity, and when nNOS expression is high. But this is only one of several overlapping mechanisms, and clearly an important issue is to determine the relative roles of nitric oxide dependent and independent forms of neurotoxicity in vivo, in animal models and in the setting of disease. Curiously, opportunities to address this issue have been missed. For example, in the studies using nNOS knockout mice, the demonstration of a reduction in infarct volume in the knockout animals compared to control was important in establishing a role of nitric oxide in infarction (Huang et al., 1994). It would have been useful to have seen in the same animals what fraction of the infarction could be blocked by NMDA antagonists. If a significant portion of the infarction observed in these animals could be blocked by NMDA antagonists, then this would have clearly made the point that multiple forms of NMDA receptor mediated neurotoxicity are at work in infarction to kill neurons.

3. *Necrosis and Apoptosis*. The fate of neurons in an infarction depends upon their vulnerability to various mechanisms of toxicity. Excitotoxicity produces reversible and irreversible forms of injury in vitro, an example of the former being cell swelling that has been shown to be calcium independent (Choi, 1987). In fact, cell swelling is aggravated by removal of calcium (Choi, 1987) due to the calcium dependence of cell volume regulation (Churchwell et al., 1996). Although cell swelling may be reversible in vitro, it can have lethal consequences in the brain due to confinement of the brain within a rigid structure. With respect to irreversible forms of injury, there is now substantial evidence showing that in addition to necrosis, apoptotic cell death is important in the neuronal loss that occurs in infarction (Hara et al., 1997a) including recent studies with transgenic mice that express a dominant negative form of interleukin-1β converting enzyme (Hara et al., 1997b; Friedlander et al., 1997). In a variety of cell types nitric oxide has been shown to cause apoptotic cell death (Albina et al., 1993; Estevez et al., 1995; Lin et al., 1995) although only in one instance so far in neurons from the central nervous system. In this instance the cell type was cerebellar granule cells, and apoptotic cell death was due to nitric oxide stimulated release of excitatory amino acids (Leist et al., 1997). It remains to be shown whether nitric oxide induces apoptotic death in forebrain neurons, either in vitro or in vivo.

What downstream mechanisms are activated by nitric oxide?

Nitric oxide generation is capable of affecting a variety of enzymes and cellular components due to its reactivity and the reactivity of peroxynitrite and its byproducts. Enzymes with iron–sulfur centers are vulnerable, including aconitase, NADH–ubiquinone oxidoreductase, succinate–ubiquinone oxidoreductase, and ribonucleotide reductase (Garthwaite and Boulton, 1995) but nitrosylation reactions may affect a great range of proteins through direct reaction with amino acid side chains (Stamler et al., 1992, 1997). In addition, because of the reactivity of the hydroxyl radical byproduct of peroxynitrite decomposition, nonspecific damage to lipids, protein, and nucleic acids may be anticipated (Beckman and Crow, 1993). It has been shown that nitric oxide induced damage to DNA causes intense activation of polyADP ribose polymerase (PARP), that inhibition of PARP results in the prevention of NMDA receptor mediated toxicity (Zhang et al., 1994), and that a PARP knockout mouse is resistant to infarction (Endres et al., 1997). This is a very important extension of the nitric oxide story, but an important question to answer will be how PARP activation relates to other downstream mechanisms implicated in neurotoxicity, including calcium dependent pathways unrelated to NOS activation (Lipton and Rosenberg, 1994) and mitochondrial dysfunction (White and Reynolds, 1996; Schinder et al., 1996).

Protective effects of nitric oxide

In the past, the primary focus for investigating protective actions of nitric oxide has been on the possibility of interaction between nitric oxide and the redox site of the NMDA receptor (Lei et al., 1992). The existence of this interaction is questionable, as is clear from the first part of this chapter, and will not be further addressed here. In the past few years evidence for other mechanisms potentially conferring protection by nitric oxide upon neurons has emerged. It is noteworthy that in the studies with the nNOS knockout mouse, it was found that although infarction size was decreased relative to controls, the use of an NOS inhibitor in this mouse increased the size of infarcted tissue (Huang et al., 1994). This was attributed to effects on blood perfusion, a hypothesis that has been supported by recent work with endothelial NOS knockout mice (Huang et al., 1996).

Protective effects of nitric oxide have been demonstrated in several cell death paradigms. Nitric oxide delays cell death in PC12 cells undergoing NGF withdrawal, and this effect is mediated by accumulation of cGMP (Farinelli et al., 1996). In endothelial cells, both endogenous and exogenous NO was shown to block TNFα induced apoptosis and CPP-32 activation by a specific nitrosylation at Cys 163, an essential amino acid conserved among ICE/CPP-32 like caspases (Dimmeler et al., 1997). Nitric oxide itself may act as an antioxidant by chelating iron (Sergent et al., 1997), and antioxidants are known to block excitotoxicity in some models (Lafon-Cazal et al., 1993). Finally, nitric oxide directly nitrosylates ras, altering its function as a GTP–GDP exchange protein, resulting in the activation of the MAP kinase pathway (Lander et al., 1997). Since this pathway has been shown to act as a "survival pathway" (Xia et al., 1995), nitrosylation of ras might have profound effects on cell survival. Thus, the largely unstudied means by which the generation of nitric oxide could enhance the survival of neurons as well as other cells might contribute to controversies concerning the action of nitric oxide in the brain, depending on the relative activation of cell death and cell survival promoting actions. More importantly, the existence of cell survival promoting actions might complicate the development and application of therapeutic interventions aimed at interfering with nitric oxide generation or action.

Conclusion

The study of excitotoxicity has seen recently the emergence of a hypothesis that NMDA receptor mediated generation of nitric oxide is an important mechanism causing the death of neurons in the brain. The controversial nature of this hypothesis lies in difficulties in reproducing the central phenomenon, namely that NMDA receptor mediated neurotoxicity in vitro is mediated by nitric oxide production. This appears to have arisen because of insufficient characterization of the conditions under which this mechanism predominates in the killing of neurons in response to exposure to NMDA agonists. The key to settling the controversy lies in pursuing this characterization as well as in the recognition of the limits of the hypothesis. Nitric oxide generation is one of multiple factors that contribute to the death of neurons following overactivation of NMDA receptors and in ischemia, and the pursuit of other mechanisms is likely to advance our knowledge of the causes of neuronal death in stroke as well as in chronic neurodegenerative diseases. In addition, recent evidence suggests that a variety of mechanisms exist by which nitric oxide generation might enhance cell survival. Understanding these mechanisms is also likely to be important in understanding the complex role nitric oxide plays in the life and death of cells in the nervous system.

References

Aghdasi, B., Zhang, J.Z., Wu, Y., Reid, M.B. and Hamilton, S.L. (1997) Multiple classes of sulfhydryls modulate the skeletal muscle Ca^{2+} release channel. *J. Biol. Chem.*, 272: 3739–3748.

Aizenman, E. (1994) Redox modulation of the NMDA receptor. In: M.G. Palfreyman, I.J. Reynolds and P. Skolnik (Eds.), Direct and allosteric control of glutamate receptors. CRC Press, *Boca Raton*, pp. 95–104.

Aizenman, E. (1995) Modulation of *N*-methyl-D-aspartate receptors by hydroxyl radicals in rat cortical neurons in vitro. *Neurosci. Letts.*, 189: 57–59.

Aizenman, E., Hartnett, K.A. and Reynolds, I.J. (1990) Oxygen free radicals regulate NMDA receptor function via a redox modulatory site. *Neuron*, 5: 841–846.

Aizenman, E., Hartnett, K.A., Zhong, C., Gallop, P.M. and Rosenberg, P.A. (1992) Interaction of the putative essential nutrient pyrroloquinoline quinone with the *N*-methyl-D-aspartate receptor redox modulatory site. *J. Neurosci.*, 12: 2362–2369.

Aizenman, E., Jensen, F.E., Gallop, P.M., Rosenberg, P.A. and Tang, L.H. (1994) Further evidence that pyrroloquinoline quinone interacts with the *N*-methyl-D-aspartate receptor redox site in rat cortical neurons in vitro. *Neurosci. Letts.*, 168: 189–192.

Aizenman, E., Lipton, S.A. and Loring, R.H. (1989) Selective modulation of NMDA responses by reduction and oxidation. *Neuron*, 2: 1257–1263.

Albina, J.E., Cui, S., Mateo, R.B. and Reichner, J.S. (1993) Nitric oxide-mediated apoptosis in murine peritoneal macrophages. *J. Immunol.*, 150: 5080–5085.

Aoki, C., Rhee, J., Lubin, M. and Dawson, T.M. (1997) NMDA-R1 subunit of the cerebral cortex co-localizes with neuronal nitric oxide synthase at pre- and postsynaptic sites and in spines. *Brain Res.*, 750: 25–40.

Beckman, J.S. and Crow, J.P. (1993) Pathological implications of nitric oxide, superoxide and peroxynitrite formation. *Biochem. Soc. Trans.*, 21: 330–334.

Bernard, C., Cannon, R.C., Ben, A.Y. and Wheal, H.V. (1997) Model of spatio-temporal propagation of action potentials in the Schaffer collateral pathway of the CA1 area of the rat hippocampus. *Hippocampus*, 7: 58–72.

Brimecombe, J.C., Boeckman, F.A. and Aizenman, E. (1997) Functional consequences of NR2 subunit composition in single recombinant *N*-methyl-D-aspartate receptors. *Proc. Natl. Acad. Sci. USA*, 94: 11019–11024.

Brorson, J.R., Sulit, R.A. and Zhang, H. (1997) Nitric oxide disrupts Ca^{2+} homeostasis in hippocampal neurons. *J. Neurochem.*, 68: 95–105.

Bruijn, L.I., Beal, M.F., Becher, M.W., Schulz, J.B., Wong, P.C., Price, D.L. and Cleveland, D.W. (1997) Elevated free nitrotyrosine levels, but not protein-bound nitrotyrosine or hydroxyl radicals, throughout amyotrophic lateral sclerosis (ALS)-like disease implicate tyrosine nitration as an aberrant in vivo property of one familial ALS-linked superoxide dismutase 1 mutant. *Proc. Natl. Acad. Sci. USA*, 94: 7606–7611.

Buchan, A.M. (1992) Do NMDA antagonists prevent neuronal injury? No. *Arch. Neurol.*, 49: 420–421.

Buchan, A.M., Gertler, S.Z., Huang, Z.G., Li, H., Chaundy, K.E. and Xue, D. (1994) Failure to prevent selective CA1 neuronal death and reduce cortical infarction following cerebral ischemia with inhibition of nitric oxide synthase. *Neuroscience*, 61: 1–11.

Butler, A.R., Flitney, F.W. and Williams, D.L. (1995) NO, nitrosonium ions, nitroxide ions, nitrosothiols and iron-nitrosyls in biology: A chemist's perspective. *Trends Pharmacol. Sci.*, 16: 18–22.

Chenard, B.L., Bordner, J., Butler, T.W., Chambers, L.K., Collins, M.A., De, C.D., Ducat, M.F., Dumont, M.L., Fox, C.B. and Mena, E.E. (1995) (1*S*,2*S*)-1-(4-hydroxyphenyl)-2-(4-hydroxy-4-phenylpiperidino)-1-propanol: A potent new neuroprotectant which blocks *N*-methyl-D-aspartate responses. *J. Med. Chem.*, 38: 3138–3145.

Choi, D.W. (1985) Glutamate neurotoxicity in cortical cell culture is calcium dependent. *Neurosci. Letts.*, 58: 293–297.

Choi, D.W. (1987) Ionic dependence of glutamate neurotoxicity. *J. Neurosci.*, 7: 369–379.

Choi, D.W. (1988) Glutamate neurotoxicity and diseases of the nervous system. *Neuron*, 1: 623–634.

Choi, D.W., Yokoyama, M. and Koh, J. (1988) Zinc neurotoxicity in cortical cell culture. *Neuroscience*, 24: 67–79.

Churchwell, K.B., Wright, S.H., Emma, F., Rosenberg, P.A. and Strange (1996) NMDA receptor activation inhibits neuronal volume regulation after swelling induced by veratridine-stimulated Na+ influx in rat cortical cultures. *J. Neurosci.*, 16: 7447–7457.

Clapp, L.H. and Gurney, A.M. (1991) Modulation of calcium movements by nitroprusside in isolated vascular smooth muscle cells. *Pflugers Archiv. - Eur. J. Physiol.*, 418: 462–470.

Crepel, V., Hammond, C., Krnjevic, K., Chinestra, P. and Ben-Ari, Y. (1993) Anoxia-induced LTP of isolated NMDA receptor-mediated synaptic responses. *J. Neurophysiol.*, 69: 1774–1778.

Dawson, V.L., Dawson, T.M., London, E.D., Bredt, D.S. and Snyder, S.H. (1991) Nitric oxide mediates glutamate neurotoxicity in primary cortical cultures. *Proc. Natl. Acad. Sci. USA*, 88: 6368–6371.

Dawson, V.L., Dawson, T.M., Bartley, D.A., Uhl, G.R. and Snyder, S.H. (1993) Mechanisms of nitric oxide-mediated neurotoxicity in primary brain cultures. *J. Neurosci.*, 13: 2651–2661.

Dawson, V.L., Kizushi, V.M., Huang, P.L., Snyder, S.H. and Dawson, T.M. (1996) Resistance to neurotoxicity in cortical cultures from neuronal nitric oxide synthase-deficient mice. *J. Neurosci.*, 16: 2479–2487.

Demerle-Pallardy, C., Lonchampt, M.O., Chabrier, P.E. and Braquet, P. (1991) Absence of implication of L-arginine/nitric oxide pathway on neuronal cell injury induced by L-glutamate or hypoxia. *Biochem. & Biophys. Res. Comm.*, 181: 456–464.

Dimmeler, S., Haendeler, J., Galle, J. and Zeiher, A.M. (1997) Oxidized low-density lipoprotein induces apoptosis of human endothelial cells by activation of CPP32-like proteases. A mechanistic clue to the 'response to injury' hypothesis. *Circulation*, 95: 1760–1763.

Endres, M., Laufs, U., Merz, H. and Kaps, M. (1997) Focal expression of intercellular adhesion molecule-1 in the human carotid bifurcation. *Stroke*, 28: 77–82.

Estevez, A.G., Radi, R., Barbeito, L., Shin, J.T., Thompson, J.A. and Beckman, J.S. (1995) Peroxynitrite-induced cytotoxicity in PC12 cells: Evidence for an apoptotic mechanism differentially modulated by neurotrophic factors. *J. Neurochem.*, 65: 1543–1550.

Faddis, B.T., Hasbani, M.J. and Goldberg, M.P. (1997) Calpain activation contributes to dendritic remodeling after brief excitotoxic injury in vitro. *J. Neurosci.*, 17: 951–959.

Fagni, L., Olivier, M., Lafon-Cazal, M. and Bockaert, J. (1995) Involvement of divalent ions in the nitric oxide-induced blockade of *N*-methyl-D-aspartate receptors in cerebellar granule cells. *Mol. Pharmacol.*, 47: 1239–1247.

Farinelli, S.E., Park, D.S. and Greene, L.A. (1996) Nitric oxide delays the death of trophic factor-deprived PC12 cells and sympathetic neurons by a cGMP-mediated mechanism. *J. Neurosci.*, 16: 2325–2334.

Friedlander, R.M., Gagliardini, V., Hara, H., Fink, K.B., Li, W., MacDonald, G., Fishman, M.C., Greenberg, A.H., Moskowitz, M.A. and Yuan, J. (1997) Expression of a dominant negative mutant of interleukin-1-β converting enzyme in transgenic mice prevents neuronal cell death induced by trophic factor withdrawal and ischemic brain injury. *J. Exper. Med.*, 185: 933–940.

Fujimori, H. and Pan-Hou, H. (1991) Effect of nitric oxide on L-[^{3}H]glutamate binding to rat brain synaptic membranes. *Brain Res.*, 554: 355–357.

Furukawa, K., Tawada, Y. and Shigekawa, M. (1988) Regulation of the plasma membrane Ca^{2+} pump by cyclic nucleotides in cultured vascular smooth muscle cells. *J. Biol. Chem.*, 263: 8058–8065.

Garg, U.C. and Hassid, A. (1991) Nitric oxide decreases cytosolic free calcium in Balb/c 3T3 fibroblasts by a cyclic GMP-independent mechanism. *J. Biol. Chem.*, 266: 9–12.

Garthwaite, J. and Boulton, C.L. (1995) Nitric oxide signaling in the central nervous system. *Ann. Rev. Physiol.*, 57: 683–706.

Garthwaite, J., Charles, S.L. and Chess-Williams, R. (1988) Endothelium-derived relaxing factor release on activation of NMDA receptors suggests role as intercellular messenger in the brain. *Nature*, 336: 385–388.

Gilbert, K.R., Aizenman, E. and Reynolds, I.J. (1991) Oxidized glutathione modulates *N*-methyl-D-aspartate- and depolarization-induced increases in intracellular Ca^{2+} in cultured rat forebrain neurons. *Neurosci. Letts.*, 133: 11–14.

Gozlan, H. and Ben-Ari, Y. (1995) NMDA receptor redox sites: Are they targets for selective neuronal protection? *Trends Pharmacol. Sci.*, 16: 368–374.

Gozlan, H., Diabira, D, Chinestra, P. and Ben-Ari, Y. (1994). Anoxic LTP is mediated by the redox modulatory site of the NMDA receptor. *J. Neurophysiol.*, 72: 3017–3022.

Gozlan, H., Khazipov, R., Diabira, D. and Ben-Ari, Y. (1995) In CA1 hippocampal neurons, the redox state of NMDA receptors determines LTP expressed by NMDA but not by AMPA receptors. *J. Neurophysiol.*, 73: 2612–2617.

Hara, H., Fink, K., Endres, M., Friedlander, R.M., Gagliardini, V., Yuan, J. and Moskowitz, M.A. (1997a) Attenuation of transient focal cerebral ischemic injury in transgenic mice expressing a mutant ICE inhibitory protein. *J. Cerebral Blood Flow & Metabol.*, 17: 370–375.

Hara, H., Friedlander, R.M., Gagliardini, V., Ayata, C., Fink, K., Huang, Z., Shimizu-Sasamata, M., Yuan, J. and Moskowitz, M.A. (1997b) Inhibition of interleukin 1β converting enzyme family proteases reduces ischemic and excitotoxic neuronal damage. *Proc. Natl. Acad. Sci. USA*, 94: 2007–2012.

Hewett, S.J., Corbett, J.A., McDaniel, M.L. and Choi, D.W. (1993) Inhibition of nitric oxide formation does not protect murine cortical cell cultures from *N*-methyl-D-aspartate neurotoxicity. *Brain Res.*, 625: 337–341.

Hewett, S.J., Csernansky, C.A. and Choi, D.W. (1994) Selective potentiation of NMDA-induced neuronal injury following induction of astrocytic iNOS. *Neuron*, 13: 487–494.

Hoyt, K.R., Tang, L.H., Aizenman, E. and Reynolds, I.J. (1992) Nitric oxide modulates NMDA-induced increases in intracellular Ca^{2+} in cultured rat forebrain neurons. *Brain Res.*, 592: 310–316.

Huang, Z., Huang, P.L., Panahian, N., Dalkara, T., Fishman, M.C. and Moskowitz, M.A. (1994) Effects of cerebral ischemia in mice deficient in neuronal nitric oxide synthase. *Science*, 265: 1883–1885.

Huang, Z., Huang, P.L., Ma, J., Meng, W., Ayata, C., Fishman, M.C. and Moskowitz, M.A. (1996) Enlarged infarcts in endothelial nitric oxide synthase knockout mice are attenuated by nitro-L-arginine. *J. Cerebral Blood Flow & Metabol.*, 16: 981–987.

Iadecola, C. (1997) Bright and dark sides of nitric oxide in ischemic brain injury. *Trends in Neurosci.*, 20: 132–139.

Iadecola, C., Yang, G., Ebner, T.J. and Chen, G. (1997) Local and propagated vascular responses evoked by focal synaptic activity in cerebellar cortex. *J. Neurophysiol.*, 78: 651–659.

Izumi, Y., Clifford, D.B. and Zorumski, C.F. (1992) Inhibition of long-term potentiation by NMDA-mediated nitric oxide release. *Science*, 257: 1273–1276.

Jensen, F.E., Gardner, G.J., Williams, A.P., Gallop, P.M., Aizenman, E. and Rosenberg, P.A. (1994) The putative essential nutrient pyrroloquinoline quinone is neuroprotective in a rodent model of hypoxic/ischemic brain injury. *Neuroscience*, 62: 399–406.

Kato, K. and Zorumski, C.F. (1993) Nitric oxide inhibitors facilitate the induction of hippocampal long-term potentiation by modulating NMDA responses. *J. Neurophysiol.*, 70: 1260–1263.

Kiedrowski, L., Manev, H., Costa, E. and Wroblewski, J.T. (1991) Inhibition of glutamate-induced cell death by sodium nitroprusside is not mediated by nitric oxide. *Neuropharmacology*, 30: 1241–1243.

Koh, J.Y. and Choi, D.W. (1988) Vulnerability of cultured cortical neurons to damage by excitotoxins: Differential susceptibility of neurons containing NADPH-diaphorase. *J. Neurosci.*, 8: 2153–2163.

Koh, J.Y. and Choi, D.W. (1994) Zinc toxicity on cultured cortical neurons: Involvement of *N*-methyl-D-aspartate receptors. *Neurosci.*, 60: 1049–1057.

Koh, J.Y., Gwag, B.J., Lobner, D. and Choi, D.W. (1995) Potentiated necrosis of cultured cortical neurons by neurotrophins. *Science*, 268: 573–575.

Koh, J.Y., Suh, S.W., Gwag, B.J., He, Y.Y., Hsu, C.Y. and Choi, D.W. (1996) The role of zinc in selective neuronal death after transient global cerebral ischemia. *Science*, 272: 1013–1016.

Kohr, G., Eckardt, S., Luddens, H., Monyer, H. and Seeburg, P.H. (1994) NMDA receptor channels: Subunit-specific potentiation by reducing agents. *Neuron*, 12: 1031–1040.

Lafon-Cazal, M., Pietri, S., Culcasi, M. and Bockaert, J. (1993) NMDA-dependent superoxide production and neurotoxicity. *Nature*, 364: 535–537.

Lander, H.M., Hajjar, D.P., Hempstead, B.L., Mirza, U.A., Chait, B.T., Campbell, S. and Quilliam, L.A. (1997) A molecular redox switch on p21(ras). Structural basis for the nitric oxide-p21(ras) interaction. *J. Biol. Chem.*, 272: 4323–4326.

Lei, S.Z., Pan, Z.H., Aggarwal, S.K., Chen, H.S., Hartman, J., Sucher, NJ and Lipton, S.A. (1992) Effect of nitric oxide production on the redox modulatory site of the NMDA receptor-channel complex. *Neuron*, 8: 1087–1099.

Leist, M., Volbracht, C., Kuhnle, S., Fava, E., Ferrandomay, E. and Nicotera, P. (1997) Caspase-mediated apoptosis in neuronal excitotoxicity triggered by nitric oxide. *Mol. Med.*, 3: 750–764.

Lin, K.T., Xue, J.Y., Nomen, M., Spur, B. and Wong, P.Y. (1995) Peroxynitrite-induced apoptosis in HL-60 cells. *J. Biol. Chem.*, 270: 16487–16490.

Lipton, S.A., Choi, Y.B., Pan, Z.H., Lei, S.Z., Chen, H.S., Sucher, N.J., Loscalzo, J., Singel, D.J. and Stamler, J.S. (1993) A redox-based mechanism for the neuroprotective and neurodestructive effects of nitric oxide and related nitroso-compounds. *Nature*, 364: 626–632.

Lipton, S.A. and Rosenberg, P.A. (1994) Excitatory amino acids as a final common pathway for neurologic disorders. *New Eng. J. Med.*, 330: 613–622.

Majewska, M.D., Bell, J.A. and London, E.D. (1990) Regulation of the NMDA receptor by redox phenomena: Inhibitory role of ascorbate. *Brain Res.*, 537: 328–332.

Manzoni, O. and Bockaert, J. (1993) Nitric oxide synthase activity endogenously modulates NMDA receptors. *J. Neurochem.*, 61: 368–370.

Manzoni, O., Prezeau, L., Marin, P., Deshager, S., Bockaert, J. and Fagni, L. (1992) Nitric oxide-induced blockade of NMDA receptors. *Neuron*, 8: 653–662.

McBain, C.J. and Mayer, M.L. (1994) *N*-methyl-D-aspartic acid receptor structure and function. *Physiol. Rev.*, 74: 723–760.

Meldrum, B. and Garthwaite, J. (1990) Excitatory amino acid neurotoxicity and neurodegenerative disease. *Trends Pharmacol. Sci.*, 11: 379–387.

Morgan, R.O. and Newby, A.C. (1989) Nitroprusside differentially inhibits ADP-stimulated calcium influx and mobilization in human platelets. *Biochem. J.*, 258: 447–454.

Nowak, L.M. and Wright, J.M. (1992) Slow voltage-dependent changes in channel open-state probability underlie hysteresis of NMDA responses in $Mg^{(2+)}$-free solutions. *Neuron*, 8: 181–187.

Paoletti, P., Ascher, P. and Neyton, J. (1997) High-affinity zinc inhibition of NMDA NR1-NR2A receptors. *J. Neurosci.*, 17: 5711–5725.

Park, J.S., Bateman, M.C. and Goldberg, M.P. (1996) Rapid alterations in dendrite morphology during sublethal hypoxia or glutamate receptor activation. *Neurobiol. Disease*, 3: 215–227.

Pasqui, A.L., Capecchi, P.L., Ceccatelli, L., Mazza, S., Gistri, A., Laghi, P.F. and Di, P.T. (1991) Nitroprusside in vitro inhibits platelet aggregation and intracellular calcium translocation. Effect of haemoglobin. *Thrombosis Res.*, 61: 113–122.

Peters, A., Palay, S.L. and Webster, H.D. (1991) In A. Peters, S.L. Palay, and H.D. Webster, (Eds.), *The Fine Structure of the Nervous System*, Oxford University Press, New York, p. 82.

Peterson, D.A., Peterson, D.C., Archer, S. and Weir, E.K. (1992) The non specificity of specific nitric oxide synthase inhibitors. *Biochem. & Biophys. Res. Comm.*, 187: 797–801.

Radi, R., Beckman, J.S., Bush, K.M. and Freeman, B.A. (1991) Peroxynitrite-induced membrane lipid peroxidation: The cytotoxic potential of superoxide and nitric oxide. *Arch. Biochem. & Biophy.*, 288: 481–487.

Rosenberg, P.A. and Aizenman, E. (1989) Hundred-fold increase in neuronal vulnerability to glutamate toxicity in astrocyte-poor cultures of rat cerebral cortex. *Neurosci. Letts.*, 103: 162–168.

Rosenberg, P.A., Amin, S. and Leitner, M. (1992) Glutamate uptake disguises neurotoxic potency of glutamate agonists in cerebral cortex in dissociated cell culture. *J. Neurosci.*, 12: 56–61.

Samdani, A.F., Dawson, T.M. and Dawson, V.L. (1997a) Nitric oxide synthase in models of focal ischemia. *Stroke*, 28: 1283–1288.

Samdani, A.F., Newcamp, C., Resink, A., Facchinetti, F., Hoffman, BE, Dawson, V.L. and Dawson, T.M. (1997b) Differential susceptibility to neurotoxicity mediated by neurotrophins and neuronal nitric oxide synthase. *J. Neurosci.*, 17: 4633–4641.

Schinder, A.F., Olson, E.C., Spitzer, N.C. and Montal, M. (1996) Mitochondrial dysfunction is a primary event in glutamate neurotoxicity. *J. Neurosci.*, 16: 6125–6133.

Sergent, O., Griffon, B., Morel, I., Chevanne, M., Dubos, M.P., Cillard, P. and Cillard, J. (1997) Effect of nitric oxide on iron-mediated oxidative stress in primary rat hepatocyte culture. *Hepatology*, 25: 122–127.

Sinor, J.D., Boeckman, F.A. and Aizenman, E. (1997) Intrinsic redox properties of *N*-methyl-D-aspartate receptor can determine the developmental expression of excitotoxicity in rat cortical neurons in vitro. *Brain Res.*, 747: 297–303.

Speliotes, E.K., Hartnett, K.A., Blitzblau, R.C., Aizenman, E. and Rosenberg, P.A. (1994) Comparison of the potency of competitive NMDA antagonists against the neurotoxicity of glutamate and NMDA. *J. Neurochem.*, 63: 879–885.

Stamler, J.S., Singel, D.J. and Loscalzo, J. (1992) Biochemistry of nitric oxide and its redox-activated forms. *Science*, 258: 1898–1902.

Stamler, J.S., Toone, E.J., Lipton, S.A. and Sucher, N.J. (1997) (*S*)NO signals: Translocation, regulation and a consensus motif. *Neuron*, 18: 691–696.

Stoyanovsky, D., Murphy, T., Anno, P.R., Kim, Y.M. and Salama, G. (1997) Nitric oxide activates skeletal and cardiac ryanodine receptors. *Cell Calcium*, 21: 19–29.

Strijbos, P.J., Leach, M.J. and Garthwaite, J. (1996) Vicious cycle involving Na^+ channels, glutamate release and NMDA receptors mediates delayed neurodegeneration through nitric oxide formation. *J. Neurosci.*, 16: 5004–5013.

Sucher, N.J., Aizenman, E. and Lipton, S.A. (1991) *N*-methyl-D-aspartate antagonists prevent kainate neurotoxicity in rat retinal ganglion cells in vitro. *J. Neurosci.*, 11: 966–971.

Sullivan, J.M., Traynelis, S.F., Chen, H.S., Escobar, W., Heinemann, S.F. and Lipton, S.A. (1994) Identification of two cysteine residues that are required for redox modulation of the NMDA subtype of glutamate receptor. *Neuron*, 13: 929–936.

Tang, L.H. and Aizenman, E. (1993a) The modulation of *N*-methyl-D aspartate receptors by redox and alkylating reagents in rat cortical neurones in vitro. *J. Physiol.*, 465: 303–323.

Tang, L.H. and Aizenman, E. (1993b) Allosteric modulation of the NMDA receptor by dihydrolipoic and lipoic acid in rat cortical neurons in vitro. *Neuron*, 11: 857–863.

Tang, L.H. and Aizenman, E. (1993c) Long-lasting modification of the *N*-methyl-D-aspartate receptor channel by a voltage-dependent sulfhydryl redox process. *Mol. Pharmacol.*, 44: 473–478.

Tokita, Y., Bessho, Y., Masu, M., Nakamura, K., Nakao, K., Katsuki, M. and Nakanishi, S. (1996) Characterization of excitatory amino acid neurotoxicity in *N*-methyl-D-aspartate receptor-deficient mouse cortical neuronal cells. *Eur. J. Neurosci.*, 8: 69–78.

Weiss, J.H., Hartley, D.M., Koh, J.Y. and Choi, D.W. (1993) AMPA receptor activation potentiates zinc neurotoxicity. *Neuron*, 10: 43–49.

White, R.J. and Reynolds, I.J. (1996) Mitochondrial depolarization in glutamate-stimulated neurons: An early signal specific to excitotoxin exposure. *J. Neurosci.*, 16: 5688–5697.

Xia, Z., Dickens, M., Raingeaud, J., Davis, R.J. and Greenberg, M.E. (1995) Opposing effects of ERK and JNK-p38 MAP kinases on apoptosis. *Science*, 270: 1326 1331.

Zhang, J., Dawson, V.L., Dawson, T.M. and Snyder, S.H. (1994) Nitric oxide activation of poly(ADP-ribose) synthetase in neurotoxicity. *Science*, 263: 687–689.

Zhong, J., Russell, S.L., Pritchett, D.B., Molinoff, P.B. and Williams (1994) Expression of mRNAs encoding subunits of the *N*-methyl-D-aspartate receptor in cultured cortical neurons. *Mol. Pharmacol.*, 45: 846–853.

R.R. Mize, T.M. Dawson, V.L. Dawson and M.J. Friedlander (Eds.)
Progress in Brain Research, Vol 118

CHAPTER 6

Redox modulation of the NMDA receptor by NO-related species

Stuart A. Lipton*, Posina V. Rayudu, Yun-Beom Choi, Nikolaus J. Sucher[†] and H.S.-Vincent Chen

The CNS Research Institute, Brigham and Women's Hospital, and Program in Neuroscience, Harvard Medical School, Boston, MA 02115, USA

Abstract

The chemical reactions of NO are largely dictated by its redox state. Increasing evidence suggests that the various redox states of the NO group exist endogenously in biological tissues. In the case of NO^+ equivalents, the mechanism of reaction often involves *S*-nitrosylation (transfer of the NO group to a cysteine sulfhydryl to form an RS-NO); further oxidation of critical thiols can possibly form disulfide bonds. We have physiological and chemical evidence that NMDA receptor activity can be modulated by *S*-nitrosylation, resulting in a decrease in channel opening. Recent data suggest that NO^-, probably in the singlet (or high-energy) state, can also react with critical sulfhydryl group(s) of the NMDA receptor to down-regulate its activity; in the triplet (lower-energy) state NO^- may oxidize these NMDA receptor sulfhydryl groups by formation of an intermediate such as peroxynitrite. It has also been reported that $NO^{\cdot}$ can react with thiol but only under specific circumstances, e.g., if an electron acceptor such as O_2 is present, as well at catalytic amounts of metals like copper, and if the conditions do not favor the kinetically preferred reaction with $O_2^{\cdot-}$ to yield peroxynitrite. Mounting evidence in many fields suggests that *S*-nitrosylation can regulate the biological activity of a great variety of proteins, perhaps analogous to phosphorylation. Thus, this chemical reaction is gaining acceptance as a newly-recognized molecular switch to control protein function via reactive thiol groups such as those encountered on the NMDA receptor.

Introduction: Redox modulation and NO-related species

In recent years, as endogenous sources of oxidizing and reducing agents have been discovered, redox modulation of protein function has been recognized to be an important mechanism for many cell types affecting a variety of proteins. For our purposes, we will confine our review of redox modulation to covalent modification of sulfhydryl (thiol) groups on protein cysteine residues. If they possess a sufficient redox potential, oxidizing agents can react to form adducts on single sulfhydryl groups or, if two free sulfhydryl groups are vicinal (in close proximity), disulfide bonds may possibly be formed. Reducing agents can regenerate free sulfhydryl (–SH) groups by donating electron(s). The redox modulatory sites

*Corresponding author. Tel.: +1 617 278 0363; fax: +1 617 264 5277; e-mail: slipton @rics.bwh. harvard. edu

[†]Present address: Department of Biology, Hong Kong University of Science and Technology.

of the NMDA receptor consist of critical cysteine residues which, when chemically reduced, increase the magnitude of NMDA-evoked responses.

In contrast, after oxidation, NMDA-evoked responses are decreased in size. Considering endogenous redox agents, in addition to the usual suspects including glutathione, lipoic acid, and reactive oxygen species, nitric oxide and its redox-related species have recently come to the fore. This has occurred largely because of the rediscovery and application to biological systems of work from the early part of this century showing the organic synthesis of nitrosothiols (RS–NO) (Reviewed by Stamler et al., 1992). NO group donors represent different redox-related species of the NO group, each with its own distinctive chemistry, which lead to entirely different biological effects. NO-related species include nitric oxide ($NO^{\bullet}$) but also the other redox-related forms of the NO group: with one less electron (NO^+ or nitrosonium ion) or one additional electron (NO^- or nitroxyl anion) (Stamler et al., 1992). Recent evidence suggests that all three of these redox-related forms or their functional equivalents are important pharmacologically and physiologically, participating in distinctive chemical reactions. NO^+ can be transferred from either endogenous or exogenous donors to thiol (a reaction termed *S*-nitrosylation) to form nitrosothiols on the NMDA receptor (NRS–NO), and the singlet (or high-energy state) of NO^-can also react with thiol groups. The reaction of the NO group with NMDA receptor thiol decreases NMDA-evoked responses similar to an oxidizing agent (Fig. 1). A consensus motif of amino acids comprised of nucleophilic residues surrounding a critical cysteine, which increase the cysteine sulfhydryl's susceptibility to *S*-nitrosylation by NO^+ donors, has been proposed (Stamler et al., 1997). Thus the NMDA receptor subunits contain typical consensus motifs for *S*-nitrosylation.

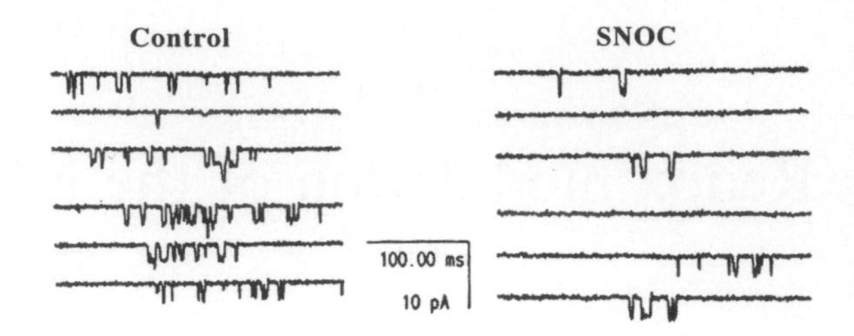

Fig. 1. Addition of *S*-nitrosocysteine (SNOC) to an outside-out patch decreases the opening frequency of NMDA-evoked single-channel events. A patch electrode was used to record NMDA-evoked currents in the outside-out configuration. A 2-min addition of SNOC (100 μM) resulted in fewer openings. This effect persisted for at least several min and was reversed by DTT (2 mM for 2 min, data not shown). Channel activity was elicited with 10 μM NMDA/5 μM glycine. The patch pipette-filling solution contained (in mM) 120 CsCl, 20 TEA-Cl, 2 $MgCl_2$, 1 $CaCal_2$, 2 EGTA, 10 Hepes, and phenol red (0.001% vol/vol), at pH 7.2. The bath (external) solution was comprised of Hanks' balanced salt solution with (in mM) 137.6 NaCl, 1 $NaHCO_3$, 0.34 Na_2HPO_4, 5.36 KCl, 0.44 $KHPO_4$, 22.2 glucose, 5 Hepes, at pH 7.2. To eliminate confounding currents, the following were added to the bath (in μM): 1 tetrodotoxin (TTX), 10 strychnine, 100 bicucculine, and 5 NBQX.

Reaction of cysteine sulfhydryls with the NO group

Free endogenous nitrosonium (NO^+) exists only at low pH. However, functional equivalents of NO^+ can be transferred to thiol, or more properly perhaps, thiolate anion (RS^-), at physiological pH. For example, transfer of NO^+ equivalents occur from one nitrosothiol to another, a reaction termed transnitrosylation, i.e., $R\text{–}SH + R'\text{–}SNO \rightleftharpoons R\text{–}SNO + R'\text{–}SH$. Since transfer of NO^+ equivalents involves thiolate anion ($R\text{–}S^-$), it is pH dependent (Arnelle and Stamler, 1995). Endogenous nitrosothiols, such as *S*-nitroso-glutathione, have been demonstrated to exist in brain and in lung and to react in this manner. The enzymatic machinery underlying the formation and breakdown of nitrosothiols is just beginning to be characterized. Recently, for example, thioredoxin reductase was shown to catalyze the homolytic cleavage of nitrosothiol (R–SNO) to nitric oxide ($NO^{\bullet} + RS^{\bullet}$) (Nikitovic and Holmgren, 1996).

Very recently, the groups of Schmidt and Feelish presented evidence that neuronal nitric oxide synthase (NOC) may produces NO^- rather than or in addition to $NO^{\bullet}$ (Schmidt et al., 1996). NO^- presents an arcane chemistry since NO^- can apparently be encountered in two different states, consisting of either a high (singlet) or low (triplet) energy state, each with distinctive chemistries. In particular, singlet NO^- can react directly with thiol while triplet NO^- does not (Bonner and Stedman, 1996). However, in the triplet state NO^- may react with O_2 to form peroxynitrite ($ONOO^-$), which in turn may oxidize free thiols to disulfide (Radi et al., 1991; Kim et al., 1996).

Classically, there was no precedent for direct reaction of $NO^•$ with thiols under anaerobic conditions (Pryor and Lightsey, 1981; Pryor et al., 1982). Recently, however, Ischiropoulos and co-workers demonstrated that under particular conditions, e.g., in the presence of an electron acceptor such as O_2 and catalytic amounts of transition metals, $NO^•$ can react with thiol to form a nitrosothiol (Gow et al., 1997). One overly simplistic but useful conceptualization of this reaction is that the intermediate formed by an electron acceptor and $NO^•$ would be effecting NO^+ transfer in the presence of thiol. Nonetheless, the reaction of $NO^•$ and $O_2^{•-}$ to form peroxynitrite would be kinetically favored over the formation of nitrosothiol if superoxide anion is also present (e.g., if $O_2^{•-}$ is not scavenged by superoxide dismutase (SOD).

Another important concept in considering the possible chemical reactions of the NO group involves our image of the local diffusion and ephemeral nature of $NO^•$. Recently, David Bredt and colleagues demonstrated that neuronal NOS is located very close to potential targets of NO by virtue of its PDZ domain (Brenman et al., 1996). For example, NOS interacts via its PDZ domain with carboxyl-terminal tail of NMDAR1, the subunit of the *N*-methyl-D-aspartate (NMDA) receptor that is essential for functional activity. Therefore, restricted diffusional constraints and the need for high local concentrations to facilitate NO reactions should not present a problem.

With some of the chemical reactions of these NO-related species in hand, we will next pay particular attention to the mechanism of *S*-nitrosylation or transfer of the NO moiety to cysteine sulfhydryl group(s) on the NMDA receptor. Of particular note, in recent months *S*-nitrosylation has also been shown to regulate the activity of various other ion channel proteins, G-proteins, growth factors, enzymes, and transcription factors. These reactions of NO-related species do not involve the well-known activation of guanylate cyclase by reaction with heme to increase cGMP formation. Rather they involve reactions with cysteine sulfhydryls on an increasing number of protein targets to provide modulation of function, analogous to phosphorylation of critical serine, threonine, or tyrosine residues.

The chemical reactivity of the NO group and its associated redox states is related to the local redox milieu and peptide environment, pH, temperature, and the presence of catalytic amounts of transition metals. Here, we will illustrate *S*-nitrosylation (transfer of NO^+ equivalents to thiol groups) to modulate protein functional activity. The first published example of this phenomenon is represented by the reaction of the NO group with regulatory sulfhydryl(s) of the NMDA receptor's redox modulatory site(s), resulting in down-regulation of receptor activity (Lipton et al., 1993). It had been known that NO donors could decrease NMDA function (Lei et al., 1992; Manzoni et al., 1992; Manzoni and Bockaert, 1993), but the exact mechanism remained in question (Fagni et al., 1995). The redox basis for this reaction will be presented below.

Most importantly, each of the NO-related species (NO^+, $NO^•$, and NO^- in singlet or triplet energy states) participates in different chemical reactions (Stamler et al., 1992; Lipton et al., 1993; Lipton and Stamler, 1994). Nowhere is this more apparent than in the reactions of $NO^•$ versus NO^+, which influence neuronal survival in a diametrically opposed fashion (Lipton et al., 1993). While $NO^•$ reacts with superoxide anion $O_2^{•-}$ to form peroxynitrite ($ONOO^-$), which in turn triggers neurotoxic reactions either by itself or via its breakdown products, NO^+ equivalents can react with a redox modulatory site(s) on the NMDA receptor to down-regulate the receptor's activity, which produces neuroprotection (Beckman et al., 1990; Dawson et al., 1991; Lipton et al., 1993)

S-Nitrosylation, NMDA receptor down-regulation, and neuroprotection

The NO group can down-regulate NMDA receptor activity (Hoyt et al., 1992; Lei et al., 1992; Manzoni et al., 1992), apparently at a redox modulatory site(s) of the receptor, consisting of critical cysteine sulfhydryl or thiol group(s) (Lei et al., 1992; Lipton et al., 1993; Kohr et al., 1994; Sullivan et al., 1994). In native neurons (e.g. cerebrocortical cells), we measured amplitude of NMDA-evoked responses, monitored by whole-cell and single-channel recording with a patch electrode or by digital calcium imaging with the Ca^{2+} sensitive dye fura-2 (Lei et al., 1992; Lipton et al., 1993). We found that sulfhydryl reducing agents, such as dithiothreitol (DTT) which promote the formation of free thiol groups, increased NMDA responses, predominantly by increasing the opening frequency of NMDA receptor-operated channels. In contrast, oxidizing agents, such 5,5′-dithio(2-bisnitrobenzoic acid) (DTNB) decreased NMDA responses, by forming thiobenzoate protein conjugates at single sulfhydryl groups or perhaps by facilitating disulfide bond formation. Additionally, taken together with the DTT and DTNB results, we knew that thiols on the NMDA receptor were involved because under our conditions *N*-ethylmaleimide (NEM), a relatively specific agent for alkylating thiols, irreversibly blocked the effects of these redox reagents while itself slightly decreasing responses to NMDA (Lei et al., 1992; Lipton et al., 1993). Importantly, under specific conditions, NEM also prevented the subsequent effects of NO donors, indicating that reactions of thiol and NO groups were involved. Recently, both our group and that of Joël Bockaert (Manzoni and Bockaert, 1993) have also demonstrated that endogenous production of NO can decrease NMDA receptor activity, indicating the potential physiological importance of this effect. In these experiments implicating the involvement of endogenous NO, inhibition of NOS was found to enhance subsequent NMDA receptor responses.

As an example of an NO^+ chemical reaction at the NMDA receptor, we found that *S*-nitrosocysteine decreases NMDA receptor activity as demonstrated by whole-cell recording or by digital calcium imaging (Lipton et al., 1993). During single-channel recording, *S*-nitrosocysteine (SNOC) decreased the opening frequency of NMDA receptor-operated channels in outside-out patches from cerebrocortical neurons (Fig. 1). In the presence of SOD, SNOC attenuated NMDA-evoked Ca^{2+} influx, a prerequisite for NMDA receptor-mediated neurotoxicity. Not surprisingly therefore, under the same conditions, application of SNOC ameliorated NMDA receptor-mediated neurotoxicity. These findings can be explained best by SNOC donating NO^+ equivalents. Thus, *S*-nitrosylation or facile transfer of an NO^+ equivalent to thiol groups of the NMDA receptor (i.e., heterolytic fission of RS-NO) results in a nitrosothiol derivative of the NMDA receptor, which down-regulates receptor activity (Fig. 2). Under these conditions, any $NO^\bullet$ produced by alternative homolytic cleavage of SNOC is prevented from entering a neurotoxic pathway of peroxynitrite formation ($ONOO^-$) via reaction with $O_2^{\bullet-}$ because of the presence of excess SOD (Lipton et al., 1993; Lipton and Stamler, 1994). Rather, NO group transfer leads to down-regulation of NMDA receptor activity, possibly through facilitation of disulfide formation, although this chemical reaction has not yet been proven definitively at the NMDA receptor's redox modulatory site (Lipton et al., 1996a) (hence the dashed line in

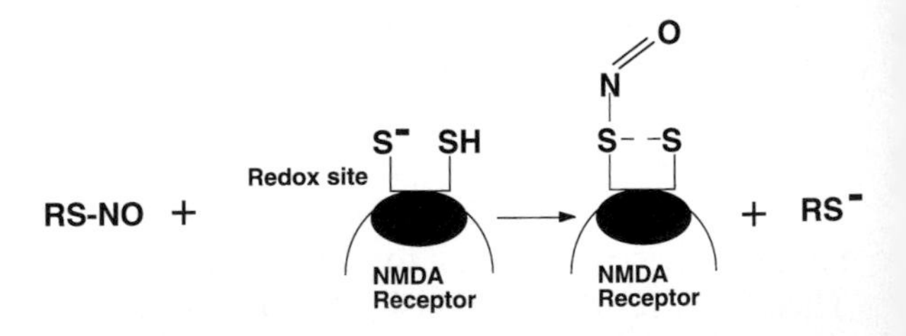

Fig. 2. *S*-Nitrosylation of critical thiol group(s) of the NMDA receptor's redox modulatory site by *S*-nitrosocysteine (SNOC), a more general example of which is RS–NO, a nitrosylated protein. The redox modulatory site of the NMDA receptor is transnitrosylated by transferring the NO group (in the NO^+ form) from RS–NO to cysteine sulfhydryl group(s) on the NMDA receptor-channel complex. This results in a decreased frequency of channel opening and hence decreased NMDA receptor activity.

Fig. 2). The fact that EDTA can prevent the effects of the NO group on NMDA receptor activity (Fagni et al., 1995) supports rather than refutes this chemistry. In particular, metals can facilitate nitrosative reactions involving $NO^{\bullet}$ (Stamler, et al., 1992; Lipton et al., 1996a). Nitrosation of redox sites is facilitated by oxygen, transition metals, and perhaps $O_2^{\bullet -}$ (superoxide anion) (Stamler, 1994; Gow et al., 1997). The common event is transfer of a NO^+ equivalent or another intermediate with NO^+-like character to form an RS–NO.

In accordance with the recent report that nNOS may produce NO^- (nitroxyl anion) (Schmidt et al., 1996), we have also tested the ability of exogenous and endogenous generation of NO^- to modulate NMDA receptor function. We have found that donors of singlet NO^- can decrease NMDA receptor activity apparently via *S*-nitrosylation because the effects can be prevented by pretreatment with NEM. Additionally, enzymatic generation of NO^-, presumably in the triplet state, can also decrease NMDA responses but apparently via formation of peroxynitrite following reaction of triplet NO^- with O_2 (Kim et al., 1996; Lipton et al., 1996b).

Nitrosylation of recombinant NMDA receptors

In order to better understand the aforementioned redox results, all of which were obtained on primary neurons with native NMDA receptors, we have turned to a recombinant system. Our work on nitrosylation and other redox reactions of recombinant NMDA receptors in the *Xenopus* oocyte expression system is instructive but also must be interpreted with a degree of caution (Sullivan et al., 1994; Sucher et al., 1996). We do not yet appreciate how to form recombinant NMDA receptors that exactly mimic native receptors and therefore conclusions based on site-directed mutagenesis studies cysteines must be viewed with tempered enthusiasm. In fact, in the course of performing PCR reactions based on primers containing the cysteines known to be unique to NMDA receptor subunits, our group discovered a new NMDA receptor subunit (originally termed NMDAR-L or χ^{-1}, but more recently named NR3A) (Ciabàrra et al., 1995; Sucher et al., 1995; Takasu et al., 1997; Rothe et al., 1997; Das et al., 1998). Additional unidentified NMDA subunits probably remain to be identified. Thus, it is not yet possible to definitively understand native NMDA receptor responses based on recombinant subunits. This statement notwithstanding, our preliminary data suggest that the cysteines at position 744 and 798 are not only important to redox reactions in general but also to the effect of NO on the NMDA receptor; however, additional cysteines on the NR2A subunit contribute most of the NO effect. This is still very much a work in progress, but our preliminary results show that (i) specific NMDA receptor subunit combinations manifest larger NO-induced decreases in activity than other receptor subunit combinations (Omerovic et al., 1995; Sucher et al., 1996), and (ii) possibly as many as seven cysteine residues on NR1 and NR2A influence NO, Zn^{2+}, or redox agent effects on the NMDA receptor. While it appears that NO effects may be predominantly due to reaction with a single cysteine sulfhydryl, other cysteine residues may also contribute producing polynitrosylation (Choi, Y.-B., Chen, H.-S.V. and Lipton, S.A. in preparation).

Other effects of NO are of course not ruled out by these findings. It is also true that the effects of Zn^{2+} on the NMDA receptor and that of redox agents can be confused because some reducing agents (such as DTT) bind Zn^{2+}, because EDTA chelates Zn^{2+}, and Zn^{2+} may also be coordinated, at least in part, by cysteine residues. It has been proposed by Joël Bockaert's group (Fagni et al., 1995) that NO can react with a Zn^{2+} site of the NMDA receptor However, Zn^{2+} can be coordinated by cysteine residues (as well as by histidine, glutamate, aspartate and possibly other amino acids), and therefore we hypothesize that the sites of NO and Zn^{2+} actions may share at least some cysteine residue(s). Experiments in progress should be able to determine if this hypothesis is correct.

Nitroglycerin down-regulates NMDA receptor activity and ameliorates neurotoxicity in vitro and in vivo

Based on the above findings, the ideal NO group donor drug would be one that reacts readily with the critical thiol group(s) of the redox modulatory site(s) of the NMDA receptor to inhibit excessive Ca^{2+} influx. We therefore studied nitroglycerin (NTG) as an exemplary compound. Specifically, this drug does not spontaneously liberate true nitric oxide ($NO^{\bullet}$) to any significant extent, and it is known to react readily with thiol groups forming derivative thionitrites (RS–NO) or thionitrates ($RS–NO_2$) (together, these are represented as $RS–NO_x$, $x = 1$ or 2) (Lei et al., 1992; Lipton et al., 1993).

Using whole-cell recording via patch clamp electrodes and digital calcium imaging with fura-2 on primary cerebrocortical neurons, we found that nitroglycerin inhibited NMDA-evoked currents and Ca^{2+} influx (Lei et al., 1992; Lipton et al., 1993). Strong evidence that this effect of nitroglycerin is mediated by its reactions with thiol in the above-illustrated manner came from a series of chemical experiments. These studies showed that specific alkylation of thiol groups with NEM completely abrogated the inhibitory effect of nitroglycerin on subsequent NMDA-evoked responses under our conditions (Lei et al., 1992).

The finding that nitroglycerin could inhibit NMDA-evoked responses was corroborated by the demonstration that nitroglycerin also significantly ameliorates NMDA-induced neuronal killing in cerebrocortical cultures (Lei et al., 1992; Lipton et al., 1993). In addition, preliminary data suggest that high doses of nitroglycerin are neuroprotective in rat models of focal ischemia under conditions of constant systemic blood pressure and modestly increased cerebral blood flow in the penumbra (Lipton and Wang, 1996). These parameters are held stable either by inducing tolerance to the systemic effects of nitroglycerin through chronic transdermal application (Sathi et al., 1993), or by intravenous infusion of a pressor agent concurrently with nitroglycerin (Lipton and Wang, 1996). Although difficult to prove in vivo, it appears likely that the decrease in stroke size observed after treatment with nitroglycerin is at least in part due to its effect on decreasing NMDA receptor activity although other beneficial actions are also possible (Lipton and Wang, 1996).

S-Nitrosylation of caspases

Another example of beneficial nitroglycerin and nitrosylation reactions that can prevent neuronal cell death involves caspases. Caspases are members of the ICE-CED-3 protease family of enzymes that play crucial roles in mammalian apoptosis during development or due to growth factor deprivation. Additionally, caspases have been recently implicated in the pathway to neuronal apoptosis from mild excitotoxic insults (Lipton, S.A. and Nicotera, P., in press). More intense excitotoxic injuries evoke rapid and irreversible energy compromise, leading to the failure of ionic homeostasis with consequent swelling and lysis (Ankarcrona et al., 1995; Bonfoco et al., 1995). This form of cell death represents necrosis and is not dependent on caspases. One new regulatory pathway of caspase activity involves *S*-nitrosylation. In a very recent development in the apoptosis field, in primary neurons (Tenneti et al., 1997) and in other cells (Dimmeler et al., 1997; Melino et al., 1997; Ogura et al., 1997) caspase activity has been found to be decreased by *S*-nitrosylation or transfer of an NO-like group to a critical cysteine sulfhydryl that is located in the active site in all caspase enzymes. This finding has important implications for the regulation of apoptosis by the NO group: under specific redox conditions that favor nitrosylation of caspases, apoptosis can be attenuated by down-regulating caspase activity, whereas under other conditions $NO^{\bullet}$ will react with $O_2^{\bullet -}$ to form peroxynitrite ($ONOO^-$) and precipitate cell death (either apoptotic or necrotic depending on the intensity of the initial insult).

S-Nitrosylation of cysteine sulfhydryls on other ion channels, enzymes, transcriptions factors and regulatory proteins

Shortly after NMDA receptor activity was shown to be regulated by NO-related species, similar data were presented for the Ca^{2+}-activated K^+ channel of cardiac muscle (Bolotina et al., 1994). In this case, donors of NO^+ equivalents were shown to activate the channel, and, similar to findings at the NMDA receptor in our laboratory, NEM blocked the effect by irreversibly alkylating thiol groups. Along similar lines, several other ion channels, enzymes, G-proteins, transcription factors, and other proteins are either up-regulated or down-regulated by similar mechanisms of *S*-nitrosylation or donation of NO^+ equivalents to regulatory sulfhydryl centers (Stamler et al., 1997). The list will undoubtedly grow just as in recent years phosphorylation, myristolation, and palmitoylation have become recognized as important biochemical processes for regulatory function. Interestingly, palmitoylation may be aimed at similar critical thiol group targets, resulting in thioester bond formation. In fact, on some proteins such as SNAP-25 it is possible that *S*-nitrosylation and palmitoylation may compete for the same sulfhydryl, possibly with different physiological outcomes (Hess et al., 1993).

In contrast to phosphorylation, however, in the case of *S*-nitrosylation evidence is accumulating that the critical cysteine residues may be located extracellularly, intracellularly, or possibly even within the putative membrane-spanning region of a protein. From this point of view *S*-nitrosylation may offer additional versatility in the mode of control that can be exerted compared to phosphorylation and other better known post-translational forms of modification.

A candidate consensus motif for S-nitrosylation

Many functionally important sites or target sites for post-translational modification of proteins are distinguished in the primary amino acid sequence by the occurrence of certain patterns or motifs. In many cases such motifs constitute only a very minor part of the entire protein primary sequence. Thus, small patterns often are not detected by overall alignment of protein sequences that are only distantly or not at all related. Such motifs, however, can be identified by the occurrence of a particular cluster of residue types in the primary sequence. A collection of such sequence fingerprints has been developed for PROSITE, a database of biologically significant sites and patterns that can be used to identify families of functionally related proteins or sites for post-translational modification. Examples of such motifs are the consensus sequence patterns required for glycosylation or phosphorylation.

In an attempt to define a possible consensus motif that might be required or at least be facilitatory for *S*-nitrosylation, we initially examined the putative target sites for redox modulation of NMDA receptors. Most importantly, cysteine residues in similar motifs to that described below for the NMDA receptor have been shown by various chemical criteria to be nitrosylated on proteins such as hemoglobin, $p21^{Ras}$, cyclooxygenase and others (Stamler et al., 1997). Two cysteines (abbreviated C in the single letter amino acid code) in the NMDAR1 subunit have been found by site directed mutagenesis to be necessary for redox modulation of that receptor (Sullivan et al., 1994). Unexpectedly, however, these cysteines, C744 and C798, appear to be conserved in all ionotropic glutamate receptors when the sequences are aligned by overall homology (Moriyoshi et al., 1991; Sucher et al., 1995). Nonetheless, among the ionotropic glutamate receptors, only NMDA receptors are exquisitely sensitive to redox modulation (Aizenman et al., 1989). Inspection of the immediate amino acid neighbors of these cysteines revealed that the NMDAR1 cysteines are distinguished from the cysteines conserved in the other ionotropic glutamate receptors in that they are preceded at position −2 by a polar amino acid (G,S,T,C,Y,N,Q), an acidic (D,E) or basic (K,R,H) amino acid at position −1, and an acidic amino acid at position +1. Based on this observation, we constructed the degenerate amino acid pattern designated (G,S,T,C,Y,N,Q)(K,R,H,

D,E)C(D,E) in standard single letter amino acid code and used it in a search of the Protein Identification Resource (PIR) and Swiss Protein (SW) databases with the program Findpatterns of the GCG software package (Program Manual for the Wisconsin Package, Version 8, September 1994, Genetics Computer Group, 575 Science Drive, Madison, Wisconsin, USA 53711). In the PIR database (Release 44.0; March 1995), 3878 sequences out of 77 573 contained this pattern at least once, in the SW database (Release 31.0; March 1995), 2383 out of 43 470 sequences contained this pattern. Viral, bacterial, plant and animal sequences contained this motif.

Among candidates for regulation by *S*-nitrosylation that were identified by the database search were ion channels (NMDA receptor, voltage sensitive Na^+ channel, cyclic nucleotide-gated channel), transporters (Ca-ATPase, K-transporter), receptors (inositol trisphosphate receptor, nerve growth factor receptor), enzymes (oxidoreductases, dehydrogenases, adenylate and guanylate cyclases, proteases, DNA topoisomerases, DNA and RNA polymerases, kinases, phosphatases), transcription factors (helix loop helix proteins, NF-κB, zinc finger proteins), small GTP binding proteins (rab, ras, sas, ypt), cell adhesion molecules (integrins, neural cell adhesion molecule), cell adhesion substrates (laminin, collagen), cyclins and coagulation factors (IXa, Xa, XIII).

In fact, 20 out of 27 proteins that had been listed in a recent review (Stamler, 1994) as bioregulatory targets of nitrogen oxides contain the full motif. The presence of the putative "nitrosylation" motif in guanylate cyclase suggests that the NO-group may regulate the functional activity of guanylate cyclase by *S*-nitrosylation in addition to the interaction with the heme group of this enzyme. It is possible, however, that *S*-nitrosylation might occur at sites other than the proposed motif or that the motif may only be evident in the tertiary rather than the primary structure of some proteins. Moreover, it appears that a certain subset of the motif may bear the highest statistical correlation to the propensity for nitrosylation (Stamler et al., 1997). While the proposed motif was defined post hoc based on our results with redox modulation of the NMDA receptor, it should allow us to identify possible target proteins for *S*-nitrosylation. Most importantly, this motif predicts a target sequence that can be subjected to site directed mutagenesis in order to experimentally verify its importance for *S*-nitrosylation.

Conclusions

In summary, the possible chemical reactions of the NO group are dictated by its redox state. In the case of NO^+ equivalents, this mechanism appears to involve *S*-nitrosylation and possibly further oxidation of critical thiols to disulfide bonds in the NMDA receptor's redox modulatory site(s) to down-regulate channel activity. Very recent data also suggest that NO^-, probably in the single state, can react with critical sulfhydryl group(s) of the NMDA receptor to down-regulate its activity. In the triplet state NO^- may oxidize these NMDA receptor sulfhydryl groups by the formation of an intermediate such as peroxynitrite (Lipton et al., 1996). NO^- can react with thiol but only under specific circumstances, e.g., if an electron acceptor such as O_2 and catalytic transition metals are present and if the conditions do not favor the kinetically preferred reaction with $O_2^{\bullet -}$ to yield peroxynitrite.

It is becoming increasingly evident that in addition to NMDA receptors, biological activities of many other proteins containing critical cysteine residues can be regulated by *S*-nitrosylation and other redox reactions, in a sense similar to the type of control exerted by phosphorylation (Lipton et al., 1993). This type of chemical reaction may represent a new and ubiquitous pathway for the molecular control of protein function by potentially reactive sulfhydryl centers.

Acknowledgements

This review is based upon work performed in a close collaboration between the laboratories of Stuart Lipton at Harvard Medical School and Jonathan Stamler at Duke Medical Center. We are grateful to the many members of the two laboratories who contributed to the work. This was a case in which the mutual ignorances of the two groups exactly complemented one another, resulting in a true collaboration. The work was presented nearly simultaneously at four meetings – in Kyoto, Berlin, Barcelona, and New Orleans – and therefore four related but in, some way, distinct versions of this manuscript were offered for the proceedings of these meetings.

References

Aizenman, E., Lipton, S.A. and Loring, R.H. (1989) Selective modulation of NMDA responses by reduction and oxidation. *Neuron*, 2: 1257–1263.

Ankarcrona, M., Dypbukt, J.M., Bonfoco, E., Zhivotovsky, B., Orrenius, S., Lipton, S.A. and Nicotera, P. (1995) Glutamate-induced neuronal death: A succession of necrosis or apoptosis depending on mitochondrial integrity. *Neuron*, 15: 961–973.

Arnelle, D.R. and Stamler, J.S. (1995) NO^+, $NO^{\cdot}$, and NO^- donation by *S*-nitrosothiols: Implications for regulation of physiological functions by *S*-nitrosylation and acceleration of disulfide formation. *Arch. Biochem. Biophys.*, 318: 279–285.

Beckman, J.S., Beckman, T.W., Chen, J., Marshall, P.A. and Freeman, B.A. (1990) Apparent hydroxyl radical production by peroxynitrite: Implications for endothelial injury from nitric oxide and superoxide. *Proc. Natl. Acad. Sci. USA*, 87: 1620–1624.

Bolotina, V.M., Najibi, S., Palacino, J.J., Pagaon, P.J. and Cohen, R.A. (1994) Nitric oxide directly activates calcium-dependent potassium channels in vascular smooth muscle. *Nature*, 368: 850–853.

Bonfoco, E., Krainc, D., Ankarcrona, M., Nicotera, P. and Lipton, S.A. (1995) Apoptosis and necrosis: Two distinct events induced respectively by mild and intense insults with NMDA or nitric oxide/superoxide in cortical cell cultures. *Proc. Natl. Acad. Sci. USA*, 92: 7162–7166.

Bonner, F.T. and Stedman, G. (1996) The chemistry of nitric oxide and redox-related species. In: M. Feelisch and J.S. Stamler (Eds.) *Methods in Nitric Oxide Research*, Wiley, Chichester, England, pp. 3–18.

Brenman, J.E., Chao, D.S., Gee, S.H., McGee, A.W., Craven, S.E., Santilliano, D.R., Wu, Z., Huang, F., Xia, H., Peters, M.F., Froehner, S.C. and Bredt, D.S. (1996) Interaction of nitric oxide synthase with the postsynaptic density protein PSD-95 and α1-syntrophin mediated by PDZ domains. *Cell*, 84: 757–767.

Ciabarra, A.M., Sullivan, J.M., Gahn, L.G., Pecht, G., Heinemann, S. and Sevarino, K.A. (1995) Cloning and characterization of χ^{-1}: A developmentally regulated member of a novel class of the ionotropic glutamate receptor family. *J. Neurosci.*, 15: 6498–6508.

Das, S., Sasaki, Y.F., Rothe, T., Premkumar, L.S., Takasu, M., Crandall, J.E., Dikkes, P., Connor, D.A., Rayudu, P.V., Cheung, W., Chen, H.-S.V., Lipton, S.A. and Nakanishi, N. (1998) Increased NMDA current and spine density in mice lacking the NMDAR subunit, NR3A *Nature*, 393: 377–381.

Dawson, V.L., Dawson, T.M., London, E.D., Bredt, D.S. and Snyder, S.H. (1991) Nitric oxide mediates glutamate neurotoxicity in primary cortical cultures. *Proc. Natl. Acad. Sci. USA*, 88: 6368–6371.

Dimmeler, S., Haendeler, J., Nehls, M. and Zeiher, A.M. (1997) Suppression of apoptosis by nitric oxide via inhibition of interleukin-1β-converting enzyme (ICE)-like and cysteine protease protein (CPP)-32-like proteases. *J. Exp. Med.*, 185: 601–607.

Fagni, L., Olivier, M., Lafon-Cazal, M. and Bockaert, J. (1995) Involvement of divalent ions in the nitric oxide-induced blockade of *N*-methyl-D-aspartate receptors in cerebellar granule cells. *Mol. Pharmacol.*, 47: 1239–1247.

Gow, A.J., Buerk, D.G. and Ischiropoulos, H. (1997) A novel reaction mechanism for the formation of *S*-nitrosothiol *in vivo*. *J. Biol. Chem.*, 272: 2841–2845.

Hess, D.T., Patterson, S.I., Smith, D.S. and Skene, J.H.P. (1993) Neuronal growth cone collapse and inhibition of protein fatty acylation by nitric oxide. *Nature*, 366: 562–565.

Hoyt, K.R., Tang, L.-H., Aizenman, E. and Reynolds, I.J. (1992) Nitric oxide modulates NMDA-induced increases in intracellular Ca^{2+} in cultured rat forebrain neurons. *Brain Res.*, 592: 310–316.

Kim, W.-K., Rayudu, P.V., Mullins, M.E., Stamler, J.S. and Lipton, S.A. (1996) Down regulation of NMDA receptor activity in cortical neurons by peroxynitrite. In: S. Moncada, J.S. Stamler, S. Gross and E.A. Higgs (Eds.), *The Biology of Nitric Oxide, part 5*, Portland Press, London, p. 26.

Kohr, G., Eckardt, S., Lddens, H., Monyer, H. and Seeburg, P.H. (1994) NMDA receptor channels: Subunit-specific potentiation by reducing agents. *Neuron*, 12: 1031–1040.

Lei, S.Z., Pan, Z.-H., Aggarwal, S.K., Chen, H.-S.V., Hartman, J., Sucher, N.J. and Lipton, S.A. (1992) Effect of nitric oxide production on the redox modulatory site of the NMDA receptor-channel complex. *Neuron*, 8: 1087–1099.

Lipton, S.A., Choi, Y.-B., Pan, Z.-H., Lei, S.Z., Chen, H.-S.V., Sucher, N.J., Loscalzo, J., Singel, D.J. and Stamler, J.S. (1993) A redox-based mechanism for the neuroprotective and neurodestructive effects of nitric oxide and related nitroso-compounds. *Nature*, 364: 626–632.

Lipton, S.A., Choi, Y.-B., Sucher, N.J., Pan, Z.-H. and Stamler, J.S. (1996a) Redox state, NMDA receptors, and NO-related species. *Trends Pharmacol. Sci.*, 17: 186–187.

Lipton, S.A., Kim, W.-K., Rayudu, P.V., Asaad, W., Arnelle, D.R. and Stamler, J.S. (1996b) Singlet and triplet nitroxyl anion (NO^-) lead to *N*-methyl-D-aspartate (NMDA) receptor downregulation and neuroprotection. In: S. Moncada, J. S. Stamler, S. Gross and E.A. Higgs (Eds.), *The Biology of Nitric Oxide – Part 5*, Portland Press, London, p. 125.

Lipton, S.A. and Nicotera, P. (1998) Excitotoxicity, free radicals, and apoptosis. *The Neuroscientist* (In press).

Lipton, S.A. and Stamler, J.S. (1994) Actions of redox-related congeners of nitric oxide at the NMDA receptor. *Neuropharmacology*, 33: 1229–1233.

Lipton, S.A. and Wang, Y.F. (1996) NO-related species can protect from focal cerebral ischemia/reperfusion. In: J. Krieglstein (Ed.) *Pharmacology of Cerebral Ischemia*, Medpharm Scientific Publishers, Stuttgart, pp. 183–191.

Manzoni, O. and Bockaert, J. (1993) Nitric oxide synthase activity endogenously modulates NMDA receptors. *J. Neurochem.*, 61: 368–370.

Manzoni, O., Prezeau, L., Marin, P., Deshager, S., Bockaert, J. and Fagni, L. (1992) Nitric oxide-induced blockade of NMDA receptors. *Neuron*, 8: 653–662.

Melino, G., Bernassola, F., Knight, R.A., Corasaniti, M.T., Nisticò, G. and Finazzi-Agrò, A. (1997) *S*-nitrosylation regulates apoptosis. *Nature*, 388: 432–433.

Moriyoshi, K., Masu, M., Ishii, T., Shigemoto, R., Mizuno, N. and Nakanishi, S. (1991) Molecular cloning and characterization of the rat NMDA receptor. *Nature*, 354: 31–37.

Nikitovic, D. and Holmgren, A. (1996) *S*-nitrosoglutathione is cleaved by the thioredoxin system with liberation of glutathione and redox regulating nitric oxide. *J. Biol. Chem.*, 271: 19180–19185.

Ogura, T., Tatemichi, M. and Esumi, H. (1997) Nitric oxide inhibits CPP32-like activity under redox regulation. *Biochem. Biophys. Res. Commun.*, 236: 365–369.

Omerovic, A., Chen, S.-J., Leonard, J.P. and Kelso, S.R. (1995) Subunit-specific redox modulation of NMDA receptors expressed in Xenopus oocytes. *J. Recep. Sign. Transduc. Res.*, 15: 811–827.

Pryor, W.A., Church, D.F., Govinden, C.K. and Crank, G. (1982) Oxidation of thiols by nitric oxide and nitrogen dioxide: Synthetic utility and toxicological implications. *J. Org. Chem.*, 47: 156–159.

Pryor, W.A. and Lightsey, J.W. (1981) Mechanisms of nitrogen dioxide reactions: Initiation of lipid peroxidation and the production of nitrous acid. *Science*, 214: 435–437.

Radi, R., Beckman, J.S., Bush, K.M. and Freeman, B.A. (1991) Peroxynitrite oxidation of sulfhydryls. The cytotoxic potential of superoxide and nitric oxide. *J. Biol. Chem.*, 266: 4244–4250.

Rothe, T., Chen, H.-S.V., Sucher, N.J., Das, S., Nakanishi, N. and Lipton, S.A. (1997) Increased NMDA currents in cerebral cortex of NMDAR-L deficient mice. *Soc. Neurosci. Abstr.*, 23: 948.

Sathi, S., Edgecomb, P., Warach, S., Manchester, K., Donaghey, T., Stieg, P.E., Jensen, F.E. and Lipton, S.A. (1993) Chronic transdermal nitroglycerin (NTG) is neuroprotective in experimental rodent stroke models. *Soc. Neurosci. Abstr.*, 19: 849.

Schmidt, H.H.H.W., Holman, H., Schindler, U., Shutenko, Z.S., Cunningham, D.D. and Feelisch, M. (1996) No NO$^{\bullet}$ from NO synthase. *Proc. Natl. Acad. Sci. USA*, 93: 14492–14497.

Stamler, J.S. (1994) Redox signaling: Nitrosylation and related target interactions of nitric oxide. *Cell*, 78: 931–936.

Stamler, J.S., Singel, D.J. and Loscalzo, J. (1992) Biochemistry of nitric oxide and its redox activated forms. *Science*, 258: 1898–1902.

Stamler, J.S., Toone, E.J., Lipton, S.A. and Sucher, N.J. (1997) (S)NO signals: Translocation, regulation, and a consensus motif. *Neuron*, 18: 691–696.

Sucher, N.J., Awobuluyi, M., Choi, Y.-B. and Lipton, S.A. (1996) NMDA receptors: From genes to channels. *Trends Pharmacol. Sci.*, 17: 348–355.

Sucher, N.J., Schahram, A., Chi, C.L., Leclerc, C.L., Awobuluyi, M. Deitcher, D.L., Wu, M.K., Yuan, J.P., Jones, E.G. and Lipton, S.A. (1995) Developmental and regional expression pattern of a novel NMDA receptor-like subunit (NMDAR-L) in the rodent brain. *J. Neurosci.*, 15: 6509–6520.

Sullivan, J.M., Traynelis, S.F., Chen, H.-S.V., Escobar, W., Heinemann, S.F. and Lipton, S.A. (1994) Identification of two cysteine residues that are required for redox modulation of the NMDA subtype of glutamate receptor. *Neuron*, 13: 929–936.

Takasu, M., Das, S., Sasaki, Y., Sucher, N.J., Lipton, S.A. and Nakanishi, N. (1997) Abnormal dendritic morphology in NMDA receptor-like (NMDAR-L) deficient mice. *Soc. Neurosci. Abstr.*, 23: 947.

Tenneti, L., D'Emilia, D.M. and Lipton, S.A. (1997) Suppression of neuronal apoptosis by *S*-nitrosylation of caspases. *Neurosci. Lett.*, 236: 139–142.

R.R. Mize, T.M. Dawson, V.L. Dawson and M.J. Friedlander (Eds.)
Progress in Brain Research, Vol 118

CHAPTER 7

The subcellular distribution of nitric oxide synthase relative to the NR1 subunit of NMDA receptors in the cerebral cortex

Chiye Aoki[1,*], David S. Bredt[2], Suzanne Fenstemaker[1] and Mona Lubin[1]

[1]*Center for Neural Science, New York University*
[2]*Department of Physiology, University of California, San Francisco*

Abstract

Results from several electrophysiological studies predict that the neuronal NO-synthesizing enzyme, nNOS, resides within spines formed by pyramid-to-pyramid axo-spinous synaptic junctions of the cortex. On the other hand, light microscopic neuroanatomical detection of nNOS within pyramidal neurons has been difficult, suggesting that these neurons contain nNOS at levels below threshold for detection. Our results obtained by electron microscopic immunocytochemistry indicate that nNOS occurs within spiny neurons, such as those of pyramidal neurons, albeit discretely within their spines. Dual electron microscopic immunocytochemistry, whereby antigenic sites to the NR1 subunit of NMDA receptors are probed simultaneously with sites immunoreactive for nNOS, reveals that some, although not all, nNOS within spines co-exist with NR1 subunits. Additionally, immunoreactivity for the NR1 subunit is detectable within nNOS-axons, indicating that NO may be generated in response to axo-axonic interactions with glutamatergic axons in the vicinity and independently of action potential propagation. Immunoreactivity for NR1 subunits within axons (with or without nNOS-immunoreactivity) may additionally serve to confer receptivity of these axons to NO generated coincidentally with activity. Analysis of the visual cortex of monocular adult animals indicates that the level of nNOS within neurites is dependent on chronic activity levels of the surrounding neuropil and independent of somatic input level. Together, these findings point to plasticity of nNOS neurons within adult brain tissue, involving regulation of subcellular nNOS distribution.

Introduction

Nitric oxide (NO) is a member of a newly recognized class of gaseous neuronal messengers with pivotal roles in synapse plasticity, excitotoxicity and neuromodulation (Schuman and Madison, 1994; Dawson and Snyder, 1994). Since NO is highly reactive (see Lipton, this volume), a fruitful approach for determining the site of action of NO has been to localize the synthetic enzyme, neuronal nitric oxide synthase (nNOS), rather than NO. Activation of this enzyme is strictly dependent on the rise, then binding, of

*Corresponding author. 4 Washington Place, Rm 809, NY, NY 10003. Tel.: +1 212 998 3929; fax: +1 212 995 4011; e-mail: chiye@cns.nyu.edu

intracellular Ca^{2+} to calmodulin, the enzyme's cofactor that is prevalent in the post-synaptic density (PSD) (Grab et al., 1980) (Bredt and Snyder, 1992; Garthwaite and Boulton, 1995). Amongst NO's various roles in synaptic transmission is its enhancement of neurotransmitter release (e.g., Montague et al., 1994). Moreover, since the majority of intercellular events involving NO is linked to NMDA receptor activation, it has been hypothesized that NO synthesis follows the influx of Ca^{2+} via activated NMDA receptors (Garthwaite and Boulton, 1995). However, since the diffusion of Ca^{2+} within cells and particularly within fine processes can be extremely limited (Gamble and Kock, 1987; Muller and Connor, 1991; Llins et al., 1992), the mere co-existence of NMDA receptors and nNOS within single cells (Price et al., 1993) does not guarantee that NMDA receptor activation will be linked to NO generation. Thus, the studies summarized in this chapter aimed to determine whether nNOS and NMDA receptors co-exist at single synapses in intact neural tissue. Earlier work attempted to localize nNOS by NADPH-diaphorase histochemistry (Hope et al., 1991; Vincent and Kimura, 1992). We have found this histochemical approach to be limited in sensitivity, particularly for ultrastructural analyses needed to determine the presence of the enzyme at synapses (unpublished observations). Fortunately, we have been able to localize the site of NO generation by electron microscopic immunocytochemistry (EM–ICC), using a highly specific antiserum directed against nNOS (Aoki et al., 1993, 1997). In this chapter, EM–ICC data pertinent to the following questions are presented:

(1) Is nNOS present in spines, such as of pyramidal neurons reported to undergo strengthening at pyramid-to-pyramid synapses, and do these spines utilize NMDA receptors for synaptic transmission?
(2) Can nNOS activation occur independently of NMDA receptor activation?
(3) Is nNOS also present pre-synaptically, where NMDA receptors are localized?
(4) Does activity-dependent generation of NO involve changes in nNOS levels?
(5) Does the localization of nNOS to the plasma membrane depend on post-synaptic peripheral membrane proteins, such as PSD-93?

Results

Does nNOS occur in spines and do these spines use NMDA receptors for synaptic transmission?

Earlier light microscopic studies showed that cortical and hippocampal nNOS occur in GABAergic interneurons co-storing NPY. These interneurons are highly reactive to the NADPH-diaphorase histochemical staining procedure (Vincent and Kimura, 1992), as neuronal NADPH-diaphorase is a NOS (Hope et al., 1991; Dawson et al., 1991). Typically, these nNOS-neurons exhibit dendritic and axonal arbors that ramify extensively throughout the neuropil and are highly varicose (Fig. 1). Since these neurons are only sparsely spiny and certainly not pyramidal (Vincent et al., 1983; Hendry et al., 1984; Aoki and Pickel, 1989; Vincent and Kimura, 1992), a question remained whether nNOS might also occur in spiny neurons, such as pyramidal neurons. Earlier published studies indicated that pyramidal neurons do not contain nNOS or that the level is too low for detection by light microscopy. One study which was able to demonstrate the presence of nNOS within pyramidal neurons (Wendland et al., 1994) indicated that the tissue had to be fixed very weakly (e.g., using $<1\%$ of paraformaldehyde) in order to retain nNOS-antigenicity. An obvious interpretation of these results is that nNOS occurs only in low levels in the pyramidal neurons or is localized too discretely (e.g., within spines) to be easily detectable by light microscopy. Assessment of the distribution pattern of nNOS within specific neuronal populations is pertinent to understanding mechanisms of synaptic plasticity, since NO released from post-synaptic sites of pyramid-to-pyramid axo-spinous junctions are reported to be responsible for activity-dependent, and more specifically, NMDA receptor-dependent synapse strengthening (reviewed by Hawkins et al. in

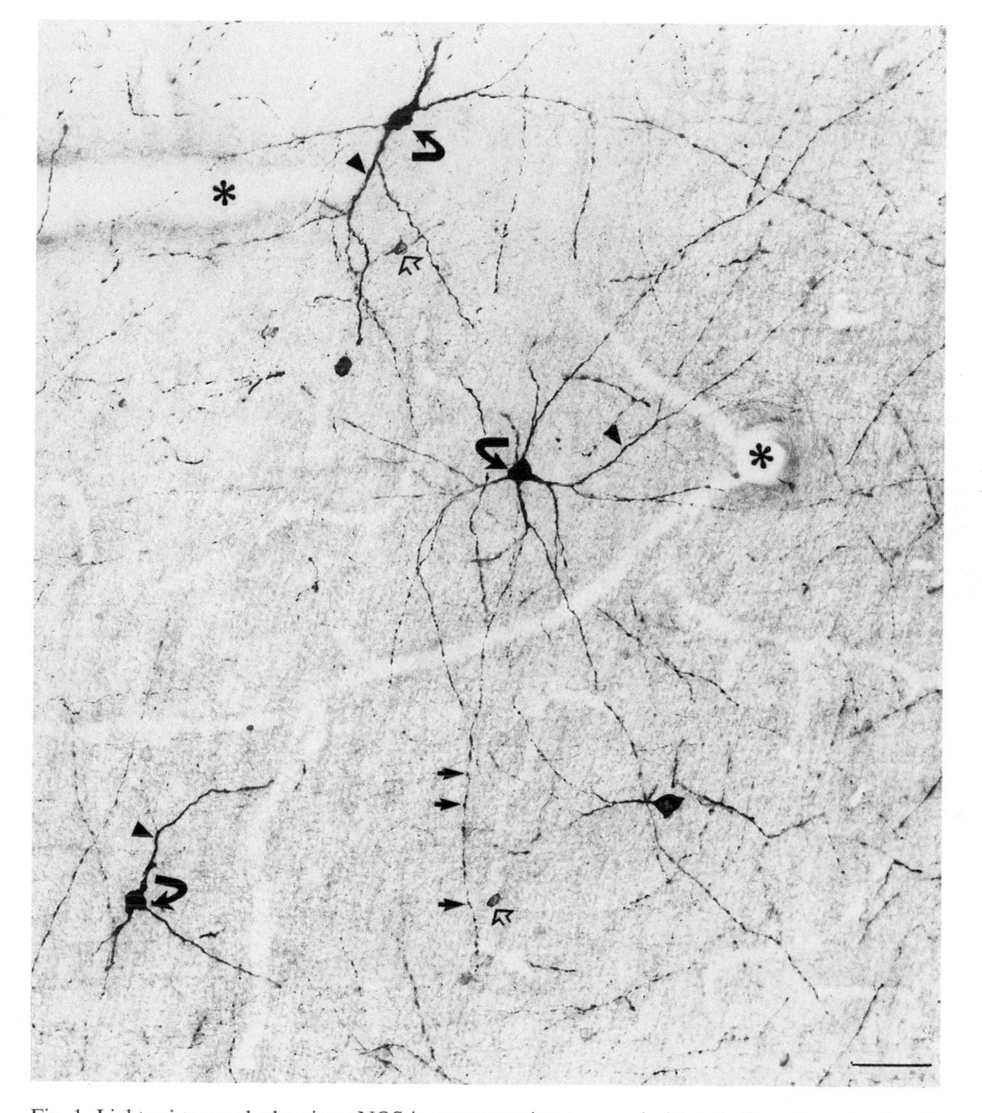

Fig. 1. Light micrograph showing nNOS-immunoreactive neurons in layer 6 of monkey visual cortex, as revealed by the presence of peroxidase-based label. nNOS-immunoreactive neurons are multipolar. Their dendrites are aspiny, varicose and extensive. Curved arrows point to large, intensely immunoreactive somata, arrowheads point to dendrites emanating from these somata, and small arrows point to varicose axonal processes. Open arrows point to smaller, weakly nNOS-immunoreactive perikarya. Asterisks point to blood vessel lumen. Calibration bar = 100 μm.

this volume). One would, therefore, expect nNOS to be contained selectively in spines of neurons.

Our first ultrastructural analysis (Aoki et al., 1993) revealed that nNOS occurs in neurons with morphological characteristics typical of GABAergic/NPY-containing neurons (Aoki and Pickel, 1990). Specifically, nNOS occurs in axon terminals forming symmetric synaptic junctions upon dendritic shafts and somata, the predominant targets for GABAergic terminals (Figs. 2A–C). The occurrence of nNOS within axons juxtaposed to axo-spinous synaptic junctions (Fig. 2D) is reminiscent of the synaptic patterns of NPY-terminals, as is the formation of synaptic junctions onto its own kind, i.e., between nNOS-terminals and nNOS-dendrites (Fig. 3B). As expected for

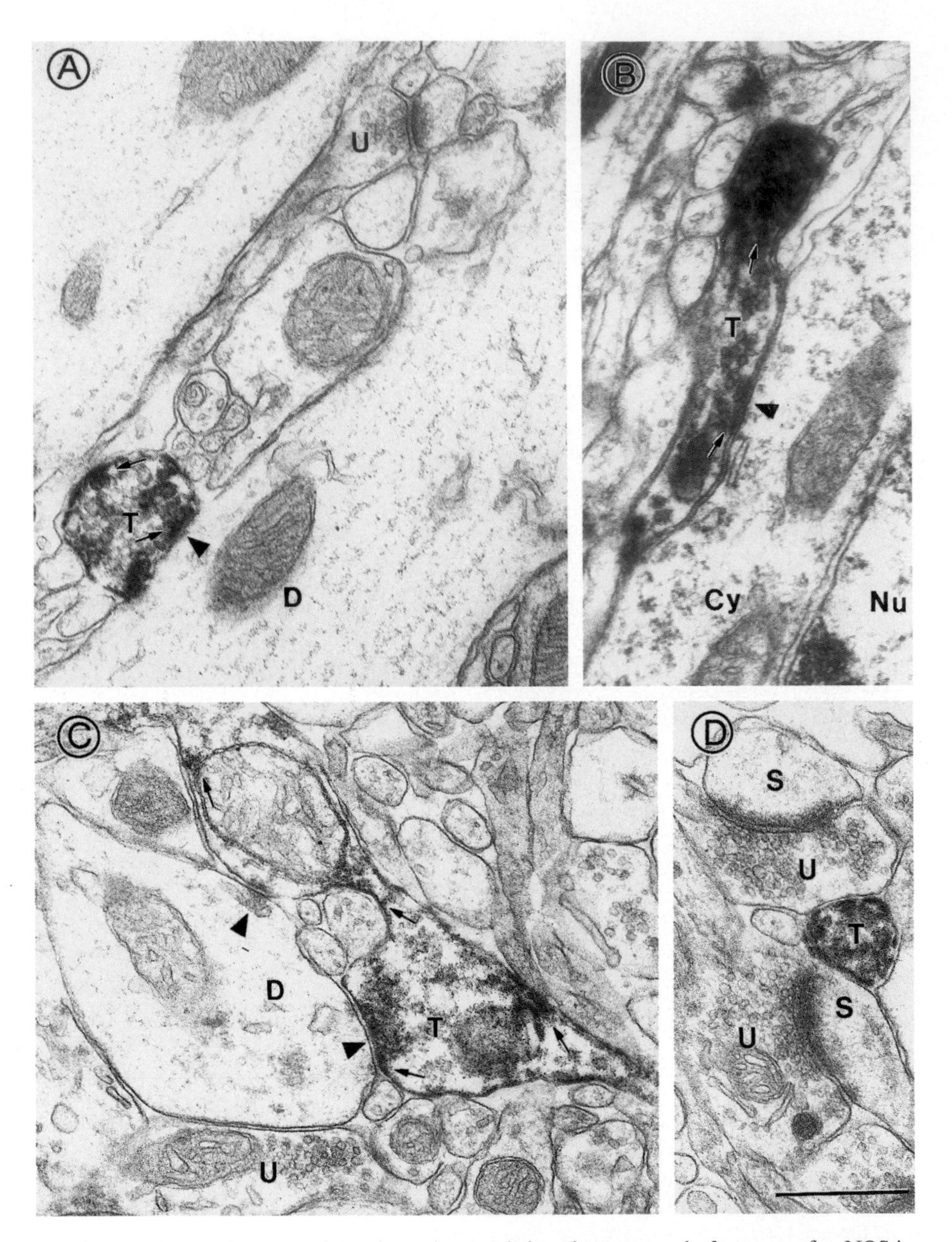

Fig. 2. Electron micrographs showing characteristic ultrastructural features of nNOS-immunoreactive terminals. nNOS-immunoreactive terminals, T, form the symmetric type of synaptic junction with dendritic shafts (D) (panels A and C) and a soma (panel B). Cy and Nu indicate the cytoplasm and nucleus of the soma. Arrowheads point to postsynaptic plasma membranes, while small arrows point to nNOS-immunoreactivity within the labeled terminals. U in panels A, C and D points to unlabeled terminals forming axo-spinous asymmetric junctions. Panel D shows a nNOS-immunoreactive terminal, T, lacking a clear target within the plane of section but is simultaneously juxtaposed to two other terminals forming asymmetric, presumably excitatory, synapses with spines. Calibration bar = 500 nm.

NPY neurons, nNOS-immunoreactivity occurs diffusely within the cytoplasm of somata exhibiting highly invaginated nuclear envelopes (Fig. 3A) and sparsely spiny dendrites (Fig. 3C).

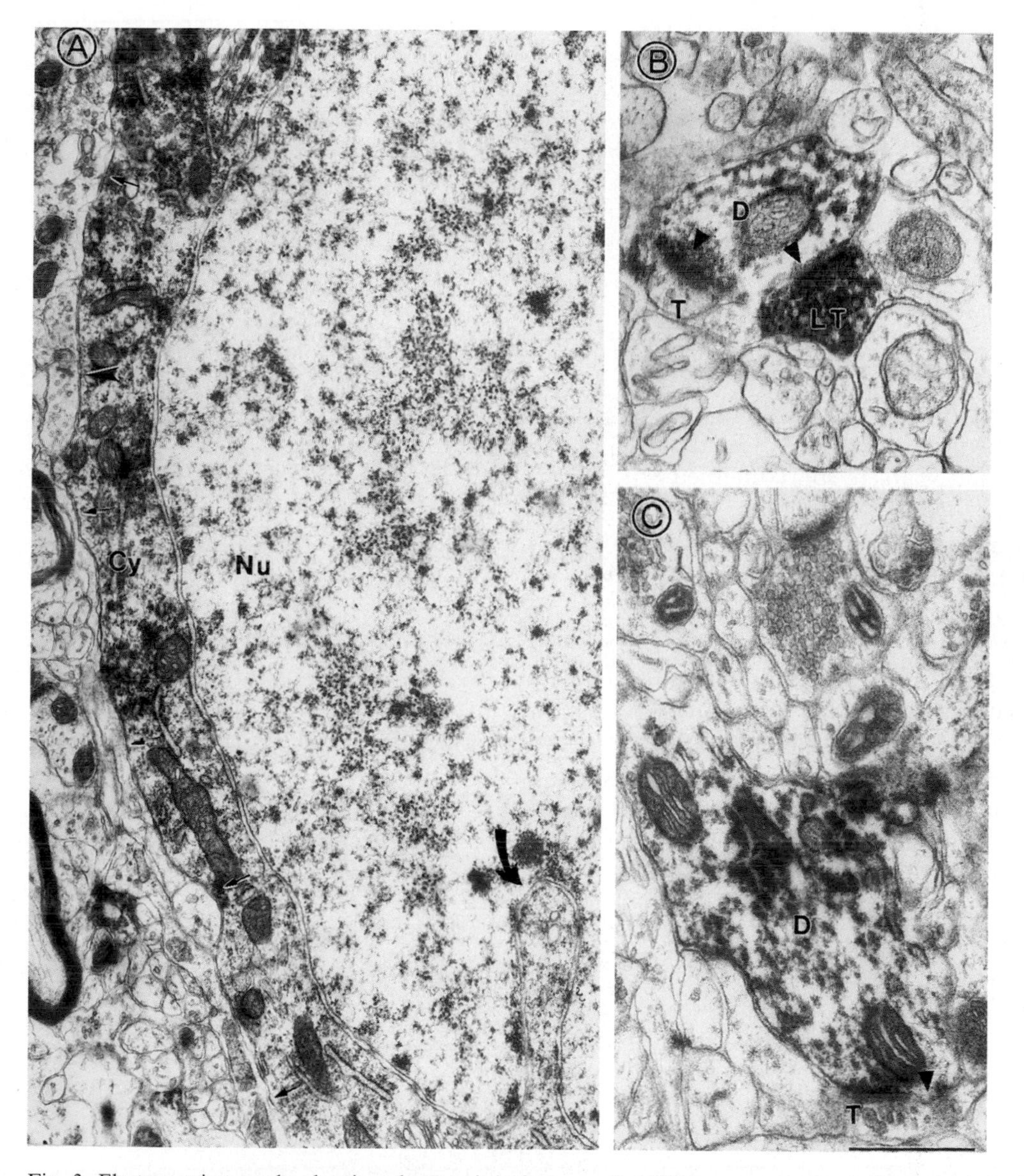

Fig. 3. Electron micrographs showing characteristic features of nNOS-immunoreactive perikarya and dendrites. Panel A: nNOS-immunoreactivity is prevalent in the perikaryal cytoplasm (Cy) and absent from the nucleoplasm (Nu). The nuclear envelopes often exhibit deep indentations (curved arrow). Small arrows point to the plasma membrane, while arrowhead points to an axo-somatic junction. Panels B and C: nNOS-immunoreactive dendrites (D) rarely exhibit dendritic spines but, instead, show local swellings and constrictions. nNOS-immunoreactivity is distributed evenly within the cytoplasm but appears accumulated over thick PSDs that are postsynaptic to unlabeled terminals (T). A labeled terminal (LT) exhibits a symmetric synaptic junction with the immunolabeled dendrite. Calibration bar = 1000 nm in panel A and 500 nm in panels B and C.

More relevant to the current topic is our observation that nNOS can occur within dendritic spines. When found within spines and shafts of spiny dendrites, such as those of pyramidal neurons, nNOS-immunoreactivity is discrete, associated primarily with post-synaptic membranes and PSDs within spine heads (Figs. 4A and 5B, C). Alternatively, nNOS can occur slightly removed from spine heads but still within a micrometer from axo-spinous asymmetric junctions, such as within the neck or base of the neck of spines (i.e., where the spine attaches to the dendritic shaft (Figs. 5A, B). The remaining cytoplasm of dendritic shafts is largely devoid of nNOS-immunoreactivity (Figs. 4B and 5). Because of this selective distribution, it is no wonder that localization of nNOS within pyramidal neurons would be difficult by light microscopy.

Together, these ultrastructural and light microscopic results indicate that nNOS is uniformly distributed and highly concentrated within the cytoplasm of aspiny and sparsely spiny interneurons. In contrast, the nNOS level within spiny neurons, such as pyramidal neurons, is low overall but highly localized to spines. Within dendritic shafts of spiny and aspiny neurons, nNOS occurs both along and away from the plasma membrane.

Spines immunoreactive for nNOS exhibit asymmetric synaptic junctions i.e., those with thick PSDs (Figs. 4 and 5). Synapses of this morphological type have been shown to be excitatory (Gray, 1959; Aoki and Kabak, 1992) and to be labeled by NMDA receptor antibodies (Aoki et al., 1994; Siegel et al., 1994; Petralia et al., 1994; Aoki, 1997). In order to visualize the localization of nNOS in relation to NMDA receptors, a study (Aoki et al., 1997) was conducted in which horseradish peroxidase (HRP)-based label was combined with pre-embedding colloidal gold label (silver-intensified 1-nm colloidal gold particles) for the simultaneous immunocytochemical detection of nNOS with the NR1 subunit of NMDA receptors within single ultrathin sections. The dual EM–ICC results obtained in this way demonstrated the co-existence of nNOS and the NR1 subunit within single spines of the cerebral cortex (Fig. 6) and the hippocampal formation (Fig. 7). However, a more specific co-localization of NR1 subunits with nNOS over single PSDs has been difficult to demonstrate (Figs. 6 and 7). This may be due to technical difficulties of localizing antigens directly over PSDs by the pre-embedding immunogold procedure. This interpretation is corroborated by results obtained using a single, HRP-based

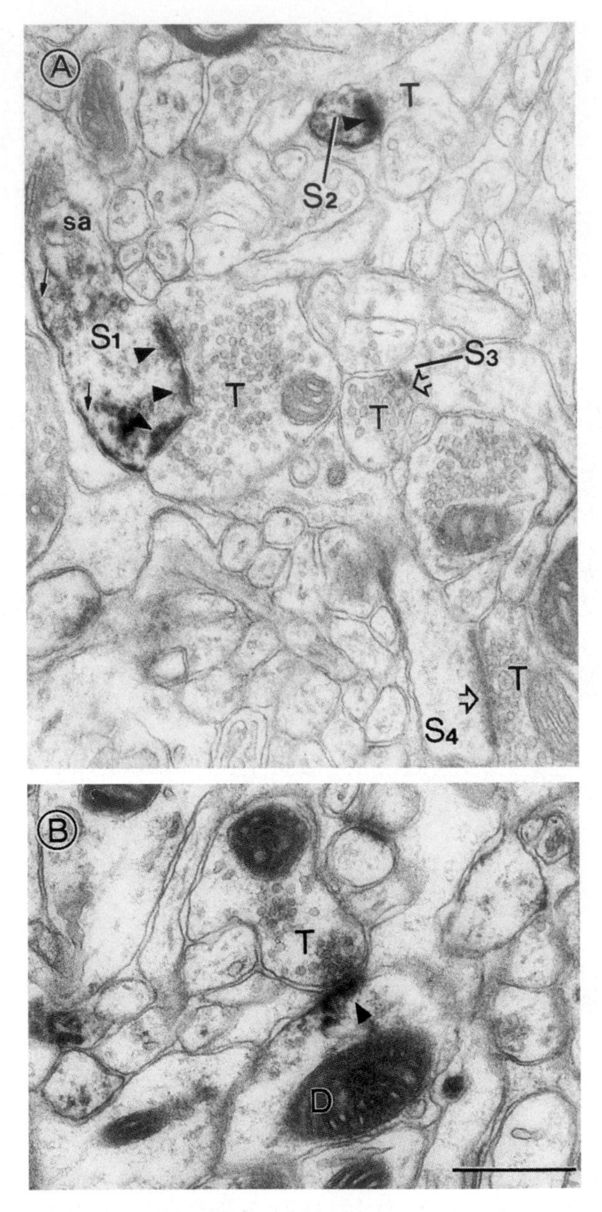

Fig. 4. Post-synaptic nNOS-immunoreactivity. Panel A: A large dendritic spine, S1, exhibits perforated PSDs with nNOS-immunoreactivity (arrowheads). nNOS-immunoreactivity is also present along the plasma membrane (small arrow) near the spine apparatus (sa). Another small spine, S2, exhibits nNOS-immunoreactivity along the plasma membrane and over the PSD. S3 and S4 are unlabeled dendritic spines, shown for comparison. T = unlabeled presynaptic terminals. Open arrows point to unlabeled thick PSDs. Panel B: nNOS-immunoreactivity is concentrated over a PSD (arrowhead) formed along a dendritic shaft (D). Calibration bar = 500 nm.

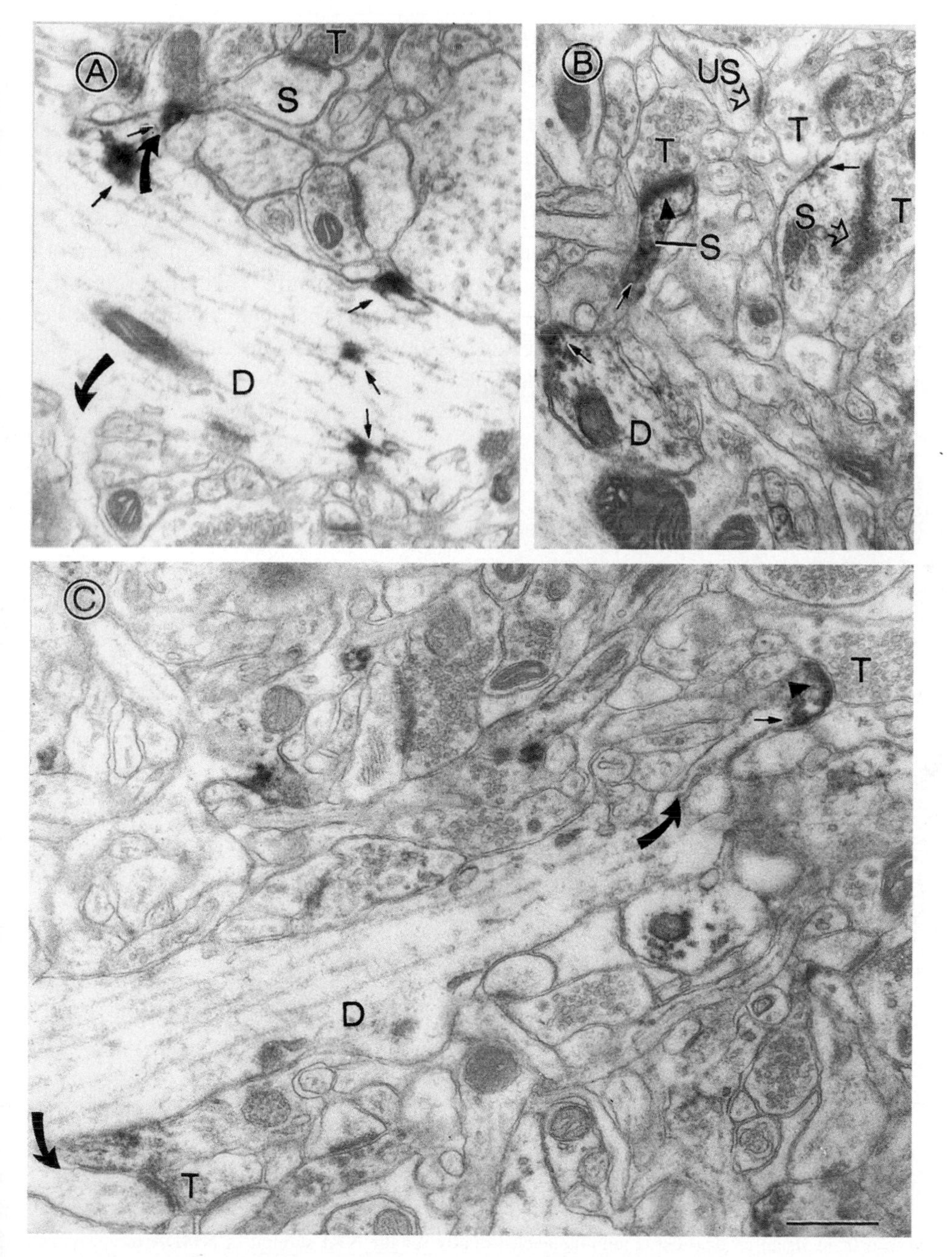

Fig. 5. Electron micrographs showing the occurrence of nNOS discretely within spines but little to none within their shafts. Small arrows point to nNOS-immunoreactivity. The spine in panel A exhibits nNOS-immunoreactivity at its base (upper curved arrow). The dendritic shaft also exhibits clumps of nNOS-immunoreactivity (small arrows). The lower curved arrow points to another, unlabeled dendritic spine emanating from the dendritic shaft. The dendrite in panel B exhibits immunoreactivity within the spine neck, plasma membrane forming the spine head, and over the PSD (arrowhead). The dendritic shaft portion (D) also exhibits limited nNOS-immunoreactivity along the plasma membrane. Another spine in its vicinity exhibits limited nNOS-immunoreactivity along the plasma membrane of the spine head but not over the PSD (open arrow). The dendrite spine in panel C shows immunoreactivity only in the spine head region (arrowhead points to the labeled PSD): the dendritic shaft and spine necks exhibit little nNOS-immunoreactivity (curved arrows). Calibration bar = 500 nm.

immunolabel in lieu of pre-embedding immunogold label. Using HRP-based labels, which are more sensitive than the immunogold labels, localization directly over PSDs of spines has been demonstrated for the NR1 subunit of NMDA receptors (Figs. 6A, B and 7) and for nNOS (Figs. 4A, B, 5B, C and 6C–E). In any case, dual EM–ICC reveals that even when co-existence over single PSDs cannot be demonstrated, the two molecules appear extremely close to one another (< 1 μm) and within single spines (Figs. 6 and 7). This would allow the two molecules to interact, following intracellular diffusion of Ca^{2+}/calmodulin.

Can nNOS activation occur independently of NMDA receptor activation?

This question was posed to consider the possibility that a rise in intracellular Ca^{2+}, leading to nNOS activation, may occur by means other than NMDA receptor activation. Although the mere presence of nNOS- or NMDA receptor-immunoreactivity does not ensure that the respective molecules are activatable, the absence of either molecule would be indication that the molecules might operate independently. Based on this rationale, the dually immunolabeled samples were

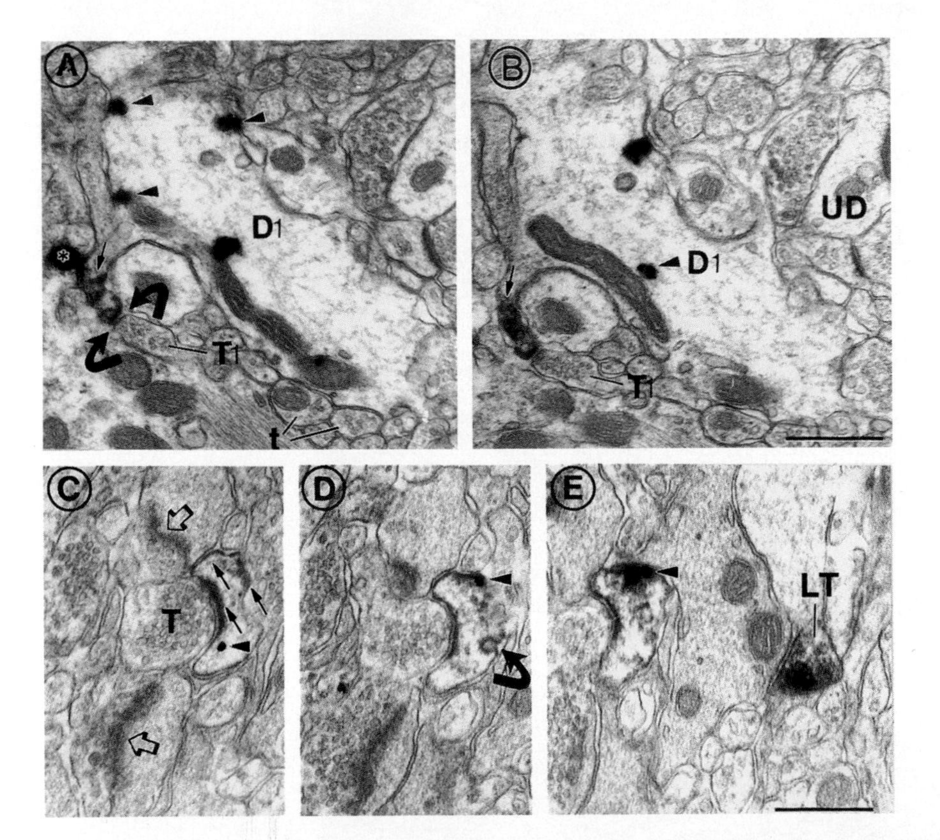

Fig. 6. Electron micrographs taken from visual cortical tissue immunolabeled dually for the NR1 subunit of NMDA receptors and nNOS. Panels A and B are from a serially collected set of ultrathin section immunolabeled for NR1 subunit by ABC-peroxidase (small arrow) and for nNOS by the silver-intensified colloidal gold label (arrowheads). A spine emanating from a dendritic shaft (D1) is immunoreactive for the NR1 subunit along its head and over the PSD (two curved arrows) that is postsynaptic to T1. nNOS-immunoreactivity is absent in the spine but occurs in the dendrite within a micrometer from the NMDA-receptor-immunoreactive spine. Panels C, D and E are taken from a serially collected set of ultrathin sections immunolabeled for the NR1 subunit by silver-intensified gold (arrowhead) and for nNOS by the ABC-peroxidase label (small arrows). The two molecules co-exist within a spine receiving an asymmetric synaptic input from an unlabeled terminal (T). LT in panel E points to another labeled terminal without recognizable synaptic targets. Calibration bar = 500 nm. (Adopted from Figs. 2 and 3 of Aoki et al., Brain Research, 750: 25–40.)

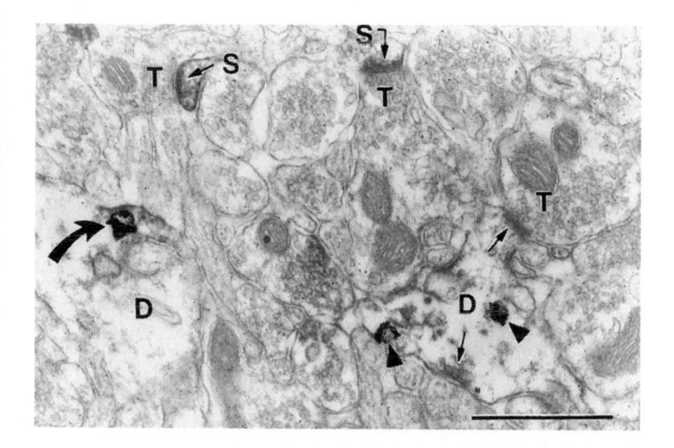

Fig. 7. Dual localization of nNOS and the NR1 subunit of NMDA receptors within the CA1 field of the hippocampal formation. Arrowheads point to silver-intensified gold particles reflecting nNOS-immunoreactivity while ABC-peroxidase/DAB reflects NR1-subunit immunoreactivity. NR1-immunoreactivity is prevalent along the plasma membrane and aggregated over PSDs (small arrows) of a dendrite, D to the left. T points to unlabeled axon terminals forming asymmetric synaptic junctions with immunolabeled dendritic profiles. Two other dendritic spines in the vicinity also exhibit NR1-immunoreactivity (small arrows along the top of the panel). The large curved arrow points to a dually labeled dendritic spine emerging from a dendritic shaft (D). An NRI-immunoreactive axon residue between the two dendrites. Calibration bar = 500 nm.

analyzed to determine whether nNOS localization occurs independently of NMDA receptor localization. Results from this analysis (Aoki et al., 1997) indicated that neuronal processes immunoreactive for nNOS do not always label for the obligatory subunits of NMDA receptors, i.e., the NR1 subunits. Since failures in immunodetection is greater when using silver-intensified gold labels than of HRP-based immunolabeles, we have tried both combinations of dual immunolabeling – i.e., using silver-intensified gold as the label for detecting nNOS, combined with the HRP-based label for detecting NR1 and visa versa. Regardless of the labels used for the two molecules, profiles immunoreactive for nNOS without immunoreactivity for NR1 subunits were observed. While some of those nNOS sites lacking NR1 subunits could reflect failures to detect the immunolabel, the presence of NR1-immunoreactivity in the vicinity (less than a few microns) of nNOS-immunoreactivity indicates that at least some nNOS-immunoreactive profiles probably do not contain NR1 subunits. These observations support the idea that nNOS and NMDA receptors can and do operate independently. Other candidates for mediating activity-linked elevation of intracellular Ca^{2+} include the voltage-activated calcium channels and a variety of neurotransmitter receptors linked to G-proteins which, when activated, cause intracellular Ca^{2+} mobilization via the phospholipase-C pathway. In addition, the α7 subunit of nicotinic acetylcholine receptors, which are similarly distributed within dendritic spine heads and neck (Fig. 8), also are permeable to Ca^{2+}, when activated (Mulle et al., 1992; Vijayaraghavan et al., 1992; Seguela et al., 1993).

NR1 subunits also occur in presynaptic terminals near nNOS-immunoreactive processes (Fig. 7). What might be the function of these axonal NR1 subunits? We are intrigued by the possibility that these NR1 subunits may serve to confer specificity for NO operating as a retrograde messenger. This idea is an outgrowth of the question, 'how NO can mediate synapse-specific changes in LTP,' particularly since NO is freely diffusible across plasma membranes and, thus, can encounter countless numbers of synapses near the site of generation. In spite of this free access to NO, NO-dependent LTP involves changes that are selective to axons that have undergone tetanus or LTP-inducing stimulation. Perhaps delivery of LTP-inducing pulses causes activation of NMDA receptors on axons as well as those receptors residing post-synaptically. This event, in turn, may make those and only those axons bearing recently activated NMDA receptors somehow receptive to NO, thereby allowing NO to induce cellular changes leading to enhanced release of glutamate. Conversely, NMDA receptors residing on axons that have not received LTP-inducing pulses would remain inactive and this state might prevent the inactive axons from becoming receptive to NO. Electrophysiological results gathered from LTP experiments fit this hypothesis, since LTP induction is prevented when NMDA receptors (presum-

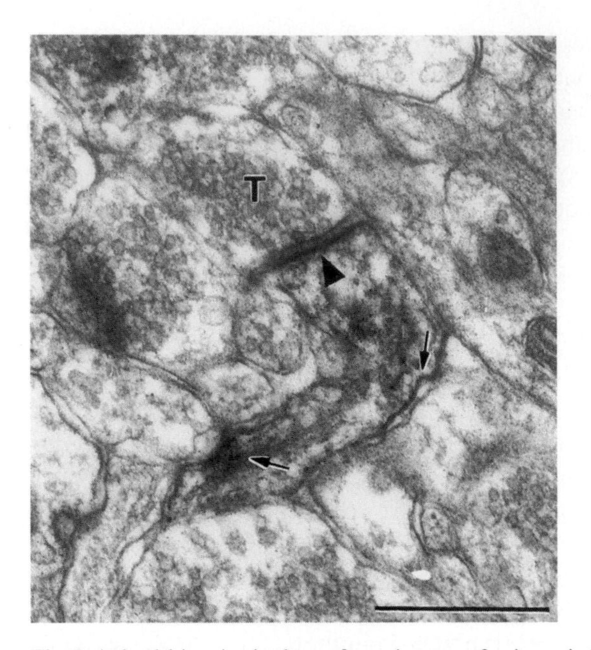

Fig. 8. A dendritic spine in the prefrontal cortex of guinea pig is immunolabeled for the $\alpha 7$ subunit of nicotinic acetylcholine receptors by a peroxidase-based label. Note the immunoreactivity along the plasma membrane and neck portion (small arrows) in addition to the thick PSD. This distribution of receptor subunit within spines resembles that for nNOS. T = unlabeled presynaptic terminal. Calibration bar = 500 nm.

ably pre-synaptic as well as post-synaptic receptors) are blocked, even when the need for activation of post-synaptic NMDA receptors is bypassed by providing exogenous NO (Murphy et al., 1994). Recent work of Hawkins and his colleagues also indicate reduction of NO-dependent LTP when NMDA receptors (presumably pre- and post-synaptic) are blocked (Son et al., 1997).

Does nNOS also occur presynaptically, where NMDA receptors are localized?

The same dual EM–ICC procedure described above revealed co-localization of nNOS with NR1 subunits of NMDA receptors within axon terminals. Our earlier ultrastructural analysis (Aoki et al., 1994; Aoki, 1997) had indicated that NR1 subunits are more prevalent within axons than in spines while another light and EM–ICC study also had indicated that nNOS occurs in high concentration within axon terminals (Aoki et al., 1993). The additional information obtained from the dual EM–ICC study was that the axons exhibiting both, nNOS and NR1 subunit immunoreactivity are a heterogeneous population, because some form asymmetric axo-spinous synaptic junctions (not shown), others form symmetric synaptic junctions onto dendritic shafts (Fig. 9) and yet others exhibit no obvious synaptic targets, at least in the plane of section (not shown). The latter observation indicates that pre-terminal portions of axons may be engaged in the generation of NO following en passant release of L-glutamate from near-by glutamatergic axons. Co-existence of the NR1 subunit of NMDA receptors with nNOS within synaptic terminals suggests that NO may be generated within terminals by two mechanisms. One would be subsequent to the influx of Ca^{2+} via voltage-activated calcium channels that become activated by the arrival of action potentials. The other would follow the release of glutamate from nearby glutamatergic terminals, independent of recent activity of the dually labeled terminal.

Does activity-dependent generation of NO involve changes in nNOS levels?

Ocular dominance columns in layer 4C of the visual cortex are a useful model for identifying activity-dependent alterations in synaptic molecules (Shatz, 1990). Ocular dominance columns represent segregated, yet immediately juxtaposed, pathways originating from the two eyes (Gilbert, 1983). While ocular dominance columns subserving the ipsilateral and contralateral eyes receive equivalent afferent activity levels in an intact visual system, the input from one eye relative to the other can be altered relatively easily by monocular eyelid suture, intraocular injection of TTX, or enucleation. This model has been used to investigate whether the cellular and ultrastructural distributions and concentration of nNOS might be regulated by afferent activity levels. To this end, the distribution of nNOS within perikarya and layer

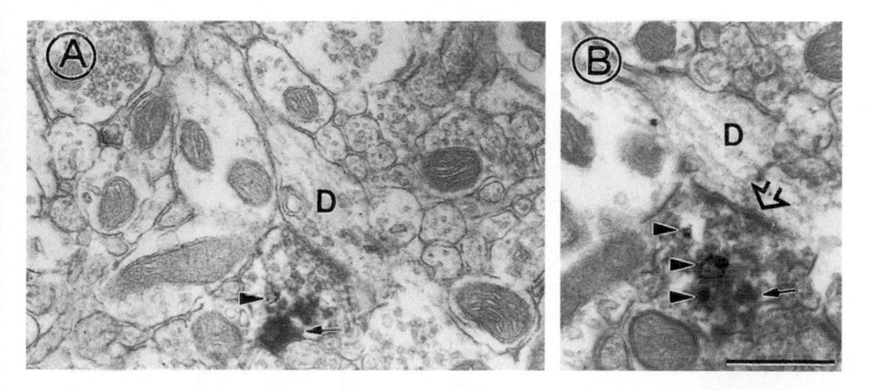

Fig. 9. Dually immunolabeled axon terminal in the rat visual cortex forming a symmetric synaptic junction onto a dendritic shaft. Peroxidase-label reflects the presence of the NR1 subunit (small arrows pointing to HRP reaction product) while silver-intensified gold particles reflect the presence of nNOS (arrowheads). The large open arrow points the postsynaptic membrane of the dendritic shaft, D. Calibration bar = 621 nm in panel A and 500 nm in panel B. (Adopted from Fig. 4 of Aoki et al., Brain Research, 750: 25–40.)

4C-neuropil of monocular adult monkeys was compared (Aoki et al., 1993).

Light microscopic results showed that neuropil content of nNOS is dramatically altered by monocular deprivation, revealing prominent, alternating stripes of darkly and lightly nNOS-immunolabeled neuropil in layer 4C (Fig. 10). These stripes register precisely with active and deprived ocular dominance columns, subserving the intact and enucleated visual pathways, respectively, as visualized by the cytochrome oxidase histochemical procedure that measures chronic levels of oxidative metabolism. Interestingly, neuropil labeling for nNOS in cortex is accompanied by changes in the laminar distribution of nNOS within the visual thalamus, i.e., the lateral geniculate nucleus (LGN). In the LGN, three of the six layers lose afferent activity due to enucleation. The layers that exhibit diminished levels of nNOS-immunoreactivity are those that have lost afferent input (Aoki et al., 1993). Although this finding is in accordance with the columnar labeling in the visual cortex, it is also somewhat contrary to expectation. The literature indicates that nNOS within the LGN is provided strictly by cholinergic afferents arising from the pontine cholinergic neurons (Bickford et al., 1993) but the inputs to these cell bodies are not retinally segregated. Our results, thus, suggest that nNOS levels within axons in the LGN must be regulated by retinally segregated neural activity within the LGN neuropil, rather than at the cell body level. The idea that nNOS level within axons can be regulated by surrounding neuropil activity and independent of input to the soma is further strengthened by the results obtained from analysis of the visual cortex. In the visual cortex, our results show that while the layer 4C neuropil exhibits strikingly different nNOS immunoreactivity in the right versus left eye's columns, the density of somata immunoreactive for nNOS is not different between the two types of columns. Thus, here, as in the LGN, nNOS level in neurites (represented by neuropil labeling) appears to be regulated independently of somatic concentrations of nNOS.

The reduced neuropil labeling for nNOS in the inactive ocular dominance columns may reflect reduced sprouting or retraction of nNOS-immunoreactive neurites (Figs. 11A, B). Alternatively, reduced neuropil labeling for nNOS may result from decremenent in the concentration of nNOS within neurites that are, otherwise, intact and functional using other neurotransmitter(s) (Fig. 11C). At present, there is no cell biological mechanism known to permit differential nNOS content within the numerous neurites that emanate from single neurons. It is not known, for that matter, whether the concentration of nNOS is the same among the multiple axon collaterals or dendrites that emanate from single cell bodies.

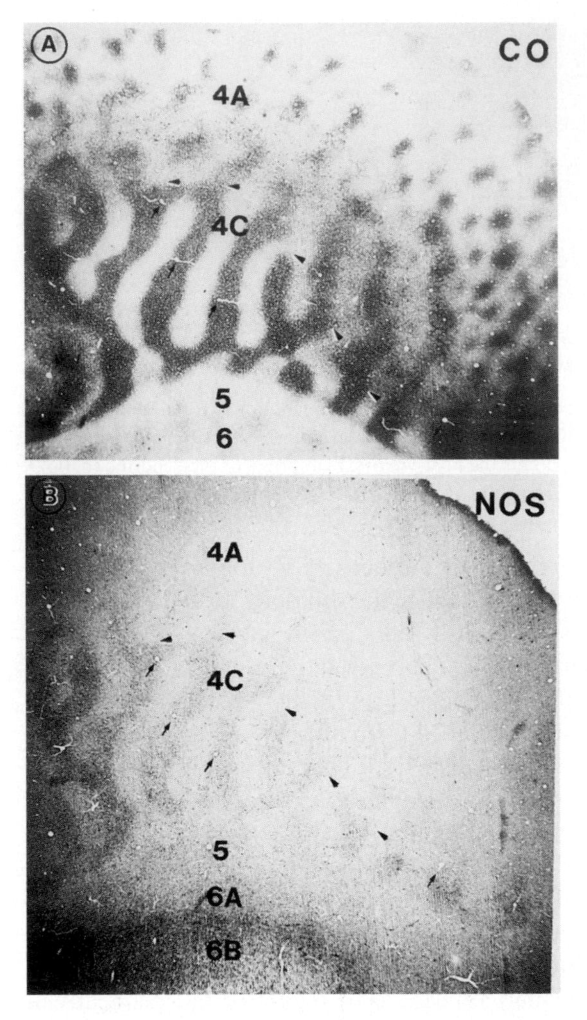

Fig. 10. Lamina 4C of a monocular monkey's visual cortical area V1 exhibits low levels of nNOS immunoreactivity within the deprived ocular dominance columns. Panel A: tangential section was reacted for cytochrome oxidase histochemistry to reveal the dark, active and light, inactive ocular dominance columns within layer 4C. Panel B: an adjacent section reacted for nNOS-immunocytochemistry reveals lower levels of nNOS within the deprived columns. Small arrows point to blood vessel lumens used as fiduciary marks while arrowheads point to transitions between derived and active columns in the two sections. (Adopted from Fig. 3 of Aoki et al., Brain Research, 620: 97–113.)

Perhaps the rate of synthesis and anterograde transport of newly synthesized nNOS or of the retrograde transport of degraded nNOS depends on neural activity in the neuropil surrounding the neurites. Future studies that allow complete visu-

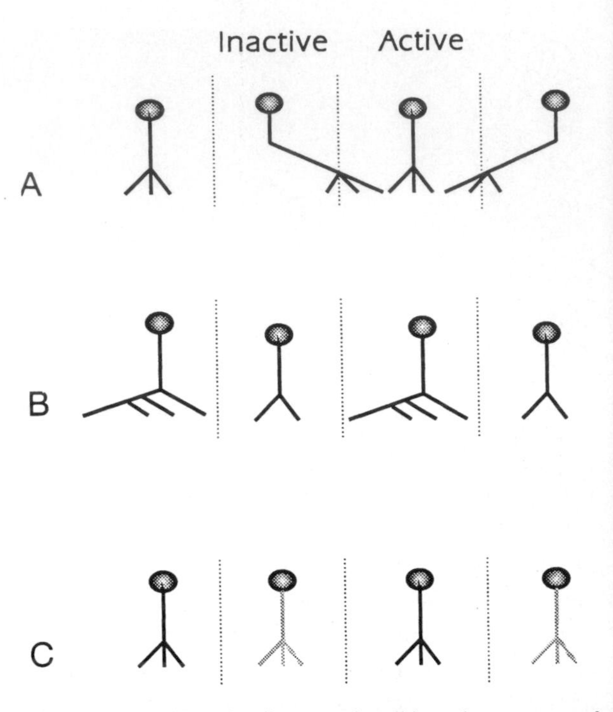

Fig. 11. A schematic diagram describing three types of subcellular changes that can result in differential distribution of nNOS-immunoreactive processes within layer 4C of the active and inactive ocular dominance columns in monkey visual cortex.

alization of all neurites from single somata, both containing and lacking nNOS-immunoreactivity, would be required to determine how, exactly, nNOS distribution within single cells is regulated.

The cortical and thalamic results, together, indicate that nNOS neurons in adult brain tissue exhibits plasticity, since they respond to reduction in afferent activity by altering the levels of nNOS protein within neurites. As reviewed by Garthwaite et al. and Friedlander et al., in this volume. NO modulates neuronal circuitry by enhancing transmitter release and blood flow. Removal of these nNOS effects would tend to decrease neural activity further within the deprived columns. Such a response by nNOS neurons may help to maximize efficiency of cortical circuitry, since noise generated by spontaneous activity within deprived columns, particularly as the receptors within these columns become supersensitized, could be reduced.

Does localization of nNOS at the plasma membrane depend on PSD-93 and related proteins?

One of the most interesting recent developments in receptor physiology is the finding that targeting of neurotransmitter receptors and nNOS to the plasma membrane and their subsequent clustering at synapses is dependent on the interaction of synaptic molecules with a class of intracellular proteins characterized by the existence of PDZ domains (Brenman et al., 1996a, b). In particular, by using the yeast two-hybrid screening procedure, two structurally related proteins, PSD-93 and PSD-95 (both reported to be prevalent at synapses), have been identified to play important roles in clustering of nNOS, NMDA receptors and other synaptic molecules along the plasma membrane (Brenman et al., 1996; Kim et al., 1996; Ehlers et al., 1996; Gomperts, 1996). These bind to nNOS as well as to the tSXV motif of NMDA receptor subunit 2B. The data obtained by EM–ICC indicate that PSD-93 and PSD-95 occur discretely over PSDs of spines (Fig. 12B) as well as in the cytoplasm of dendritic shafts (Fig. 12A). Additionally, dual EM–ICC, performed as described above for nNOS and NR1 subunits of the NMDA receptors, indicate that the two occur co-localized within single dendritic spines (Fig. 12B). However, the two molecules also occur frequently dissociated from one another and away from dendritic plasma membranes (Fig. 12A). This latter observation supports the idea that even though localization of nNOS to synapses may depend on PSD-93/95, additional cues may be required for their translocation from the cytosol in dendritic shafts to the plasma membrane of spines and dendritic shafts. One possibility is that the association of PSD-93/95 and nNOS to plasma membranes may depend on synaptic activity, promoting the addition of a fatty acid group (palmitoylation) to PSD-93/95. This idea can be tested in the future by using antisera that can differentiate PSD-93/95 with and without fatty acid modification. Furthermore, cell physiological studies should be performed to elucidate how nNOS activation may differ when anchored along the plasma membrane near synapses as opposed to remaining in the dendritic cytosol.

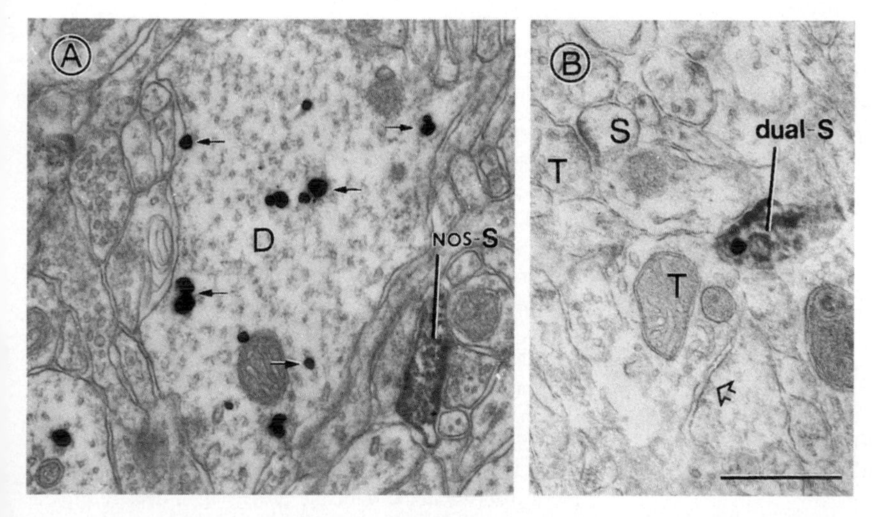

Fig. 12. Dual EM–ICC of nNOS and PSD-93/95. Panel A shows a large dendrite, D, immunolabeled for PSD-93 by the silver-intensified gold method and a small dendritic spine in its vicinity immunolabeled for nNOS by a peroxidase-based label (NOS-S). Note that the two immunolabels do not co-localize. Moreover, immunoreactivity is prevalent in the cytosol. Panel B: An example of a small spine labeled dually for nNOS (peroxidase) and PSD-95 (silver-intensified gold particle). This spine is post-synaptic to a terminal, T in the center of the panel, that forms a synapse with another spine (open arrow points to the postsynaptic membrane). The panel shows another axo-spinous junction that is devoid of immunoreactivity pre- or post-synaptically (T and S, upper left). Calibration bar = 500 nm.

Conclusion

Our results obtained from EM–ICC provides morphological support for the idea that activation of nNOS and NMDA receptors is coordinated. Specifically, since nNOS occurs within and near spines that also contain NR1-immunoreactivity, influx of Ca^{2+} via activated NMDA channels is likely to cause NO generation. In addition, proximity of the two molecules would facilitate NMDA receptor inhibition brought about by interaction between NO and the NMDA receptor's redox modulatory site (Lei et al., 1992; Lipton et al., 1993).

Our EM–ICC results also revealed the presence of nNOS and NMDA receptors at sites that would not be predicted to contain these molecules, based upon electrophysiological and neurochemical studies. In particular, the presence of nNOS with NMDA receptors within pre-synaptic axon terminals indicates that nNOS activation can occur following axo-axonic interactions mediated by en passant release of glutamate. Alternatively, should the processes immunoreactive for nNOS (with or without NMDA receptors) also harbor voltage-activated calcium channels or other receptor channels capable of altering the influx of Ca^{2+}, such as the $\alpha7$ subunit of nicotinic acetylcholine receptors, then NO generation could also occur by means other than glutamatergic synaptic transmission, such as the arrival of depolarizing pulses or cholinergic volume transmission. The presence of NMDA receptors in axons suggests that these pre-synaptic receptors may confer selective receptivity of axons to NO. Finally, results obtained from the ultrastructural localization of nNOS at sites that are immunoreactive for PSD-93/95 indicate that further studies are needed to determine the cellular mechanisms regulating the translocation of cytosolic PSD-93/95 to the plasma membrane, leading to effective anchoring of nNOS at synapses.

Acknowledgements

We are grateful to Dr. Ted M. Dawson for providing the nNOS and NMDAR1 antisera. This work was supported by the NIH grants EY08055 and NS30944, the NSF Presidential Faculty Fellowship RCD92-53750 and the Human Frontiers Science Program RG-16/93 to CA and the NIH Grant NS36017 to DSB. We thank Zak Shusterman, Byron Taylor and Mian Hou for the photographic reproductions and Alice Elste, C.G. Go, and X.-Z. Song for their technical assistance.

References

Aoki, C. (1997) Postnatal changes in the laminar and subcellular distribution of NMDA-R1 subunits in the cat visual cortex as revealed by immunoelectron microscopy. *Brain Res.*, 98: 41–59.

Aoki, C., Fenstemaker, S., Lubin, M. and Go, C.-G. (1993) Nitric oxide synthase in the visual cortex of monocular monkeys as revealed by light and electron microscopic immunocytochemistry. *Brain Res.*, 620: 97–113.

Aoki, C. and Kabak, S. (1992) Cholinergic terminals in the cat visual cortex: Ultrastructural basis for interaction with glutamate-immunoreactive neurons and other cells. *Visual Neurosci.*, 8: 177–191.

Aoki, C., and Pickel, V.M. (1989) Neuropeptide Y in the cerebral cortex and the caudate-putamen nuclei: Ultrastructural basis for interactions with GABAergic and non-GABAergic neurons. *J. Neurosci.*, 9: 4333–4354.

Aoki, C., Rhee, J., Lubin, M. and Dawson, T.M. (1997) NMDA-R1 subunit of the cerebral cortex co-localizes with neuronal nitric oxide synthase at pre- and postsynaptic sites and in spines. *Brain Res.*, 750: 25–40.

Aoki, C., Venkatesan, C., Go, C.-G., Mong, J.A. and Dawson, T.M. (1994) Cellular and subcellular localization of NMDA-R1 subunit immunoreactivity in the visual cortex of adult and neonatal rats. *J. Neurosci.*, 14: 5202–5222.

Bickford, M.E., Gunluk (Erisir), A.E., Guido, W. and Sherman, S.M. (1993) Evidence that cholinergic axons from the parabrachial region of the brainstem are the exclusive source of nitric oxide in the lateral geniculate nucleus of the cat. *J. Comp. Neurol.*, 334: 410–430.

Bredt, D.S. and Snyder, S.H. (1992) Nitric oxide, a novel neuronal messenger. *Neuron*, 8: 3–11.

Brenman, J.E., Chao, D.S., Gee, S.H., McGee, A., Craveri, S.E., Santillano, D.R., Wu, Z., Huang, F., Xia, H., Peters, M.F., Froehner, S.C. and Bredt, D.S. (1996a) Interaction of nitric oxide synthase with the postsynaptic density protein PSD-95 and $\alpha1$-syntrophin mediated by PDZ domains. *Cell*, 84: 757–767.

Brenman, J.E., Christopherson, K.S., Craven, S.E., McGee, A.W. and Bredt, D.S. (1996b) Cloning and characterization of postsynaptic density 93, a nitric oxide synthase interacting protein. *J. Neurosci.*, 16: 7407–7415.

Dawson, T.M. and Snyder, S.H. (1994) Gases as biological messengers: nitric oxide and carbon monoxide in the brain. *J. Neurosci*, 14: 5147–5159.

Dawson, T.M., Bredt, D.S., Fotuh, I.M., Hwang, P.M. and Snyder, S.H. (1991) Nitric oxide synthase and neuronal NADPH diaphorase are identical in brain and peripheral tissues. *Proc. Natl. Acad. Sci.*, USA, 88: 7797–7801.

Ehlers, M.D., Mammen, A.L., Lau, L-F. and Huganir, R.L. (1996) Synaptic targeting of glutamate receptors. *Current Opinion in Cell. Biol.*, 8: 484–489.

Gamble, E. and Koch, C. (1987) The dynamics of free calcium in dendritic spines in response to repetitive synaptic input. *Science*, 236: 1311–1315.

Garthwaite, J. and Boulton C.L. (1995) Nitric oxide signaling in the central nervous system. *Ann. Rev. Physiol.*, 57: 683–706.

Gilbert, C.D. (1983) Microcircuitry of the visual cortex. *Annu. Rev. Neurosci.*, 6: 217–247.

Gomperts, S.N. (1996) Clustering membrane proteins: It's all coming together with the PSD-95/SAP90 protein family. *Cell*, 84: 659–602.

Grab, D.J., Carlin, R.K. and Siekevitz, P. (1980) The presence and functions of calmodulin in the postsynaptic density. *Ann. NY Acad. Sci.*, 356: 55–71

Gray, E.G. (1959) Axo-somatic and axo-dendritic synapses of the cerebral cortex. *J. Anatomy*, 93: 420–433.

Hendry, S.H.C., Jones, E.G. and Emson, P.C. (1984) Morphology, distribution, and synaptic relations of somatostatin- and neuropeptide γ-immunoreactive neurons in rat and monkey neocortex, *J. Neurosci.*, 4: 2497–2517.

Hope, B.T., Michael, G.J., Knigge, K.M. and Vincent, S.R. (1991) Neuronal NADPH diaphorase is a nitric oxide synthase. *Proc. Natl. Acad. Sci. USA*, 88: 2811–2814.

Kim, E., Cho, K.-O., Rothchild, A. and Sheng, M. (1996) Heteromultimerization and NMDA receptor-clustering activity of Chapsyn-110, a member of the PSD-95 family of proteins. Neuron, 17: 103–113.

Lei, S.Z., Pan, Z.-H., Aggarwal, S.K., Chen, H.-S.V., Hartman, J., Sucher, N.J. and Lipton, S.A. (1992) Effect of nitric oxide production on the redox modulatory site of the NMDA receptor-channel complex. *Cell*, 8: 1087–1099.

Lipton, S.A., Choi, Y.B., Pan, Z.H., Lei, S.Z., Chen, H.S., Sucher, N.J., Loscalzo, J., Singel, D.J. and Stamler, J.S. (1993) A redox-based mechanism for the neuroprotective and neurodestructive effects of nitric oxide and related nitroso-compounds, *Nature*, 364: 626–632.

Llins, R., Sugimori, M. and Silver, R.B. (1992) Microdomains of high calcium concentration in a presynaptic terminal. *Science*, 256: 677–679.

Montague, P.R., Gancayco, C.D., Winn, M.J., Marchase, R.B. and Friedlander, M.J. (1994) Role of NO production in NMDA receptor mediated neurotransmitter release in cerebral cortex. *Science*, 263: 973–977.

Mulle, C., Choquet, D., Korn, H. and Changeux, J.P. (1992) Calcium influx through nicotinic receptor in rat central neurons: Its relevance to cellular regulation. *Neuron*, 8: 135–143.

Mller, W. and Connor, J.A. (1991) Dendritic spines as individual neuronal compartments for synaptic Ca^{2+} responses. *Nature*, 354: 73–79.

Murphy, K.P.S.J., Williams, J.H., Bettache, N. and Bliss, T.V.P. (1994) Photolytic release of nitric oxide modulates NMDA receptor-mediated transmission but does not induce long-term potentiation at hippocampal synapses. *Neuropharmacol.*, 33: 1375–1385.

Petralia, R.S., Yokotani, N. and Wenthold, R.J. (1994) Light and electron microscope distribution of the NMDA receptor subunit NMDAR1 in the rat nervous system using a selective anti-peptide antibody. *J. Neurosci.*, 14: 667–696.

Price, R.H., Mayer, B. and Beitz, A.J. (1993) Nitric oxide synthase neurons in rat brain express more NMDA receptor mRNA than non-NOS neurons. *NeuroReport*, 4: 807–810.

Schuman, E.M. and Madison, D.V. (1994) Nitric oxide and synaptic function. *Ann. Rev. Neurosci.*, 17: 153–83.

Seguela, P., Wadiche, J., Dineley-Miller, K., Dani, J.A. and Patrick, J.W. (1993) Molecular cloning, functional properties, and distribution of rat brain $\alpha 7$: A nicotinic cation channel highly permeable to calcium. *J. Neurosci.*, 13: 596–604.

Shatz, C.J. (1990) Impulse activity and the patterning of connections during CNS development. *Neuron*, 5: 745–756.

Siegel, S.J., Brose, N., Janssen, W.G., Gasic, G.P., Jahn, R., Heinemann, S.F. and Morrison, J.H. (1994) Regional, cellular, and ultrastructural distribution of *N*-methyl-D-aspartate receptor subunit 1 in monkey hippocampus. *Proc. Natl. Acad. Sci. USA*, 91: 564–568.

Son H., Zhuo M., Arancio O., Kandel E.R. and Hawkins, R.D. (1997) Further tests on the role of cGMP in hippocampal LTP. *Soc for Neurosci Abstract*, 23: 1393.

Vijayaraghavan, S., Pugh, P.C., Zhang, Z.W., Rathouz, M.M. and Berg, D.K. (1992) Nicotinic receptors that bind α-bungarotoxin on neurons raise intracellular free Ca^{2+}. *Neuron*, 8: 353–362.

Vincent, S.R. and Kimura, H. (1992) Histochemical mapping of nitric oxide synthase in the rat brain. *Neurosci.*, 46: 755–84.

Vincent, S.R., Johansson, O., Hokfelt, T., Skirboll, L., Elde, R.P., Terenius, L., Kimmel, J. and Goldstein, M. (1983) NADPH-diaphorase: A selective histochemical marker for striatal neurons containing both somatostatin- and avian pancreatic polypeptide (APP)-like immunoreactivities. *J. Comp. Neurol.*, 217: 252–263.

Wendland, B., Schweizer, F.E., Ryan, T., Nakane, M., Murad, F., Scheller, R.H. and Tsien, R.W. (1994) Existence of nitric oxide synthase in rat hippocampal pyramidal cells. *Proc. Natl. Acad. Sci. USA*, 91: 2151–2155.

SECTION III

Nitric oxide in brain development

R.R. Mize, T.M. Dawson, V.L. Dawson and M.J. Friedlander (Eds.)
Progress in Brain Research, Vol 118

CHAPTER 8

Nitric oxide as a signaling molecule in visual system development

Karina S. Cramer*,†, Catherine A. Leamey and Mriganka Sur

Department of Brain and Cognitive Sciences, Massachusetts Institute of Technology, Cambridge, MA 02139, USA

Abstract

The lateral geniculate nucleus (LGN) of the ferret is characterized by the readily discernible anatomical patterning of afferent terminations from the retina into both eye-specific layers and On/Off sublaminae. The eye-specific layers form during the first post-natal week, and On/Off sublaminae become apparent during the third to fourth post-natal weeks. The post-natal appearance of these patterns thus provides an advantageous model for the study of the mechanisms of activity-dependent development. The second phase of pattern formation, the appearance of On/Off sublaminae, involves the elaboration of appropriately placed axonal terminals and the restriction (or retraction) of inappropriately placed terminals. Previous work has demonstrated that this process is dependent on the activation of NMDA-receptors. Other studies have provided strong evidence that nitric oxide, a diffusible gas which is produced downstream of NMDA-receptor activation, acts as a retrograde messenger molecule to induce changes in pre-synaptic structures. In this article we review the evidence that nitric oxide plays a role in activity-dependent synaptic plasticity in the developing retinogeniculate pathway. The role of nitric oxide in other aspects of visual system development is also discussed.

Introduction

During nervous system development, neurons initially project broadly to target regions. In the subsequent refinement of synaptic projections, patterns of neuronal activity further improve the specificity of connections, resulting in the removal of inappropriate connections and the strengthening and growth of appropriate connections. The mechanisms underlying this activity-dependent refinement of connections are the subject of intense study in the field of nervous system development. Activity-dependent mechanisms are not only involved in specifying connections during development, but also appear to be involved in specifying synaptic strength in adulthood. Both activity-dependent development and synaptic plasticity require detection of correlated activity using post-synaptic mechanisms, and both processes may result in changes in pre-synaptic structure or function. Thus, both anterograde and retrograde signaling molecules are required to transmit information between pre- and post-synaptic elements.

* Corresponding author. Tel.: +1 206 616 4652; fax: +1 206 616 1828; e-mail: kcramer@u.washington.edu

† Present address: Virginia Merrill Bloedel Hearing Research Center, University of Washington, Box 357923 Seattle, WA 98195, USA

We use the development of the retinogeniculate projection in the ferret to study the activity-dependent refinement of connections. This system is particularly well suited to this purpose as the retinogeniculate projection patterns are robust and discernible, form post-natally, and the activity-dependence and activity patterns in the system have been well characterized. In this system, retinal ganglion cell axons from both eyes project to the lateral geniculate nucleus (LGN) of the thalamus. At birth, these axons are present and inputs from both eyes are intermingled. By one post-natal week, axons from the two eyes have segregated and thus eye-specific layers within the LGN are evident (Linden et al., 1981). The largest and most clearly discernible are the A layers (Fig. 1). Layer A receives input from the contralateral retina, while layer A1 receives input from the ipsilateral retina. The eye-specific layers A and A1 are later subdivided into sublaminae that correspond to the electrophysiological response properties of retinal ganglion cells. The inner sublamina receives input from On-center retinal ganglion cells, while the outer sublamina receives input from Off-center retinal ganglion cells (Stryker and Zahs, 1983). Sublamination, based on anterograde labeling of afferents, is evident by three to four post-natal weeks (Fig. 1). The On/Off sublamine are evident within eye-specific layers because retinal ganglion cell axons form terminations in either the inner (On) or outer (Off) sublamina. Axonal branching is very sparse in the intersublaminar region (Hahm et al., 1991). Thus the pattern of orderly connections and selective axon arborization is evident when the entire projection is labeled (e.g., with tracers injected intraocularly).

Segregation of retinogeniculate afferents is dependent on neuronal activity. When tetrodotoxin (TTX), a sodium channel blocker, is superfused over the optic chiasm in pre-natal cats, eye-specific layer formation is disrupted (Shatz and Stryker, 1988). The formation of On/Off sublaminae in the ferret is also dependent on afferent activity. Sublaminae fail to form properly when TTX is applied intraocularly during the third and fourth post-natal weeks (Cramer and Sur, 1997a). Activity is also required for specific connections between On- and Off-center retinal ganglion cells and corresponding cell types in the LGN of the cat (Archer et al., 1982). In this species the connec-

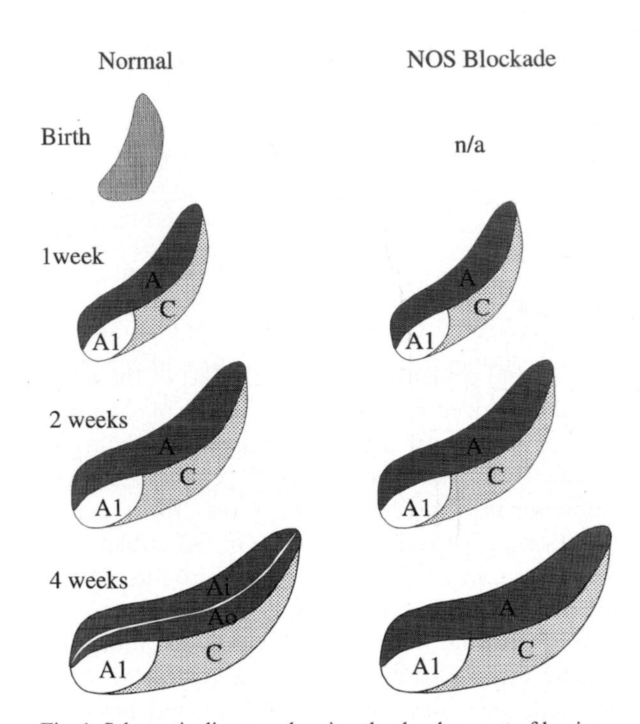

Fig. 1. Schematic diagram showing the development of laminae and sublaminae in the LGN of the ferret as seen in horizontal sections. All sections are of the right LGN, with stippling denoting location and density of projections after injection of label into the contralateral (left) eye. The left column illustrates normal development. At P0, axons from the two eyes are intermingled in the LGN. By 1 week post-natal, the fibres from the two eyes have segregated to form the A and A1 layers which receive contralateral and ipsilateral inputs respectively. The C layer is also apparent at this stage. Between 1 and 2 weeks, the eye-specific layers become more well defined. By 4 weeks however, the A layer has become subdivided into an inner sublamina (Ai) which receives input from On-center retinal ganglion cells and an outer sublamina which receives input from Off-center retinal ganglion cells. The right-hand column shows how normal development of the LGN is perturbed by NOS blockade. NOS blockade between birth and one week post-natal has no effect on the formation of eye-specific layers: they form as in the normal animal. However, NOS blockade between 2 and 4 weeks prevents the formation of On-Off sublaminae in the LGN.

tions are specific but not anatomically sublaminated.

Interestingly, several studies suggest that the neuronal activity required for On/Off segregation is spontaneous rather than visually evoked, as photoreceptors and bipolar synapses are relatively late to mature (Greiner and Weidman, 1981; Maslim and Stone, 1986), and the eyes do not open until after four post-natal weeks in ferrets. Spontaneous activity has been recorded in the retina, and occurs in spatially correlated waves (Galli and Maffei, 1988; Meister et al., 1991; Wong et al., 1993). This correlation is due to electrical coupling between ganglion cells in the retina (Penn et al., 1994) and suggests that inputs from one eye are better correlated than inputs from two different eyes. Additionally, On-center and Off-center retinal ganglion cell dendrites are differentially stratified within the retina (Bodnarenko et al., 1995; Maslim and Stone, 1988; Wingate, 1996), suggesting that spontaneous activity from two inputs from a single cell class are correlated better with each other than inputs from two different classes. This selective correlation of activity within retinal ganglion cell types is supported by the fact that spontaneous bursting patterns of On and Off retinal ganglion cells become distinct during early post-natal development (Wong and Oakley, 1996). The patterns of activity in retinal ganglion cells are thus a reflection of the positions and types of these cells within the retina, and may serve to organize the eye-specific and On/Off termination patterns within the LGN.

Retinogeniculate synaptic transmission is mediated by glutamate and, in the developing animal, largely via *N*-methyl-D-aspartate (NMDA) receptors (Esguerra et al., 1992; Kwon et al., 1991; Mooney et al., 1993; Ramoa and McCormick, 1994). These receptors may function as detectors of coincident activity in LGN cells because they allow Ca^{2+} influx only when they are depolarized. Consistent with this postulated role for NMDA receptors, the formation of sublaminae is disrupted when NMDA receptors are inhibited during the third post-natal week (Hahm et al., 1991). NMDA receptors are also required for the formation of numerous other patterns of axonal projections in the developing brain, including the refinement of topographic maps in the superior colliculus of the rat (Simon et al., 1992), the formation of auditory maps in the superior colliculus of the ferret (King et al., 1996), and the formation of eye-specific stripes in the tectum of the "three-eyed" frog (Cline and Constantine-Paton, 1990). NMDA receptors are also involved in the critical period of developmental plasticity for the formation of ocular dominance columns in the cat (Bear et al., 1990) and barrels in the somatosensory cortex of the rat (Schlaggar et al., 1993). They are critical for the induction of long-term potentiation (LTP) in the CA1 region of the hippocampus (Gustafsson and Wigstrom, 1990; Harris et al., 1984; Murphy et al., 1997). In these systems, correlated pre- and post-synaptic activity is detected by the post-synaptic cell using NMDA receptors. This correlated activity however, also results in changes in pre-synaptic terminals. During development, blocking NMDA receptors or adding NMDA can influence the size and position of pre-synaptic axon arbors (Hahm et al., 1991; Yen et al., 1995). In addition, a possible mechanism for LTP involves upregulation of neurotransmitter release in pre-synaptic terminals. In these cases, a retrograde messenger is required to transmit information about post-synaptic NMDA receptor activation to pre-synaptic terminals.

Nitric oxide (NO), a diffusible gas, has been proposed as a retrograde messenger that acts downstream of NMDA receptor activation (Bredt and Snyder, 1992; Gally et al., 1990; Montague et al., 1991). Numerous studies have provided evidence that NO is required for LTP in CA1 of the hippocampus (Arancio et al., 1996; Bohme et al., 1991; O'Dell et al., 1991; Schuman and Madison, 1993; Son et al., 1996). In this chapter, we will review the evidence for a role for NO in visual system development, and examine the nature of the influence of NO on developing axons and dendrites. In addition, we will evaluate whether NO is a retrograde messenger, and whether it has other developmental roles as well.

Expression of nitric oxide synthase

In order to determine whether NO has a role in retinogeniculate development, it is first necessary to demonstrate that NO is present in the LGN at the appropriate periods of development. Nitric oxide synthase (NOS) is the synthetic enzyme for NO, and is present in several forms in the brain. It is Ca^{2+}/calmodulin-dependent and uses NADPH and arginine to synthesize NO. Regions of NOS activity can therefore be revealed using NADPH-diaphorase histochemistry (Dawson et al., 1991; Hope et al., 1991). We used this method to study the developmental expression of NOS in the LGN (Cramer et al., 1995). At birth, NADPH-diaphorase labeling was present in blood vessels and in neuropil but not in LGN cell bodies. At three weeks post-natal, some LGN cells stained positive for NADPH-diaphorase. The density of NADPH-positive cell bodies had increased by four post-natal weeks. The density then began to decline, so that by six weeks post-natal no cell bodies were labeled, but staining of neuropil and blood vessels was retained.

In the cat, the neuropil staining with NADPH-diaphorase in the LGN is localized within cholinergic brainstem afferents and not within retinal afferents (Bickford et al., 1993); thus it appears that NO in the LGN is not produced in retinal ganglion cell axon terminals. The stained cell bodies in the ferret LGN are of various types; some (but not most) NADPH-diaphorase positive cells are also stained by immunohistochemistry for γ-amino-butyric acid (GABA) (Polonsky, A, unpublished data). The remainder of the cells have a wide range of diameters and morphologies, and most likely include the thalamic relay neurons that project to visual cortex (Cramer et al., 1995).

In another study, we examined the developmental expression of the neuronal form of NOS (nNOS), using immunohistochemistry (Cramer and Sur, 1996). As for NADPH-diaphorase, an increase followed by a decrease in expression was found, but in contrast to NADPH-diaphorase staining, the peak staining in cell bodies was at about three weeks post-natal. By four weeks post-natal, when NADPH-diaphorase had the most dense cell body staining, nNOS immunohistochemical staining had diminished and was present only in a few cell bodies. Thus the NADPH-diaphorase staining probably represents several different forms of NOS that are developmentally regulated in LGN cells.

Regulation of NOS expression

Transient NOS expression is present in several areas of the vertebrate nervous system during development. In the pre-natal rat, this includes the dorsal root ganglia (Wetts and Vaughn, 1993; Bredt and Snyder, 1994), the embryonic and early post-natal cortical plate (Bredt and Snyder, 1994), embryonic spinal cord (Wetts et al., 1995), olfactory bulb (Samama and Boehm, 1996), and the olfactory epithelium (Bredt and Snyder, 1994; Roskams et al., 1994; Arnhold et al., 1997). Transient NOS expression is also present in the mouse whisker-barrel pathway (Mitrovic and Schachner, 1996), and the chick optic tectum (Williams et al., 1994). The mechanisms by which NOS is regulated have been examined in a number of systems. One interesting possibility is that NOS expression is itself regulated by neuronal activity. This type of regulation has been shown in a number of systems using deafferentation paradigms. In the chick optic tectum, the expression of NOS coincides with the arrival of retinal afferents, and removal of these afferents results in the reduction of NOS in several tectal cell types (Williams et al., 1994). Monocular enucleation in the rat results in reduced NOS in both the LGN (Zhang et al., 1996) and the superior colliculus contralateral to the removed input (Vercelli and Cracco, 1994; Zhang et al., 1996). These experiments seem to support a positive correlation between expression of NOS and afferent input to a target. However, in LGN cells of the adult cat, in which NAPDH-diaphorase is not normally expressed, monocular lid suture leads to increased expression of NADPH-diaphorase activity in the layers receiving input from the *intact* eye (Gunluk et al., 1994), and not to down-regulation of NADPH-diaphorase in the deprived layers. This study suggests that expression of NADPH-diaphorase in LGN cells uses cues that are dependent on levels of *relative* activity between the two eyes. While manipulating afferent input seems to have an effect in many cases, we found that when post-natal ferrets were treated monocularly with TTX during the third and fourth post-natal weeks, there was no difference in the density of cells labeled with NADPH-diaphorase between deprived and normal LGN layers (unpublished observations). Other experimental results suggest a negative correlation between afferent activity and NOS expression. In cultures of cerebellar granule cells, application of TTX results in increased expression of nNOS, and depolarization causes reduced expression of nNOS (Baader and Schilling, 1996).

Regulation of NOS expression may also depend on neurotrophin levels. Nerve growth factor (NGF) increases nNOS mRNA levels and the level of nNOS activity in basal forebrain neurons of the developing and adult animal (Holtzman et al., 1994; Holtzman et al., 1996). In cultures of embryonic rat spinal cord, brain-derived neurotrophic factor (BDNF) and neurotrophins NT-3 and NT-4 increase the number of NADPH-diaphorase neurons. The regulation of NOS expression by neuronal activity may be related to regulation of NOS expression by neurotrophins, because neurotrophin expression is modified by neuronal activity (Bonhoeffer, 1996; Castren et al., 1993; Lindholm et al., 1994). Several molecular pathways may thus work together to regulate the expression of NOS during development (Baader et al., 1997).

NO is activated by Ca^{2+} influx through NMDA receptors, and by the association of Ca^{2+} with calmodulin (Bredt and Snyder, 1989; Bredt and Snyder, 1990; Garthwaite et al., 1988; Garthwaite et al., 1989). Activation of NMDA receptors can lead to high levels of Ca^{2+} concentrated in dendritic spines (Regehr and Tank, 1990). The subcellular localization of nNOS is consistent with a role for NO downstream of NMDA receptor activation. Neuronal NOS is membrane-associated (Brenman et al., 1996a; Hecker et al., 1994), and binds to post-synaptic density proteins PSD-93 and PSD-95, which are coupled to the NMDA receptor 2B subunit (Brenman et al., 1996b). PSD-93/95 may have a role in clustering receptors and ion channels, and may thus couple Ca^{2+} influx to activation of nNOS (Brenman et al., 1996b). In addition, electron microscopy studies support the colocalization of NMDA receptors and nNOS (Aoki et al., 1997). This coupling is not strict, however, and NMDA receptors and nNOS are also found unassociated with each other.

A role for NO in development

Regulation of axonal terminations

Several studies have examined the role of NO in axon outgrowth and in pattern formation in nervous system development. We have examined the effect of NOS blockade on the formation of eye-specific layers and On/Off sublaminae in the ferret LGN (Cramer et al., 1996). To examine On/Off segregation, we administered daily intraperitoneal injections of 4.0 to 40 mg/kg of the arginine analog N^{ω}-nitro-L-arginine (L-NoArg), which blocks NOS (Dwyer et al., 1991), or used osmotic minipump application of 0.5 mM L-NoArg to superfuse the drug over the LGN on the right side of the brain. Application of L-NoArg began at post-natal day (P) 14, and continued through P26; this period of development coincides with the formation of On/Off sublaminae in the ferret LGN. On P24, animals were anesthetized and received an intraocular injection of horseradish peroxidase coupled to wheat germ agglutinin (WGA–HRP) in the left eye to anterogradely label axon terminations in the right LGN. We assessed On/Off sublamination by blind-scoring sections of the right LGN on a scale from 0 to 3, where 0 signifies no evidence of sublamination, and 3 signifies a clear intersublaminar pale staining region throughout the stained A layer. Sublamination (Fig. 1), and corresponding scores (Fig. 2), were significantly reduced with focal or systemic

L-NoArg treatment. In addition, systemic treatment with L-NoArg produced a dose-dependent inhibitory effect on sublamination scores (Fig. 2). Animals that received 40 mg/kg/day of the inactive isomer N^G-nitro-D-arginine methyl ester (D-NAME) during the same treatment period had normal sublamination. In addition, some animals received 40 mg/kg/day L-NoArg together with 40 mg/kg/day L-arginine; in these animals sublamination of the LGN was normal, suggesting that the drug acts at the arginine binding site on NOS.

One potential problem associated with systemic blockade of NOS is that the treatment increases blood pressure. We verified that blood pressure was increased by our treatment (Cramer et al., 1996). As a control to examine whether this increase in blood pressure contributed to the effect of NOS inhibition on sublamination, we administered 40 mg/kg/day L-NoArg together with 5 mg/kg/day verapamil, an antihypertensive Ca^{2+} channel blocker, to a separate set of animals. Blood pressure in these animals was normal, but On/Off sublamination was significantly disrupted (Fig. 2), suggesting that hypertension is unrelated to the effect of NOS blockade on retinogeniculate sublamination. In addition, we attempted to rule out other indirect effects of systemic drug application by using focal treatment with osmotic minipumps. Focal application disrupted sublamination only on the operated side; the contralateral LGN appeared normal.

We also examined the role of NOS blockade on the development of individual axonal arborizations. To compare our results with those obtained in NMDA blockade experiments, we blocked NOS systemically with 40 mg/kg/day L-NoArg from P14 to P21. Control animals received 40 mg/kg/day D-NAME. At P21, axons were stained in acute thalamic preparations. For these, animals were anesthetized, and perfused with artificial cerebrospinal fluid (ACSF) saturated with oxygen. The diencephalon was dissected out and small deposits of HRP were placed in the optic tract overlying the LGN. Thalamic preparations were maintained for 3–5 h in ACSF to allow transport of label to axon terminals. They were then fixed and sectioned and axons were reconstructed using *camera lucida* tracings. In L-NoArg treated animals, individual axon arbors were positioned inappropriately in the center of the eye-specific layers, thus obscuring the pale staining intersublaminar region that normally separates the On- and Off-sublaminae. These findings are similar to those obtained following NMDA receptor blockade (Hahm et al., 1991), and are consistent with the hypothesis that NMDA receptor-mediated processes in this system act through NOS to influence development of sublaminae. At present it is unclear whether

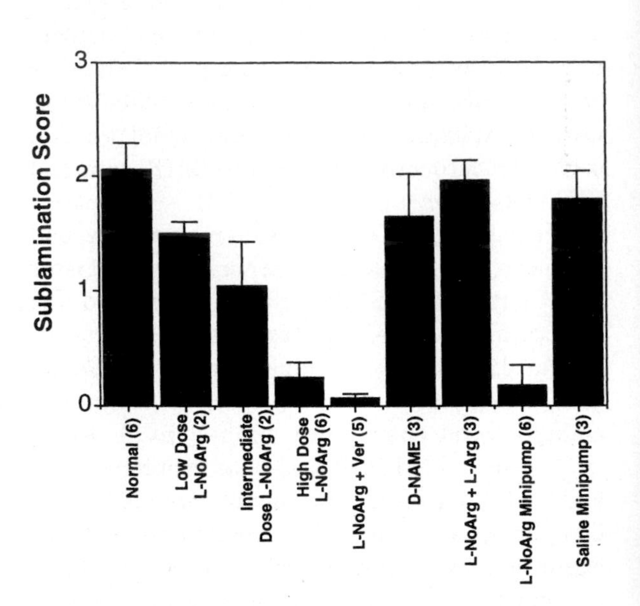

Fig. 2. Bar plot summarising the sublamination scores (see text) in the ferret LGN following systemic or focal application of the NOS inhibitor L-NoArg. The numbers in parentheses indicate the number of animals in each treatment group. There was a dose dependent decrease in sublamination following systemic blockade of NOS. In animals receiving a high dosage, sublamination was significantly reduced compared to normal animals. In animals where the pressor effect of L-NoArg was blocked by the application of verapamil (ver), sublamination was also disrupted, indicating that the effects of NOS inhibitors were not due to hypertensive effects. Control animals treated with the inactive isomer of L-NoArg, D-NAME, or with L-NoArg together with L-Arginine, had normal levels of sublamination. Focal application of L-No-Arg via osmotic minipumps also disrupted sublamination, whereas sublamination was normal in control animals with saline minipumps. (Reprinted from Cramer et al., 1996; with permission).

treatment with NOS inhibitors produces an increase in the number of synaptic contacts with LGN neurons. However, it has been shown that treatment with NMDA significantly decreases the number of synaptic contacts in the frog retinotectal system (Yen et al., 1995). This occurs as a consequence of the presence of fewer high order axonal branches. Since in this system NO induces growth cone collapse (Renteria and Constantine Paton, 1996) (see below), it is possible that NMDA receptor activation regulates synapse number during development through an NO-mediated pathway.

To examine whether NO plays a role in the formation of eye-specific layers, we treated animals from P1 to P7 systemically with 20 mg/kg/day L-NoArg. Control animals received 20 mg/kg/day D-NAME during the same period. On P7, animals were anesthetized and WGA-HRP was injected into one eye and the anterograde tracer cholera toxin B (CTB) into the other eye. At P8, brains were fixed and sectioned, and alternate sections were processed histologically to reveal HRP and CTB. Treated, control, and untreated animals all had similar segregation of eye-specific layers, suggesting that NO is not required for the segregation of these layers (Fig. 1). HPLC analysis of brain [arginine]/[citrulline] ratios suggest that NOS activity blockade in the treated animals was at least as great as in the older animal group in which sublamination was disrupted; thus it is likely that treated animals in the early age group received a dose of L-NoArg sufficient to inhibit most NOS activity.

These results are similar to those seen with blockade of NMDA receptors in that On/Off sublamination is disrupted (Hahm et al., 1991), but the formation of eye-specific layers is not (Smetters et al., 1994). Together, these studies suggest that pattern formation in the retinogeniculate pathway of the LGN occurs in two stages that use distinct sets of molecular mechanisms within a single population of axons to mediate two phases of activity-dependent refinement. In the LGN of the ferret, NO is part of this activity-dependent mechanism only when NMDA receptor activation is required, and only when high levels of NADPH-diaphorase are present in LGN cell bodies.

Other studies have similarly found a role for NO in visual system development. In the chick retinotectal projection, a transient ipsilateral projection is removed in an activity-dependent, NMDA receptor-dependent manner (Ernst et al., 1997). Inhibition of NOS prevents the removal of this projection (Wu et al., 1994). In this system, expression of NOS is also developmentally modulated so that the peak in expression coincides with the requirement for NO during developmental refinement (Williams et al., 1994). A role for NO has also been found in the development of the central projections of photoreceptors in Drosophila melanogaster (Gibbs and Truman, 1998). Exogenously applied NO stimulates guanylyl cyclase activity, and blocking NOS during metamorphosis results in disorganized projections from photoreceptors to the optic lobe in a cyclic GMP-dependent manner.

It is significant to note that many projection patterns develop in an apparently NO-independent manner. As noted above, the formation of eye-specific layers does not require NO. These layers form during the first post-natal week, when NADPH-diaphorase staining in the LGN includes only the neuropil; the source of this stained neuropil is probably cholinergic brainstem afferents (Bickford et al., 1993). In the superior colliculus of the cat and rat, cholinergic innervation is organized into patches (Beninato and Spencer, 1986; Jeon and Mize, 1993). Although these cholinergic fibers express NADPH-diaphorase, inhibition of NOS during the period of patch formation in the rat does not interfere with the organization of patches (Mize et al., 1997). The authors postulate that NO selectively influences the developmental refinement of pathways involving glutamatergic neurotransmission and NMDA receptors. Consistent with this view, cholinergic patches in the superior colliculus are not glutamatergic, and while retinogeniculate connections are glutamatergic, the formation of eye-specific layers does not require NMDA receptor activation (Smetters et al., 1994). However, glutamatergic pathways whose development requires NMDA receptor activation do not necessarily require NO for developmental refinement. For example, NO appears not to be involved in the formation of ocular dominance columns, even though NOS is present during the appropriate period of development (Daw et al., 1994; Finney and Shatz, 1994; Finney and Shatz, 1997; Ruthazer et al., 1996).

It is not clear why NO regulates refinement of some axonal projections and not others. One possibility is that there are several biochemical pathways that use neuronal activity patterns to set up specific projections during development. The use of an NO-mediated pathway may be favored when glutamatergic transmission and NMDA receptors are involved because of the coupling of the NMDA response with activation of NOS. However, these conditions may be insufficient in some circumstances – for example, if NOS is not present in a suitable distribution or quantity or if a more appropriate mechanism is in place. Future experiments comparing several elements of activity-dependent pathways will be required to elucidate the conditions under which NO is involved in developmental refinement of projections.

Regulation of neurite outgrowth

We have described above the effects of NOS blockade on the refinement of synaptic connections. In addition, experiments on primary cultures of neurons have shown that NO may influence neurite outgrowth. In cultures of dorsal root ganglia (Hess et al., 1993), addition of NO donors is associated with growth cone collapse. Growth cone collapse was also seen in studies in cultures of frog retinal ganglion cells exposed to NO donors (Renteria and Constantine Paton, 1996). In these studies, growing retinal ganglion cell axons responded to exogenously applied NO, and not to other components of the NO donors. NOS blockade had no effect, however, suggesting that these growth cones do not normally respond to endogenously produced NO, and that growing retinal ganglion cell axons respond to NO produced in the tectum. However, nitric oxide donors promote neurite outgrowth and branching in primary hippocampal neuron cultures and enhance NGF mediated neurite outgrowth in cultures of PC12 cells (Hindley et al., 1997). The effects of NO on neurite outgrowth in culture may depend on several other factors, including the tissue type and the developmental age of the tissue.

Regulation of dendritic morphology

Neuronal activity can regulate dendritic morphology in several regions of the developing nervous system. The effects of neuronal activity on dendritic morphology have been shown in several systems to be mediated by NMDA receptor activation. In the visual system, TTX treatment during development increases the density of dendritic spines in cat LGN cells, but does not appear to alter dendritic branching or complexity (Dalva et al., 1994). Chronic treatment of the ferret LGN with inhibitors of NMDA receptors significantly increases the extent of dendritic branching and the number of dendritic spines (Rocha and Sur, 1995).

The time scale of the effect of NMDA receptor blockade on dendritic spine density was examined in living slices of ferret LGN (Rocha and Sur, 1995). Horizontal slices of LGN were prepared from ferrets of different ages and maintained in oxygenated ACSF. Small crystals of a lipophilic dye, diI, were placed within the LGN to label dendritic segments. Confocal images of labeled dendritic segments were made initially, and segments were imaged again after treatment with NMDA receptor inhibitors for several hours. Over the course of several hours, new spines could be observed on labeled segments, and there was a net increase in spine density in slices from P14 to P21 ferrets. The effects of NMDA receptor blockade may be mediated by NO in this system. We have begun to accumulate evidence that an increase in spine density in acute slice preparations is similarly seen using inhibitors of NOS (Cramer and Sur, 1997b) during the same developmental period.

In the retinogeniculate pathway, NMDA receptor activation is correlated with a restriction in dendritic branching and spine density. The stratification of retinal ganglion cell dendrites in the retina is also blocked by the inhibition of NMDA receptors, but this treatment does not seem to alter the extent of dendritic arbors (Bodnarenko and Chalupa, 1993; Bodnarenko et al., 1995). In other systems, NMDA receptor activation is associated with an *increase* in growth and branching. For example, in spinal cord motor neurons, NMDA receptor blockade *decreases* the extent of dendritic branching during a growth phase in spinal cord motor neurons (Kalb, 1994). In this system, NO is required for neuronal maturation (Kalb and Agostini, 1993), and NO seems to act downstream of NMDA receptors in the regulation of dendritic branching (R. Kalb, personal communication).

NMDA receptors and NO

During development, NMDA receptor activation is required for the normal formation of several pathways, and NO also seems to be necessary in at least a small subset of these. When NMDA receptors and NO are both involved in a system, does NO act as a retrograde messenger? Several lines of evidence suggest that retrograde signaling downstream of NMDA receptor activation is a function of NO. A recent study of hippocampal neurons (Arancio et al., 1996) elegantly showed that during the induction of LTP, NO is produced in the post-synaptic cell, diffuses through the extracellular space, and acts at the pre-synaptic terminal. In the developing retinotectal projection, NOS is present in the frog tectum (Bruning and Mayer, 1996; Munoz et al., 1996), and NO acts on growing processes of cultured frog retinal ganglion cells (Renteria and Constantine Paton, 1996), which are not believed to produce NO independently (Renteria and Constantine Paton, 1996). In the retinogeniculate projection of the ferret, NOS is produced in post-synaptic neurons (Cramer et al., 1995), and most likely not in retinal ganglion cell axons (Bickford et al., 1993). Blocking NOS in the LGN results in changes in the position of retinal ganglion cell axon arbors (Cramer et al., 1996), suggesting that NO made in LGN neurons diffuses to retinal ganglion cell axons as a retrograde messenger during development.

While these studies support a role for NO as a retrograde signaling molecule, other evidence suggests that NO has other roles as well. In the regulation of dendritic maturation, the effects of NO are on post-synaptic cells. In dendrites of LGN cells in the ferret, it is likely that NMDA receptor activation leads to production of NO, which then acts in the same cells to regulate dendritic morphology (Cramer and Sur, 1997b; Rocha and Sur, 1995). In the regulation of dendritic maturation in the spinal cord, NOS is made in premotor interneurons, and seems to affect motor neurons post-synaptic to these interneurons in an orthograde manner (Kalb and Agostini, 1993). The diffusible nature of NO and its colocalization with NMDA receptors make it well suited to regulate pre- and post-synaptic aspects of activity-dependent development. Interestingly, a small fraction of electron dense profiles that are positive for both nNOS and NMDA R1 subunit are pre-synaptic (Aoki et al., 1997). This finding suggests that the NMDA/NO pathway may in some cases regulate pre-synaptic terminals independently of signaling to or from the post-synaptic cell. In order to understand the roles of NO during development, it is essential to identify the molecular targets for NO.

Molecular targets for NO

Several molecules have been proposed as targets for NO. One target is soluble guanylyl cyclase (sGC), which in turn stimulates production of cyclic GMP (cGMP) (Bredt and Snyder, 1989; East and Garthwaite, 1991). Cyclic GMP activates several cation conductances, including a cGMP-gated nonselective cation current in retinal ganglion cells (Ahmad et al., 1994) and potassium-selective channels (Yao et al., 1995). There is evidence for a role for sGC downstream of NO in hippocampal LTP (Arancio et al., 1995), in fly retinal development (Gibbs and Truman, 1998; Truman et al., 1996), and in NGF-induced neurite outgrowth in PC12 cells (Hindley et al., 1997). However, production of cGMP does not appear to be involved in mediating the effects of NO in neurite outgrowth in cultured retinal ganglion cells (Renteria and Constantine Paton, 1996) or cultured dorsal root ganglia (Hess et al., 1993).

In LTP, there is evidence that NO acts on ADP ribosyltransferase in pre-synaptic cells (Schuman et al., 1994). Targets for ADP ribosylation include the growth associated protein GAP-43 (Coggins et al., 1993) and actin (Aktories, 1990), both of which influence development. Another mechanism by which NO may exert its effects is through modification of protein fatty acylation (Hess et al., 1993), which may regulate expression of growth associated proteins SNAP-25 and GAP-43. SNAP-25 is involved in axon growth and synapse formation, as well as synaptic vesicle release (Osen-Sand et al., 1996), and thus may be involved in both LTP and activity-dependent refinement of connections during development.

NO may also act to inhibit its own production. NMDA receptors are inhibited by NO donors (Manzoni et al., 1992), although some evidence suggests that NO donors act directly on NMDA receptors independently of NO (Kiedrowski et al., 1992). In addition, the activity of NOS is inhibited by NO (Hu and El-Fakahany, 1996; Vickroy and Malphurs, 1995). These studies suggest that NO limits the period of time over which it can exert an effect at a synapse, and thus presents an interesting mechanism for increasing specificity in NO signaling.

Concluding remarks

Activity-dependent refinement of synaptic connections during development is similar to some forms of LTP in that correlated input activity is detected by NMDA receptor activation in the post-synaptic cell, resulting in changes in pre-synaptic structure or function (Cramer and Sur, 1995). In development, these pre-synaptic changes include strengthening or weakening of synapses and expansion or retraction of axon terminals and branches. During some forms of LTP, pre-synaptic changes may include increases in neurotransmitter release (Bliss and Collingridge, 1993; Hawkins et al., 1993). In both systems, a retrograde messenger is required to signal coincidence detection in the post-synaptic cell and effect changes in pre-synaptic terminals. Thus, in both systems NO has been studied as a potential retrograde messenger.

Several studies support a role for NO in hippocampal LTP (Arancio et al., 1996; Bohme et al., 1991; O'Dell et al., 1991; Schuman and Madison, 1993; Son et al., 1996). NO plays a role in some aspects of development as well. During development, NOS is expressed in several different areas within the visual system (Cramer et al., 1995; Finney and Shatz, 1994; Williams et al., 1994). Blockade of NOS seems to interfere with normal pattern formation LGN of the ferret (Cramer et al., 1996), the chick retinotectal system (Wu et al., 1994), and retinal projections in Drosophila (Gibbs and Truman, 1998). Application of NO donors induces growth cone collapse in retinal ganglion cells (Renteria and Constantine Paton, 1996) and cultured dorsal root ganglia (Hess et al., 1993). However, NO does not appear to be involved in several other aspects of nervous system development; perhaps other molecules act as retrograde messengers in these systems.

While NO has a role as a retrograde signaling molecule in some systems, there are other roles for NO as well, including regulation of dendritic morphology during development. The molecular pathways by which NO acts are not fully understood. Future experiments will address potential targets for NO in visual system development, and will examine how these targets ultimately lead to changes in synaptic structure.

References

Ahmad, I., Leinders-Zufall, T., Kocsis, J.D., Shepherd, G.M., Zufall, F. and Barnstable, C.J. (1994) Retinal ganglion cells express a cGMP-gated cation conductance activatable by nitric oxide donors. *Neuron*, 12: 155–165.

Aktories, K. (1990) ADP-ribosylation of actin. *J. Muscle Res. Cell Motil.*, 11: 95–97.

Aoki, C., Rhee, J., Lubin, M. and Dawson, T.M. (1997) NMDA-R1 subunit of the cerebral cortex co-localizes with neuronal nitric oxide synthase at pre- and post-synaptic sites and in spines. *Brain Res.*, 750: 25–40.

Arancio, O., Kandel, E.R. and Hawkins, R.D. (1995) Activity-dependent long-term enhancement of transmitter release by pre-synaptic 3′,5′-cyclic GMP in cultured hippocampal neurons. *Nature*, 376: 74–80.

Arancio, O., Kiebler, M., Lee, C.J., Lev-Ram, V., Tsien, R.Y., Kandel, E.R. and Hawkins, R.D. (1996) Nitric oxide acts directly in the pre-synaptic neuron to produce long-term potentiation in cultured hippocampal neurons. *Cell*, 87: 1025–1035.

Archer, S., Dubin, M. and Stark, L. (1982) Abnormal retinogeniculate connectivity in the absence of action potentials. *Science*, 217: 743–745.

Arnhold, S., Andressen, C., Bloch, W., Mai, J.K. and Addicks, K. (1997) NO synthase-II is transiently expressed in embryonic mouse olfactory receptor neurons. *Neurosci. Lett.*, 229: 165–168.

Baader, S.L., Bucher, S. and Schilling, K. (1997) The developmental expression of neuronal nitric oxide synthase in cerebellar granule cells is sensitive to GABA and neurotrophins. *Dev. Neurosci.*, 19: 283–290.

Baader, S.L. and Schilling, K. (1996) Glutamate receptors mediate dynamic regulation of nitric oxide synthase expression in cerebellar granule cells. *J. Neurosci.*, 16: 1440–1449.

Bear, M.F., Kleinschmidt, A., Gu, Q. and Singer, W. (1990) Disruption of experience-dependent synaptic modifications in striate cortex by infusion of an NMDA receptor antagonist. *J. Neurosci.*, 10: 909–925.

Beninato, M. and Spencer, R.F. (1986) Cholinergic projections to the rat superior colliculus demonstrated by retrograde transport of horseradish peroxidase and choline acetyltransferase innumohistochemistry. *J. Comp. Neurol.*, 253: 525–538.

Bickford, M.E., Gunluk, A.E., Guido, W. and Sherman, S.M. (1993) Evidence that cholinergic axons from the parabrachial region of the brainstem are the exclusive source of nitric oxide in the lateral geniculate nucleus of the cat. *J. Comp. Neurol.*, 334: 410–430.

Bliss, T.V.P. and Collingridge, G.L. (1993) A synaptic model of memory: Long-term potentiation in the hippocampus. *Nature*, 361: 31–39.

Bodnarenko, S.R. and Chalupa, L.M. (1993) Stratification of ON and OFF ganglion cell dendrites depends on glutamate-mediated afferent activity in the developing retina. *Nature*, 364: 144–146.

Bodnarenko, S.R., Jeyarasasingam, G. and Chalupa, L.M. (1995) Development and regulation of dendritic stratification in retinal ganglion cells by glutamate-mediated afferent activity. *J. Neurosci.*, 15: 7037–7045.

Bohme, G.A., Bon, C., Stutzmann, J.-M., Doble, A. and Blanchard, J.-C. (1991) Possible involvement of nitric oxide in long-term potentiation. *Eur. J. Pharmacol.*, 199: 379–381.

Bonhoeffer, T. (1996) Neurotrophins and activity-dependent development of the neocortex. *Curr. Opin. Neurobiol.*, 6: 119–126.

Bredt, D.S. and Snyder, S.H. (1989) Nitric oxide mediates glutamate-linked enhancement of cGMP levels in the cerebellum. *Proc. Natl. Acad. Sci. USA*, 86: 9030–9033.

Bredt, D.S. and Snyder, S.H. (1990) Isolation of nitric oxide synthetase, a calmodulin-requiring enzyme. *Proc. Natl. Acad. Sci. USA*, 87: 682–685.

Bredt, D.S. and Snyder, S.H. (1992) Nitric oxide, a novel neuronal messenger. *Neuron*, 8: 3–11.

Bredt, D.S. and Snyder, S.H. (1994) Transient nitric oxide synthase neurons in embryonic cerebral cortical plate, sensory ganglia, and olfactory epithelium. *Neuron*, 13: 301–313.

Brenman, J.E., Chao, D.S., Gee, S.H., McGee, A.W., Craven, S.E., Santillano, D.R., Wu, Z., Huang, F., Xia, H., Peters, M.F., Froehner, S.C. and Bredt, D.S. (1996a) Interaction of nitric oxide synthase with the post-synaptic density protein PSD-95 and α1-syntrophin mediated by PDZ domains. *Cell*, 84: 757–767.

Brenman, J.E., Christopherson, K.S., Craven, S.E., McGee, A.W. and Bredt, D.S. (1996b) Cloning and characterization of post-synaptic density 93, a nitric oxide synthase interacting protein. *J. Neurosci.*, 16: 7407–7415.

Bruning, G. and Mayer, B. (1996) Localization of nitric oxide synthase in the brain of the frog, Xenopus laevis. *Brain Res.*, 741: 331–343.

Castren, E., da Penha Berzaghi, M., Lindholm, D. and Thoenen, H. (1993) Differential effects of MK-801 on brain-derived neurotrophic factor mRNA levels in different regions of the rat brain. *Exp. Neurol.*, 122: 244–252.

Cline, H.T. and Constantine-Paton, M. (1990) NMDA receptor agonist and antagonists alter retinal ganglion cell arbor structure in the developing frog retinotectal projection. *J. Neurosci.*, 10: 1197–1216.

Coggins, P.J., McLean, K., Nagy, A. and Zwiers, H. (1993) ADP-ribosylation of the neuronal phosphoprotein B-50/GAP-43. *J. Neurochem.*, 60: 368–371.

Cramer, K.S., Angelucci, A., Hahm, J.O., Bogdanov, M.B. and Sur, M. (1996) A role for nitric oxide in the development of the ferret retinogeniculate projection. *J. Neurosci.*, 16: 7995–8004.

Cramer, K.S., Moore, C.I. and Sur, M. (1995) Transient expression of NADPH-diaphorase in the lateral geniculate nucleus of the ferret during early post-natal development. *J. Comp. Neurol.*, 353: 306–316.

Cramer, K.S. and Sur, M. (1995) Activity-dependent remodeling of connections in the mammalian visual system. *Curr. Opin. Neurobiol.*, 5: 106–111.

Cramer, K.S. and Sur, M. (1996) On/off sublamination in the ferret LGN is independent of systemic changes in blood pressure and requires neuronal NOS. *Soc. Neurosci. Abs.*, 22: 761.

Cramer, K.S. and Sur, M. (1997a) Blockade of afferent impulse activity disrupts on/off sublamination in the ferret lateral geniculate nucleus. *Dev. Brain Res.*, 98: 287–290.

Cramer, K.S. and Sur, M. (1997b) Blockade of nitric oxide synthase alters dendritic morphology in the ferret lateral geniculate nucleus. *Soc. Neurosci. Abs.*, 23: 1159.

Dalva, M.B., Ghosh, A. and Shatz, C.J. (1994) Independent control of dendritic and axonal form in the developing lateral geniculate nucleus. *J. Neurosci.*, 14: 3588–3602.

Daw, N.W., Reid, S.N.M., Czepita, D. and Flavin, H. (1994) A nitric oxide synthase (NOS) inhibitor reduces the ocular dominance shift after monocular deprivation. *Soc. Neurosci. Abs.*, 20: 1428.

Dawson, T.M., Bredt, D.S., Fotuhi, M., Hwang, P.M. and Snyder, S.H. (1991) Nitric oxide synthase and neuronal NADPH-diaphorase are identical in brain and peripheral tissues. *Proc. Natl. Acad. Sci. USA*, 88: 7797–7801.

Dwyer, M.A., Bredt, D.S. and Snyder, S.H. (1991) Nitric oxide synthase: Irreversible inhibition by L-NG-nitroarginine in brain *in vitro* and *in vivo*. *Biochem. Biophys. Res. Commun.*, 176: 1136–1141.

East, S.J. and Garthwaite, J. (1991) NMDA receptor activation in rat hippocampus induces cyclic GMP formation through the L-arginine-nitric oxide pathway. *Neurosci. Lett.*, 123: 17–19.

Ernst, A.F., Wu, H.H., El-Fakahany, E.E. and McLoon, S.C. (1997) NMDA receptor mediated refinement of a transient retinotectal projection is dependent on nitric oxide synthesis. *Soc. Neurosci. Abs.*, 23: 1975.

Esguerra, M., Kwon, Y.H. and Sur, M. (1992) Retinogeniculate EPSP's recorded intracellularly in the ferret lateral geniculate nucleus in vitro: Role of NMDA receptors. *Vis. Neurosci.*, 8: 545–555.

Finney, E.M. and Shatz, C.J. (1994) Onset of nitric oxide synthase expression in ferret visual cortex correlates with formation of ocular dominance columns. *Soc. Neurosci. Abs.*, 20: 466.

Finney, E.M. and Shatz, C.J. (1997) Normal ferret ocular dominance column formation and mouse barrel field plasticity after inhibition of nitric oxide synthase. *Soc. Neurosci. Abs.*, 23: 305.

Galli, L. and Maffei, L. (1988) Spontaneous impulse activity of rat retinal ganglion cells in pre-natal life. *Science*, 242: 90–91.

Gally, J.A., Montague, P.R., Reeke, G.N., Jr. and Edelman, G.M. (1990) The NO hypothesis: Possible effects of a short-lived, rapidly diffusible signal in the development and function of the nervous system. *Proc. Natl. Acad. Sci.*, 87: 3547–3551.

Garthwaite, J., Charles, S.L. and Chess-Williams, R. (1988) Endothelium-derived relaxing factor release on activation of NMDA receptors suggests role as intercellular messenger in the brain. *Nature*, 336: 385–388.

Garthwaite, J., Garthwaite, G., Palmer, R.M. and Moncada, S. (1989) NMDA receptor activation induces nitric oxide synthesis from arginine in rat brain slices. *Eur. J. Pharmacol.*, 172: 413–416.

Gibbs, S.M. and Truman, J.W. (1998) Nitric oxide and cyclic GMP regulate retinal patterning in the optic lobe of *Drosophila* melanogaster. *Neuron*, 20: 83–93.

Greiner, J.V. and Weidman, T.A. (1981) Histogenesis of the ferret retina. *Exp. Eye Res.*, 33: 315–322.

Gunluk, A.E., Bickford, M.E. and Sherman, S.M. (1994) Rearing with monocular lid suture induces abnormal NADPH-diaphorase staining in the lateral geniculate nucleus of cats. *J. Comp. Neurol.*, 350: 215–218.

Gustafsson, B. and Wigstrom, H. (1990) Long-term potentiation in the hippocampal CA1 region: Its induction and early temporal development. *Prog. Brain Res.*, 83: 223–232.

Hahm, J., Langdon, R.B. and Sur, M. (1991) Disruption of retinogeniculate afferent segregation by antagonists to NMDA receptors. *Nature*, 351: 568–570.

Harris, E.W., Ganong, A.H. and Cotman, C.W. (1984) Long-term potentiation in the hippocampus involves activation of *N*-methyl-D-aspartate receptors. *Brain Res.*, 323: 132–137.

Hawkins, R.D., Kandel, E.R. and Siegelbaum, S.A. (1993) Learning to modulate transmitter release: Themes and variations in synaptic plasticity. *Annu. Rev. Neurosci.*, 16: 625–665.

Hecker, M., Mulsch, A. and Busse, R. (1994) Subcellular localization and characterization of neuronal nitric oxide synthase. *J. Neurochem.*, 62: 1524–1529.

Hess, D.T., Patterson, S.I., Smith, D.S. and Skene, J.H. (1993) Neuronal growth cone collapse and inhibition of protein fatty acylation by nitric oxide. *Nature*, 366: 562–565.

Hindley, S., Juurlink, B.H., Gysbers, J.W., Middlemiss, P.J., Herman, M.A. and Rathbone, M.P. (1997) Nitric oxide donors enhance neurotrophin-induced neurite outgrowth through a cGMP-dependent mechanism. *J. Neurosci. Res.*, 47: 427–439.

Holtzman, D.M., Kilbridge, J., Bredt, D.S., Black, S.M., Li, Y., Clary, D.O., Reichardt, L.F. and Mobley, W.C. (1994) NOS induction by NGF in basal forebrain cholinergic neurons: Evidence for regulation of brain NOS by a neurotrophin. *Neurobiol. Dis.*, 1: 51–60.

Holtzman, D.M., Lee, S., Li, Y., Chua-Couzens, J., Xia, H., Bredt, D.S. and Mobley, W.C. (1996) Expression of neuronal-NOS in developing basal forebrain cholinergic neurons: Regulation by NGF. 21: 861–868.

Hope, B.T., Michael, G.J., Knigge, K.M. and Vincent, S.R. (1991) Neuronal NADPH-diaphorase is a nitric oxide synthase. *Proc. Natl. Acad. Sci.*, 88: 2811–2814.

Hu, J. and El-Fakahany, E.E. (1996) Intricate regulation of nitric oxide synthesis in neurons. *Cell Signal*, 8: 185–189.

Jeon, C.-J. and Mize, R.R. (1993) Choline acetyltransferase immunoreactive patches overlap specific efferent cell groups in the cat superior colliculus. *J. Comp. Neurol.*, 337: 127–150.

Kalb, R.G. (1994) Regulation of motor neuron dendrite growth by NMDA receptor activation. *Development*, 120: 3063–3071.

Kalb, R.G. and Agostini, J. (1993) Molecular evidence for nitric oxide-mediated motor neuron development. *Neuroscience*, 57: 1–8.

Kiedrowski, L., Costa, E. and Wroblewski, J. T. (1992) Sodium nitroprusside inhibits *N*-methyl-D-aspartate-evoked calcium influx via a nitric oxide- and cGMP-independent mechanism. *Mol. Pharmacol.*, 41: 779–84.

King, A.J., Schnupp, J.W., Carlile, S., Smith, A.L. and Thompson, I.D. (1996) The development of topographically-aligned maps of visual and auditory space in the superior colliculus. *Prog. Brain Res.*, 112: 335–350.

Kwon, Y.H., Esguerra, M. and Sur, M. (1991) NMDA and non-NMDA receptors mediate visual responses of neurons in the cat's lateral geniculate nucleus. *J. Neurophysiol.*, 66: 414–428.

Linden, D.C., Guillery, R.W. and Cucchiaro, J. (1981) The dorsal lateral geniculate nucleus of the normal ferret and its post-natal development. *J. Comp. Neurol.*, 203: 189–211.

Lindholm, D., Castren, E., Berzaghi, M., Blochl, A. and Thoenen, H. (1994) Activity-dependent and hormonal regulation of neurotrophin mRNA levels in the brain: Implications for neuronal plasticity. *J. Neurobiol.*, 25: 1362–1372.

Manzoni, O., Prezeau, L., Marin, P., Deshager, S., Bockaert, J. and Fagni, L. (1992) Nitric oxide-induced blockade of NMDA receptors. *Neuron*, 8: 653–662.

Maslim, J. and Stone, J. (1986) Synaptogenesis in the retina of the cat. *Brain Res.*, 373: 35–48.

Maslim, J. and Stone, J. (1988) Time course of stratification of the dendritic fields of ganglion cells in the retina of the cat. *Dev. Brain Res.*, 44: 87–93.

Meister, M., Wong, R.O., Baylor, D.A. and Shatz, C.J. (1991) Synchronous bursts of action potentials in ganglion cells of the developing mammalian retina. *Science*, 252: 939–943.

Mitrovic, N. and Schachner, M. (1996) Transient expression of NADPH-diaphorase activity in the mouse whisker to barrel field pathway. *J. Neurocytol.*, 25: 429–437.

Mize, R.R., Scheiner, C.A., Salvatore, M.F. and Cork, R.J. (1997) Inhibition of nitric oxide synthase fails to disrupt the development of cholinergic fiber patches in the rat superior colliculus. *Dev. Neurosci.*, 19: 260–273.

Montague, P.R., Gally, J.A. and Edelman, G.M. (1991) Spatial signaling in the development and function of neural connections. *Cereb. Cortex*, 1: 199–220.

Mooney, R., Madison, D.V. and Shatz, C.J. (1993) Enhancement of transmission at the developing retinogeniculate synapse. *Neuron*, 10: 815–825.

Munoz, M., Munoz, A., Marin, O., Alonso, J.R., Arevalo, R., Porteros, A. and Gonzalez, A. (1996) Topographical distribution of NADPH-diaphorase activity in the central nervous system of the frog, *Rana perezi*. *J. Comp. Neurol.*, 367: 54–69.

Murphy, K.P., Reid, G.P., Trentham, D.R. and Bliss, T.V. (1997) Activation of NMDA receptors is necessary for the induction of associative long-term potentiation in area CA1 of the rat hippocampal slice. *J. Physiol (Lond)*, 504: 379–385.

O'Dell, T.J., Hawkins, R.D., Kandel, E.R. and Arancio, O. (1991) Tests of the roles of two diffusible substances in long-term potentiation: Evidence for nitric oxide as a possible early retrograde messenger. *Proc. Natl. Acad. Sci.*, 88: 11285–11289.

Osen-Sand, A., Staple, J.K., Naldi, E., Schiavo, G., Rossetto, O., Petitpierre, S., Malgaroli, A., Monecucco, C. and Catsicas, S. (1996) Common and distinct fusion proteins in axonal growth and transmitter release. *J. Comp. Neurol.*, 367: 222–234.

Penn, A.A., Wong, R.O. and Shatz, C.J. (1994) Neuronal coupling in the developing mammalian retina. *J. Neurosci.*, 14: 3805–3815.

Ramoa, A.S. and McCormick, D.A. (1994) Enhanced activation of NDA receptor responses at the immature retinogeniculate synapse. *J. Neurosci.*, 14: 2098–2105.

Regehr, W.G. and Tank, D.W. (1990) Postsynaptic NMDA receptor-mediated calcium accumulation in hippocampal CA1 pyramidal cell dendrites. *Nature*, 345: 807–810.

Renteria, R.C. and Constantine Paton, M. (1996) Exogenous nitric oxide causes collapse of retinal ganglion cell axonal growth cones in vitro. *J. Neurobiol.*, 29: 415–428.

Rocha, M. and Sur, M. (1995) Rapid acquisition of dendritic spines by visual thalamic neurons after blockade of *N*-methyl-D-aspartate receptors. *Proc. Natl. Acad. Sci.*, 92: 8026–8030.

Roskams, A.J., Bredt, D.S., Dawson, T.M. and Ronnett, G.V. (1994) Nitric oxide mediates the formation of synaptic connections in developing and regenerating olfactory receptor neurons. *Neuron*, 13: 289–299.

Ruthazer, E.S., Gillespie, D.C., Dawson, T.M., Snyder, S.H. and Stryker, M.P. (1996) Inhibition of nitric oxide synthase does not prevent ocular dominance plasticity in kitten visual cortex. *J. Physiol.(Lond).*, 494: 519–527.

Samama, B. and Boehm, N. (1996) Ontogenesis of NADPH-diaphorase activity in the olfactory bulb of the rat. *Dev. Brain Res.*, 96: 192–203.

Schlaggar, B.L., Fox, K. and O'Leary, D.D.M. (1993) Post-synaptic control of plasticity in developing somatosensory cortex. *Nature*, 364: 623–626.

Schuman, E.M. and Madison, D.V. (1993) A requirement for the intercellular messenger nitric oxide in long-term potentiation. *Science*, 254: 1503–1506.

Schuman, E.M., Meffert, M.K., Schulman, H. and Madison, D.V. (1994) An ADP-ribosyltransferase as a potential target for nitric oxide action in hippocampal long-term potentiation. *Proc. Natl. Acad. Sci.*, 91: 11958–11962.

Shatz, C.J. and Stryker, M.P. (1988) Prenatal tetrodotoxin infusion blocks segregation of retinogeniculate afferents. *Science*, 242: 87–89.

Simon, D.K., Prusky, G.T., O'Leary, D.D.M. and Constantine-Paton, M. (1992) *N*-methyl-D-aspartate receptor antagonists disrupt the formation of a mammalian neural map. *Proc. Natl. Acad. Sci. USA*, 89: 10593–10597.

Smetters, D.K., Hahm, J. and Sur, M. (1994) An *N*-Methyl-D-Aspartate receptor antagonist does not prevent eye-specific segregation in the ferret retinogeniculate pathway. *Brain Res.*, 658: 168–178.

Son, H., Hawkins, R.D., Martin, K., Keibler, M., Huang, P.L., Fishman, M.C. and Kandel, E.R. (1996) Long-term potentiation is reduced in mice that are doubly mutant in endothelial and neuronal nitric oxide synthase. *Cell*, 87: 1015–1023.

Stryker, M.P. and Zahs, K.R. (1983) ON and OFF sublaminae in the lateral geniculate nucleus of the ferret. *J. Neurosci.*, 3: 1943–1951.

Truman, J.W., DeVente, J. and Ball, E.E. (1996) Nitric oxide-sensitive guanylate cyclase activity is associated with the maturational phase of neuronal development in insects. *Development*, 122: 3949–3958.

Vercelli, A.E. and Cracco, C.M. (1994) Effects of eye enucleation on NADPH-diaphorase positive neurons in the superficial layers of the rat superior colliculus. *Dev. Brain Res.*, 83: 85–98.

Vickroy, T.W. and Malphurs, W.L. (1995) Inhibition of nitric oxide synthase activity in cerebral cortical synaptosomes by nitric oxide donors: Evidence for feedback autoregulation. *Neurochem. Res.*, 20: 299–304.

Wetts, R., Phelps, P.E. and Vaughn, J.E. (1995) Transient and continuous expression of NADPH-diaphorase in different neuronal populations of developing rat spinal cord. *Dev. Dyn.*, 202: 215–228.

Wetts, R. and Vaughn, J.E. (1993) Transient expression of β-NADPH-diaphorase in developing rat dorsal root ganglia neurons. *Dev. Brain Res.*, 76: 278–282.

Williams, C.V., Nordquist, D. and McLoon, S.C. (1994) Correlation of nitric oxide synthase expression with changing pattern of axonal projections in the developing visual system. *J. Neurosci.*, 14: 1746–1755.

Wingate, R.J. (1996) Retinal ganglion cell dendritic development and its control. Filling the gaps. *Mol. Neurobiol.*, 12: 133–144.

Wong, R.O.L., Meister, M. and Shatz, C.J. (1993) Transient period of correlated bursting activity during development of the mammalian retina. *Neuron*, 11: 923–938.

Wong, R.O.L. and Oakley, D.M. (1996) Changing patterns of spontaneous bursting activity of On and Off retinal ganglion cells during development. *Neuron*, 16: 1087–1095.

Wu, H.H., Williams, C.V. and McLoon, S.C. (1994) Involvement of nitric oxide in the elimination of a transient retinotectal projection in development. *Science*, 265: 1593–1596.

Yao, X., Segal, A.S., Welling, P., Zhang, X., McNicholas, C.M., Engel, D., Boulpaep, E.L. and Desir, G.V. (1995) Primary structure and functional expression of a cGMP-gated potassium channel. *Proc. Natl. Acad. Sci. USA*, 92: 11711–11715.

Yen, L., Sibley, J.T. and Constantine-Paton, M. (1995) Analysis of synaptic distribution within single retinal axonal arbors after chronic NMDA treatment. *J. Neurosci.*, 15: 4712–4725.

Zhang, C., Granstrom, L. and Wong-Riley, M.T. (1996) Deafferentation leads to a down-regulation of nitric oxide synthase in the rat visual system. *Neurosci. Lett.*, 211: 61–64.

R.R. Mize, T.M. Dawson, V.L. Dawson and M.J. Friedlander (Eds.)
Progress in Brain Research, Vol 118

CHAPTER 9

Mechanisms involved in development of retinotectal connections: Roles of Eph receptor tyrosine kinases, NMDA receptors and nitric oxide

Alan F. Ernst, William M. Jurney and Steven C. McLoon*

Department of Cell Biology and Neuroanatomy, University of Minnesota, 4-144 Jackson Hall, 321 Church Street, SE, Minneapolis, MN 55455, USA

Abstract

Axons of retinal ganglion cells exhibit a specific pattern of connections with the brain. Within each visual nucleus in the brain, retinal connections are topographic such that axons from neighboring ganglion cells have neighboring synapses. Research is beginning to shed light on the mechanisms responsible for development of topographic connections in the visual system. Much of this research is focused on the axonal connections of the retina with the tectum. *In vivo* and *in vitro* experiments indicate that the pattern of retinotectal connections develops in part due to positional labels carried by the growing retinal axons and by the tectal cells. Evidence suggests that gradients of Eph receptor tyrosine kinases serve as positional labels on the growing retinal axons, and gradients of ligands for these receptors serve as positional labels in the tectum. Blocking expression of EphA3, a receptor tyrosine kinase, in the developing retina resulted in disruption of the topography of the retinotectal connections, further supporting the role of these molecules. Although positional labels appear to be important, other mechanisms must also be involved. The initial pattern of retinotectal connections lacks the precision seen in the adult. The adult pattern of connections arises during development by activity dependent refinement of a roughly ordered prepattern. The refinement process results in elimination of projections to the wrong side of the brain, to non-visual nuclei and to inappropriate regions within a nucleus. Blocking NMDA receptors during the period of refinement preserved anomalous retinotectal projections, which suggests that elimination of these projections is mediated by NMDA receptors. Furthermore, tectal cells normally express high levels of nitric oxide synthase (NOS) during the period of refinement, and blocking nitric oxide (NO) synthesis also preserved inappropriate projections. Thus, both NMDA receptors and NO appear to be involved in refinement. Blocking NMDA receptor activation reduced NOS activity in tectal cells, which suggests the possibility that NO is the downstream mediator of NMDA function related to refinement. A quantitative comparison of blocking NMDA receptors, NO synthesis or both showed that all three treatments have comparable effects on refinement. This indicates that the role of NMDA receptor activation relative to refinement may be completely mediated through nitric oxide. Quantitative analysis also

*Corresponding author. Tel.: +1 612 624 9182; fax: +1 612 624 8118; e-mail: mcloons@lentimed.umn.edu

suggests that other mechanisms not involving NMDA receptors or NO must be involved in refinement. Other mechanisms appear to include cell death.

Introduction

Axonal connections between functionally related groups of neurons have precise, stereotypic patterns. For example, in the motor and somatosensory systems, the patterns of connections recreate a two dimensional map of the body surface. The auditory system is organized in a frequency map, and the visual system uses a map of visual space. These patterns of connections are essential for normal function of the nervous system. In the case of the visual system, when this pattern of connections was altered, animals were incapable of responding to visual information in a meaningful manner (So et al., 1981). The mechanisms by which these precise patterns of connections develop are incompletely understood.

The axonal connections between the retina and tectum serve as an important model system for study of the mechanisms involved in development of specific patterns of neuronal connections. Visual information is relayed from the retina to the brain via axons of the retinal ganglion cells. With regard to the retinotectal connections, the retina can be considered a two-dimensional sheet of ganglion cells. The visual image is mapped onto the sheet of ganglion cells, with each ganglion cell receiving a signal representing a discrete region of visual space. In non-mammalian vertebrates, most ganglion cell axons synapse in the optic tectum, which forms from the roof of the midbrain. The homologous structure in mammals is the superior colliculus. Other pretectal and diencephalic nuclei receive axons from the retina, particularly the lateral geniculate nucleus in the thalamus. The spatial arrangement of the ganglion cells across the retina is approximately recreated in the pattern of connections in the brain, so that the neighboring ganglion cells have neighboring connections (Hamdi and Whitteridge, 1954). This is often referred to as a topographic or retinotopic pattern of connections. The retinotectal projection is particularly advantageous for study because, like the retina, the tectum has an organization that can be reduced to a two-dimensional sheet of cells, and both the retina and the tectum are readily accessible for experimental manipulation and analysis in the adult and the embryo.

The right and left sides of the visual world are also reflected in the pattern of the retinal axon connections in the brain (Mason et al., 1996). Generally, information related to each side of the visual world is carried to the opposite side of the brain in adult animals. Since mammals have eyes on the front of the head, at least a portion of both sides of the visual world falls onto each retina. The projection from a retina splits at the optic chiasm so that axons connect to the side of the brain appropriate for the side of the visual world served by each axon. In most non-mammalian vertebrates, the eyes are on the side of the head so that each retina receives information representing only one side of the visual world. The entire retina in these species projects to the opposite side of the brain. A considerable body of work has begun to shed some light on how the proper laterality and topography of the retinal projections to the brain develop.

The axons of the retinal ganglion cells grow from the eye to the brain during embryonic development. The axons grow through the optic fiber layer of the retina, the optic nerve, the chiasm and the optic tract to reach the various visual centers distributed along the tract. Clearly, accurate guidance of the growing retinal axons to the visual centers in the brain is an essential antecedent to development of the proper pattern of connections. Although a significant degree of order is maintained among the axons as they grow through the pathway (Bunt et al., 1983; Fawcett and Gaze, 1982; Thanos and Bonhoeffer, 1983), this order apparently is not essential for development of patterned connections (Finlay et al., 1979; Harris, 1982; Thanos and Dutting, 1987). The mechanisms responsible for guiding the growth of the retinal axons to the visual centers in the brain are complicated and beyond the scope of this review (Tessier-Lavigne and Goodman, 1996). This review will focus on the mechanisms that determine the position in which the retinal axons terminate and form synapses upon reaching the visual centers, particularly the tectum.

The pioneering studies by Sperry form a cornerstone of our current hypothesis on how topographic connections develop. Sperry observed that surgically cut retinal axons in newts regenerated to their original termination site in the tectum, and they did so even if the eye was rotated 180° (Sperry, 1943). In the case of the eye rotation, the reformed connections were behaviorally maladaptive in that animals would seek prey in the wrong position when it was presented visually to the affected eye. A later study from Sperry's laboratory showed that after removing a portion of the retina in goldfish, the remaining retinal axons still regenerated to their original termination sites in the tectum, initially leaving a portion of the tectum without retinal innervation (Attardi and Sperry, 1963). Based on these studies, Sperry suggested that neurons carry positionally dependent chemical labels and that the pattern of connections between two groups of neurons is due to the interaction between the labels on the growing axons and the cells to which they connect (Sperry, 1963). This has come to be known as the chemospecificity hypothesis.

Although a strong case can be made for chemospecificity having a role in development of patterned neuronal connections, other mechanisms must also be involved. The initial pattern of connections formed by retinal ganglion cells with visual nuclei, particularly in warm-blooded vertebrates, lacks the topographic precision of mature connections. During development, many retinal axons project to inappropriate positions within the primary visual centers, to inappropriate nuclei or to the inappropriate side of the brain (Cowan et al., 1984; Holt and Harris, 1993). These targeting errors are eventually corrected during a discrete period of development by a process of refinement. During refinement, axons with inappropriate connections retract and axons in appropriate positions arborize and form more synapses (Fraser and O'Rourke, 1990; Simon et al., 1994). The mechanisms involved in refinement are poorly understood. It is widely believed that refinement requires neuronal activity and is independent of mechanisms associated with chemospecificity.

This review will examine in more detail the cellular mechanisms underlying chemospecificity and activity dependent refinement in the retinotectal system during development. Evidence is presented suggesting that Eph receptor tyrosine kinases serve as positional labels. Evidence is also presented that shows activity dependent refinement is in part mediated by nitric oxide synthesis linked to NMDA receptor activation.

Chemospecificity

A considerable body of evidence supports the chemospecificity hypothesis. As well as Sperry's original experiments, a variety of *in vivo* studies showed that retinal axons recognize the position of cells in their target centers (reviewed by Holt and Harris, 1993; Udin and Fawcett, 1988). In general, these experiments involved surgically altering the position or number of retinal or tectal cells and then characterizing the pattern of retinotectal connections after development or regeneration of the retinal axons. An example of a particularly compelling study involved transplanting small numbers of dye labeled retinal cells to a heterotopic position of the retina in developing tadpoles (Fraser and O'Rourke, 1990). The axons from the transplanted cells grew to sites in the tectum appropriate for their original position rather than for their location following transplantation. Several studies also showed that retinal axons forced to enter the tectum via abnormal routes still formed projections with the proper orientation (Finlay et al., 1979; Harris, 1982). Since retinal axons generally found their appropriate target cells regardless of the position of either cell population, it has been widely concluded that the axons and target cells carry positional labels.

Various *in vitro* studies provided further evidence for differences among cells from different regions of the developing retina that could reflect positional labels. Early *in vitro* studies described differences in the ability of cells from dorsal and ventral regions of the retina to adhere to each other (Gottlieb et al., 1976) and to different regions of the optic tectum (Barbera, 1975). The retinal cells adhered to tectal tissue in a pattern that matched the normal pattern of axonal connections from the retina to the tectum. Other tissue culture studies showed that growing retinal axons from the nasal and temporal sides of the retina differ in their interactions with each other (Bonhoeffer and Huf, 1985; Fan and Raper, 1995; Raper and Grunewald, 1990) and with cell membranes from different positions of the tectum (Baier and Bonhoeffer, 1992; Halfter et al., 1981). These differences in cell–cell interactions exhibited *in vitro* by cells from different retinal positions are believed to reflect mechanisms active in development of patterned axonal connections *in vivo*.

Further insight into the mechanisms that may be involved in development of topographically organized connections came from the *in vitro* 'stripe assay' developed by Bonhoeffer and colleagues (Walter et al., 1987a,b). Retinal explants were grown adjacent to a substrate of alternating stripes of cell membranes prepared from anterior or posterior tectum. Neurites from temporal retina grew selectively on membranes from anterior tectum, the region to which they would normally have connected *in situ*. This preference appeared to stem from a repellent property of the posterior tectal cell membranes rather than selective affinity for the anterior membranes, as heating the posterior membranes eliminated the activity. Furthermore, the repellent activity appeared to be in a gradient across the tectum with lowest activity in anterior tectum and highest activity in posterior tectum. It was also shown that the growth cones of temporal but not nasal retinal axons collapsed when they encountered membranes from posterior tectum (Cox et al., 1990). These findings suggest that part of the mechanism active in development of topographic connections may involve repulsion of axons from inappropriate regions within their target tissue. This repellent activity is most likely mediated by an interaction of one or more molecules distributed in gradients across the tectum with specific receptors for those molecules on the growing retinal axons.

The molecules that serve as positional labels need to be identified in order to validate the chemospecificity hypothesis and to understand the mechanisms underlying development of patterned axonal connections. It was recognized very early that the genome is too small for the chemical labels to be unique genes expressed by every ganglion cell. As an alternative, it was suggested that two or more molecules distributed across the retina in concentration gradients perpendicular to one another could impart unique positional labels to each cell (Fraser, 1980; Gierer, 1983). Investigators have been searching for over 20 years for molecules that mark position in the developing retina. Over 10 molecules have now been found that are distributed asymmetrically across the developing retina (reviewed by Kaprielian and Patterson, 1994; Sanes, 1993). Unfortunately, there is no definitive evidence showing that any specific molecule functions as a positional label for development of retinotectal connections.

There is a growing body of evidence, however, that suggests the Eph family of receptor tyrosine kinases and their ligands are involved in development of retinotectal connections. Receptor tyrosine kinases are a diverse group of transmembrane receptors, which include growth factor and neurotrophin receptors, that have a high degree of homology in their cytoplasmic catalytic domains and considerable variability in their extracellular ligand-binding domains (reviewed by van der Geer et al., 1994). Several properties of the Eph family set them apart from other receptor tyrosine kinase families. In particular, Eph ligands must be anchored to the cell membrane to activate their receptors, suggesting a requirement for cell–cell contact (Davis et al., 1994). Many of the Eph family members are expressed exclusively in the nervous system and show restricted patterns of expression (Gale et al., 1996). In the retinotectal system, two Eph family ligands, ephrin-A2 and ephrin-A5, are expressed in gradients across the anterior/posterior axis of the developing tectum. As diagramed in Fig. 1, both ligands have their highest level of expression in the posterior pole of the tectum and the lowest in the anterior pole. Ephrin-A5, however, appears to be expressed in a steeper gradient with expression confined to the posterior half of the tectum (Cheng et al., 1995; Drescher et al., 1995; Monschau et al., 1997). Thus, the expression patterns suggest the possibility that these ligands for Eph receptors serve as positional labels in the developing tectum.

The response of growing retinal axons to ephrin-A2 and ephrin-A5 *in vitro* and *in vivo* further supports the possibility that these molecules have a role in development of the pattern of retinotectal connections. Ephrin-A5 was expressed in COS cells, and membranes isolated from these cells repelled temporal retinal axons in the stripe assay similar to membranes prepared from posterior tectum (Drescher et al., 1995). At low concentrations, Ephrin-A5 caused collapse of axons from the temporal side of the retina and had no effect on axons from the nasal side; at higher concentrations, ephrin-A5 caused collapse of axons from both sides of the retina (Monschau et al., 1997). Ephrin-A2 at all concentrations tested caused axons from the temporal side of the retina to collapse with no effect on axons from the nasal side. Ephrin-A2 was overexpressed in patches of tectum using a retroviral vector (Nakamoto et al., 1996). Axons from the temporal side of the retina avoided regions of ectopic ephrin-A2 expression, and there was no detectable change in behavior of axons from the nasal side of the retina. Both ephrin-A5 and ephrin-A2 were overexpressed indirectly in anterior tectum via retroviral overexpression of *engrailed* (Friedman and O'Leary, 1996; Itasaki and Nakamura, 1996; Logan et al., 1996). This caused premature termination of growth of axons from the temporal side of the retina where they enter the tectum at its anterior pole, while axons from the nasal side were either unaffected or showed abnormal arborizations. This evidence suggests the intriguing possibility that during normal development axons from the temporal side of the retina are restricted to

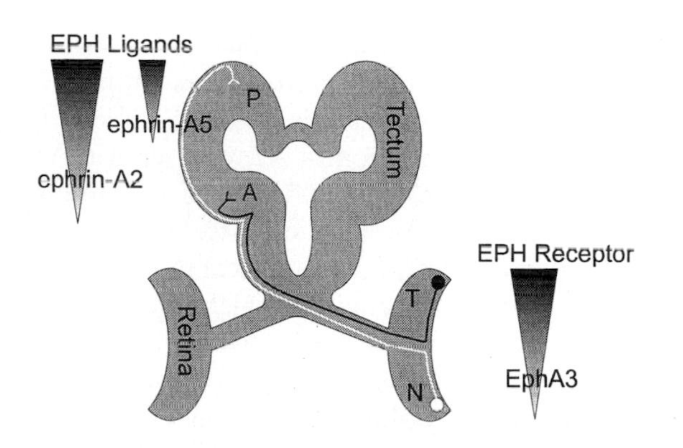

Fig. 1. Schematic of the retinotectal projection and the expression patterns of the Eph receptor, EphA3, in the retina and the Eph ligands, ephrin-A2 and ephrin-A5, in the tectum. Retinal ganglion cells from the temporal side of the retina express high levels of EphA3 (shown in black) and form connections in the anterior tectum. Retinal ganglion cells from the nasal side of the retina express low levels of EphA3 (shown in white) and grow through anterior tectum to form connections in posterior tectum. N = nasal retina, T = temporal retina, A = anterior tectum, and P = posterior tectum.

anterior regions of the tectum by ephrin-A2. Axons from the nasal side of the retina may be topographically organized in the rostral–caudal axis of the posterior tectum by a gradient of ephrin-A5.

One complication in the proposed role of ephrin-A5 in retinotectal specificity relates to the specific cells that express ephrin-A5 in the developing tectum. Based on immunostaining, the ephrin-A5 protein is on glia cell processes but not on neurons in the tectal layers in which retinal axons grow and form synapses (Drescher et al., 1995). However, in an *in vitro* assay that examined retinal axon responses to interactions with dissociated tectal cells, retinal axons were not repelled by contact with posterior tectal glia (Davenport et al., 1996). Axons from temporal retina but not from nasal retina were repelled by posterior tectal neurons. If ephrin-A2 is expressed by tectal neurons, then its proposed role in restricting temporal axons to anterior tectum remains viable. The failure of glia cells to repel axons from nasal retina underscores the need to investigate further the role of Eph ligands in development of the retinotectal system.

If Eph ligands serve as positional labels in the developing tectum, then Eph receptors should be expressed by the growing retinal axons. Three Eph receptors that bind ephrin-A2 and ephrin-A5 normally are expressed in the developing retina (Cheng et al., 1995). One of these, EphA3, is expressed in a concentration gradient in the nasal–temporal axis of the retina with high levels on the temporal side and low levels on the nasal side (Figs. 1 and 2). Among the Eph receptors expressed in the retina, EphA3 has the highest binding affinity for ephrin-A2 and ephrin-A5 (Monschau et al., 1997). Axons that express low levels of EphA3 (i.e. those from the nasal side of the retina) grow into areas of the tectum that express high levels of ephrin-A2 and ephrin-A5 (i.e. the posterior tectum).

It may be that the level of EphA3 expressed by the ganglion cells determines the position in which their axons terminate in the anterior–posterior axis of the tectum. If this is the case, then reducing expression of EphA3 in the developing retina should allow retinal axons, particularly those from the temporal side of the retina, to grow further back in the tectum than they would normally. We recently tested this in developing chick embryos (Jurney et al., 1997). Antisense oligonucleotides were injected into the developing eye to reduce expression of EphA3. Following the antisense treatment, the topography of the retinotectal projection was examined using retrograde axonal tracing techniques. As shown in Fig. 3, there was a clear difference in the pattern of labeling between EphA3 antisense oligonucleotide and control oligonucleotide treatments. A small injection of tracer into the posterior tectum in normal or control embryos retrogradely labeled a tight cluster of ganglion cells on the nasal side of the contralateral retina. In retinas treated with antisense oligonucleotides, retrogradely labeled cells were broadly distributed across the nasal–temporal axis of the retina. There was a significant increase in the number of labeled ganglion cells on the temporal side of the retina in antisense treated embryos. Therefore, reduction in EphA3 expression in the retina resulted in an inappropriate invasion of axons from the temporal side of the retina into posterior regions of the tectum. This indicates that the EphA3 receptor expressed by the growing retinal axons mediates a positionally

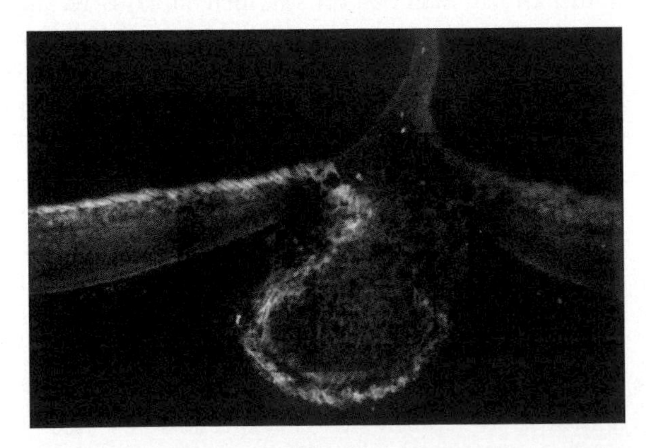

Fig. 2. Micrograph of EphA3 expression in the developing retina. A cross-section of embryonic day nine chick retina was processed for immunohistochemistry with an antibody to EphA3. Axons in the optic fiber layer on the temporal side of the retina (left side) are positive for EphA3, while the axons on the nasal side of the retina (right side) are negative. The optic nerve is in the center of the micrograph.

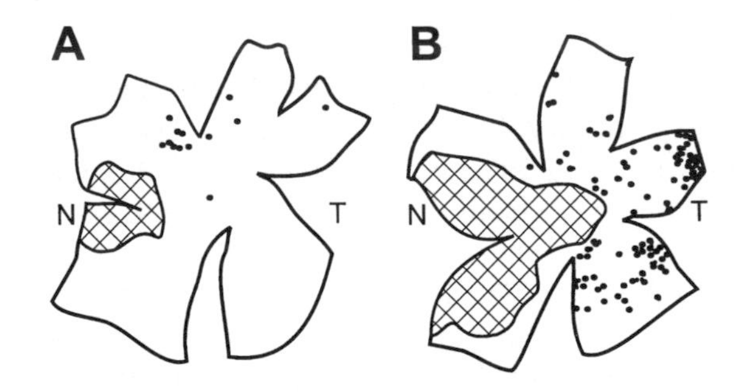

Fig. 3. Loss of retinotectal topography following reduction in EphA3 expression in the developing retina. The enlarged outlines of retinal whole-mounts from day 11 chick embryos show the position of retrogradely labeled ganglion cells following a DiI injection into the posterior pole of the contralateral tectum. The dots indicate individual labeled cells, and the cross-hatching indicates areas with a high density of retrogradely labeled cells. Eyes were injected with (A) a missense control oligonucleotide or (B) an antisense EphA3 oligonucleotide beginning on embryonic day 6. Compared to the control, the antisense treatment resulted in an increased number of cells in temporal retina projecting to posterior tectum. N = nasal retina, T = temporal retina.

appropriate stop-growing signal activated by specific levels of Eph ligands in the tectum.

Relatively little is known about the downstream effectors of Eph receptor activation. Association of EphA2 with PI 3-kinase was demonstrated using the yeast two-hybrid system, and EphA2 activation by ephrin-A1 increased PI 3-kinase activity (Pandey et al., 1994). It was also shown that a novel Src-like adapter protein (SLAP) binds to activated EphA2 receptors (Pandey et al., 1995). The signal transduction pathways of PI 3-kinase and SLAP relative to development of topographic neuronal connections remain to be studied.

While the evidence is compelling that Eph receptors and ligands are important players in guiding development of retinal topography in the anterior/posterior axis of the tectum, there remain many unanswered questions. A one-dimensional labeling system cannot completely account for the connections between two-dimensional sheets of cells. Another axis of labels appears to be necessary, possibly in the dorsal–ventral axis. The Eph receptor, EphB2, is expressed in a dorsal–ventral concentration gradient in the developing retina (Holash and Pasquale, 1995). The expression patterns of putative ligands for EphB2 have yet to be described in the tectum. A dorsal–ventral gradient of a ganglioside also was identified in the developing retina (Constantine-Paton et al., 1986), and gangliosides have been implicated in adhesion of retinal cells to tectum in a pattern that matches the topography of retinotectal connections in the dorsal–ventral axis (Marchase and Roth, 1978). At least four other Eph receptors, EphA4, EphA5, EphB3 and EphB5, are expressed in the developing retina (Drescher et al., 1997; Holash and Pasquale, 1995; Monschau et al., 1997; Soans et al., 1996). Although none of these appear to be expressed in concentration gradients, they could be involved in determining the laminar distribution of retinotectal connections or in some other aspect of development of neuronal connections.

It seems likely that molecules other than EphA3/ephrin-A2/ephrin-A5 also are involved in development of the topography of retinal connections in the anterior–posterior axis of the tectum. It is not clear that a topographic map could be ordered by repulsive mechanisms alone. Growing axons would be expected to stop at the first encounter with a repulsive molecule unless a counter-acting positive force drives their growth. Such a positive force could also be in a gradient in the same pattern as the repulsive force. A balance of attractive and repulsive signals could determine where a retinal axon stops growing in the anterior–posterior axis of the tectum. There is some evidence that attractive forces are involved in development of the pattern of retinotectal connections. Axons from the nasal side of the retina exhibited a slight preference for posterior tectal membranes in the stripe assay (von Boxberg et al., 1993). Certainly, chemospecificity requires a number of molecular cues, only some of which are beginning to be understood.

Refinement

The precision of the early retinotectal projection varies substantially among different species. In cold-blooded animals, the projection has a high degree of order from the very beginning (Holt and Harris, 1993). In chick, the initial pattern of retinotectal connections lacks the precision seen in the adult, and many axons project incorrectly (McLoon, 1982; Nakamura and O'Leary, 1989). In rodents, the early projection is so rough that it is difficult to detect the adult topography (Cowan et al., 1984; Land and Lund, 1979; Simon et al., 1994). The precision of the retinotectal projection improves, and transient projections are eliminated during a discrete period of development.

A large body of work demonstrated that developmental refinement of the retinotectal projection is activity-dependent. The nature of the visual activity influences the improvement in the topographic precision that normally takes place during development (Brickley et al., 1998; Schmidt and Buzzard, 1993). Blocking action potentials in the retinal axons by treatment with TTX during the period of refinement prevented the normal elimination of inappropriate retinal connections in the brain (e.g. Kobayashi et al., 1990; Reh and

Constantine-Paton, 1985; Shatz and Stryker, 1988; Stryker and Harris, 1986). Since action potentials are rare in the retinal axons during development of the early pattern of connections (Galli and Maffei, 1988; Rager, 1980), it seems likely that chemospecificity works independently of activity. This leads to the conclusion that refinement, which is activity-dependent, and chemospecificity are separate mechanisms. Although it is still untested experimentally, it is likely that chemospecificity establishes a rough blueprint in the early projection pattern from which refinement mechanisms work to create the adult pattern of connections. The mechanisms involved in activity-dependent refinement are poorly understood, but recent research is beginning to shed light on this process.

Mounting evidence suggests that post-synaptic activity, as well as pre-synaptic activity, is required for remodeling neuronal connections during the refinement period. It is currently thought that activation of *N*-methyl-D-aspartic acid (NMDA) type glutamate receptors on post-synaptic neurons is required for activity-dependent refinement of the retinotectal projection (reviewed by Constantine-Paton et al., 1990). Retinal axons are known to be glutamatergic (Dye and Karten, 1996; Kalloniatis et al., 1994), and their targets express NMDA receptors (Cline et al., 1994; Fohr et al., 1995; Guido et al., 1997). Blocking NMDA receptors in the tectum during this period of development disrupted refinement of retinal connections (Cline et al., 1987; Cline and Constantine-Paton, 1989; Hahm et al., 1991; Simon et al., 1992). In chick, there is a transient ipsilateral retinotectal projection, and this projection is completely eliminated during the period of refinement (McLoon and Lund, 1982; O'Leary et al., 1983; Thanos and Bonhoeffer, 1984). Elimination of this projection in chick is believed to be analogous to segregation of the projection from the two eyes in the superior colliculus of mammals. We recently demonstrated that NMDA receptors are required for elimination of the chick ipsilateral retinotectal projection (Ernst et al., 1997). As shown in Fig. 4, increasing doses of MK-801, an NMDA receptor antagonist, rescued increasing numbers of retinal ganglion cells with projections to the ipsilateral tectum at an age by which this projection would normally have disappeared. Since the number of cells comprising the chick ipsilateral retinotectal projection is few, it is easy to quantify changes in the projection that result from various experimental manipulations. By quantifying the number of ipsilaterally projecting retinal ganglion cells that remained following NMDA receptor blockade, it was determined that NMDA receptor activity accounts for approximately 30% of the refinement of the ipsilateral retinotectal projection. This implies that mecha-

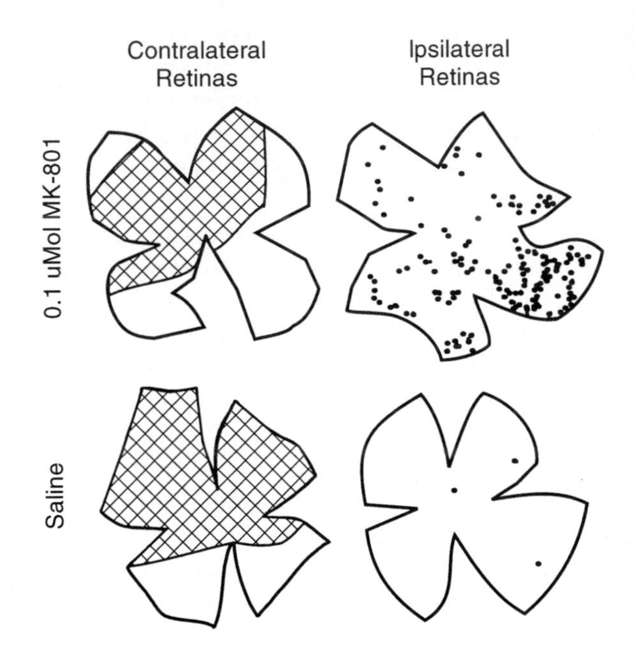

Fig. 4. Rescue of the ipsilateral retinotectal projection following blockade of NMDA receptors in embryonic chick. The enlarged outlines of retinal whole-mounts from day 17 embryos show the position of retrogradely labeled ganglion cells following a Fast Blue injection throughout one tectum. The dots indicate individual labeled cells in retinas ipsilateral to the injected tectum, and the cross-hatching indicates areas with a high density of retrogradely labeled cells in retinas contralateral to the injected tectum. Embryos were treated daily during the period of refinement from E9 to E16 with either 0.1 μmol MK-801 (top row) or saline (bottom row). Compared to the saline treated embryos, the MK-801 treatment rescued a significant number of ipsilaterally projecting retinal ganglion cells that would normally have been eliminated during the period of refinement.

nisms not involving NMDA receptors are also important in refinement of this projection.

Cell death appears to be involved in refinement of the retinotectal projection. Approximately half of all ganglion cells die during the developmental period in which the retinal projection is refined (Finlay et al., 1979; Hughes and McLoon, 1979; O'Leary et al., 1986). Many of the ganglion cells that die have connections in inappropriate positions of the brain (Cowan et al., 1984). Approximately 60% of the ganglion cells that give rise to the ipsilateral retinotectal projection in chick are normally eliminated by cell death (Williams and McLoon, 1991). The number of ganglion cells that die was not altered by blocking NMDA receptors or nitric oxide synthesis (Wu et al., 1996). This suggests that cell death eliminates transient projections independently of NMDA receptor activation or nitric oxide synthesis.

Though activation of post-synaptic NMDA receptors appears to be involved in the activity-dependent refinement of retinotectal connections, it is unclear what downstream signal transduction pathways are involved. NMDA receptors are cation channels, which are highly permeable to Ca^{2+} in the activated state (Scatton, 1993). By regulating internal Ca^{2+} levels, NMDA receptors could in principle modulate numerous signal transduction pathways. For example, rises in intracellular Ca^{2+} can activate at least two Ca^{2+}/calmodulin-dependent enzymes: CamKII and nitric oxide synthase (NOS). In the frog retinotectal system, overexpression of a constitutively active form of CamKII in tectal neurons prevented retinal axons from developing normal arborization patterns (Zou and Cline, 1996). CamKII is also capable of phosphorylating AMPA receptors, which may sensitize post-synaptic target cells to future glutamate release from pre-synaptic axons (Barria et al., 1997). Thus, CamKII could modulate activity in the post-synaptic cell, but no studies have shown that it acts directly on retinal axons.

The fact that activation of NMDA receptors on tectal cells results in changes in the pre-synaptic connections of retinal axons strongly suggests that tectal cells communicate retrogradely with the retinal axons. The nature of this retrograde signal is not known, but recent studies suggest that nitric oxide (NO) is involved. NO has characteristics that could allow it to function as a retrograde messenger (Gally et al., 1990; Jessell and Kandel, 1993; Kandel and O'Dell, 1992). In some brain cells, when NMDA receptors were activated by the neurotransmitter glutamate, NO was synthesized (Bredt and Snyder, 1989; Garthwaite et al., 1988). NO is a soluble gas that can diffuse freely across cell membranes. In certain cells, NO causes an increase in cyclic guanosine 3′,5′-monophosphate levels (cGMP) (Bredt and Snyder, 1989; Garthwaite et al., 1988). Retinal ganglion cells express a cGMP-gated calcium channel that is modulated by NO, which suggests that ganglion cells may respond to NO with an increase in intracellular calcium levels (Ahmad et al., 1994). An increase in intracellular calcium concentration can influence the extension and retraction of axons through molecules such as gelsolin that can interact with the cytoskeleton (Furukawa et al., 1997; Kater et al., 1988; Lankford and Letourneau, 1989).

Nitric oxide synthase (NOS), the enzyme that synthesizes nitric oxide, is expressed by tectal cells during the period of refinement in many species (Gonzalez-Hernandez et al., 1993; Tenorio et al., 1995; Williams et al., 1994). In embryonic chick tectum, NOS expression first appears in the superficial layers in concert with the arrival of retinal axons (Williams et al., 1994). This suggests that the retinal axons may regulate NOS expression in the tectum. This hypothesis is supported by experiments in which one eye was removed early in development, which resulted in reduced NOS expression in the retinorecipient layers of the tectum (Fig. 5). The correlation between the arrival of retinal axons and the first expression of NOS and the dependence of NOS expression on the presence of retinal axons leads to the suggestion that retinal axons synapse on tectal cells that express NOS. Peak levels of NOS expression in the retinorecipient layers of the tectum coincide with the period during which transient retinotectal projections are eliminated, consistent with NO having a role in refinement.

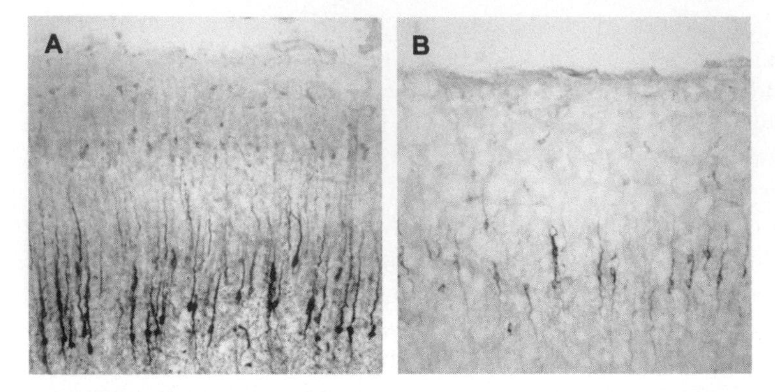

Fig. 5. Micrograph of NOS expression in embryonic chick tectum. A section of embryonic day 13 chick tectum was processed for NADPH diaphorase histochemistry. Diaphorase staining in formaldehyde fixed tissue is coextensive with nitric oxide synthase (Bredt et al., 1991; Dawson et al., 1991; Hope et al., 1991). The micrograph shows the superficial layers of the tectum (A) contralateral to a normal eye and (B) contralateral to the side of the embryo in which the eye was removed on the second day of incubation. The tectum receiving retinal innervation has significantly more diaphorase staining than does the uninnervated tectum.

NMDA receptor activity also regulates NOS activity in tectal cells. Blocking NMDA receptors in chick embryos with MK-801 during the period of refinement resulted in a significant reduction in NOS activity in the tectum (Ernst et al., 1997). These results are consistent with the possibility that NO is important in activity-dependent refinement of the retinotectal projection.

The strongest evidence that nitric oxide acts downstream of NMDA receptors relative to refinement comes from studies on refinement of the chick ipsilateral retinotectal projection. Similar to NMDA receptor blockade, inhibition of NO synthesis during the period of refinement prevented complete elimination of this transient projection (Wu et al., 1994). Embryos were treated systemically with L-NAME or L-NoArg, which are arginine analogues that inhibit NO synthesis. Each of these drugs rescued the ipsilateral retinotectal projection in a dose dependent manner. In a similar study, blocking NO synthesis also blocked the refinement in the topographical precision of the contralateral retinotectal projection (Wu et al., 1996). Since the chick ipsilateral retinotectal projection can be quantified, the effect of blocking NMDA receptors could be compared to the effect of blocking NO synthesis. As shown in Fig. 6, doses of MK-801 or L-NoArg that reduced tectal NOS activity to comparable levels resulted in the same number of ipsilaterally projecting retinal ganglion cells, which suggests that the effect of NMDA receptors on refinement is mediated by NO (Ernst et al., 1997). The finding that no additive effects were observed when embryos were treated simultaneously with both NMDA receptor blockers and NOS activity blockers further supports this conclusion. It is likely then, that NMDA receptor mediated refinement of the chick retinotectal projection requires nitric oxide.

There are systems, however, in which NMDA receptor mediated refinement does not appear to require NO. The segregation of geniculocortical afferents into eye-specific ocular dominance columns in mammalian cortex is believed to require activation of NMDA receptors on cortical cells (Fonta et al., 1997; Gu et al., 1989; Rauschecker et al., 1990). Infusing local regions of cortex with NOS activity inhibitors, however, failed to prevent the shift in ocular dominance associated with monocular deprivation (Ruthazer et al., 1996). Infusion of the visual cortex with excess BDNF or NT4/5, on the other hand, disrupted the normal segregation of geniculocortical afferents into ocular dominance columns (Cabelli et al., 1995). Thus, NMDA mediated refinement in some systems may work through pathways that do not involve NO but that may use neurotrophins.

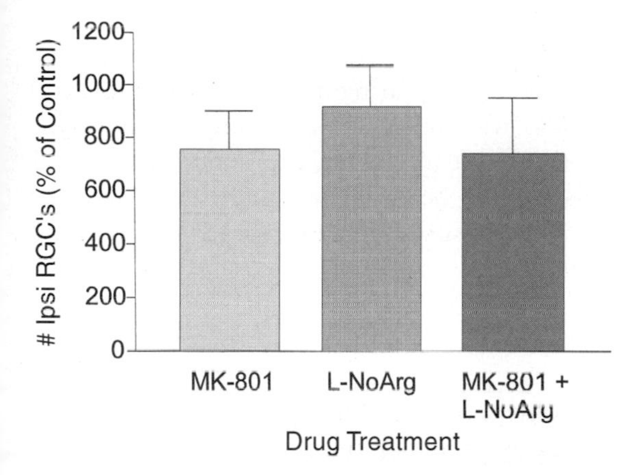

Fig. 6. Effect of MK-801, L-NoArg and MK-801 plus L-NoArg on the persistence of retinal ganglion cells projecting to the ipsilateral tectum. For each treatment condition, the number of labeled retinal ganglion cells in retinas ipsilateral to a Fast-Blue injected tectum was quantified. All values are expressed as a percentage of the number of ipsilaterally projecting ganglion cells present in retinas of E17 control embryos. Embryos were treated with 0.01 μmol MK-801, 1.0 μmol L-NoArg or 0.01 μmol MK-801 plus 1.0 μmol L-NoArg during the period of refinement of the ipsilateral retinotectal projection from E9 to E16. There was no statistically significant difference in the number of ipsilaterally projecting retinal ganglion cells rescued by any of the treatment conditions (Anova; $F = 3.24$; $P = 0.92$).

It is not entirely clear how NO production could lead to refinement of neuronal projections during development. One possibility is that NO leads to the strengthening of synapses made by "appropriately connected" axons in a manner that is similar to the synaptic strengthening associated with LTP. LTP is a sustained increase in the strength of a synaptic connection induced by tetanic stimulation of the pre-synaptic input. Post-synaptically, LTP induction appears to require activation of NMDA receptors and influx of Ca^{2+} ions (Collingridge et al., 1992; Daw et al., 1993; Malenka, 1994; Otani and Ben-Ari, 1993; Rison and Stanton, 1995). Changes in pre-synaptic transmitter release are believed to accompany LTP as well. It has been suggested that pre-synaptic changes associated with LTP require the release of a retrograde messenger from the post-synaptic cell upon activation of NMDA receptors (Medina and Izquierdo, 1995; Williams, 1996). Nitric oxide may be the retrograde messenger responsible for induction of LTP as blockade of NO synthesis or sequestration of extracellular NO with the scavenger, hemoglobin, prevented LTP in some paradigms (Arancio et al., 1996; Boulton et al., 1995; Haley et al., 1992; Izumi et al., 1992; O'Dell et al., 1994; Schuman and Madison, 1991; Son et al., 1996). Thus, both NMDA receptor activation and nitric oxide are important for the plasticity associated with LTP and may be involved as well in the selective strengthening of certain synaptic connections during development. With respect to the retinotectal system of the chick, NMDA receptor mediated NO release might induce LTP-like synaptic enhancement in the synapses made by ganglion cells that have more appropriate connections.

It is also possible that NMDA receptor mediated NO release destabilizes the synapses made by axons that have made "inappropriate connections" during development, such as the ipsilateral retinotectal projection in chick. It was demonstrated *in vitro* that retinal ganglion cell axons extending from explants of both rat and chick retina retract in response to nitric oxide (Renteria and Constantine-Paton, 1996; Ernst et al., 1998). Nitric oxide released from post-synaptic cells, therefore, might cause retraction of exposed fibers. This does not explain, however, why NO causes retraction of only selected axons. NO is understood to be released globally from cells in which it is synthesized. Thus, both appropriate and inappropriate synapses on a given tectal cell should be exposed to NO simultaneously. One possibility is that nitric oxide specifically causes the retraction of afferent inputs that are not firing in synchrony with the post-synaptic cell. This notion is supported by *in vitro* experiments with myocyte-motor neuron co-cultures (Lo and Poo, 1991). When single myocytes were innervated by two motor neurons, the synapse made by the motor neuron that fired out of synchrony with the myocyte was gradually reduced in strength, whereas the synapse made by the motor neuron that fired in synchrony with the myocyte was unaffected. Importantly, this effect was shown to be mediated by NO (Wang et al., 1995). In the retinotectal system, NO might selectively cause weakening and retraction of pre-synaptic terminals that were not recently active and might have no effect on pre-synaptic terminals that were active at the time of NO release from the post-synaptic cell.

An inappropriately connected neuron most likely will be inactive when NO is released from its target cell, and its connection will be weakened. This suggestion is based on a number of observations. An individual tectal cell receives synapses from many retinal ganglion cells. NMDA channels are voltage dependent and require a cell to be sufficiently depolarized before they can be activated by glutamate (Collingridge et al., 1992; Mayer et al., 1984). Therefore, a number of ganglion cells must be active at the same time to sufficiently depolarize a tectal cell so that the NMDA receptor can be opened. Individual tectal cells are likely to receive their major input from ganglion cells that neighbor each other in the retina due to chemospecificity, as discussed above. The activity of neighboring ganglion cells appears to be synchronous (Galli and Maffei, 1988; Maffei and Galli-Resta, 1990; Meister et al., 1991; Wong et al., 1997). Synchronous firing in neighboring ganglion cells that project to a common target cell could depolarize the target cell enough to activate NMDA receptors (Mayer and Westbrook, 1985; Jahr and Stevens, 1987; Ascher et al., 1988). Minor projections to a given tectal cell, i.e. the inappropriate projections, would probably be few and thus too weak to activate NMDA receptors on post-synaptic target cells. Thus, the major projection, with synchronized activity, would be more likely than the minor projection to activate NMDA receptors on a common target cell. Ganglion cells in the developing retina fire spontaneous bursts of action potentials followed by substantial periods of inactivity (Galli and Maffei, 1988; Maffei and Galli-Resta, 1990; Meister et al., 1991). Thus, it would be rare for the major and minor projections to a common target cell to be active at the same time. Since NO is synthesized by tectal cells in response to activation of NMDA receptors, NO is probably released in conjunction with activity in the major (i.e. appropriate) connections, and the minor (i.e. inappropriate) connections are likely to be exposed to NO during their period of inactivity. Since NO was linked to weakening inactive synapses (Wang et al., 1995), the minor retinotectal synapses could be weakened and ultimately eliminated by exposure to NO released from the tectal cells.

Conclusion

Development of the normal pattern of retinotectal connections is a complicated process involving multiple mechanisms and a variety of molecules. Two processes appear to be involved, formation of a prepattern and refinement. Formation of the prepattern is guided by multiple positional labels carried by the growing axons and the tectal cells. One set of positional labels appears to involve 'stop signals' mediated by gradients of an Eph receptor tyrosine kinase expressed across one axis of the retina and ligands for this receptor expressed across a complementary axis of the tectum. Some type of 'growth signal' with positional properties is probably required in this axis as well. Presumably, positional labels in another axis of the retinotectal system also are required. Refinement of the pre-pattern appears to involve multiple independent mechanisms. These include cell death and signalling mediated by NMDA receptors and nitric oxide. Further studies are needed to determine whether the initial pattern of connections that develop as a result of positional labels is an essential antecedent to the refinement process. The universality of these mechanisms needs to be determined by examination of multiple species and neuronal systems. The conclusions reached here were derived largely from study of the developing chick retinotectal system. Other species show obvious differences. For example, the prepattern in rodents is much more poorly organized than in chick (Simon et al., 1994). Other systems also appear to work differently. For example, development of ocular dominance columns, another type of refinement, appears to involve activation of NMDA receptors but not nitric oxide (Ruthazer et al., 1996). It seems likely that evolution has provided a menu of mechanisms that can be used to develop precise patterns of axonal connections between groups of neurons, and that different menu items are selected for use by different systems.

References

Ahmad, I., Leinders-Zufall, T., Kocsis, J.D., Shepherd, G.M., Zufall, F. and Barnstable, C.J. (1994) Retinal ganglion cells express a cGMP-gated cation conductance activatable by nitric oxide donors. *Neuron*, 12: 155–165.

Arancio, O., Kiebler, M., Lee, C.J., Lev-Ram, V., Tsien, R.Y., Kandel, E.R. and Hawkins, R.D. (1996) Nitric oxide acts directly in the presynaptic neuron to produce long-term potentiation in cultured hippocampal neurons. *Cell*, 87: 1025–1035.

Ascher, P., Bregestovski, P. and Nowak, L. (1988) *N*-Methyl-D-aspartate-activated channels of mouse central neurones in magnesium-free solutions. *J. Physiol.*, 399: 207–226.

Attardi, D.G. and Sperry, R.W. (1963) Preferential selection of central pathways by regenerating optic fibers. *Exp. Neurol.*, 7: 46–64.

Baier, H. and Bonhoeffer, F. (1992) Axon guidance by gradients of a target-derived component. *Science*, 255: 472–475.

Barbera, A.J. (1975) Adhesive recognition between developing retinal cells and the optic tecta of the chick embryo. *Dev. Biol.*, 46: 167–191.

Barria, A., Muller, D., Derkach, V., Griffith, L.C. and Soderling, T.R. (1997) Regulatory phosphorylation of AMPA-type glutamate receptors by CaMKII during long-term potentiation. *Science*, 276: 2042–2045.

Bonhoeffer, F. and Huf, J. (1985) Position-dependent properties of retinal axons and their growth cones. *Nature*, 315: 409–410.

Boulton, C.L., Southam, E. and Garthwaite, J. (1995) Nitric oxide-dependent long-term potentiation is blocked by a specific inhibitor of soluble guanylyl cyclase. *Neuroscience*, 3: 699–703.

Bredt, D.S. and Snyder, S.H. (1989) Nitric oxide mediates glutamate-linked enhancement of cGMP levels in the cerebellum. *Proc. Natl. Acad. Sci. USA*, 86: 9030–9033.

Bredt, D.S., Glatt, C.E., Hwang, P.M., Fotuhi, M., Dawson, T.M. and Snyder, S.H. (1991) Nitric oxide synthase protein and mRNA are discretely localized in neuronal populations of the mammalian CNS together with NADPH diaphorase. *Neuron*, 7: 615–624.

Brickley, S.G., Dawes, E.A., Keating M.J. and Grant S. (1998) Synchronizing retinal activity in both eyes disrupts binocular map development in the optic tectum. *J. Neurosci.*, 18: 1491–1504.

Bunt, S.M., Lund, R.D. and Land, P.W. (1983) Prenatal development of the optic projection in albino and hooded rats. *Dev. Brain Res.*, 6: 149–168.

Cabelli, R.J., Hohn, A. and Shatz, C.J. (1995) Inhibition of ocular dominance column formation by infusion of NT-4/5 or BDNF. *Science*, 267: 1662–1666.

Cheng, H., Nakamoto, M., Bergemann, A.D. and Flanagan, J.G. (1995) Complementary gradients in expression and binding of ELF-1 and Mek4 in development of the topographic retinotectal projection map. *Cell*, 82: 371–381.

Cline, H.T. and Constantine-Paton, M. (1989) NMDA receptor antagonists disrupt the retinotectal topographic map. *Neuron*, 3: 413–426.

Cline, H.T., Debski, E.A. and Constantine-Paton, M. (1987) NMDA receptor antagonist desegregates eye-specific stripes. *Proc. Natl. Acad. Sci. USA*, 84: 4342–4345.

Cline, H.T., McDonald, J.W. and Constantine-Paton, M. (1994) Glutamate receptor binding in juvenile and adult *Rana pipiens* CNS. *J. Neurobiol.*, 25: 488–502.

Collingridge, G.L., Randall, A.D., Davies, C.H. and Alford, S. (1992) The synaptic activation of NMDA receptors and Ca^{2+} signalling in neurons. *Ciba Found. Sym.*, 164: 162–175.

Constantine-Paton, M., Blum, A.S., Mendez, O.R. and Barnstable, C.J. (1986) A cell surface molecule distributed in a dorsoventral gradient in the perinatal rat retina. *Nature*, 324: 459–462.

Constantine-Paton, M., Cline, H.T. and Debski, E. (1990) Patterned activity, synaptic convergence, and the NMDA receptor in developing visual pathways. *Annu. Rev. Neurosci.*, 13: 129–154.

Cowan, W.M., Fawcett, J.W., O'Leary, D.D.M. and Stanfield, B.B. (1984) Regressive events in neurogenesis. *Science*, 225: 1258–1265.

Cox, E.C., Muller, B. and Bonhoeffer, F. (1990) Axonal guidance in the chick visual system: Posterior tectal membranes induce collapse of growth cones from the temporal retina. *Neuron*, 4: 31–37.

Davenport, R.W., Thies, E. and Nelson, P.G. (1996) Cellular localization of guidance cues in the establishment of retinotectal topography. *J. Neurosci.*, 16: 2074–2085.

Davis, S., Gale, N.W., Aldrich, T.H., Maisonpierre, P.C., Lhotak, V., Pawson. T. et al. (1994) Ligands for Eph-related receptor tyrosine kinases that require membrane attachment or clustering for activity. *Science*, 266: 816–819.

Daw, N.W., Stein, P.S. and Fox, K. (1993) The role of NMDA receptors in information processing. *Annu. Rev. Neurosci.*, 16: 207–222.

Dawson, T.M., Bredt, D.S., Fotuhi, M., Hwang, P.M. and Snyder, S.H. (1991) Nitric oxide synthase and neuronal NADPH diaphorase are identical in brain and peripheral tissues. *Proc. Natl. Acad. Sci. USA*, 88: 7797–7801.

Drescher, U., Bonhoeffer, F. and Muller, B.K. (1997) The Eph family in axon guidance. *Curr. Opin. Neurobiol.*, 7: 75–80.

Drescher, U., Kremoser, C., Handwerker, C., Loschinger, J., Noda, M. and Bonhoeffer, F. (1995) In vitro guidance of retinal ganglion cell axons by RAGS, a 25 kDa tectal protein related to ligands for Eph receptor tyrosine kinases. *Cell*, 82: 359–370.

Dye, J.C. and Karten, H.J. (1996) An *in vitro* study of retinotectal transmission in the chick: Role of glutamate and GABA in evoked field potentials. *Vis. Neurosci.*, 13: 747–758.

Ernst, A.F., Gallo, G., Letourneau, P. and McLoon, S.C. (1998) BONF protects against nitric oxide induced retinal growth cone collapse. *Soc. Neurosci. Abstr.*, 24: in press.

Ernst, A.F., Wu, H.H., El-Fakahany, E.E. and McLoon, S.C. (1997) NMDA receptor mediated refinement of a transient retinotectal projection is dependent on nitric oxide synthesis. *Soc. Neurosci. Abstr.*, 23: 1975.

Fan, J. and Raper, J.A. (1995) Localized collapsing cues can steer growth cones without inducing their full collapse. *Neuron*, 14: 263–274.

Fawcett, J.W. and Gaze, R.M. (1982) The retinotectal fibre pathways from normal and compound eyes in *Xenopus*. *J. Embryol. Exp. Morphol.*, 72: 19–37.

Finlay, B.L., Wilson, K.G. and Schneider, G.E. (1979) Anomalous ipsilateral retinotectal projections in Syrian hamsters with early lesions: Topography and functional capacity. *J. Comp. Neurol.*, 183: 721–740.

Fohr, K.J., Schirm, T. and Finger, W. (1995) NMDA-induced whole-cell currents and single channel conductances in tectal neurons during two stages of early development of chicken. *Neurosci. Lett.*, 183: 87–90.

Fonta, C., Chappert, C. and Imbert, M. (1997) *N*-methyl-D-aspartate subunit R1 involvement in the post-natal organization of the primary visual cortex of *Callithrix jacchus*. *J. Comp. Neurol.*, 386: 260–276.

Fraser, S.E. (1980) A differential adhesion approach to the patterning of nerve connections. *Dev. Biol.*, 79: 453–464.

Fraser, S.E. and O'Rourke, N.A. (1990) In situ analysis of neuronal dynamics and positional cues in the patterning of nerve connections. *J. Exp. Biol.*, 153: 61–70.

Friedman, G.C. and O'Leary, D.D.M. (1996) Retroviral misexpression of *engrailed* genes in the chick optic tectum perturbs the topographic targeting of retinal axons. *J. Neurosci.*, 16: 5498–509.

Furukawa, K., Fu, W., Li, Y., Witke, W., Kwiatkowski, D.J. and Mattson, M.P. (1997) The actin-severing protein gelsolin modulates calcium channel and NMDA receptor activities and vulnerability to excitotoxicity in hippocampal neurons. *J. Neurosci.*, 17: 8178–8186.

Gale, N.W., Holland, S.J., Valenzuela, D.M., Flenniken, A., Pan, L., Ryan, T.E. et al. (1996) Eph receptors and ligands comprise two major specificity subclasses and are reciprocally compartmentalized during embryogenesis. *Neuron*, 17: 9–19.

Galli, L. and Maffei, L. (1988) Spontaneous impulse activity of rat retinal ganglion cells in pre-natal life. *Science*, 242: 90–91.

Gally, J.A., Montague, P.R., Reeke, G.N. and Edelman, G.M. (1990) The NO hypothesis: Possible effects of a short-lived, rapidly diffusible signal in the development and function of the nervous system. *Proc. Natl. Acad. Sci. USA*, 87: 3547–3551.

Garthwaite, J., Charles, S.L. and Chess-Williams, R. (1988) Endothelium-derived relaxing factor release on activation of NMDA receptors suggests role as intercellular messenger in the brain. *Nature*, 336: 385–388.

Gierer, A. (1983) Model for the retino-tectal projection. *Proc. R. Soc. Lond. B*, 218: 77–93.

Gonzalez-Hernandez, T., Conde-Sendin, M., Gonzalez-Gonzalez, B., Mantolan-Sarmiento, B., Perez-Gonzalez, H. and Meyer, G. (1993) Post-natal development of NADPH-diaphorase activity in the superior colliculus and the ventral lateral geniculate nucleus of the rat. *Dev. Brain Res.*, 76: 141–145.

Gottlieb, D.I., Rock, K. and Glaser, L. (1976) A gradient of adhesive specificity in developing avian retina. *Proc. Natl. Acad. Sci. USA*, 73: 410–414.

Gu, Q.A., Bear, M.F. and Singer, W. (1989) Blockade of NMDA-receptors prevents ocularity changes in kitten visual cortex after reversed monocular deprivation. *Dev. Brain Res.*, 47: 281–288.

Guido, W., Lo, F.S. and Erzurumlu, R.S. (1997) An *in vitro* model of the kitten retinogeniculate pathway. *J. Neurophysiol.*, 77: 511–516.

Hahm., J.-O., Langdon, R.B. and Sur, M. (1991) Disruption of retinogeniculate afferent segregation by antagonists to NMDA receptors. *Nature*, 351: 568–570.

Haley, J.E., Wilcox, G.L. and Chapman, P.F. (1992) The role of nitric oxide in hippocampal long-term potentiation. *Neuron*, 8: 211–216.

Halfter, W., Claviez, M. and Schwarz, U. (1981) Preferential adhesion of tectal membranes to anterior embryonic chick retina neurites. *Nature*, 292: 67–70.

Hamdi, F.A. and Whitteridge, D. (1954) The representation of the retina on the optic tectum of the pigeon. *Q. J. Exp. Physiol.*, 39: 111–119.

Harris, W.A. (1982) The transplantation of eyes to genetically eyeless salamanders: Visual projections and somatosensory interactions. *J. Neurosci.*, 2: 339–353.

Holash, J.A. and Pasquale, E.B. (1995) Polarized expression of the receptor protein tyrosine kinase Cek5 in the developing avian visual system. *Dev. Biol.*, 172: 683–693.

Holt, C.E. and Harris, W.A. (1993) Position, guidance and mapping in the developing visual system. *J. Neurobiol.*, 24: 1400–1422.

Hope, B.T., Michael, G.J., Knigge, K.M. and Vincent, S.R. (1991) Neuronal NADPH diaphorase is a nitric oxide synthase. *Proc. Natl. Acad. Sci. USA*, 88: 2811–2814.

Hughes, W.F. and McLoon, S.C. (1979) Ganglion cell death during normal retinal development in the chick: Comparisons with cell death induced by early target field destruction. *Exp. Neurol.*, 66: 587–601.

Itasaki, N. and Nakamura, H. (1996) A role for gradient *en* expression in positional specification of the optic tectum. *Neuron*, 16: 55–62.

Izumi, Y., Clifford, D.B. and Zorumski, C.F. (1992) Inhibition of long-term potentiation by NMDA-mediated nitric oxide release. *Science*, 257: 1273–1276.

Jahr, C.E. and Stevens, C.F. (1987) Glutamate activates multiple single channel conductances in hippocampal neurons. *Nature*, 325: 522–525.

Jessell, T.M. and Kandel, E.R. (1993) Synaptic transmission: A bidirectional and self-modifiable form of cell-cell communication. *Cell*, 72 (Suppl.): 1–30.

Jurney, W.M., Selski, D.J. and McLoon, S.C. (1997) Reduction of Cek4 expression in developing chick retina alters topography of retinotectal projections. *Soc. Neurosci. Abstr.*, 23: 1975.

Kalloniatis, M., Tomisich, G. and Marc, R.E. (1994) Neurochemical signatures revealed by glutamine labeling in the chicken retina. *Vis. Neurosci.*, 11: 793–804.

Kandel, E.R. and O'Dell, T.J. (1992) Are adult learning mechanisms also used for development? *Science*, 258: 243–245.

Kaprielian, Z. and Patterson, P.H. (1994) The molecular basis of retinotectal topography. *BioEssays*, 16: 1–11.

Kater, S.B., Mattson, M.P., Cohan, C.S. and Connor, J.A. (1988) Calcium regulation of the neuronal growth cone. *Trends Neurosci.*, 11: 315–321.

Kobayashi, T., Nakamura, H. and Yasuda, M. (1990) Disturbance of refinement of retinotectal projection in chick embryos by tetrodotoxin and grayanotoxin. *Dev. Brain Res.*, 57: 29–35.

Land, P.W. and Lund, R.D. (1979) Development of the rat's uncrossed retinotectal pathway and its relation to plasticity studies. *Science*, 205: 698–700.

Lankford, K.L. and Letourneau, P.C. (1989) Evidence that calcium may control neurite outgrowth by regulating the stability of actin filaments. *J. Cell Biol.*, 109: 1229–1243.

Lo, L.J. and Poo, M.M. (1991) Activity-dependent synaptic competition *in vitro*: Heterosynaptic suppression of developing synapses. *Science*, 254: 1019–22.

Logan, C., Wizenmann, A., Drescher, U., Monschau, B., Bonhoeffer, F. and Lumsden, A. (1996) Rostral optic tectum acquires caudal characteristics following ectopic *engrailed* expression. *Curr. Biol.*, 6: 1006–1014.

Maffei, L. and Galli-Resta, L. (1990) Correlation in the discharges of neighboring rat retinal ganglion cells during pre-natal life. *Proc. Natl. Acad. Sci. USA*, 87: 2861–2864.

Malenka, R.C. (1994) Synaptic plasticity in the hippocampus: LTP and LTD. *Cell* , 78: 535–538.

Marchase, R.B. and Roth, S. (1978) Properties of a double gradient model for retinotectal specificity. *Prog. Clinl. Biol. Res.*, 23: 637–645.

Mason, C.A., Marcus, R.C. and Wang, L.C. (1996) Retinal axon divergence in the optic chiasm: Growth cone behaviors and signaling cells. *Prog. Brain Res.*, 108: 95–107.

Mayer, M.L. and Westbrook, G.L. (1985) The action of *N*-methyl-D-aspartic acid on mouse spinal neurons in culture. *J. Physiol.*, 361: 65–90.

Mayer, M.L., Westbrook, G.L. and Guthrie, P.B. (1984) Voltage-dependent block by Mg^{2+} of NMDA responses in spinal cord neurones. *Nature*, 309: 261–263.

McLoon, S.C. (1982) Alterations in precision of the crossed retinotectal projection during chick development. *Science*, 215: 1418–1420.

McLoon, S.C. and Lund, R.D. (1982) Transient retinofugal pathways in the developing chick. *Exp. Brain. Res.*, 45: 277–284.

Medina, J.H. and Izquierdo, I. (1995) Retrograde messengers, long-term potentiation and memory. *Brain Res. Rev.*, 21: 185–194.

Meister, M., Wong, R.O.L., Baylor, D.A. and Shatz, C.J. (1991) Synchronous bursts of action potentials in ganglion cells of the developing mammalian retina. *Science*, 252: 939–943.

Monschau, B., Kremoser, C., Ohta, K., Tanaka, H., Kaneko, T., Yamada, T. et al. (1997) Shared and distinct functions of RAGS and ELF-1 in guiding retinal axons. *EMBO J.*, 16: 1258–1267.

Nakamoto, M., Cheng, H.J., Friedman, G.C., McLaughlin, T., Hansen, M.J., Yoon, C.H., O'Leary, D.D. and Flanagan, J.G. (1996) Topographically specific effects of ELF-1 on retinal axon guidance *in vitro* and retinal axon mapping *in vivo*. *Cell*, 86: 755–766.

Nakamura, H. and O'Leary, D.D.M. (1989) Inaccuracies in initial growth and arborization of chick retinotectal axons followed by course corrections and axon remodeling to develop topographic order. *J. Neurosci.*, 9: 3776–3795.

O'Dell, T.J., Huang, P.L., Dawson, T.M., Dinerman, J.L., Snyder, S.H., Kandel, E.R. and Fishman, M.C. (1994) Endothelial NOS and the blockade of LTP by NOS inhibitors in mice lacking neuronal NOS. *Science*, 265: 542–546.

O'Leary, D.D.M., Fawcett, J.W. and Cowan, W.M. (1986) Topographic targeting errors in the retinocollicular projections and their elimination by selective ganglion cell death. *J. Neurosci.*, 6: 3692–3705.

O'Leary, D.D.M., Gerfen, C.R. and Cowan, W.M. (1983) The development and restriction of the ipsilateral retinofugal projection in the chick. *Dev. Brain Res.*, 10: 93–109.

Otani, S. and Ben-Ari, Y. (1993) Biochemical correlates of long-term potentiation in hippocampal synapses. *Intl. Rev. Neurobio.*, 35: 1–41.

Pandey, A., Duan, H. and Dixit, V.M. (1995) Characterization of a novel Src-like adapter protein that associates with the Eck receptor tyrosine kinase. *J. Biol. Chem.*, 270: 19201–19204.

Pandey, A., Lazar, D.F., Saltiel, A.R. and Dixit, V.M. (1994) Activation of the Eck receptor tyrosine kinase stimulates phosphatidylinositol 3-kinase activity. *J. Biol. Chem.*, 269: 30154–30157.

Rager, G.H. (1980) Development of the retinotectal projection in the chicken. *Adv. Anat. Embryol. Cell Biol.*, I–VIII: 1–90.

Raper, J.A. and Grunewald, E.B. (1990) Temporal retinal growth cones collapse on contact with nasal retinal axons. *Exp. Neurol.*, 109: 70–74.

Rauschecker, J.P., Egert, U. and Kossel, A. (1990) Effects of NMDA antagonists on developmental plasticity in kitten visual cortex. *Intl. J. Dev. Neurosci.*, 8: 425–435.

Reh, T.A. and Constantine-Paton, M. (1985) Eye-specific segregation requires neural activity in three-eyed *Rana pipiens*. *J. Neurosci.*, 5: 1132–1143.

Renteria, R.C. and Constantine-Paton, M. (1996) Exogenous nitric oxide causes collapse of retinal ganglion cell axonal growth cones *in vitro*. *J. Neurobiol.*, 29: 415–428.

Rison, R.A. and Stanton, P.K. (1995) Long-term potentiation and *N*-methyl-D-aspartate receptors: Foundations of memory and neurologic disease? *Neurosci. Behav. Rev.*, 19: 533–552.

Ruthazer, E.S., Gillespie, D.C., Dawson, T.M., Snyder, S.H. and Stryker, M.P. (1996) Inhibition of nitric oxide synthase does not prevent ocular dominance plasticity in kitten visual cortex. *J. Physiol.*, 494: 519–527.

Sanes, J.R. (1993) Topographic maps and molecular gradients. *Curr. Opin. Neurobiol.*, 3 (1): 67–74.

Scatton, B. (1993) The NMDA receptor complex. *Fund. Clinl. Pharm.*, 7: 389–400.

Schmidt, J.T. and Buzzard, M. (1993) Activity-driven sharpening of the retinotectal projection in goldfish: Development under stroboscopic illumination prevents sharpening. *J. Neurobiol.*, 24: 384–399.

Schuman, E.M. and Madison, D.V. (1991) A requirement for the intercellular messenger nitric oxide in long-term potentiation. *Science*, 254: 1503–1506.

Shatz, C.J. and Stryker, M.P. (1988) Prenatal tetrodotoxin infusion blocks segregation in retinogeniculate afferents. *Science,* 242: 87–89.

Simon, D.K., Prusky, G.T., O'Leary, D.D.M. and Constantine-Paton, M. (1992) *N*-Methyl-D-aspartate receptor antagonists disrupt the formation of a mammalian neural map. *Proc. Natl. Acad. Sci. USA*, 89: 10593–10597.

Simon, D.K. Roskies, A.L. and O'Leary, D.D.M. (1994) Plasticity in the development of topographic order in the mammalian retinocollicular projection. *Dev. Biol.*, 162: 384–393.

So, K.-F., Schneider, G.E. and Ayres, S. (1981) Lesions of the brachium of the superior colliculus in neonate hamsters: Correlation of anatomy with behavior. *Exp. Neurol.*, 72: 379–400.

Son, H., Hawkins, R.D., Kiebler, M.K., Huang, P.L., Fishman, M.C. and Kandel, E.R. (1996) Long-term potentiation is reduced in mice that are doubly mutant in endothelial and neuronal nitric oxide synthase. *Cell*, 87: 1015–1023.

Soans, C., Holash, J.A., Pavlova, Y. and Pasquale, E.B. (1996) Developmental expression and distinctive tyrosine phosphorylation of the Eph-related receptor tyrosine kinase Cek9. *J. Cell Biol.*, 135: 781–795.

Sperry, R.W. (1943) Effect of 180° rotation of the retinal field on visuomotor coordination. *J. Exp. Zool.*, 92: 263–279.

Sperry, R.W. (1963) Chemoaffinity in the orderly growth of nerve fiber patterns and connections. *Proc. Natl. Acad. Sci. USA*, 50: 703–710.

Stryker, M.P. and Harris, W.A. (1986) Binocular impulse blockade prevents the formation of ocular dominance columns in cat visual cortex. *J. Neurosci.*, 6: 2117–2133.

Tessier-Levigne, M. and Goodman, C.S. (1996) The molecular biology of axon guidance. *Science*, 274: 1123–1133.

Tenorio, F., Giraldi-Guimaraes, A. and Mendez-Ortero, R. (1995) Developmental changes of nitric oxide synthase in the rat superior colliculus, *J. Neurosci. Res.*, 42: 633–637.

Thanos, S. and Bonhoeffer, F. (1983) Investigations on the development and topographic order of retinotectal axons: Anterograde and retrograde staining of axons and perikarya with rhodamine *in vivo*. *J. Comp. Neurol.*, 219: 420–430.

Thanos, S. and Bonhoeffer, F. (1984) Development of the transient ipsilateral retinotectal projections in the chick embryo: A numerical fluorescence-microscopic analysis. *J. Comp. Neurol.*, 224: 407–414.

Thanos, S. and Dutting, D. (1987) Outgrowth and directional specificity of fibers from embryonic retinal transplants in the chick optic tectum. *Dev. Brain Res.*, 32: 161–179.

Udin, S.B. and Fawcett, J.W. (1988) Formation of topographic maps. *Annu. Rev. Neurosci.*, 11: 289–327.

van der Geer, P., Hunter, T. and Lindberg, R.A. (1994) Receptor protein-tyrosine kinases and their signal transduction pathways. *Annu. Rev. Cell Biol.*, 10: 251–337.

von Boxberg, Y., Deiss, S. and Schwarz, U. (1993) Guidance and topographic stabilization of nasal chick retinal axons on target-derived components *in vitro*. *Neuron*, 10: 345–357.

Walter, J., Kern-Veits, B., Huf, J., Stolze, B., Bonhoeffer, F. (1987a) Recognition of position-specific properties of tectal cell membranes by retinal axons *in vitro*. *Development*, 101: 685–696.

Walter, J., Henke-Fahle, S. and Bonhoeffer, F. (1987b) Avoidance of posterior tectal membranes by temporal retinal axons. *Development*, 101 (4): 909–913.

Wang, T., Xie, Z. and Lu, B. (1995) Nitric oxide mediates activity-dependent synaptic suppression at developing neuromuscular synapses. *Nature*, 374: 262–266.

Williams, C.V. and McLoon, S.C. (1991) Elimination of the transient ipsilateral retinotectal projection is not solely achieved by cell death in the developing chick. *J. Neurosci.*, 11: 445–453.

Williams, C.V., Nordquist D. and McLoon, S.C. (1994) Correlation of nitric oxide synthase expression with changing patterns of axonal projections in the developing visual system. *J. Neurosci.*, 14: 1746–1755.

Williams, J.H. (1996) Retrograde messengers and long-term potentiation: A progress report. *J. Lipid Med. Cell Signal.*, 14: 331–339.

Wong, W.T., Sanes, J.R. and Wong, R.O.L. (1997) Spontaneous activity patterns in the ganglion cell layer of the embryonic chick retina. *Soc. Neurosci. Abstr.*, 23: 641.

Wu, H.H., Waid, D.K. and McLoon, S.C. (1996) Nitric oxide and the developmental remodeling of retinal connections in the brain. *Prog. Brain Res.*, 108: 273–286.

Wu, H.H. Williams, C.V. and McLoon, S.C. (1994) Involvement of nitric oxide in the elimination of a transient retinotectal projection in development. *Science*, 265: 1593 1596.

Zou, D.-J. and Cline, H.T. (1996) Expression of constituitively active CamKII in target tissue modifies presynaptic axon arbor growth. *Neuron*, 16: 529–539.

R.R. Mize, T.M. Dawson, V.L. Dawson and M.J. Friedlander (Eds.)
Progress in Brain Research, Vol 118

CHAPTER 10

The role of nitric oxide in development of the patch–cluster system and retinocollicular pathways in the rodent superior colliculus

R. Ranney Mize*, Hope H. Wu, R. John Cork and Christopher A. Scheiner

Department of Cell Biology and Anatomy and the Neuroscience Center, Louisiana State University Medical Center, New Orleans, LA 70112, USA

Abstract

Nitric oxide (NO) has been implicated as a retrograde signal in the process of refining axonal pathways during brain development. To determine some of the factors involved in this process, we have used two model pathway systems in the rat and mouse superior colliculus (SC). The first, the patch–cluster system, consists of clusters of neurons in the intermediate gray layer (igl) which transiently express NO during development and which receive input from a cholinergic pathway from the parabrachial brainstem as well as from other pathways containing different transmitters. The second system, the retinocollicular pathway, consists of glutamatergic fibers that project to the superficial gray layer. We have used both nitric oxide synthase inhibition (nw-nitro-L-arginine, NoArg) and single (nNOS) and double (nNOS and eNOS) gene knockout mice to examine the effect that reduction in NOS has upon the development of these two systems.

The onset of NOS expression in rat, as revealed by nicotinamide adenine dinucleotide phosphate diaphorase (NADPH-d) labeling, occurred in igl cells as early as postnatal day P5, with clusters being well-established by P14. Cholinergic fibers were first visible at P10 and formed obvious patches and tiers by P14. Intraperitoneal injections of NoArg from P1–P22 had no effect upon the development of these cholinergic patches. The pathway also developed normally in both single and double-knockout mice. In contrast, the ipsilateral retinocollicular pathway was altered in the double, but not in the single knockout mouse. This pathway is exuberant during the first week of life, being distributed across much of the mediolateral axis of the rostral SC. By P8–P15, this pathway has retracted to the most mediorostral SC. This refinement was delayed substantially in the double NOS gene knockout mouse. Ipsilateral fibers were found within 3–5 separate medio-lateral patches within the rostral 600 μm of SC at P15, and patches of abnormal size and extent were also seen at P18.

We conclude from these results that NO plays a role in pathway development in the rodent SC, but only in glutamatergic pathways and only when both endothelial and neuronal forms of NOS have been deleted. The mechanism of this effect must involve pathway elimination in situations where there is non-correlated electrical activity. It is likely that NO promotes fiber retraction rather than fiber stabilization in these developing nerve fibers.

* Corresponding author. Tel.: +1 504 568 4012; fax: +1 504 568 4392; e-mail: rmize@isumc.edu

Introduction

Beginning with the seminal work of Garthwaite (Garthwaite et al., 1988, Garthwaite, 1991) and Snyder (Bredt and Snyder, 1989, 1990, Snyder, 1992), nitric oxide (NO) has been shown to be an important messenger molecule in the central nervous system (Bohme et al., 1993, Dawson et al., 1992, Williams et al., 1993, Jaffery, 1995, Nelson, 1995). NO is a free radical gas that is known to diffuse freely across membranes and to affect pre-synaptic neurotransmitter release. It thus can serve as a retrograde messenger that signals to pre-synaptic neurons that an event has occurred in the post-synaptic neuron. Evidence suggests that NO is capable of enhancing the activity of the presynaptic neuron and potentially reinforcing the synaptic connection. This hypothesis has been best tested in models where plasticity occurs in the adult, including long term potentiation (Bohme et al., 1991; Schuman and Madison, 1991; O'Dell et al., 1991; Izumi et al., 1992; Zhuo et al., 1993; Hawkins, 1994; Son et al., 1996) and long term depression (Daniel et al., 1993; Shibuki, 1993; Zhuo et al., 1994; but see Malen and Chapman, 1997).

In 1990, Gally and co-workers suggested that nitric oxide could also serve as a retrograde messenger in the developing nervous system to refine developing synaptic connections (Gally et al., 1990). A more detailed report of the hypothesis and a model to test it was reported by Montague et al., 1991. According to the Gally et al., hypothesis: (1) "postsynaptic elevation of Ca^{2+} concentration after synaptic activity causes NO production and release from postsynaptic sites in such a fashion that NO alters the synaptic efficacy both of its synapses of origin and of the synapses in surrounding space"; (2) "synapses that fire in a coordinated manner in the same region of neural space are strengthened and those that are not so coordinated are weakened". These authors stated further that: (3) "the presence of such a signal provides a means of matching afferent axonal arbors and their recipient dendritic arbors...in such a way that correlations present in the afferent input can be reflected in the final anatomy of the network"; and (4) "this conclusion is consistent with glutamate-induced production and release of NO and the disruption of this effect by agents that block glutamatergic transmission".

This hypothesis predicts explicitly the role that NO plays in pathway refinement in several respects. First, it predicts that NO is active not only in strengthening synapses that show correlated electrical activity but also in weakening synapses that are not firing in a coordinated manner. Second, the hypothesis suggests that the effects of NO are produced by glutamate-induced production of NO that involves the *n*-methyl-*d*-aspartate (NMDA) receptor, although it does not exclude the possibility that other transmitters could also play a role.

Although the hypothesis is reasonable, a number of questions remain to be answered regarding it. Thus, for example, we do not know if glutamatergic transmission and the NMDA receptor are essential to the release of NO, or whether increases in cytoplasmic calcium via other mechanisms (i.e. inositol trisphosphate (IP3)-dependent release after activation of the acetylcholine (ACh) receptor or voltage-gated calcium channels) can also release NO from post-synaptic cells. In addition, we do not know whether the effect of NO is confined to the presynaptic terminal (i.e. a guanylate cyclase-cGMP enhancement of transmitter release) or whether it can also directly affect the postsynaptic NMDA receptor or other receptor channel ion permeability. Finally, there is no direct evidence as to whether NO affects only axons contacting the NO releasing neuron or whether it can affect synapses in surrounding space.

The studies that we review in this chapter attempt to address the first of these questions. Although it is traditionally stated that the increase in cytoplasmic calcium ($[Ca^{2+}]i$) that activates nitric oxide synthase (NOS) is due to glutamate-dependent calcium influx through the NMDA receptor/channel, there are many other ways in which suitable $[Ca^{2+}]i$ increases could be produced. Some of those that might be operating in the superior colliculus (SC) are illustrated in Fig. 1. Glutamate from presynaptic glutamatergic terminals (Fig. 1(1)) could trigger calcium influx by several mechanisms. One likely channel is the NMDA receptor (Fig. 1A) (Garthwaite et al., 1988; Bredt and Snyder, 1989). Calcium could also flow in through voltage-gated calcium channels (VGCCs) in response to AMPA-channel mediated depolarization, or directly through the AMPA-channel itself (see Muller and Bicker, 1994; Wang et al., 1996). Cholinergic synaptic transmission (Fig. 1(2)) could activate muscarinic acetylcholine receptors (e.g. M1 or M3) that are linked to the IP3/DAG signal transduction pathway leading to calcium release from internal calcium stores (Fig. 1B). Even GABAergic synaptic transmission (Fig. 1(3)) has been reported to increase $[Ca^{2+}]i$ (Owens et al., 1996). In some immature neurons GABAa receptor activation results in a chloride efflux which depolarizes the cell and could open VGCCs (Fig. 1C) (see Baader et al., 1997).

We have addressed this issue of transmitter specificity using two model pathways in the SC. The first is an anatomical circuit within the intermediate gray layer (igl) of SC which we call the patch–cluster system. This system includes an acetylcholine-containing pathway from the para-

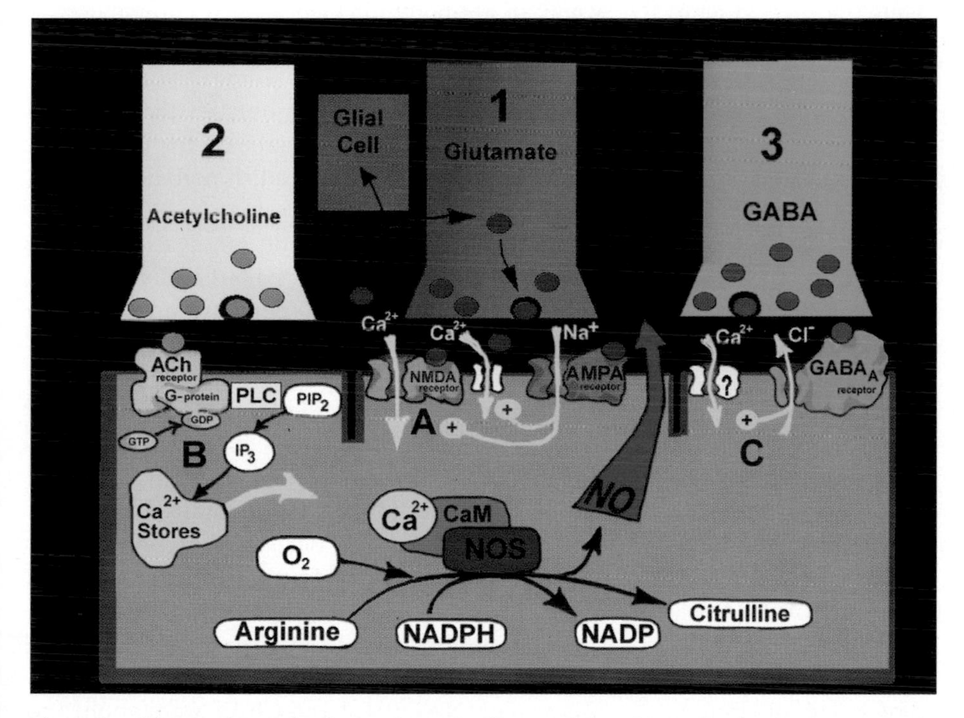

Fig. 1. Possible signal transduction pathways leading to nitric oxide production in the cat superior colliculus (SC). Synaptic inputs thought to contact nitric oxide synthase (NOS) containing cells in the SC include: (1) Glutamatergic axons, some of which arise from the retina and project to the superficial layers of SC and others from the cortical frontal eye fields (FEF) which project to the patch–cluster system in the intermediate gray layer of SC; (2) cholinergic axons, which arise from the pedunculopontine tegmental and lateral dorsal tegmental nuclei and project to the patch–cluster system of SC. ACh fibers also terminate in the region of NOS containing cells in the sgl; (3) GABAergic axons which arise from the substantia nigra (SN) and which innervate the patch–cluster system. Other GABA inputs arise from other sources and from intrinsic interneurons. All three sources could increase intracellular Ca^{2+} levels and trigger release of nitric oxide (NO). Possible mechanisms of calcium increase include: (A) influx of Ca^{2+} through the NMDA receptor, with consequent binding to NOS and production of NO; (B) increase in intracellular Ca^{2+} triggered by the G-protein linked acetylcholine receptor which activates an inositol trisphosphate linked release of calcium stores to initiate NO release; (C) influx of calcium through a voltage-gated Ca^{2+} channel depolarized by chloride efflux through the GABAa receptor. Further detail is in the text.

brachial brainstem which terminates in patches that precisely overlap 'clustered' neurons in the igl that are known to project to regions of the cuneiform nucleus of the brainstem (Jeon and Mize, 1993). This system appears to be an appropriate model to test the role of NO in development because the clustered neurons express NOS transiently at about the time at which afferent fibers from the parabrachium reach the igl and establish patches.

The other model system is the retinocollicular pathway which terminates in the upper superficial gray layer of SC. This pathway is glutamatergic and also projects to a region of SC in which cells begin to express NOS at about the time that the pathway is being refined. Thus, the two systems allow us to determine whether manipulations of NO expression can alter the development of a cholinergic or glutamatergic pathway, or both.

The patch–cluster system in the mammalian superior colliculus

The patch–cluster system consists of clusters of projection neurons that precisely overlap several afferent pathways to this region of the mammalian SC. The clustered neurons were first described in cat by Jeon and Mize (1993) who used retrograde tracers to show that these cells project to the cuneiform nucleus within the mesencephalic tegmentum. In cat, the clusters consist of 3–20 closely apposed neurons that are separated by intercluster intervals that have few or no cells. Similar clusters of neurons are labeled by NADPH-d in the developing cat SC (Scheiner et al., 1995; Mize et al., 1996) and in both rat and mouse SC (see below). The clustering pattern is less obvious in rodent than in cat, but the NADPHd labeled neurons do largely overlap the patch-like ACh fiber pathway in rodent as they do in cat (Mize et al., 1996, 1997). Although we do not as yet have direct evidence that these clusters are the same neurons that project to the cuneiform region (CFR) in cat, their morphology, position within the igl, and clustered distribution strongly argue in favor of this homology.

These neuron clusters receive at least three inputs: the first is a GABAergic input that arises from the substantia nigra (SN) (Illing and Graybiel, 1985; Ficalora and Mize, 1989). The second input is from the frontal eye fields (Illing and Graybiel, 1985) and is probably glutamatergic (Dori et al., 1992). The third arises from the pedunculopontine tegmental (PPTN) and lateral dorsal tegmental (LDTN) nuclei within the mesencephalic tegmentum (Beninato and Spencer, 1986; Hall et al., 1989, Harting and Leishout, 1991). In cat, all three inputs form distinctive fiber patches that overlap the clustered neurons (Illing and Graybiel, 1985; Harting and Leishout, 1991). The acetylcholine (ACh) patches are readily visible when an antibody directed against choline acetyltransferase (ChAT), the synthetic enzyme of ACh, is applied to sections of SC (Illing 1990), and they can be shown to perfectly overlap the clustered neurons in sections double-labeled for ChAT and the retrograde tracer horseradish peroxidase (Jeon and Mize, 1993).

In rodent, the pattern of innervation of the cholinergic fibers from the PPTN/LDTN is slightly different from that in cat (Beninato and Spencer, 1986). ChAT labeled fibers in the igl of both rat and mouse form two separate tiers, one within the dorsal most igl, the other within a more ventral tier (Fig. 2A). These two tiers have regions that contain dense concentrations of fiber terminals interspersed with regions that contain many fewer fiber terminals. The two tiers are also connected by bridges or streams of fibers that are dispersed vertically between the two tiers (Fig. 2A).

From double-labeling studies in cat we know that these ACh containing fibers also contain NO. When both NADPH-d and ChAT were applied to the same section, both labels could be visualized in the same fiber. In rodent, adjacent sections labeled by ChAT or NADPH-d show a pattern of fiber staining that is virtually identical, where the two tiers and interconnecting fiber streams can be seen with both labels (Fig. 2A,B). In addition, both NADPHd and ChAT colocalize in most of the neurons in the PPTN and LDTN, the nuclei of origin for these fibers (see Cork et al., this volume).

In summary, the patch–cluster model offers several advantages for study of the role of NO in development: (1) the system has well-defined and segregated afferents and target neurons, both of which contain NOS; (2) it receives multiple afferent inputs, each of which uses a different neurotransmitter; (3) the synaptic connections between the patch-like fibers and efferent neurons are formed at about the time that NOS appears to be expressed in these cells. We can thus test whether NOS inhibition can alter the development of a non-glutamatergic, cholinergic pathway.

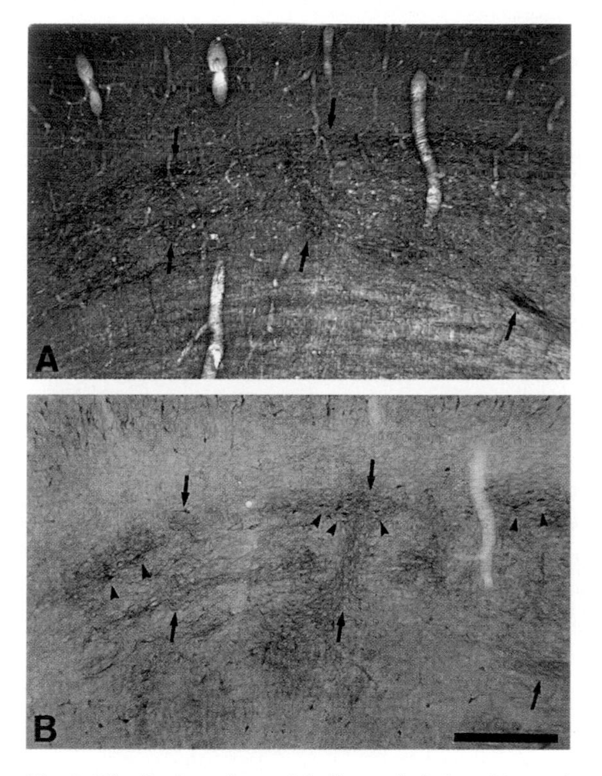

Fig. 2. Distribution of acetylcholine and nitric oxide synthase within the patch–cluster system of the rodent SC: (A) Distribution of acetylcholine containing fibers within the intermediate gray layer (igl) of mouse SC, visualized using an antibody to choline acetyltransferase (ChAT). Note the dorsal and ventral tiers of labeled fibers connected by intertier fiber streams. High densities of fibers that form patches are indicated by arrows. (B) Distribution of fibers in the mouse igl that are labeled by NADPH-d, a marker of NOS containing fibers and cells. Note the close correspondence of the labeling pattern with that seen in A. Note also the presence of some NADPH-d labeled neurons within the fiber tiers (arrowheads). Scale bar, A, B = 300 μm.

Development of the patch–cluster system

As a first step in examining the role that nitric oxide may play in the development of the patch–cluster system, we studied the onset of expression of NOS, NO's synthesizing enzyme, in the SC of cats, rats, and mice. Previous studies in cat had shown that the clustered neurons that project to the CFR have migrated to the igl by about embryonic day E41 and form clear clusters by E51. However, they do not yet express NOS at this age (see below). Axons that form patches in the igl from the three input sources – the PPTN/LDTN, the frontal eye fields, and the SN – have apparently not reached the igl by E51, and cholinergic fiber patches from PPTN have not been identified prior to about postnatal day P10 in cat (McHaffee et al., 1991). Thus, the refinement of the afferents into patches that match the cell clusters appears to occur between P7–P21.

In cat, the most intense expression of NOS in cells in the igl, as determined by NADPH-d histochemistry, occurs in this same time frame. NADPHd labeling can first be found in individual cells in the igl between E51–E58. Closely grouped clusters of NOS expressing neurons are seen clearly by P3 and are well developed by P14 (Scheiner et al., 1995; Mize et al., 1996). The intensity of labeling and the number of neurons is reduced after about P42 and in the adult cat (Scheiner et al., 1995; Mize et al., 1996). In rat, NADPH-d labeled neurons in SC are first seen in the igl as early as P5. They increase dramatically in number by P8–9, where their distribution is largely confined to the tiers of ACh fibers. Quantitative plots of the distribution of NADPHd positive cells at different ages showed that the number of cells averaged about 40–45 between P8 and P21 and then increased further with a mean of 56.3 at P35 (Mize et al., 1997). The number of NADPH-d labeled cells then decreased in juvenile and adult animals. This decrease occurred despite the fact that the igl has increased in volume in the adult. Thus, the peak expression of NOS occurs at approximately the time that the ACh fibers from the PPTN/LDTN are forming synapses with cells in the igl in both cat and rat.

The development of NOS expression in the mouse is similar. There is a ventral to dorsal progression in expression. The earliest cells labeled by NADPH-d appear within a dorsal lateral segment of the periaquaductal gray (PAG), with a few lightly labeled cells also present within the deep layers of SC at P5 (Fig. 3A). By P8, NADPH-d positive neurons within the deep layers have well-labeled cell bodies and dendrites. Some cells have reached the presumptive dorsal tier of the igl (Fig. 3C, P8), but these are lightly labeled and few in number. Little or no fiber labeling is apparent at this age. Between P10 and P12, a number of NADPHd positive cells are present in the igl and these are scattered within the two tiers of labeled fibers (Fig. 4A). By P18, many cells are embedded within the fiber tiers (Fig. 4B, arrows), and the fibers themselves are more darkly stained and clearly organized into dorsal and ventral tiers that have obvious dense patches interspersed with intervals containing fewer fibers.

Computer generated maps of the distribution of NADPH-d labeled cells in SC confirm this developmental sequence. Computer plots (Fig. 5) illustrating the distribution of NADPH-d cells in SC at ages P5, P8, P12, P15, and adult show that the number of neurons in the igl increases between P8 and P15 and is reduced in number in adult animals. These data suggest, that as in rat, the largest number of NOS neurons is present in the igl tiers during the time that fibers are forming synapses with these neurons. In summary, NOS (and presumably NO) begins to appear in igl neurons as fibers are growing into that region of SC; and that the highest expression of NOS occurs at the time when the patches are in the process of refinement.

Inhibition of NOS fails to alter the development of the ACh fiber pathway

If NOS shows a maximal expression during the time of patch–cluster formation, then it is reasonable to speculate that NO plays a role in the refinement process. To test experimentally whether the development of the patch–cluster system is altered by reduction of NO, we examined the distribution of the ACh containing fibers in rats in which NOS had been inhibited, and in NOS

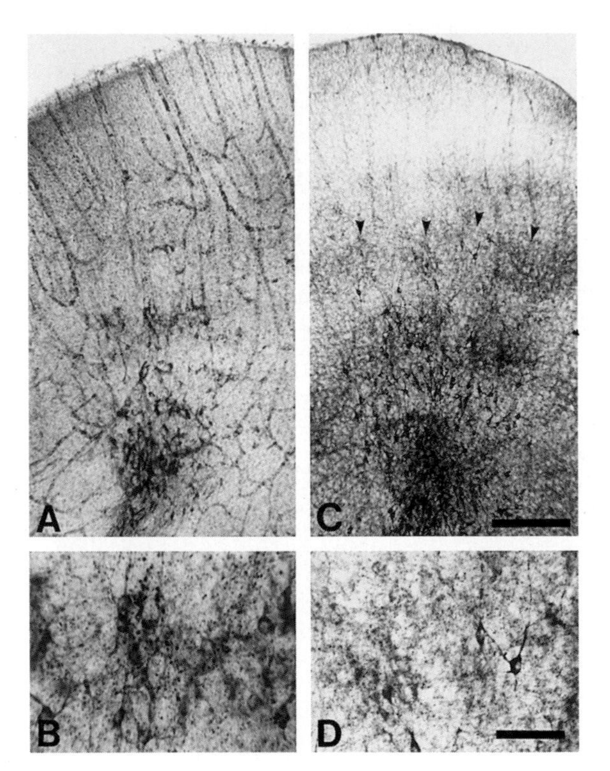

Fig. 3. Early development of NOS expression in the mouse SC: (A) P5. Scattered neurons are seen within the dorsolateral wedge of the periaquaductal gray as well as within the deep layers of SC. There are very few labeled neurons within the intermediate layers. Virtually no fiber labeling is present at this age; (B) higher magnification of labeled neurons within the deep gray layer of SC at P5; (C) P8. A number of labeled neurons are visible in the intermediate gray layer. Note that they have a clustered pattern with groups of neurons (arrowheads) separated by regions that contain few labeled cells; (D) higher magnification view of two labeled neuron clusters. Note labeling of primary and some secondary dendrites of one of these cells. Scale bar, A, C = 200 μm; B, D = 50 μm.

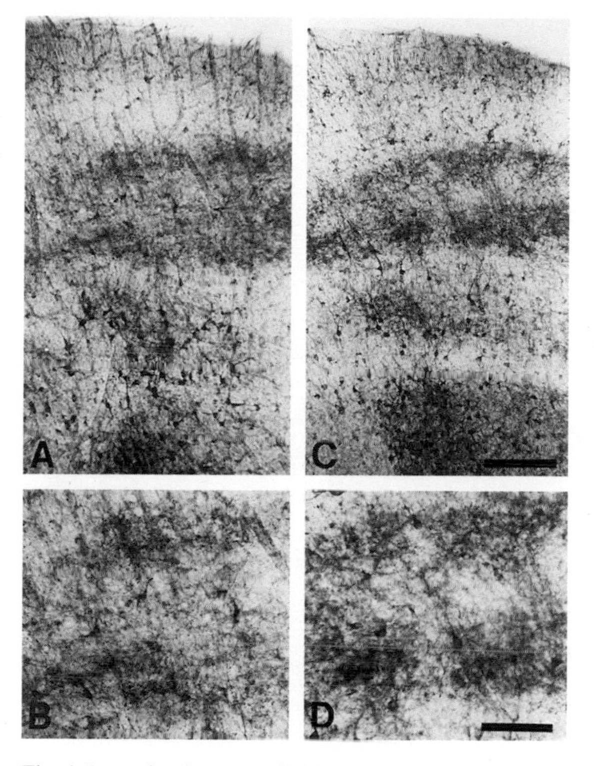

Fig. 4. Later development of NOS expression in the mouse SC. (A) P12. Many labeled neurons are present within the periaquaductal gray and deep gray layer (dgl) of SC. Some darkly labeled cells are also seen within the intermediate gray layer and labeled fibers now form two tiers within that layer; (B) higher magnification view of labeled fibers and cells within the igl at P12; (C) P15. Many dark NADPH-d labeled cells are seen within the igl. The labeled fiber tiers are prominent, including the bridge fibers between the tiers. (D) higher magnification view of labeled fibers and cells within the igl at P15. Scale bar, A, C = 200 μm; B, D = 100 μm.

deficient mice. NOS was inhibited in rats by daily systemic injection of the NOS inhibitor nw-nitro-L-arginine (NoARG). One group received a dose of 1 μmol/g b.w. of NoArg, the other a dose of 3 μmol/g b.w. Injections were made daily from P1 (defined as the first full day after birth) until perfusion at P8, P10–11, P14, P18, or P21. Control animals were given injections of comparable vol-

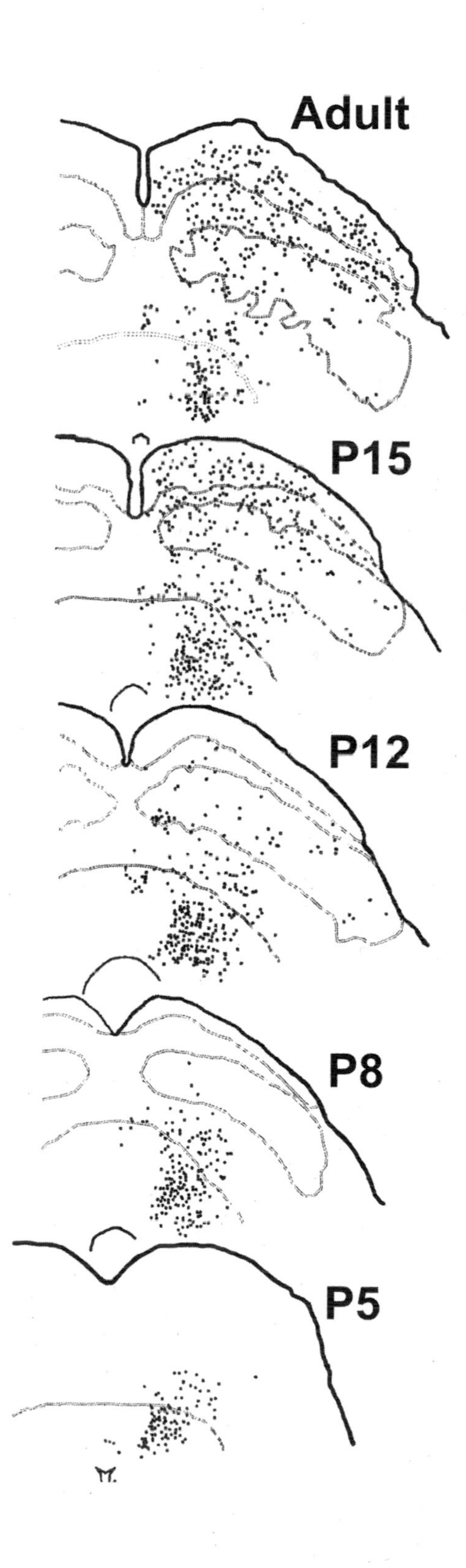

Fig. 5. Development of NADPHd labeled neurons in mouse SC. Computer-generated microscope plots of the distribution of NADPH-d labeled neurons in the mouse SC at P5, P8, P12, P15, and Adulthood.

umes of saline. The degree of NOS inhibition was determined in some animals at ages P14, P18, and P21 by using a [^{3}H] arginine–[^{3}H]citruline conversion assay described in detail previously (Mize et al., 1997).

ChAT is well expressed in fibers in both control and experimental animals by P14. At this age, the two-tiered labeling seen in the adult is beginning to appear, but there is little evidence of patchiness in these fibers. The two-tiered pattern is better developed at P18, and some bridges between the two tiers are also visible at this age. By P21 the distinctive patch-like pattern is clearly visible and it remains so at P35 and in the adult. Careful comparison of the labeling intensity, distribution, and spatial density of ChAT labeled fibers revealed virtually no qualitative differences between normal and inhibited animals in this pathway (Mize et al., 1997).

We also failed to observe any differences in the labeling of these fibers by NADPHd. NADPHd is a highly sensitive marker of ChAT fiber distribution in the igl because these fibers contain both ACh and NOS. Fig. 6 summarizes our findings using NADPH-d staining in control and experimental rats at ages P14 and P18. Control data are shown on the left, data from experimental animals on the right. NADPH-d labeled fibers are clearly visible by P14 in both groups. By P18, the two tiers of fibers are clearly segregated and vertical streams of fibers bridge the two tiers (Fig. 6). Again, no qualitative differences could be detected between the control and experimental animals.

This normal refinement of the ACh/NOS fiber tiers occurred despite a significant reduction in NOS activity that was produced by the NoArg inhibitor. This inhibition was dose dependent (see Mize et al., 1997). Reductions in L-citrulline ranged 72.1–88.8% after doses of 1 μmol/g body weight in rats aged P14–P22. A dosage of 3 μmol/g body weight further reduced L-citrulline to between 80.4–91.0% in rats aged P14–P18. This percent inhibition as measured by citrulline assay is known to be sufficient to alter the refinement of retinotectal (Wu et al., 1994) and retinogeniculate fibers (Cramer and Sur, 1996) in mammals. Thus,

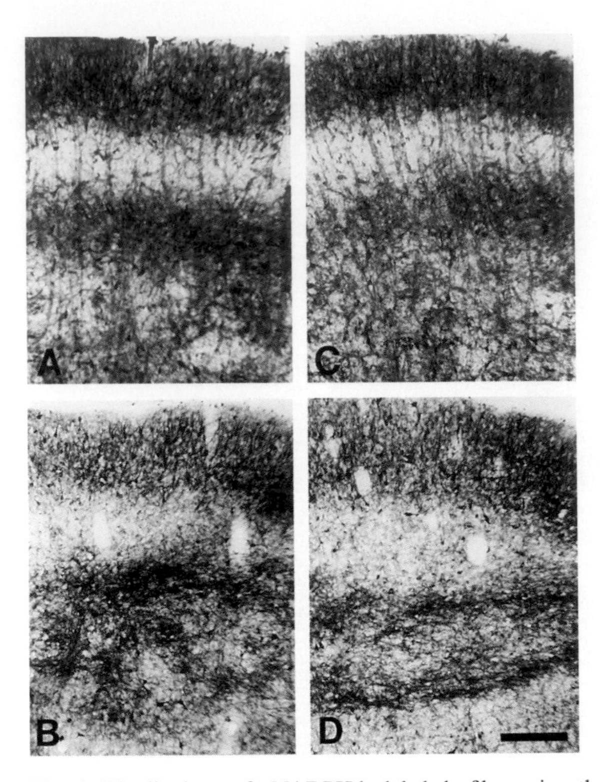

Fig. 6. Distribution of NADPHd labeled fibers in the intermediate gray layer (igl) of the rat SC after inhibition of NOS during development. NADPH-d labeling in fibers within the igl of control (A,B) and experimental (inhibited) animals (C, D) at two ages: P14 (A, B) and P21 (C, D). Note that there are virtually no differences in distribution or density of the labeled fibers between the two groups. Scale bar, A–D = 200 μm.

we conclude that NOS inhibition has no qualitative effect upon the pattern of distribution or density of the cholinergic pathway in the igl of the rat SC.

The ACh fiber pathway is also unaltered in single and double NOS gene knockout mice

The development of the ACH fiber distribution in the igl of SC was also examined in NOS deficient mice. We used both single neuronal (nNOS) and double neuronal and endothelial (n,eNOS) knockout mice (provided by Dr. Paul Huang, Massachusetts General Hospital, Harvard University). The nNOS mutant mouse was produced by deletion of the exon-2 and has been shown to have produced a reduction in NOS activity in excess of 95% (Huang et al., 1993; Huang et al, this volume). The double knockout was produced by crossbreeding the single nNOS knockout mice with mice in which the eNOS gene had been disrupted (Huang et al., 1995; Son et al., 1996).

We failed to find any alteration in the development of the ACh fiber system in the igl of single nNOS or double e,nNOS knockout mice. Fig. 7 shows the pattern of labeling in the igl of normal, single, and double knockouts at P14 and P18. As in rats, there is virtually no detectable difference in the labeling pattern of ACh fibers in the three groups, as visualized by ChAT immunocytochemistry. At P14, the full width of the two tiers is apparent in all three groups, and the distinctive patchiness is not yet visible in either normal or knockout animals. By P18, the patchiness is apparent in both the dorsal and ventral tiers, and again there are no visible differences among the three groups of animals.

Thus, we conclude that neither inhibition of NOS nor disruption of the NOS genes is able to alter the refinement of the cholinergic fibers into tiers and patches within the igl of the rodent SC. This is so despite evidence that the neurons express NOS during the time that this process is taking place.

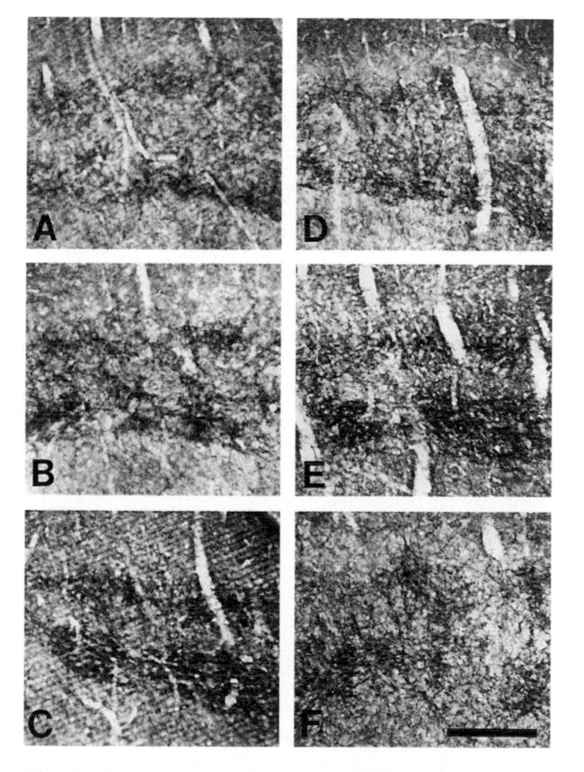

Fig. 7. Comparison of the distribution of acetylcholine containing fibers in the intermediate gray layer (igl) of normal vs single vs double-knockout mouse SC visualized used choline acetyltransferase immunocytochemistry at two ages. P14 (A, B, C) and P18 (C, D, E). (A, D) Normal mouse; (B, E) single nNOS knockout mouse; (C, F) double n and eNOS knockout mouse. As with the inhibition experiments, gene deletion had no effect upon either the density or distribution of the ChAT labeled fibers in the igl. Scale bar: 200 μm.

The retinocollicular pathway in the mammalian superior colliculus

The retinocollicular pathway in mammals includes inputs from both the contralateral and ipsilateral eyes to each tectum. The extent of these pathways varies in different species, depending upon the degree of binocularity. In cat, a frontal eyed species, the overlap is significant with the ipsilateral pathway overlapping the contralateral projection throughout much of the SC, although it is partially segregated from the contralateral input in that it forms patches of input that lie ventral to the contralateral input (Graybiel, 1975, 1976; Harting and Guillery, 1976; Williams and Chalupa, 1982; Behan, 1982; Mize, 1983). In lateral eyed species, such as rodents, the ipsilateral input is much less extensive. Indeed, the two pathways are largely segregated with the ipsilateral pathway limited to the rostral medial colliculus and the contralateral pathway spreading throughout the superficial gray layer (sgl) (Land and Lund, 1979; Godement et al., 1984; Cowan et al., 1984; O'Leary et al., 1986; Simon et al., 1992a). As in cats, the inputs are also segregated dorso-ventrally, with the ipsilateral projection lying ventral to the contralateral input (Godement et al., 1987). The pathways are thus partially segregated in all species examined, and the extent of their overlap varies dramatically with species.

Development of the retinocollicular pathway in rodents

The adult distribution of the contralateral and ipsilateral pathways develops gradually in rodents. In mice, the contralateral projection appears to extend across the entire rostral–caudal and mediolateral axes of SC before birth (Godement et al., 1984, 1987; Edwards et al., 1986; Colello and Guillery, 1990). At this time the ipsilateral pathway is much more extensive than it is in the adult, spreading across the medial–lateral extent of the rostral one-half of SC (Godement et al., 1984). During development the ipsilateral pathway is refined so that by P10–P15 it has retracted to its adult position within the rostromedial segment of SC (Godement et al., 1984).

Previously, Wu et al. (1994) had demonstrated that the ipsilateral retinotectal pathway in chicks, which is a transient projection that disappears during later embryonic stages, is partially spared if NOS is inhibited during this embryonic period of development. Preliminary data from their laboratory also showed that the ipsilateral retinotectal pathway in rats was more extensive after inhibition of NOS (Wu et al., 1996). Thus, there is evidence that the retinotectal pathway, which is glutamatergic (Kvale and Fonnum, 1983; Kvale et al., 1983; Cline et al., 1987; Cline and Constantine-Paton, 1989; Fosse et al., 1989; Simon et al., 1992b), may be influenced by nitric oxide release during development. As a first step in determining whether NO also plays a role in this refinement process in the mouse, we examined the development of the contralateral and ipsilateral pathways in C57/BL6 mice.

Our results show a similar progression to that seen in rats. As Fig. 8 illustrates, the contralateral pathway (Fig. 8, left SC) is densely distributed throughout the SC at all ages examined (P8, P10, P15, and P42). The pathway extends from the surface of SC to a depth of approximately 500 μm in the dorsoventral plane, and from the medial to the lateral edges of SC throughout the rostrocaudal extent of SC (Fig. 8). By contrast, the ipsilateral pathway (Fig. 8, right SC) is confined to a

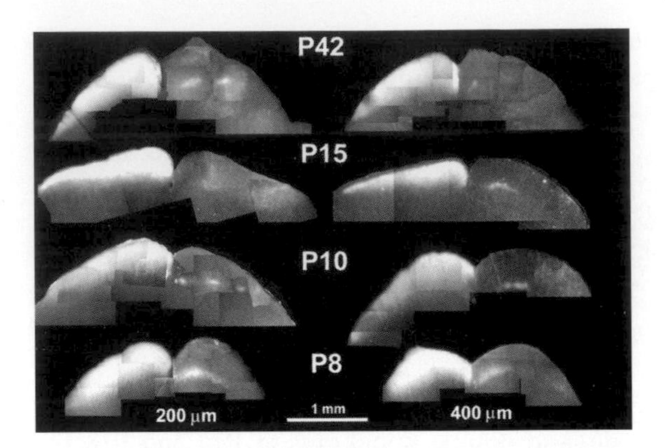

Fig. 8. Montage of contralateral and ipsilateral retinotectal projections in the normal mouse at different ages and different locations within SC. Left: 200 μm from the rostral pole; Right: 400 μm from the rostral pole. The contralateral projection (left SC) is densely and evenly distributed across the entire medial–lateral superficial gray layer. The ipsilateral retinal projection (right SC) is distributed in non-uniform patches across the medial–lateral extent of SC at P8 and P10, but is confined principally to a single medial patch in the rostral SC by P15–P42. Scale bar = 1 mm.

narrow tier of fibers that lie along the ventral border of the contralateral input. In rostral sections (rostral 200 μm, Fig. 8, left), this tier is continuous within the medial SC, with patches of label found more laterally at early ages (P8–P15, Fig. 8, right). More caudally, the ipsilateral input is confined to a rostromedial patch of fibers (400 μm, Fig. 8, right). There is thus a substantial refinement of the ipsilateral pathway to the rostromedial segment of SC by age P8–P10.

This refinement of the ipsilateral pathway occurs prior to the onset of expression of NO. The development of NOS, as revealed using NADPHd histochemistry, shows that neurons that lie within the field of retinal fibers in the sgl of SC first begin to express NOS between P12–P15 (Fig. 5). At P12 there are few cells, and most of these are found at the ventral border of the sgl. By P15, NADPH-d positive cells are found throughout the sgl (Fig. 5). Fig. 9 shows that cells within the sgl are very well labeled by P16, particularly within the caudal SC. These neurons have a variety of morphologies, including cells with a horizontal fusiform morphology (Fig. 9B), and others with a

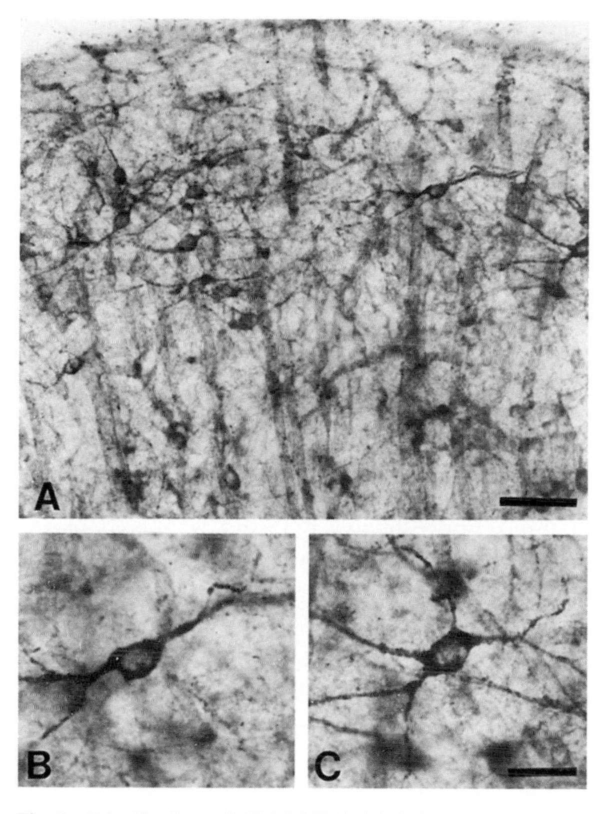

Fig. 9. Distribution of NADPH-d labeled neurons in the superficial gray layer of the mouse superior colliculus at P16. Note the large number of labeled neurons of varying morphology; (B, C) higher magnification views illustrating the morphology of two cells labeled in A. Note the intense labeling of primary and secondary dendrites at this age. Scale bar, A = 50 μm; B, C = 20 μm.

multipolar, stellate-like morphology (Fig. 9C). NADPH-d labeling is retained in adulthood, suggesting that if there is a transient expression of NOS, it is subtle. What is clear is that the onset of expression in the sgl occurs after the contralateral fibers have densely innervated the SC and also well beyond the time at which the ipsilateral pathway has started to retract to the rostromedial SC.

NOS does play a role in the refinement of the ipsilateral retinotectal pathway in mice

Preliminary results from our laboratory show that the refinement of the ipsilateral retinocollicular pathway is nevertheless delayed in knock-out mice during this period of development. This effect occurred in the double eNOS/nNOS knockout but not in the single nNOS knockout model.

We studied the development of the ipsilateral and contralateral retinal pathways by injecting the fluorescent axon tracer DiI into the right eye approximately 48 h before sacrifice. Normal (C57/BL6) and knockout mice were sacrificed at P8, P10, P12, P15, and P18, P21, and P42. Series of sections through the SC of these mice were examined with a fluorescence microscope, and digital photographs of the labeling pattern were taken using a Kodak DCS420 digital camera. From these images, we produced montages of the labeling pattern found in both the ipsilateral and contralateral SC. Adobe Photoshop 4.0 was used for image processing and production of the montages. Sections at approximate 100 μm intervals throughout the SC were examined. In some animals, we also measured the total area occupied by the ipsilateral and contralateral pathways in different sections using a digitizing tablet and MacMeasure software. We traced the outline of the label and expressed the area of the ipsilateral projection as a percent of the total area occupied by the contralateral pathway in order to normalize the data across animals.

Our results to date show no differences in the topographic distribution or total distributional area of the ipsilateral retinocollicular pathway in the normal vs single nNOS knockout mouse. By contrast, the retraction of the ipsilateral pathway was substantially altered in the double knockout model. This alteration is illustrated in Fig. 10 (compare with the pattern in normal mice illustrated in Fig. 8). In the example illustrated (a P15 double knockout), the ipsilateral pathway still extends for considerable distance across the medial–lateral extent of SC for as far caudal as 700 μm. At 500 μm four patches of ipsilateral label can be seen extending at least 60% across the medial–lateral plane of the ventral sgl. The ipsilateral pathway is also more extensive at 700 μm, where at least three separate patches of label can be observed (Fig. 10). This more extensive distribu-

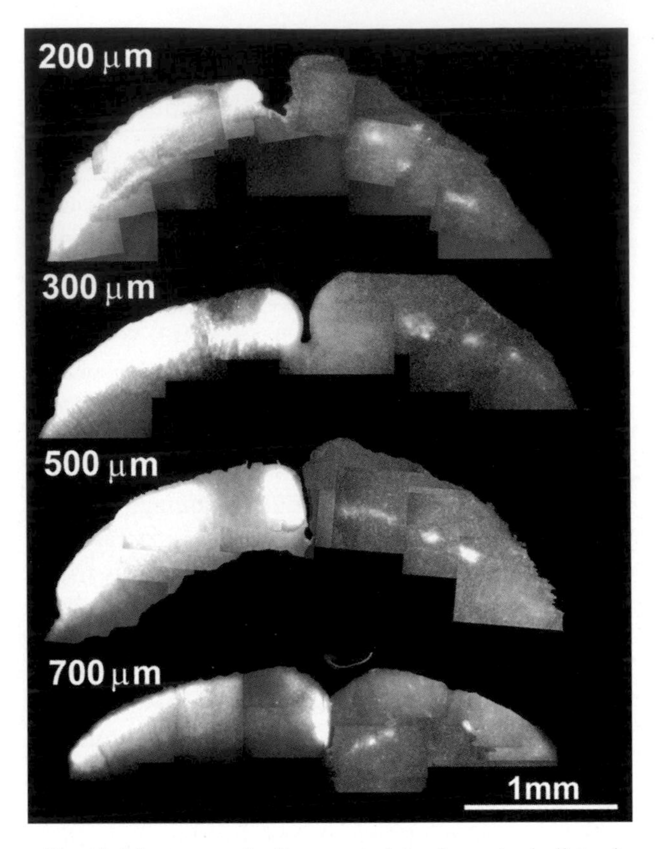

Fig. 10. Montage of the contralateral and ipsilateral retinocollicular projection in double n and eNOS knockout mouse at P15. Dense contralateral retinal input can be seen filling the medial SC as in normal animals. The ipsilateral retinal projection extends broadly across the mediolateral extent of SC throughout the rostral 500 μm and includes at least three patches 700 μm caudal to the rostral pole. The patchy pattern is similar to that seen in normal animals at earlier ages. Rostral–caudal sequence is top to bottom. Scale bar = 1 mm.

tion of label was also observed at P18. Quantitative analysis of the areal distribution of label in one P15 normal and one P15 double-knockout animal confirmed this effect.

The alteration in the double knockout seems to be a delay in the refinement of the ipsilateral pathway rather than a permanent alteration in its distribution. Data that we have collected from older age groups show that the ipsilateral pathway is less extensive at P21 than at earlier ages in the double knockout, and that it is essentially identical to that of normals by P42.

Functional implications

Our results have shown that nitric oxide can play a role in the refinement of axonal pathways in the developing mammal. This role appears to depend both on the type of refinement involved and on the neurotransmitter utilized by the pathway. Thus, the cholinergic fibers that innervate the patch–cluster system in the intermediate gray layer of the rodent SC form fiber tiers even after a substantial reduction of NOS activity has been produced by the inhibitor n-w-L-nitro-arginine or disruption of the genes for both endothelial and neuronal isoforms of NOS have been deleted. By contrast, the development of the glutamate-containing ipsilateral retinocollicular pathway is altered in the double-knockout mouse, suggesting that nitric oxide does play an important role in the refinement of that pathway in rodents.

Because nitric oxide is expressed in the postsynaptic cluster neurons at about the time that the ACh containing fibers arrive in the igl, it remains possible that the NO contained in these postsynaptic neurons does affect the development of other pathways in the system, most probably the projection from the frontal eye fields (FEF) that is thought to be glutamatergic (Dori et al., 1992). This is a difficult hypothesis to test because the FEF is more difficult to manipulate and label than is the cholinergic pathway from the PPTN/LDTN. It is nevertheless an important experiment to perform because it will allow us to confirm the transmitter selectivity of the NO effect.

The mechanism by which we thought NO might direct fiber refinement in the patch–cluster system is illustrated in Fig. 11A. In panel 1, we show the axons from the three sources (GABAergic SN, cholinergic PPTN/LDTN, glutamatergic FEF) growing towards the igl at some arbitrary early age in development. At or before this time, the clustered neurons (illustrated as red circles, Fig. 11A(1)) are already defined by their selective projection to the cuneiform region (Banfro and Mize, 1996). At some arbitrary later time in development (Fig. 11A(2)), the cells begin to release NO which stabilizes the synaptic connec-

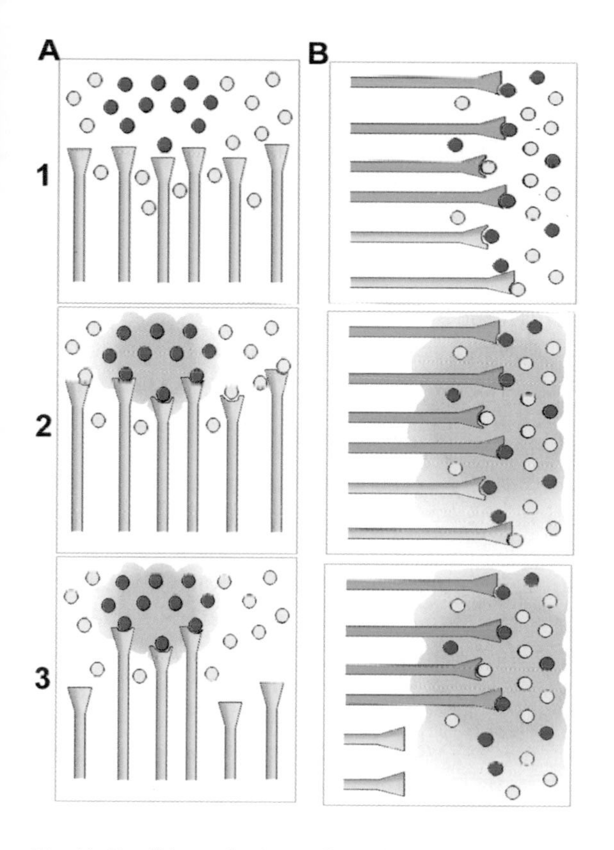

Fig. 11. Possible mechanisms of NO in axon refinement: (A) Synapse stabilization. In this schema, (1) fibers grow diffusely towards patch–cluster neurons (red) and other neurons within the intermediate gray layer (igl) of SC. (2) Patch–cluster neurons contain and release NO, facilitating synaptic stabilization. (3) Those fibers that contact or are near NO releasing cells form stable synapses. Those that do not retract or die. This mechanism is consistent with the developmental sequence of the patch–cluster system in the igl of SC, although our results show that NO is not required for the ACh pathway to form patches. (B) Synapse elimination. In this schema, (1) fibers grow exuberantly towards and reach many cells in the superficial gray layer (sgl), some of which contain NOS; (2) these cells express NOS transiently and release NO throughout the sgl. Fibers with correlated activity (green) form stable synapses on both NOS containing and other cell types. Fibers with uncorrelated activity (yellow) retract, a phenomenon also promoted by NO.

tions of the axons in the vicinity of the clustered neurons. The axons that do not encounter NO are eliminated because they have not been exposed to this substance. In this scheme, NO plays a facilitatory role in strengthening the appropriate synapses. The absence of NO in neurons outside the clusters prevents synapse formation in these regions of SC because NO is assumed to have a finite diffusion (Wood and Garthwaite, 1994).

This model requires that nitric oxide be topographically constrained, i.e. that NO be contained in target neurons that have a discrete distribution. Because of this selective distribution it is possible for NO to influence axons in a spatially constrained space (i.e. the patches) without affecting those axons that are remote from the NO releasing cells. In such a model, NO can serve to facilitate the synaptic connections of axons that contact NOS neurons without the need for NO to weaken other synapses. The fibers not contacting NOS containing cells simply die away because they lack the agent necessary to strengthen the synapse. This model does not require correlated electrical activity in the afferents, although some spontaneous electrical activity would presumably be required in order to promote Ca^{2+} influx and production of NO.

The refinement of the ipsilateral retinocollicular pathway requires more complex assumptions. A possible role for NO in retinocollicular development is illustrated in Fig. 11B. The retinocollicular system differs from that of the patch–cluster system in several respects. For one, the postsynaptic NOS containing neurons are not spatially constrained. NADPHd labeled neurons are scattered throughout the region of the sgl in which both the contralateral and ipsilateral retinal afferents terminate. There is no selective distribution which could provide a spatial cue as to which axons connect to which cells. This fact is illustrated in Fig. 11B(1) where NOS containing cells (red circles, Fig. 11B1) are shown to be scattered amongst non-NOS containing cells throughout the region in which both contralateral (green axons, Fig. 11B) and ipsilateral (yellow axons, Fig. 11B) afferents terminate.

One implication of this arrangement is that both contralateral and ipsilateral afferents (i.e. those that stay and those that retract) are exposed to diffusible NO, as shown in Fig. 11B(2). An additional factor is therefore required to explain why the contralateral afferents stay while the ipsilateral afferents retract. Gally et al. (1990) hypothesized that the additional factor is the presence of correlated activity in some of the presynaptic fibers (the contralateral pathway) and uncorrelated activity in others (the ipsilateral pathway). This effect is illustrated in Fig. 11B(3) where correlated activity (green axons) strengthens synaptic contacts while uncorrelated activity weakens them (yellow axons), resulting in axon elimination.

Evidence that NO can facilitate synaptic transmission (and thus stabilize synapses) comes largely from models of associative plasticity, including long term potentiation (LTP) in the hippocampus and long term depression (LTD) in cerebellum. Several studies have shown that NO donors can facilitate LTP while drugs that inhibit NOS or scavengers which bind NO (such as haemoglobin) can block LTP in the CA1 region of the hippocampus (Haley et al., 1992; Williams et al., 1993; Arancio et al., 1996a, b). LTD has also been shown to be altered by either application of NO donors or by inhibition of NOS (Izumi and Zorumski, 1993; Gage et al., 1997). The mechanism for facilitation of LTP by NO has been shown to involve guanylate cyclase and 3–5 cyclic GMP as the presynaptic mediator in some but not all LTP paradigms. Thus, in hippocampal pyramidal cell cultures, inhibitors of either guanylate cyclase or cGMP-dependent protein kinase can block LTP at some stimulation frequencies, while application of cGMP analogs (8-Br-cGMP) or direct injection of cGMP into presynaptic neurons can promote LTP facilitation (Arancio et al., 1995). It is generally thought that this presynaptic effect results in increases in the release of glutamate from the presynaptic terminal (Bliss and Collingridege, 1993; Hawkins et al., 1994). However, other forms of LTP appear to be independent of cGMP mediated events (Schuman et al., 1994), suggesting that other signal transduction pathways may also be involved in this form of plasticity.

There is much less evidence that NO serves the same facilitatory role in synaptic plasticity in the developing brain. NO has been implicated in the segregation of "on" and "off" retinal afferents into sublaminae in the ferret lateral geniculate nucleus (LGN), based upon the following evidence: (1) LGN cells transiently express NOS during the time at which these two pathways form distinct sublaminae; (2) inhibition of NOS blocks the development of these segregated sublaminae (Cramer et al., 1996); (3) NOS inhibition alters the distribution of individual axon arbors within the sublaminae (Cramer et al., 1996, Cramer and Sur, 1996). Although the mechanism of NO in this segregation is unknown, it likely involves the NMDA receptor, because NMDA receptor inhibition also blocks "on" and "off" pathway segregation in this species (Hahm et al., 1991). Correlated electrical activity in "on" fibers in the "on" sublamina and in "off" fibers in the "off" sublamina could promote this selective synapse stabilization. Consistent with this idea, differential waves of spontaneous electrical activity in "on" vs "off" retinal ganglion cells has been suggested to be the mediator of this effect in ferret LGN (see Cramer et al., this volume). If the segregation is due to correlated electrical activity, this could only occur if there is an initial preponderance of "on" and "off" fibers in the appropriate sublaminae. Whether this occurs is not yet established.

NO has also been shown to facilitate early stages of neurite outgrowth in both invertebrates and mammals. In *Drosophila*, evidence that NO modulates growth of the central projections of photoreceptors includes: (1) the optic lobes express both NADPH and nNOS between 24–40 h after metamorphosis; (2) NOS inhibitors and NO scavengers at this age cause photoreceptor axons to become disorganized and project beyond their normal target; (3) inhibitors of soluble guanylate cyclase also disrupt the growth of these axons, while the cyclic GMP analog 8-Br-cGMP prevents the disruption (Gibbs and Truman, 1998). A similar guanylate cyclase-cGMP mediated effect has been reported in mouse hippocampal cultures and in PC12 cells in the presence of NGF (Hindley et al., 1997). Evidence for this includes: (1) NO donors promote neurite outgrowth in both hippocampus and in PC12 cells; (2) NO scavengers block neurite outgrowth in PC 12 cells, but only in the presence of NGF; (3) NO donors increase intracellular cGMP in PC12 cells as do cGMP analogs; and (4) inhibitors of guanylate cyclase inhibit outgrowth. Thus, evidence from both invertebrate and mammalian models shows that the facilitatory effect of NO is likely mediated through a presynaptic guanylate cyclase-cGMP signal transduction pathway.

NO has also been shown to induce retraction or elimination of pathways in the developing visual system. The best studied of these models is the retinotectal pathway in chicks. In this species, the ipsilateral retinotectal pathway is transient during embryonic development. It is initially exuberant, projecting to large portions of the ipsilateral tectum, but disappears in the later embryonic stage. Evidence that NO is involved in this elimination is: (1) NOS is transiently expressed in cells in the chick optic tectum at the time that the ipsilateral retinal afferents innervate the tectum (Williams et al., 1994); (2) monocular enucleation reduces NOS activity in these neurons, suggesting that NO in these cells is regulated by retinal input (Williams et al., 1994; see also Vercelli and Cracco, 1994; Vercelli et al., 1995); (3) NOS inhibitors partially block the retraction of the ipsilateral retinotectal pathway (Wu et al., 1994, 1996). The sparing of the ipsilateral pathway is only partial, and it is not known whether the sparing is permanent or whether the retraction is merely delayed. The mechanism for this effect is unknown, but probably involves the NMDA receptor because pharmacological block of the NMDA receptor also limits the retraction of the ipsilateral pathway (Ernst et al., 1997; Ernst et al., this volume).

That NO plays a role in pathway retraction and synapse elimination has received substantial recent experimental support. Hess et al. (1993) have shown that the outgrowth of neurites in cultured xenopus dorsal root ganglion cells is inhibited by NO. Thus: (1) NO donors produce an initial arrest of elongation and growth cone collapse, followed by actual retraction of the neurite; (2) the effect is reversed by the binding of released NO with hemoglobin or its exhaustion prior to application of the donor; (3) the retraction is not dependent on guanylate cyclase-cGMP because neurites grow normally after treatment with cGMP analogues. The effect appears instead to be due to the inhibition of palmytoylation by NO by directly affecting protein long-chain fatty acetylation.

NO also promotes fiber retraction in xenopus retinal ganglion cells (RGCs) (Renteria and Constantine-Paton, 1996). (1) Application of a variety of NO donors in whole retinal explant cultures produces RGC growth cone collapse and retraction; (2) binding of NO with the scavenger oxyhemoglobin blocks the effect; (3) ferricyanide, peroxynitrite, nitrite, and cysteine do not produce collapse, suggesting that the active agent is NO and not a by-product of its decomposition. (4) Cyclic GMP analogs also failed to induce growth cone collapse, suggesting that the effect of NO is not guanylate cyclase-cGMP dependent. Whether this effect represents fiber elimination or mere cessation of growth is uncertain because the initial step in both synapse elimination and stabilization requires that axons stop the process of outgrowth.

The relationship between neurite retraction and synaptic activity has recently been elegantly demonstrated in nerve muscle cultures by Wang et al. (1995). In these experiments: (1) NO donors produced a significant decrease in both spontaneous and evoked synaptic currents at myocyte synapses; (2) asynchronous firing produced by depolarizing the myocyte also induced suppression of the amplitude of the evoked synaptic current; (3) NO synthase inhibitors abolished the suppression, an effect reversed by application of hemoglobin; (4) the same effect could be obtained by intracellular myocyte injection of an NOS inhibitor, suggesting that the source of NO was postsynaptic; (5) other experiments showed that active terminals were less susceptible to the suppression since the NO donor SNAP had no effect on the synaptic currents of the most active presynaptic terminals. The authors concluded from these studies that NO can serve as a retrograde messenger that suppresses presynaptic currents in fibers and cells that are firing out of synchrony while preserving those that are firing in synchrony. The effect is probably cGMP dependent because a cGMP-dependent protein kinase activator also decreased the spontaneous and evoked currents in this study.

That NO could induce retraction of fibers in addition to or instead of stabilization is an attractive hypothesis because it can explain our results in the double-knockout mouse where NO appears to promote retraction of the ipsilateral retinocollicular pathway while the contralateral fibers are strengthened. Whether NO promotes retraction in mammals other than mouse is as yet unclear, although McLoon and colleagues (Wu et al., 1996) have pilot data showing that NO inhibitors can block the segregation of ipsilateral and contralateral retinal afferents in the rat SC, a result similar to that shown in our studies in double-knockout mice. As in the results with chick (Wu et al., 1994, 1996), it is difficult to explain the retraction of the ipsilateral retinotectal pathway without positing an effect that promotes retraction and elimination of axons. Whether NO also serves to stabilize synapses in the contralateral retinocollicular pathway in which there is correlated activity must still be determined.

Although evidence shows that NO helps regulate pathway refinement, it is clearly not the only molecule to do so. Thus, in chick, significant inhibition of NOS results in only a 20% sparing of the transient ipsilateral pathway in the tectum. This partial effect is not sufficient to explain the complete elimination of this pathway in this species (Wu et al., 1994). Other factors that may play a role include the neurotrophins (Cabelli et al., 1995; McAllister et al., 1995; Riddle et al., 1995). In addition, NO appears to play no role in the refinement of some pathways. Thus, NOS inhibition does not prevent the ocular dominance shift that occurs when one eye is occluded during development in cat cortex (Ruthazer et al., 1996). The variety of other molecules that participate in pathway refinement is an area of active investigation in our and other laboratories.

Summary and conclusions

Our studies began with the intent to test whether nitric oxide plays a role in pathway refinement and synapse stabilization in the developing rodent superior colliculus, and whether that role is specific to the neurotransmitter utilized by the pathway. We used two model systems, one in which NOS is expressed in a spatially restricted cell group that receives afferents from cholinergic neurons (the patch–cluster system), the other in which the NOS expressing target neurons are not spatially restricted and in which the afferents utilize glutamate as a neurotransmitter (the retinocollicular pathway). Our results have shown that neither NOS inhibition in rats nor endothelial and neuronal gene deletion in double-knockout mice alter the characteristic pattern of patches and tiers formed by cholinergic fibers in rodent SC. This failure occurred despite the fact that NOS containing neurons first express NOS at about the time that the ACh fibers establish connections in the igl. By contrast, our results show that the development of the ipsilateral retinocollicular pathway in SC is affected by NO because retraction of the ipsilateral pathway to the rostromedial segment of SC is delayed for at least a week in the double-knockout model.

Although our results demonstrate an effect in a glutamatergic but not in a cholinergic pathway, we must test further whether the effect of NOS is neurotransmitter specific. In particular, it will be important to determine whether the glutamatergic pathway from the frontal eye fields to the patch–cluster system develops normally or is disrupted after NOS inhibition or in the double-knockout mouse. Our results also provide another example in which NO may serve to eliminate an exuberant pathway in which correlated activity is absent. It will be useful to extend these experiments to determine whether NO also plays a facilitatory role in stabilizing connections in which correlated activity is present, such as in the contralateral retinocollicular projection.

Acknowledgements

We thank Brett Wilson and Marly Perrone for invaluable assistance in some of the histochemistry experiments. Cathy Vial and Michelle St Onge assisted in preparation of the figures. Drs. John Haycock and Robert Roskoski and Michael Salvatore were instrumental in completing the biochemical assays of NOS. Marly Perrone, Deborah Schumann, and Brett Wilson also assisted in the computer microscope mapping of the distribution of NOS neurons and in producing the computer montages of DiI labeled SC. This research was supported by USPHS grant NS-36000 from the National Institute of Neurological Disorders and Stroke, EY-02973 from the National Eye Institute, and US Army Research and Development Grant DAMD 17-93-V3013-P-40001. Christopher Scheiner was supported by a Superior Graduate Student Fellowship from the Board of Regents, State of Louisiana (LEQSF Grant (95-00)-GF-12).

References

Arancio, O., Kandel, E.R. and Hawkins, R.D. (1995) Activity-dependent long-term enhancement of transmitter release by presynaptic 3′,5′-cyclic GMP in cultured hippocampal neurons. *Nature*, 376: 74–80.

Arancio, O., Kiebler, M., Lee, C.J., Lev-Ram, V., Tsien, R.Y. and Kandel, E.R. (1996a) Nitric oxide acts directly in the presynaptic neuron to produce long-term potentiation in cultured hippocampal neurons. *Cell*, 87: 1025–1035.

Arancio, O. Lev-Ram, V., Tsien, R.Y., Kandel, E.R. and Hawkins, R.D. (1996) Nitric oxide acts as a retrograde messenger during long-term potentiation in culture hippocampal neurons. *J. Physiol. Paris.*, 90: 321–322.

Baader, S.L., Bucher, S. and Schilling, K. (1997) The development expression of neuronal nitric oxide synthase in cerebellar granule cells is sensitive to GABA and neurotrophins. *Dev Neurosci.*, 19: 283–290.

Banfro, F.T. and Mize, R.R. (1996) The clustered cell system is present before formation of the Ach patches in the intermediate gray layer of the cat superior colliculus. *Brain Res.*, 733: 273–283.

Behan, M. (1982) A quantitative analysis of the ipsilateral retinocollicular projection in the cat: An EM degeneration and EM autoradiographic study. *J. Comp. Neurol.*, 206: 253–258.

Beninato, M. and Spencer, R.F. (1986) Cholinergic projections to the rat superior colliculus demonstrated by retrograde transport of horseradish peroxidase and choline acetyltransferase immunocytochemistry. *J. Comp. Neurol.*, 253: 525–538.

Bliss, T.V.P. and Collingridge, L.M. (1993) A synaptic model of memory, long-term potentiation in the hippocampus. *Nature*, 361: 31–39.

Bohme, G.A., Bon, C., Lemarie, M., Reibaud, M., Piot, O., Stutzman, J.M., Doble, A. and Blanchard, J.C. (1993) Altered synaptic plasticity and memory formation in nitric oxide inhibitor-treated rats. *Proc. Natl. Acad. Sci.*, 90: 9191–9194.

Bohme, G.A., Bon, C., Stutzmann, J.M., Doble, A. and Blanchard, J.C. (1991) Possible involvement of nitric oxide in long-term potentiation. *Eur. J. Pharmacol.*, 199: 379–381.

Bredt, D.S. and Snyder, S.H. (1989) Nitric oxide mediates glutamate-linked enhancement of cGMP levels in the cerebellum. *Proc. Natl. Acad. Sci. USA*, 86: 9030–9033.

Bredt, D.S. and Snyder, S.H. (1990) Isolation of nitric oxide synthase, a calmodulin-requiring enzyme. *Proc. Natl. Acad. Sci. USA*, 87: 682–685.

Cabelli, R.J., Hohn, A. and Shatz, C.J. (1995) Inhibition of ocular dominance column formation by Infusion of NT-4/5 or BDNF. *Science*, 267: 1662–1666.

Cline, H.T., Debski, E.A. and Constantine-Paton, M. (1987) NMDA receptor antagonist desegregates eye-specific stripes. *Proc. Natl. Acad. Sci. USA*, 84: 4342–4345.

Cline, H.T. and Constantine-Patton, M. (1989) NMDA receptor antagonists disrupt the retinotectal topographic map. *Neuron*, 3: 413–426.

Colello, R.J. and Guillery, R.W. (1990) The early development of retinal ganglion cells with uncrossed axons in the mouse: Retinal position and axonal course. *Development*, 108: 513–253.

Cowan, W.M., Fawcett, J.W., O'Leary, D.D.M. and Stanfield, B.B. (1984) Regressive events in neurogenesis. *Science*, 225: 1258–1265.

Cramer, K.S., Angelucci, A., Hahm, J., Bogdanov, M.B. and Sur, M. (1996) A role of nitric oxide in the development ferret retinogeniculate projection. *J. Neurosci.*, 16: 7995–8004.

Cramer, K.S. and Sur, M. (1996) The role of NMDA receptors and nitric oxide in retinogeniculate development. *Prog. Brain Res.*, 108: 235–244.

Daniel, H., Hemart, N., Jaillard, D. and Crepel, F. (1993) Long-term depression requires nitric oxide and quanosine 3′:5′ cyclic monophosphate production in rate cerebellar Purkinje cells. *Eur. J. Neurosci.*, France, 5(5): 1079–1082.

Dawson, T.M., Dawson, V.L. and Snyder, S.H. (1992) A novel neuronal messenger molecule in brain: The free radical, nitric oxide. *Ann. Neurol.*, 32: 297–311.

Dori, I., Dinopoulos, A., Cavanaugh, M.E. and Parnavelas, J.G. (1992) Proportion of glutamate- and aspartate-immunoreactive neurons in the efferent pathways of the rat visual cortex varies according to the target. *J. Comp. Neurol.*, 319: 191–204.

Edwards, M.A., Caviness, V.S. and Schneider, G.E. (1986) Development of cell and fiber lamination in the mouse superior colliculus. *J. Comp. Neurol.*, 248: 395–409.

Ernst, A.F., Wu, H.H., El-Fakahany, E.E. and McLoon, S.C. (1997) NMDA receptor mediated refinement of a transient retinotectal projection is dependent on nitric oxide synthase. *Soc. Neurosci. Abstr.*, 23: 1975.

Ficalora, A.N. and Mize, R.R. (1989) The neurons of the substantia nigra and zona incerta which project to the cat superior colliculus are GABA immunoreactive: A double-label study using GABA immunocytochemistry and lectin retrograde transport. *Neuroscience*, 29: 567–581.

Fosse, V.M., Heggelund, P. and Fonnum, P. (1989) Postnatal development of glutamatergic, GABAergic, and cholinergic neurotransmitter phenotypes in the visual cortex, lateral geniculate nucleus, pulvinar, and superior colliculus in cats. *J. Neurosci.*, 9: 426–435.

Gage, A.T., Reyes, M. and Stanton, P.K. (1997) Nitric-oxide-guanylyl-cyclase-dependent and -independent components of multiple forms of long-term synaptic depression. *Hippocampus*, 7: 286–295.

Gally, J.A., Montague, P.R., Reeke, Jr., G.N. and Edelman, G.M. (1990) The NO hypothesis: Possible effects of a short-lived, rapidly diffusible signal in the development and function of the nervous system. *Proc. Natl. Acad. Sci. USA*, 87: 3547–3551.

Garthwaite, J. (1991) Glutamate, nitric oxide and cell–cell signaling in the nervous system. *Trends Neurosci.*, 14: 60–67.

Garthwaite, J., Charles, S.L. and Chess-Williams, R. (1988) Endothelium-derived relaxing factor release on activation of NMDA receptors suggests role as intracellular messenger in brain. *Nature*, 336: 385–388.

Gibbs, S.M. and Truman, J.W. (1998) Nitric oxide and cyclic GMP regulate retinal patterning in the optic lobe of Dosophila melanogaster. *Neuron*, 20: 83–93.

Godement, P. Salaun, J. and Imbert, M. (1984) Prenatal and postnatal development of retinogeniculate and retinocollicular projections in the mouse. *J. Comp. Neurol.*, 230: 552–575.

Godement, P., Salaun, J. and Metin, C. (1987) Fate of uncrossed retinal projections following early or late prenatal monocular enucleation in the mouse. *J. Comp. Neurol.*, 255: 97–109.

Graybiel, A.M. (1975) Anatomical organization of retinotectal afferents in the cat: An autoradiographical study. *Brain Res.*, 96: 1–23.

Graybiel, A.M. (1976) Evidence for banding of the cat's ipsilateral retinotectal connection. *Brain Res.*, 114: 318–327.

Hahm., J.-O., Langdon, R.B. and Sur, M. (1991) Disruption of retinogeniculate afferent segregation by antagonists to NMDA receptors. *Nature*, 351: 568–570.

Haley, J.E., Wilcox, G.L., and Chapman, P.F. (1992) The role of nitric oxide in hippocampal long-term potentiation. *Neuron*, 8: 211–216.

Hall, W.C., Fitzpatrick, D., Klatt, L.L. and Rackowski, D. (1989) Cholinergic innervation of the superior colliculus in the cat. *J. Comp. Neurol.*, 287: 495–514.

Harting, J.K. and Guillery, R.W. (1976) Organization of retinocollicular pathways in the cat. *J. Comp. Neurol.*, 166: 133–144.

Harting, J.K. and Lieshout, D.P.V. (1991) Spatial relationship of axons arising from the substantia nigra, spinal trigeminal nucleus and pedunculopontine tegmental nucleus within the intermediate gray of the cat superior colliculus. *J. Comp. Neurol.*, 305: 543–558.

Hawkins, R.D., Zhuo M. and Arancio, O. (1994) Nitric oxide and carbon monoxide as possible retrograde messengers in hippocampal long-term potentiation. *J. Neurobiol.*, 25: 652–655.

Hess, D.T., Patterson, S.I., Smith, D.S. and Skene J.H.P. (1993) Neuronal growth cone collapse and inhibition of protein fatty acylation by nitric oxide. *Nature*, 366: 562–565.

Hindley, S., Juurlink, B.H.J., Gysbers, J.W., Middlemiss, P.J., Herman, M.A.R.and Rathbone, M.P. (1997) Nitric oxide donors enhance neurotrophin-induced neurite outgrowth through a cGMP-dependent mechanism. *J. Neurosci. Res.*, 47: 427–439.

Huang, P.L., Dawson, T.M., Bredt, D.S., Snyder, S.H. and Fishman, M.C. (1993) Targeted disruption of the nitric oxide synthase gene. *Cell*, 75: 1273–1286.

Huang, P.L., Huang, Z., Mashimo, H., Bloch, K.D., Moskowitz, M.A., Bevan, J.A. and Fishman, M.C. (1995) Hypertension in mice lacking the gene for endothelial nitric oxide synthase. *Nature*, 377: 239–242.

Illing, R.B. (1990) Choline acetyltransferase-like immunoreactivity in the superior colliculus of the cat and its relation to the pattern of acetylcholinetransferase staining. *J. Comp. Neurol.*, 296: 32–46.

Illing, R.B. and Graybiel, A.M. (1985) Convergence of afferents from frontal cortex and substantia nigra onto acetylchoilinesterase-rich patches of the cat's superior colliculus. *Neuroscience*, 14: 455–482.

Izumi Y., Clifford, D.B. and Zorumski, C.F. (1992) Inhibition of long-term potentiation by NMDA-mediated nitric oxide release. *Science*, 257: 1273–1276.

Izumi, Y. and Zorumski, C.F. (1993) Nitric oxide and long-term synaptic depression in rat hippocampus. *Neuroreport*, 4: 1131–1134.

Jaffery, S.R. and Snyder, S.H. (1995) Nitric Oxide: A neural messenger. *Ann. Rev. Cell Dev. Biol.*, 11: 417–440.

Jeon, C.-J. and Mize, R.R. (1993) Choline acetyltransferase immunoreactive patches overlap specific efferent cell groups in the cat superior colliculus. *J. Comp. Neurol.*, 337: 127–150.

Kvale, I. and Fonnum, F. (1983) The effects of unilateral neonatal removal of visual cortex on transmitter parameters in the adult superior colliculus and lateral geniculate body. *Brain Res.*, 313: 261–266.

Kvale, I., Fosse, V.M. and Fonnum, P. (1983) Development of neurotransmitter parameters in lateral geniculate body, superior colliculus, and visual cortex of the albino rat. *Brain Res.*, 283: 137–145.

Land, P.W. and Lund, R.D. (1979) Development of the rat's uncrossed retinotectal pathway and its relation to plasticity studies. *Science*, 205: 698–700.

Lund, R.D. and Lund, J.S. (1971) Synaptic adjustment after deafferentation of the superior colliculus of the rat. *Science*, 171: 804–807.

Malen, P.L. and Chapman, P.F. (1997) Nitric oxide facilitates long-term potentiation, but not long-term depression. *J. Neuroscience*, 17: 2645–2651.

McAllister, A.K., Lo, D.C. and Katz, L.C. (1995) Neurotrophins regulate dendritic growth in developing visual cortex. *Neuron*, 15(4): 791–803.

McHaffie, J., Beninato, M., Stein, B.E. and Spencer, R.F. (1991) Postnatal development of acetylcholinesterase and cholinergic projections to the cat superior colliculus. *J. Comp. Neurol.*, 313: 113–131.

Mize, R.R. (1983) Patterns of convergence and divergence of retinal and cortical synaptic terminals in the cat superior colliculus. *Exp. Brain Res.*, 51: 88–96.

Mize, R.R., Banfro, F.T. and Scheiner, C.A. (1996) Pre- and postnatal expression of amino acid neurotransmitters, calcium binding proteins, and nitric oxide synthase in the developing superior colliculus. *Prog. Brain Res.*, 108: 313–332.

Mize, R.R., Scheiner, C.A., Salvatore, M.F. and Cork, R.J. (1997) Inhibition of nitric oxide synthase fails to disrupt the development of cholinergic fiber patches in the rat superior colliculus. *Dev. Neurosci.*, 19: 260–273.

Montague, P.R., Gally, J.A. and Edelman, G.M. (1991) Spatial signaling in the development and function of neural connections. *Cereb. Cortex*, 1: 199–220.

Muller, U. and Bicker, G. (1994) Calcium-activated release of nitric oxide and cellular distribution of nitric oxide-synthesizing neurons in the nervous system of the locust. *J. Neurosci.*, 14: 7521–7528.

Nelson, R.J., Demas, G.E., Huang, P.L., Fisnman, M.C., Dawson, V.L., Dawson, T.M. and Snyder, S.H. (1995) Behavioural abnormalities in male mice lacking neuronal nitric oxide synthase. *Nature*, 378: 383–386.

O'Dell, T.J., Hawkins, R.D., Kandel, E.R. and Arancio, O. (1991) Tests of the roles of two diffusible substances in long-term potentiation: Evidence for nitric oxide as a possible early retrograde messenger. *Proc. Natl. Acad. Sci.*, 88: 11285–11289.

O'Leary, D.D.M., Fawcett, J.W. and Cowan, W.M. (1986) Topographic targeting errors in the retinocollicular projections and their elimination by selective ganglion cell death. *J. Neurosci.*, 6: 3692–3705.

Owens, D.F., Boyce, L.H., Davis, M.B.E. and Kriegstein, A.R. (1996) Excitatory GABA responses in embryonic and neonatal cortical slices demonstrated by gramicidin perforated-patch recordings and calcium imaging. *J. Neurosci.*, 16(20): 6414–6423.

Renteria, R.C. and Constantine-Patron, M. (1996) Exogenous nitric oxide causes collapse of retinal ganglion cell axonal growth cones in vitro. *J. Neurobiol.*, 29: 415–428.

Riddle, D.R., Lo, D.C. and Katz, L.C. (1995) NT-4-Mediated rescue of lateral geniculate neurons from effects of monocular deprivation. *Nature*, 378(6553): 189–191.

Ruthhazer, E.S., Gillespie, D.C., Dawson, T.M., Synder, S.H. and Stryker, J.P. (1996) Inhibition of nitric oxide synthase does not prevent ocular dominance plasticity in kitten visual cortex. *J. Physiol.*, 494.2: 519–527.

Scheiner, C.A., Banfro, F.T., Arceneaux, R.D., Kratz, K.E. and Mize, R.R. (1995) Nitric oxide is expressed early in the prenatal development of the cat superior colliculus. *Soc. Neurosci. Abstr.*, 21: 817.

Scheiner, C.A. and Mize, R.R. (1998) Cholinergic fiber patches in the mouse superior colliculus develop normally in e,n nitric oxide synthase knockout mice. *FASEB J. Abstr.*, 12: A757.

Schuman, E.M. and Madison, D.V. (1991) A requirement for the intracellular messenger nitric oxide in long-term potentiation. *Science*, 254: 1503–1506.

Schuman, E.M., Meffert, M.K., Schulman, H. and Madison, D.V. (1994) An ADP-ribosyltransferase as a potential target for nitric oxide action in hippocampal long-term potentiation. *Proc. Natl. Acad. Sci.*, 91: 11958–11962.

Simon, D.K. and O'Leary, D.D.M. (1992a) Development of topographic order in the mammalian retinocollicular projection. *J. Neurosci.*, 12: 1212–1232.

Simon, D.K. and O'Leary, D.D.M. (1992b) Influence of position along the medial–lateral axis of the superior colliculus on the topographic targeting and survival of retinal axons. *Dev. Brain Res.*, 69: 167–172.

Shibuki, K. (1993) Cerebellar long-term depression enabled by nitric oxide, a diffusible intercellular messenger. *Ann NY Acad Sci.*, 707: 521–523.

Snyder, S.H. (1992) Nitric oxide and neurons. *Curr. Opin. Neurobiol.*, 2: 323–327.

Son, H.D., Martin, K., Kiebler, M., Huang, P.L., Fishman, M.C. and Kandel, E.R. (1996) Long-term potentiation is reduced in mice that are doubly mutant in endothelial and neuronal nitric oxide synthase. *Cell*, 87: 1015–1023.

Vercelli, A., Biasiol, S. and Jhaveri, S. (1995) Development of NADPH-diaphorase activity in the superficial layers of the rat superior colliculus (SC): Effect of eye enucleation and of activity. *Soc. Neurosci. Abstr.*, 21: 817.

Vercelli, A.E. and Cracco, C.M. (1994) Effects of eye enucleation on NADPH-diaphorase positive neurons in the superficial layers of the rat superior colliculus. *Brain Res. Dev.*, 83: 85–98.

Wang, S.Z., Lee, S.Y., Zhu, S.Z. and El-Fakahany, E.E. (1996) Differential coupling of m_1, m_3, and m_5 muscarinic receptors to activation of neuronal nitric oxide synthase. *Pharmacology*. 53: 271–280.

Wang, T., Xie, Z. and Lu, B. (1995) Nitric oxide mediates activity-dependent synaptic suppression at developing neuromuscular synapses. *Nature*, 374: 262–266.

Williams, C.V., Nordquist, D. and McLoon, S.C. (1994) Correlation of nitric oxide synthase expression with changing patterns of axonal projections in the developing visual system. *J. Neurosci.*, 14(3): 1746–1755.

Williams, J.H., Li, Y.G., Nayak, A., Erington, M.L., Murphy, K.P., Bliss, T.V. (1993) The suppression of long-term potentiation in rat hippocampus by inhibitors of nitric oxide synthase is temperature and age dependent. *Neuron*, 11: 877–884.

Williams, W.R. and Chalupa, L.M. (1982) Prenatal development of retinocollicular projections in the cat: An anterograde tracer transport study. *J. Neurosci.*, 2: 604–622.

Wood, J. and Garthwaite, J. (1994) Models of the diffusional spread of nitric oxide: Implications for neural nitric oxide signalling and its pharmacological properties. *Neuropharmacology*, 33(11): 1235–1244.

Wu, H.H., Williams, C.V. and McLoon, S.C. (1994) Involvement of nitric oxide in the elimination of a transient retinotectal projection in development. *Science*, 265: 1593–1596.

Wu, H.H., Waid, D.K. and McLoon, S.C. (1996) Nitric Oxide and the developmental remodeling of retinal connections in brain. *Prog. Brain Res.*, 108: 273–286.

Wu, H.H., Wilson, B., Cork, R.J. and Mize, R.R. (1998) Refinement of the ipsilateral retinocollicular projection is unaffected in neuronal NOS gene knockout mice. *FASEB J. Abstr.*, 12: A578.

Zhuo, M., Small, S.A., Kandel, E.R. and Hawkins, R.D. (1993) Nitric oxide and carbon dioxide produce activity dependent long-term synaptic enhancement in hippocampus. *Science*, 260: 1946–1950.

Zhou, M., Kandel, E.R. and Hawkins, R.D. (1994) Nitric oxide and cGMP can produce either synaptic depression or potentiation depending on the frequency of presynaptic stimulation in the hippocampus. *NeuroReport*, 5: 1033–1036.

SECTION IV

Nitric oxide in synaptic plasticity

R.R. Mize, T.M. Dawson, V.L. Dawson and M.J. Friedlander (Eds.)
Progress in Brain Research, Vol 118

CHAPTER 11

Nitric oxide as a retrograde messenger during long-term potentiation in hippocampus

Robert D. Hawkins[1,2,*], Hyeon Son[1,†] and Ottavio Arancio[1]

[1]*Center for Neurobiology and Behavior, College of Physicians and Surgeons, Columbia University, and* [2]*New York State Psychiatric Institute, New York, NY 10032, USA*

Abstract

Nitric oxide (NO) is widespread in the nervous system and is thought to play a role in a variety of different neuronal functions, including learning and memory (see other chapters, this volume). A number of behavioral studies have indicated that NO is involved in several types of learning such as motor learning (Yanagihara and Kondo, 1996), avoidance learning (Barati and Kopf, 1996; Myslivecek et al., 1996), olfactory learning (Okere et. al., 1996; Kendrick et al., 1997), and spatial learning (Holscher et al., 1995; Yamada et al., 1996) (for review of earlier papers see Hawkins, 1996). Moreover, NO is thought to be involved in neuronal plasticity contributing to these different types of learning in different brain areas including the cerebellum (chapter by R. Tsien, this volume) and hippocampus. In this chapter we review evidence on the role of NO in long-term potentiation (LTP), a type of synaptic plasticity in hippocampus that is believed to contribute to declarative forms of learning such as spatial learning.

* Corresponding author. Tel.: +1 212 543 5244; fax: +1 212 543 5474; e-mail: rhawkins@nypimail.cpmc.columbia.edu
† Present address: Department of Biochemistry, College of Medicine, Hanyang University, 17 Haengdang-dong, Seongdong-Gu, Seoul 133-791, Korea.

Hippocampal long-term potentiation

Since the pioneering work of Scoville and Milner (1957), the hippocampus has been known to be important for the initial storage of declarative memory (memory for people, places, and things) in humans and other mammals (Squire, 1992). These studies have shown that the hippocampus may be essential for initially storing long-term memory for a period of days to weeks before the memory trace is consolidated elsewhere, perhaps in different areas of the cerebral cortex (Zola-Morgan and Squire, 1990). The hippocampus has three, well-studied types of intrinsic feed-forward excitatory synaptic connections (Fig. 1): the perforant pathway from entorhinal cortex synapses onto the granule cells in the dentate gyrus, the mossy fiber pathway from the granule cells synapses onto the pyramidal cells in the CA3 region of the hippocampus, and the Schaffer collateral pathway from pyramidal cells in CA3 synapses onto the pyramidal neurons in the CA1 region of the hippocampus, which project back to entorhinal cortex. Entorhinal cortex, in turn, has reciprocal connections with many areas of association cortex.

The hippocampus was also the first brain area in which long-term potentiation (LTP), a cellular mechanism particularly suitable for memory, was described. Bliss and Lomo (1973) demonstrated

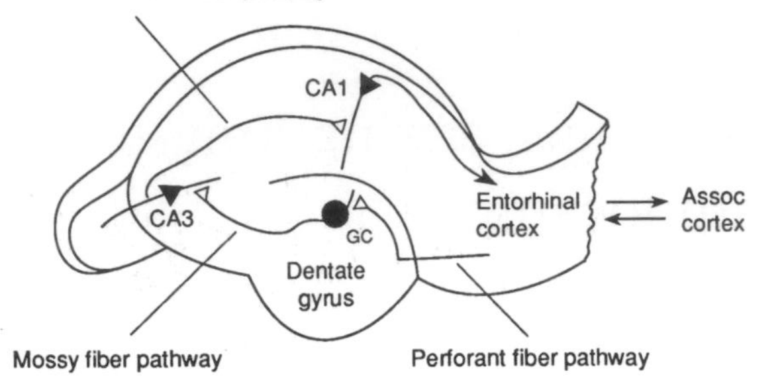

Fig. 1. Principal neural circuit connections in the hippocampus. Fibers in the perforant path from the entorhinal cortex synapse on granule cells in the dentate gyrus. The mossy fiber axons of the granule cells synapse on pyramidal cells in the CA3 region of hippocampus. The Schaffer collateral axons of the CA3 pyramidal cells synapse on pyramidal cells in the CA1 region, which project back to the entorhinal cortex. The entorhinal cortex, in turn, has reciprocal connections with many areas of association cortex.

that a brief high-frequency train of action potentials in the perforant path produces an increase in the excitatory synaptic potential in the granule cells, which can last for hours, and even, under some circumstances, for days or weeks. Later studies showed that such LTP occurs at each of the three major types of synaptic connections in the hippocampus, but that the potentiation has different properties (and perhaps different mechanisms) at different synapses (Zalutsky and Nicoll, 1990). In this chapter, we will concentrate on LTP in the CA1 region. One property of LTP in CA1 which will be relevant is pathway specificity, which means that the potentiation is restricted to the tetanized pathway (Andersen et al., 1977). This and other properties of LTP in the CA1 region can be explained by the fact that the synapses have a type of plasticity first postulated by Hebb (1949): coincident activity in the post-synaptic pyramidal neuron and pre-synaptic neurons is both necessary and sufficient for the induction of the potentiation. Blocking post-synaptic firing blocks the induction of LTP (Malinow and Miller, 1986), and intracellularly produced post-synaptic firing can induce LTP when it is paired with weak pre-synaptic stimulation (Kelso et al., 1986; Sastry et al., 1986; Wigström et al., 1986).

The Hebbian property of LTP in CA1, in turn, apparently derives from the properties of the *N*-methyl-D-aspartate (NMDA) glutamate receptor and its channel. The major excitatory transmitter in the hippocampus is thought to be glutamate, which exerts its action through two major classes of receptors: NMDA receptors and non-NMDA receptors. The ion channels associated with the NMDA receptors allow Ca^{2+} influx, which is thought to be essential for the induction of LTP in CA1 (Lynch et al., 1983; Malenka et al., 1988, 1992). The NMDA-receptor channel is blocked at the resting membrane potential by extracellular Mg^{2+} ions. The blockade of this channel is voltage dependent; when the membrane is depolarized, Mg^{2+} is expelled from the channel. Thus, Ca^{2+} influx through the channel requires the coincidence of (1) post-synaptic depolarization, and (2) activation of the NMDA receptors by glutamate. Both events are normally produced by strong, high-frequency stimulation of pre-synaptic fibers, which causes sufficient activation of non-NMDA receptor channels to depolarize the post-synaptic cell, removing the Mg^{2+} blockade of the NMDA receptor channels and allowing Ca^{2+} to enter the cell. Blocking activation of the NMDA receptor with selective inhibitors or gene knockout blocks LTP in CA1 (Collingridge et al., 1983; Tsien et al., 1996). Infusion of NMDA receptor blockers into the hippocampus or regionally restricted knockout also blocks spatial learning, suggesting that an NMDA receptor mechanism in the hippocampus, perhaps LTP, is involved in that type of learning (Morris et al., 1986; Davis et al., 1992; Tsien et al., 1996).

LTP in CA1 is thus thought to be induced by post-synaptic mechanisms, but the evidence for maintenance is less clear, and the site of expression remains controversial. Perhaps the safest conclusion at this time is that there is evidence for both pre- and post-synaptic effects during the maintenance of LTP, and both may occur (for reviews see Bliss and Collingridge, 1993; Hawkins et al., 1993; Kullman and Siegelbaum, 1995). If the induction of LTP requires post-synaptic activation of NMDA receptors and maintenance involves a pre-synaptic increase in transmitter release, then some message may be sent from the post-synaptic to the presynaptic neurons. The putative retrograde messenger is most likely released not from the cell body but from dendrites of the post-synaptic cell, perhaps from dendritic spines. Because the dendritic spines do not have the conventional machinery for the release of transmitter, the retrograde messenger may be membrane permeable and reach the presynaptic terminals by free diffusion. In this chapter we shall consider the possible roles in LTP of two molecules that are freely diffusible and might serve as retrograde messengers: nitric oxide (NO) and carbon monoxide (CO).

Nitric oxide as a candidate retrograde messenger

(i) Inhibition of nitric oxide synthase

NO is a gas that is generated by the enzyme NO synthase (NOS) from the amino acid L-arginine by splitting off stoichiometric amounts of citrulline. Activation of the neuronal (type I) and endothelial (type III) isoforms of NOS requires Ca^{2+}/calmodulin and the coenzyme NADPH. O'Dell et al. (1991), Bohme et al. (1991), Schuman and Madison (1991), and Haley et al. (1992) first investigated the possible role of NO as a retrograde messenger in LTP. All four groups found that inhibitors of NOS block the induction of LTP in hippocampal slices and that the inhibition is reversed by giving excess of the amino acid L-arginine, suggesting that the inhibitors are relatively specific. As would be expected for a retrograde messenger, O'Dell et al. (1991) and Schuman and Madison (1991) found that NO is likely to be produced post-synaptically, because injecting the inhibitors into the post-synaptic cell blocks LTP. Moreover, perfusion of the slice with the NO scavenging protein, hemoglobin (which does not penetrate cells) also blocks LTP, indicating that NO must diffuse out of the post-synaptic cell to produce its action, consistent with its possible role as a retrograde messenger in the induction of LTP.

However, subsequent studies with inhibitors of NOS have produced mixed results. Some studies have confirmed that NOS inhibitors block the induction of LTP in CA1 and dentate gyrus (Mizutani et al., 1993; Boulton et al., 1995; Doyle et al., 1996; Wu et al., 1997), but other studies have found that inhibitors of NOS either do not block LTP (Kato and Zorumski, 1993; Bannerman et al., 1994; Cummings et al., 1994) or block LTP only under some experimental circumstances and not others (Gribkoff and Lum-Ragan, 1992; Haley et al., 1993, 1996; Chetkovich et al., 1993; Williams et al., 1993; O'Dell et al., 1994; Malen and Chapman, 1997). Moreover, LTP was found to be normal in mice with a targeted mutation of the predominant neuronal isoform of NOS (type I or nNOS), suggesting that another isoform of NOS may be responsible for generating NO during the induction of LTP (O'Dell et al., 1994).

(ii) Targeted mutations of nNOS and eNOS

Another area of controversy has been whether pyramidal neurons in CA1 contain NOS. nNOS was originally detected in the CA1 area only in interneurons (Bredt et al., 1991), but it has now been reported to be localized in CA1 pyramidal neurons by using relatively gentle methods of fixation (Wendland et al., 1994). The endothelial isoform of NOS (type III or eNOS) has also been found to be expressed in the brain and to be present in the pyramidal cells of the CA1 region (Dinerman et al., 1994; O'Dell et al., 1994). This observation made it plausible that eNOS, rather than or in addition to nNOS, might play an important role in LTP. eNOS was initially thought to be more strategically located because it is predominantly membrane bound, whereas nNOS is cytoplasmic, but subsequent studies have shown that nNOS is associated with the post-synaptic density by protein–protein interactions (Brenman et al., 1996). Moreover, LTP can be blocked by either adenovirus-mediated inhibition of eNOS (Kantor et al., 1996) or inhibitors that are thought to be specific for nNOS (Doyle et al., 1996; Wu et al., 1997), suggesting that both isoforms might be involved in LTP.

To try to clarify the possible roles of the different isoforms of NOS, Son et al. (1996) tested LTP in hippocampal slices from mice in which nNOS, eNOS or both nNOS and eNOS were genetically disrupted. In agreement with O'Dell et al. (1994), they found that mice with a targeted mutation of nNOS ($nNOS^-$) have normal LTP in the stratum radiatum of CA1. Moreover, LTP in the $nNOS^-$ mice was reduced by the NOS inhibitor N^G-nitro-L-arginine (NOArg), suggesting that another isoform of NOS might be involved. Similarly, mice with a targeted mutation of eNOS ($eNOS^-$) have normal LTP that was reduced by NOArg. However, doubly mutant mice ($nNOS^-$/ $eNOS^-$) have LTP that was significantly reduced, and NOArg produced no further reduction of LTP in those animals. These results suggest that both nNOS and eNOS are involved in LTP, and either one can compensate for the other when it is disrupted. Furthermore, the block of LTP in the doubly mutant animals was not complete, suggesting that there is also an NOS-independent component of LTP that might involve other retrograde messengers or purely post-synaptic mechanisms. This result may provide an explanation for the finding that NOS inhibitors can block LTP under some experimental circumstances (which presumably favor a role of NOS) but not others (see above).

The reduction of LTP in the stratum radiatum of CA1 in the $nNOS^-$/$eNOS^-$ animals might be due to non-specific effects such as developmental abnormalities. However, those animals had no gross anatomical abnormalities in the hippocampus and they had normal hippocampal physiology including paired-pulse facilitation and long-term depression, suggesting that the defect in LTP was not due to non-specific effects of the mutation. As an additional control, Son et al. (1996) examined LTP in the stratum oriens of CA1. eNOS immunoreactivity is observed in stratum radiatum but not in stratum oriens (O'Dell et al., 1994) and LTP in stratum oriens is not reduced by NOS inhibitors, although it is NMDA-dependent (Haley et al., 1996; Son et al., 1996). Similarly, LTP in stratum oriens was not reduced in $nNOS^-$, $eNOS^-$, or doubly mutant mice. These results demonstrate that other aspects of the mechanisms involved in LTP are normal in the doubly mutant mice, and suggest that the defect in stratum radiatum LTP in those mice is due to their lack of NOS.

(iii) Activity-dependent long-term enhancement by NO

As another way of testing the involvement of NO in LTP, Zhuo et al. (1993) applied dissolved NO gas to hippocampal slices. They anticipated that NO by itself might have no effect because of an inherent problem with the retrograde messenger hypothesis, which is how to maintain the observed pathway specificity of LTP. LTP occurs only at synapses from active pre-synaptic fibers (Andersen et al., 1977), whereas a diffusible retrograde messenger might also spread to synapses from neighboring, inactive fibers. A possible solution would be for the message to be effective only at recently active pre-synaptic fibers, as during activity-dependent pre-synaptic facilitation in *Aplysia* (Hawkins et al., 1983; Walters and Byrne, 1983) (Fig. 2). Zhuo et al. (1993) tested this possibility by applying NO to hippocampal slices either alone or paired with weak tetanic stimulation of the pre-synaptic fibers that by itself produced little or no LTP. Alone, NO had no consistent effect on the field excitatory post-synaptic potential (EPSP) in the CA1 region, and neither did weak tetanic stimulation (50 Hz for 0.5 s). However, when NO was applied at the same time as weak tetanic stimulation (paired training), the EPSP was rapidly enhanced and remained enhanced for at least 1 h. When applied 5 min after weak tetanic stimulation (unpaired training), NO produced no significant long-term effect, which demonstrates that the effects of NO and weak tetanic stimulation are synergistic and not simply additive during paired training. Other studies have shown that NO donors such as sodium nitroprusside can also produce long-term enhancement (Bohme et al., 1991; Malen and Chapman, 1997), but some studies have reported no enhancement by NO donors (Boulton et al., 1994; Murphy et al., 1994).

A possible explanation for these negative results is toxic side effects of some of the NO donors or of higher concentrations of NO itself.

Zhou et al. (1993) went on to examine whether the properties of the long-term enhancement by NO were consistent with a role in LTP. First, if that enhancement has properties similar to LTP, then it should exhibit pathway specificity (Andersen et al., 1977). Zhuo et al. (1993) tested this possibility by placing stimulation electrodes on two different pre-synaptic pathways in the same slice. When NO was paired with weak tetanic stimulation of one pathway, the EPSP in that pathway was rapidly enhanced and remained so for at least 1 h. In contrast, the EPSP in the control pathway, which received weak tetanic stimulation 5 min before application of NO, showed no significant long-term effect. Therefore, long-term enhancement by NO is spatially restricted to synapses from active pre-synaptic pathways. This result also indicates that the enhancement does not involve a generalized post-synaptic effect but rather must involve either pre-synaptic enhancement or a localized post-synaptic effect.

If long-term enhancement by NO contributes to normal LTP, then that enhancement should occlude LTP. Zhuo et al. (1993) tested this possibility with strong tetanic stimulation at the end of the experiments described above. After long-term enhancement by NO paired with weak tetanic stimulation, the strong tetanic stimulation produced decrementing potentiation but no significant long-term effect. In contrast, the strong tetanic stimulation did produce significant long-term potentiation after perfusion with NO unpaired with weak tetanic stimulation (which does not produce long-term enhancement).

If NO acts as a retrograde messenger, then its effects should not require NMDA receptor activation because supplying a retrograde messenger exogenously would presumably bypass the NMDA receptor step in the induction of LTP. Zhuo et al. (1993), therefore, repeated these experiments in the presence of 2-amino-5-phosphonovaleric acid (APV), which blocks NMDA receptors and induction of LTP by normal tetanic stimulation. Paired training with NO still produced significant long-term enhancement of the synaptic potential in the presence of APV. Paired training with NO also still produced significant long-term enhancement in the presence of both APV and the voltage-dependent Ca^{2+} channel blocker nifedipine, indicating that NO does not simply act by enhancing post-synaptic Ca^{2+} influx through NMDA receptor channels or L-type voltage-dependent Ca^{2+} channels during

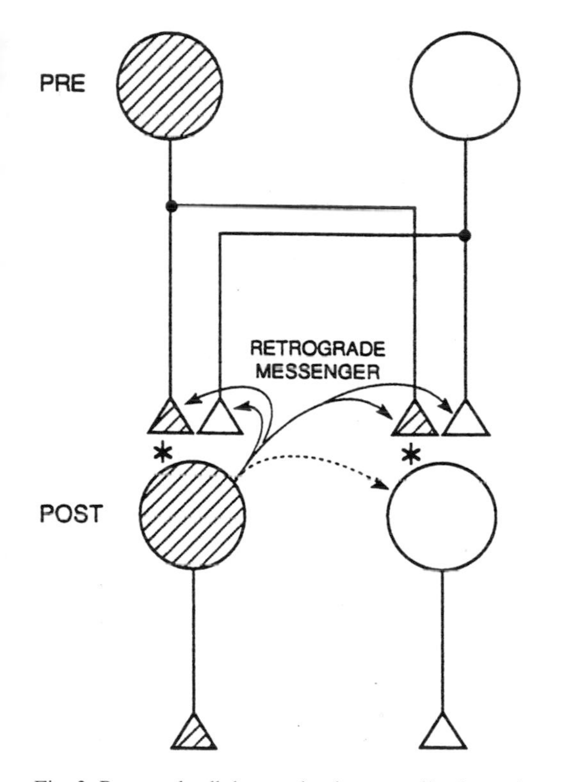

Fig. 2. Proposed cellular mechanism contributing to long-term potentiation in the CA1 region of the hippocampus. Shading indicates spike activity. Active post-synaptic neurons release a diffusible retrograde messenger that produces a long-lasting increase in transmitter release from recently active presynaptic fibers. This mechanism could account for Hebbian potentiation at synapses between active presynaptic cells and active postsynaptic cells, as well as non-Hebbian potentiation at synapses between active presynaptic cells and neighboring postsynaptic cells that are not active (asterisks). Both types of potentiation occur in hippocampus. This mechanism could also account for the pathway specificity of LTP, because there would be no potentiation at synapses from inactive presynaptic cells.

the weak tetanic stimulation. In addition, the long-term enhancement was not blocked in the presence of both APV and the GABA receptor blocker picrotoxin, suggesting that it does not importantly involve post-synaptic disinhibition. These results are consistent with the possibility that NO has a pre-synaptic locus of action.

(iv) Role of NO in LTP in cultured hippocampal neurons

The results of these experiments support the involvement of NO in LTP, but the inaccessibility of the pre-synaptic terminal in hippocampal slices or in vivo prevented a definitive test of whether NO acts as a retrograde messenger – that is whether it is produced in the post-synaptic cell and acts in the pre-synaptic cell. Because both sides of the synapse are accessible in cultured hippocampal neurons, Arancio et al. (1995, 1996) performed a series of experiments on this preparation to test the role of NO in LTP.

Bekkers and Stevens (1990) first demonstrated LTP at synapses between hippocampal neurons in culture, but their success rate was low. Arancio et al. (1995) found that LTP could be induced reliably in monosynaptically connected pairs of cultured hippocampal neurons by high frequency trains of depolarization (tetani) of the pre-synaptic neuron during temporary removal of Mg^{2+} from the bath, to allow Ca^{2+} influx through post-synaptic NMDA receptor channels. This potentiation was blocked by the NMDA receptor antagonist APV or injection of the Ca^{2+} chelator BAPTA into the post-synaptic cell, demonstrating that it has some of the key features of LTP in the CA1 region of hippocampal slices.

Arancio et al. (1996) found that like LTP in slices, LTP in culture was also blocked by delivery of an NO scavenger (oxymyoglobin) to the bath, whereas a related compound with a much lower affinity for NO (metmyoglobin) had no effect. These results suggest that NO must travel across the extracellular space during the induction of LTP. Arancio et al. (1996) next took advantage of the culture system by injecting various substances intracellularly into either neuron through the patch pipette. They first confirmed that fluorescently labeled myoglobin injected into the cell body reached the dendrites and synaptic terminals within 10 min. They then found that injection of oxymyoglobin into either the pre- or post-synaptic neuron blocked LTP, whereas injection of met-myoglobin had no effect. These results suggest that NO must travel through both the pre- and post-synaptic neurons during the induction of LTP, consistent with the retrograde messenger hypothesis.

However, the results of these experiments do not indicate where NO is synthesized or where it acts. Furthermore, oxymyogoblin is not specific for NO, because it can also bind other molecules such as CO. Arancio et al. (1996) therefore next injected an inhibitor of NO synthase, N^{G}-monomethyl-L-arginine (L-NMMA). As in slices (O'Dell et al., 1991; Schuman and Madison, 1991), injection of L-NMMA into the post-synaptic neuron blocked LTP in culture. In contrast, injection of L-NMMA into the pre-synaptic neuron had no effect. These results suggest that NO is involved in LTP in culture, and that NO synthase is activated in the post-synaptic neuron during the induction of LTP.

As an additional test of the involvement of NO, Arancio et al. (1996) applied exogenous NO to see if it induced long-lasting potentiation. As in slices (Zhuo et al., 1993), brief bath application of NO paired with a weak tetanus produced rapid and long-lasting potentiation. These experiments were conducted with normal Mg^{2+} and APV in the bath to avoid potentiation by the weak tetanus alone. Moreover, the potentiation by NO paired with weak tetanus was blocked by injection of oxymyoglobin into the pre-synaptic neuron, whereas injection of oxymyoglobin into the post-synaptic neuron had no effect. These results suggest that NO acts in the pre-synaptic neuron to produce potentiation of transmitter release. Consistent with that idea, NO produced a long-lasting increase in the frequency of spontaneous miniature synaptic currents in cultured hippocampal neurons, suggesting that it acts on aspects of transmitter release that are independent of pre-synaptic action potentials (O'Dell et al., 1991; Sporns and Jenkinson, 1997).

As a complementary test of the site of action of NO, Arancio et al. (1996) used a membrane-impermeant NO donor, CNO-4, that releases NO only upon photolysis with UV light. Intracellular injection of CNO-4 into the post-synaptic neuron, followed by photolytic release of NO paired with weak tetanus (in the presence of APV) caused rapid and long-lasting potentiation. Like tetanic LTP, this potentiation was blocked by the addition of oxymyoglobin to the bath, consistent with the hypothesis that NO produced in the post-synaptic neuron must cross the extracellular space to produce potentiation. In contrast, intracellular injection of CNO-4 into the pre-synaptic neuron, followed by photolytic release of NO paired with weak tetanus (in the presence of APV) caused rapid and long-lasting potentiation that was not blocked by oxymyoglobin in the bath. These results are consistent with the results obtained with exogenous NO, and suggest that the site of action of NO is in the pre-synaptic neuron.

Taken together, the results of these experiments indicate that NO is produced in the post-synaptic neuron, travels through the extracellular space, and acts directly in the pre-synaptic neuron to produce long-lasting potentiation, strongly supporting the hypothesis that NO acts as a retrograde messenger during LTP.

Carbon monoxide as a candidate retrograde messenger

Carbon monoxide (CO) is another membrane permeable gas that has been suggested to play a signaling role in the nervous system. It is produced during conversion of heme to biliverdin by the enzyme heme oxygenase. Because CO has actions similar to NO in other systems and heme oxygenase is present in hippocampal pyramidal neurons (Marks et al., 1991; Ewing and Maines, 1992; Verma et al., 1993), Zhuo et al. (1993) also examined a possible role of CO in LTP in hippocampal slices. Zinc protoporphyrin IX (ZnPP), which inhibits heme oxygenase, blocked the induction of LTP in a dose-dependent manner. ZnPP also blocked LTP in the presence of picrotoxin, a gamma-aminobutyric acid (GABA) receptor blocker, suggesting that ZnPP does not act by enhancing inhibition. In contrast with its effect on LTP, ZnPP did not greatly affect the decrementing potentiation in the first 10 min after the tetanus, the baseline EPSP, or the NMDA component of the EPSP. Although there are some concerns about the specificity of ZnPP (Ignarro et al., 1984; Meffert et al., 1994a; but see Zakhary et al., 1996), these results suggest that the drug's effects are relatively specific. A similar block of LTP by ZnPP was observed by Stevens and Wang (1993) and Ikegaya et al. (1994). In contrast, Poss et al. (1995) found no effect on LTP of a targeted mutation of one isoform of heme oxygenase. A possible explanation for this negative result is compensation by other isoforms of heme oxygenase or other messengers.

Zhuo et al. (1993) found that like NO, CO applied alone had no consistent effect on the EPSP in the CA1 region of slices, but when CO was paired with weak tetanic stimulation, the EPSP was rapidly enhanced and remained enhanced for at least 1 h. When applied 5 min after weak tetanic stimulation (unpaired training), CO produced no significant long-term effect. Like activity-dependent potentiation by NO, the potentiation by CO paired with weak tetanus was pathway specific, occluded tetanic LTP, and was not blocked by APV, nifedipine, or picrotoxin, consistent with a pre-synaptic locus of action.

LTP can be induced by either strong tetanic stimulation of the pre-synaptic fibers or low-frequency stimulation paired with strong post-synaptic depolarization (Wigström et al., 1986; Colino et al., 1992). If NO or CO acts as a retrograde messenger for both types of LTP, then one might expect that pairing the gas with low-frequency stimulation would also produce long-lasting synaptic enhancement. When CO was paired with low-frequency stimulation (0.25 Hz for 100 s), the EPSP was enhanced and remained enhanced for at least 1 h (Zhuo et al., 1993). Low-frequency stimulation alone produced no significant long-term effect, and no post-tetanic or decrementing potentiation, making it unlikely that CO acts by enhancing the effect of the low-frequency stimulation. In contrast, NO paired with low-frequency stimulation did not produce long-lasting synaptic enhancement but rather produced long-lasting depression (Zhuo et al., 1994b). Some other studies have also supported the possible involvement of NO in long-lasting depression (Gage et al., 1997; Wu et al., 1997), but other studies have not replicated those results (Cummings et al., 1994; Malen and Chapman, 1997). A possible explanation for these mixed results is that NO is involved in potentiation as well as depression, with the balance depending on experimental parameters such as the frequency of the paired activity.

The results of these studies suggest that CO could be involved in LTP. However, evidence for CO as a retrograde messenger is still incomplete, and it might play another role either during the induction of LTP or as a constitutively required substance. Also, if NO and CO are both involved in LTP, it is not clear what their respective roles might be. Tetanic stimulation and NMDA are known to activate NO synthase in the hippocampus (East and Garthwaite, 1991; Chetkovich et al., 1993), but such data are currently lacking for heme oxygenase. Thus, one possibility is that CO provides a widespread basal level of stimulation, whereas NO provides local, phasic stimulation during the induction of LTP. This possibility would be consistent with the fact that NO has a very short half life, whereas CO is more stable. Alternatively, NO and CO might be activated under somewhat different circumstances that could engage different receptors and second messengers. Consistent with that possibility, Zhuo et al. (1994c) found that perfusion with the metabotropic glutamate receptor (mGluR) agonist tACPD paired with weak tetanus produced long-lasting potentiation that was blocked by inhibitors of heme oxygenase but not NO synthase. These results suggest that mGluR activation might selectively stimulate heme oxygenase activity during the induction of LTP, whereas NMDA receptor activation might selectively stimulate NO synthase.

The idea that both NO and CO might act as retrograde messengers raises the possibility that there could be a family of retrograde messengers, just as there is a family of second messengers. In addition to NO and CO, there is evidence that arachidonic acid (Lynch et al., 1989; Williams and Bliss, 1989; Williams et al., 1989), platelet-activating factor (del Cerro et al., 1990; Clark et al., 1992; Wieraszko et al., 1993; see chapter by N. Bazan, this volume), and several neurotrophins (Kang and Schuman, 1995; Thoenen, 1995; Korte et al., 1996) may also act as retrograde messengers during LTP.

Guanylyl cyclase and cGMP-dependent protein kinase as possible pre-synaptic targets of NO and CO

(i) Tests of a role of cGMP in LTP

Nitric oxide, carbon monoxide, and arachidonic acid all can activate soluble guanylyl cyclase and increase cGMP levels in other tissues (Snider et al., 1984; Garthwaite et al., 1988; Verma et al., 1993). Tetanic stimulation, NMDA, or tACPD activate guanylyl cyclase in hippocampus (East and Garthwaite, 1991; Chetkovich et al., 1993; Zhuo et al., 1994c), and the activation by tetanic stimulation or NMDA is reduced by NO synthase inhibitors or hemoglobin, which scavenges CO and NO extracellularly. In situ hybridization demonstrates that soluble guanylyl cyclase is located in pyramidal cells in the CA1 and CA3 regions of hippocampus (Matsuoka et al., 1992; Verma et al., 1993), and immunocytochemistry demonstrates that NO elevates cGMP levels in fibers in hippocampus (deVente et al., 1988). Zhuo et al. (1994a), therefore, investigated a possible role for guanylyl cyclase and cGMP in LTP in hippocampal slices. If guanylyl cyclase is one of the target proteins for CO and NO during LTP, then inhibition of soluble guanylyl cyclase might block LTP. Zhuo et al. (1994a) found the LY83583, an inhibitor of soluble guanylyl cyclase, blocked the induction of

LTP in a dose-dependent manner, but did not affect its maintenance. LY83583 also did not affect the baseline EPSP or the NMDA component of EPSP. Although there are some concerns about the specificity of LY83583, these results suggest that the drug's effects are relatively specific. LY83583 also blocked long-lasting potentiation by tACPD paired with weak tetanus, suggesting that guanylyl cyclase might be stimulated by activation of metabotropic glutamate receptors as well as NMDA receptors during the induction of LTP (Zhuo et al., 1994c).

Experiments with a more specific inhibitor of soluble guanylyl cyclase, ODQ, have produced mixed results. Gage et al. (1997) found no effect of ODQ on LTP, but Boulton et al. (1995) and Son et al. (1998) found that ODQ significantly reduced LTP in the stratum radiatum of CA1. In addition, Son et al. (1998) found that, like inhibitors or knockout of NO synthase (Haley et al., 1996; Son et al., 1996), ODQ had no significant effect on LTP in stratum oriens, as might be expected if guanylyl cyclase acts downstream of NO synthase.

If cGMP is involved in LTP, then pairing membrane permeable cGMP analogs with weak pre-synaptic stimulation might produce long-term enhancement. Application of 8-Br-cGMP at the same time as weak pre-synaptic stimulation (50 Hz, 0.5 s) (paired training) produced a rapid potentiation of the synaptic potential that lasted for at least 1 h (Zhuo et al., 1994a). Neither 8-Br-cGMP alone nor weak pre-synaptic stimulation alone produced any significant enhancement. Furthermore, application of 8-Br-cGMP 5 min after weak pre-synaptic stimulation (unpaired training) did not produce significant enhancement. Like the synaptic enhancement by NO or CO, the long-lasting enhancement produced by 8-Br-cGMP paired with weak tetanic stimulation was spatially restricted to the stimulated pathway, and occluded subsequent LTP by strong tetanic stimulation. It also was not blocked by APV, nifedipine, or picrotoxin, suggesting that 8-Br-cGMP does not act by enhancing calcium influx through either NMDA receptor channels or L-type calcium channels, or by disinhibition.

Similarly, Haley et al. (1992) found that cGMP analogs paired with tetanic stimulation produced long-lasting potentiation that was not blocked by an NO synthase inhibitor, consistent with the idea that cGMP acts downstream of NO. However, some other studies have failed to replicate potentiation by 8-Br-cGMP paired with tetanic stimulation (Schuman et al., 1994; Selig et al., 1996). Son et al. (1998) have identified two experimental variables that might account for some of the different results in the different studies. First, they found that brief perfusion with 8-Br-cGMP prior to weak tetanic stimulation produced long-lasting potentiation (replicating Zhuo et al., 1994a), but surprisingly more prolonged perfusion with 8-Br-cGMP prior to the tetanus produced no potentiation. Second, Son et al. (1998) found that although potentiation by cGMP analogs paired with tetanic stimulation was not blocked by APV, it was reduced. This result is not predicted by the simple retrograde messenger hypothesis, and suggests that the hypothesis should be modified to include some involvement of pre- or post-synaptic NMDA receptors.

(ii) Tests of a role of cGMP-dependent protein kinase in LTP

In various tissues cGMP produces many of its effects by stimulating cGMP-dependent protein kinase (PKG). PKG type I (which is a soluble dimer) is expressed in CA1 and CA3 pyramidal neurons (Kingston et al., 1996), and PKG type II (which is a membrane-associated monomer) is also expressed in hippocampus (El-Husseini et al., 1995). In addition, Zhuo et al. (1994a) demonstrated cGMP-dependent protein kinase activity in homogenized hippocampal tissue by measuring phosphorylation of a PKG peptide substrate.

To test the involvement of cGMP-dependent protein kinase in LTP, Zhuo et al. (1994a) pretreated slices with three different cGMP-dependent protein kinase inhibitors. Rp-8-Br-cGMPS, which is membrane-permeable, blocked normal LTP in a dose-dependent manner. Rp-cGMPS, a weaker and less permeable inhibitor, also reduced

LTP, and KT5823, a structurally different cGMP-dependent protein kinase inhibitor, blocked it completely. The inhibitory effect of cGMP-dependent protein kinase inhibitors was selective for induction of LTP: established LTP was not affected by Rp-cGMPS or KT5823.

Other experiments with PKG inhibitors have produced mixed results. Schuman et al. (1994) found no effect on LTP of a broad spectrum kinase inhibitor, H-8, that is thought to block PKG. However, Son et al. (1998) found that either Rp-8-Br-cGMPS or KT5823 blocked LTP in the stratum radiatum of CA1, replicating Zhuo et al. (1994a). In addition, Son et al. (1998) found that, like inhibitors of NO synthase or soluble guanylyl cyclase (Haley et al., 1996; Son et al., 1998), Rp-8-Br-cGMPS and KT5823 had no significant effect on LTP in stratum oriens, consistent with the idea that PKG acts downstream of NO and cGMP. Blitzer et al. (1995) also found that LTP in stratum radiatum was reduced by bath application of Rp-8-Br-cGMPS but not by injection of Rp-8-Br-cGMPS into the post-synaptic neuron, suggesting that PKG acts pre-synaptically during LTP.

As an additional test of a role of cGMP-dependent protein kinase, Zhuo et al. (1994a) substituted for 8-Br-cGMP a selective cGMP-dependent protein kinase activator, 8-(4-chlorophenylthio)-cGMP 3′,5′-cyclic monophosphate (8-pCPT-cGMP), that does not significantly affect cGMP-regulated phosphodiesterases. Pairing 8-pCPT-cGMP with weak pre-synaptic stimulation produced a rapid and long-lasting enhancement of the synaptic potential. In contrast, 8-pCPT-cGMP applied 5 min after the weak pre-synaptic stimulation (unpaired training) failed to produce long-lasting enhancement. The long-term enhancement produced by 8-pCPT-cGMP occluded normal LTP: subsequent strong tetanus (2×100 Hz, 1 s) after 8-pCTP-cGMP paired training failed to produce significant further potentiation, but strong tetanus did produce a significant enhancement after 8-pCTP-cGMP unpaired training.

A particularly interesting test of activity dependence is pairing with low-frequency stimulation. Like NO, 8-Br-cGMP paired with low-frequency stimulation (0.25 Hz, 100 s) produced depression of the EPSP that lasted for at least 1 h (Zhuo et al., 1994b). In contrast, the cGMP-dependent protein kinase activator 8-pCPT-cGMP paired with low-frequency stimulation induced long-term enhancement of the EPSP (Zhuo et al., 1994a). These results are consistent with a role for cGMP-dependent protein kinase in LTP, and suggest that under some circumstances NO and cGMP analogs can also induce other mechanisms that are involved in long-term depression. Experiments by Gage et al. (1997) also support the involvement of cGMP in long-term depression.

(iii) Role of cGMP and PKG in LTP in cultured hippocampal neurons

The results of these experiments complement the previous suggestions that NO and CO may serve as retrograde messengers in LTP by providing evidence that two target proteins for NO and CO, soluble guanylyl cyclase and cGMP-dependent protein kinase, are involved in the induction of LTP in the hippocampus. According to this hypothesis, cGMP and cGMP-dependent protein kinase should have pre-synaptic actions. Arancio et al. (1995, 1997) have tested this possibility in dissociated cultures of hippocampal neurons. They first found that, like LTP in slices (Zhuo et al., 1994a), LTP in culture is blocked by an inhibitor of soluble guanylyl cyclase, LY83583 (Arancio et al., 1995). Conversely, brief perfusion with 8-Br-cGMP produced rapid and long-lasting potentiation of the EPSC. This result differs from the results in slices where 8-Br-cGMP alone had no effect (Zhuo et al., 1994a), perhaps due to the poorer accessibility of the neurons in slices. However, in culture, as in slices, pairing 8-Br-cGMP with weak tetanic stimulation produced significantly greater long-lasting potentiation than 8-Br-cGMP alone. These experiments were carried out in the presence of normal Mg^{2+} and APV to insure that the weak tetanus by itself produced no potentiation.

Arancio et al. (1995) next took advantage of the culture system by injecting cGMP intracellularly into the pre- or post-synaptic neuron to determine its site of action. They found that injection of cGMP into the pre-synaptic neuron paired with weak tetanic stimulation (in APV) produced rapid and long-lasting potentiation of the EPSC, whereas injection of cGMP into the post-synaptic neuron had no effect. As a control, injection of cAMP into either neuron also had no effect. Furthermore, the increase in the amplitude of the EPSC by either 8-Br-cGMP or pre-synaptic cGMP correlated significantly with an increase in the inverse of the coefficient of variation squared ($1/CV^2$), consistent with a pre-synaptic site of expression (DelCastillo and Katz, 1954).

As another way of distinguishing between pre- and post-synaptic effects of cGMP, Arancio et al. (1995) examined spontaneous miniature synaptic currents (m.e.p.s.c.s). They found that, like NO (O'Dell et al., 1991), brief perfusion with 8-Br-cGMP produced a rapid and long-lasting increase in the frequency of m.e.p.s.c.s. In contrast, 8-Br-cGMP produced no change in m.e.p.s.c. amplitude, and injection of cGMP into the post-synaptic neuron had no effect on m.e.p.s.c.s. Collectively, these results suggest that cGMP acts directly in the pre-synaptic neuron to produce an increase in transmitter release that is, at least in part, independent of any changes in the pre-synaptic action potential.

LTP in culture, as in slices, (Zhuo et al., 1994a; Blitzer et al., 1995), was blocked by bath application of an inhibitor of cGMP-dependent protein kinase, Rp-8-Br-cGMPS, suggesting that cGMP acts through PKG. However, Rp-8-Br-cGMPS might have non-specific effects, and bath application does not discriminate between pre- and post-synaptic actions. Arancio et al. (1997) therefore injected a new and highly specific peptide inhibitor of PKG intracellularly into either the pre- or post-synaptic neuron. They found that injection of the inhibitor into the pre-synaptic neuron blocked LTP, whereas injection into the post-synaptic neuron had no effect. As a control, injection of a peptide inhibitor of cAMP-dependent protein kinase into either neuron also had no significant effect. These results support the idea that cGMP acts through PKG in the pre-synaptic neuron during the induction of LTP.

Conclusions – A molecular model and remaining questions

(i) Possible targets of NO, CO, cGMP, and PKG

The results of the experiments we have reviewed are consistent with the model illustrated in Fig. 3. Tetanic stimulation activates NO synthase and perhaps also heme oxygenase in the post-synaptic neurons, leading to production of NO and CO. NO and CO then act as retrograde messengers, producing long-lasting enhancement of transmitter release from recently active pre-synaptic terminals by stimulating guanylyl cyclase and cGMP-dependent protein kinase. We should emphasize that this is a working model that is not meant to be a complete account of the mechanisms contributing to LTP, and that a number of points still need to be addressed. Are both nNOS and eNOS activated in the post-synaptic cells during LTP? Is heme oxygenase activated? If so, then how? If NO and CO are retrograde messengers, then what are their molecular targets in the pre-synaptic cell? We have presented evidence supporting a presynaptic role for soluble guanylyl cyclase, which is a common target of both NO and CO. However, NO is known to have other targets that might also play a role in LTP. NO stimulates ADP ribosyl transferase, which has been implicated in hippocampal LTP (Duman et al., 1993; Schuman et al., 1994; Sullivan et al., 1997). NO can also interact with sulfhydryl groups, and may directly modify synaptic vesicle associated proteins by this route to enhance Ca^{2+}-independent vesicle release (Meffert et al., 1994b, 1996).

Is soluble guanylyl cyclase located in the pre-synaptic terminals? If so, what are its molecular targets? Again, we have presented evidence supporting a role for cGMP-dependent protein kinase, but cGMP might also have other actions. cGMP-stimulated phosphodiesterase is present in hippocampal pyramidal cells (Repasko et al., 1993) and may participate in synaptic depression

(Doerner and Alger, 1988; Zhuo et al., 1994b). Cyclic nucleotide gated calcium channels similar to those found in olfactory receptors and photoreceptors are also expressed in hippocampal pyramidal cells (Kingston et al., 1996; Bradley et al., 1997), and cGMP stimulates Ca^{2+} influx through these channels in hippocampal neurons (Leinders-Zufall et al., 1995; Bradley et al., 1997).

Is either type I or type II cGMP-dependent protein kinase located in the pre-synaptic terminals? If so, what are its molecular targets? PKG specifically phosphorylates more than 40 proteins in brain, most of which have not been identified (Wang and Robertson, 1995). A particularly intriguing possibility is that one of the substrates of PKG could be a protein involved in exocytosis of synaptic vesicles. One synaptic vesicle associated protein that is known to be phosphorylated by PKG is rabphilin 3A (Fyske et al., 1995), which binds rab3A. Mice with a targeted mutation of rab3A have normal LTP in the CA1 region of hippocampus (Geppert et al., 1994) but greatly reduced LTP in the mossy fiber pathway (Castillo et al., 1997).

Alternatively, PKG may affect transmitter release indirectly. One known mechanism by which PKG might play such a role is that it phosphorylates and activates DARPP-32 and inhibitor 1, which inhibit protein phosphatase 1 (Tsou et al, 1993; Tokui et al., 1996). Inhibition of protein phosphatase 1 is thought to remove a brake on the actions of other kinases including CamKII, thereby facilitating the induction of LTP (Blitzer et al., 1995). This cascade is believed to operate in the post-synaptic cells but it might also occur in the pre-synaptic terminals, possibly leading to phosphorylation of synaptic vesicle associated proteins (Hirling and Scheller, 1996; Nayak et al. 1996).

Another known mechanism by which PKG might contribute is that it activates ADP ribosyl cyclase leading to the production of cyclic ADP ribose, which stimulates ryanodine receptors causing release of Ca^{2+} from intracellular stores (Galione et al., 1993). Ryanodine receptors are expressed in hippocampal pyramidal cells (Furuichi et al., 1994; Gianni et al., 1995), and ryanodine produces activity-dependent long-lasting potentiation in hippocampus (Wang et al., 1996).

PKG may also contribute to the late phase of LTP by causing induction of immediate early genes through phosphorylation of CREB (Peuno-

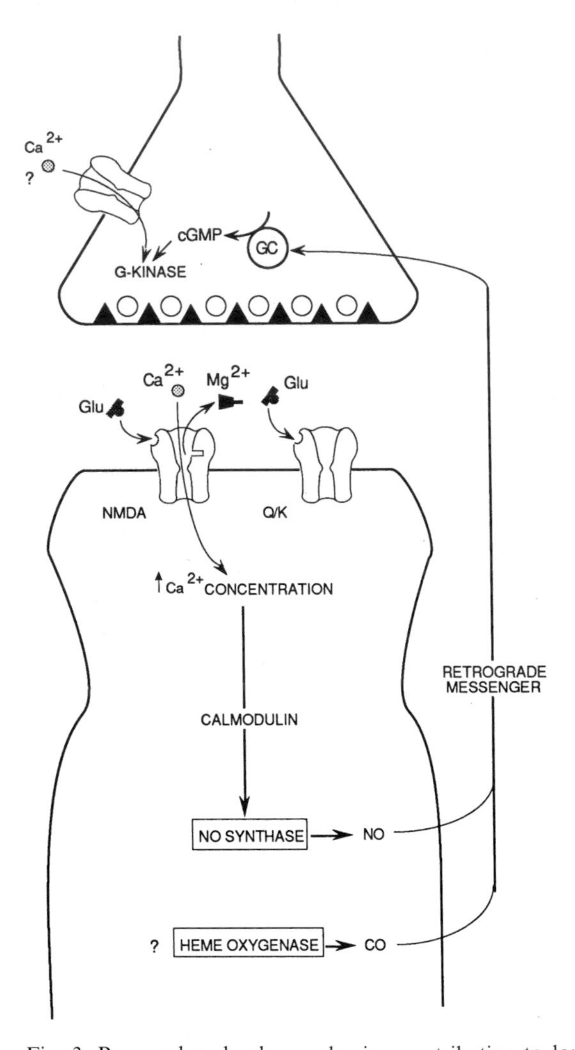

Fig. 3. Proposed molecular mechanism contributing to long-term potentiation. Tetanic stimulation activates NO synthase and perhaps also heme oxygenase in the postsynaptic neurons, leading to production of NO and CO. Either or both gases then act(s) as a retrograde messenger, causing a long-lasting increase in transmitter release from the presynaptic terminals by stimulating soluble guanylyl cyclase and cGMP-dependent protein kinase. This molecular cascade is amplified or enabled by concurrent spike activity in the presynaptic neurons, perhaps due to a synergistic effect of Ca^{2+} that enters the cell during the activity.

va and Enlkolopov, 1993; Haby et al, 1994; Gudi et al., 1996). CREB phosphorylation and gene induction are thought to contribute to the late, protein synthesis dependent phase of hippocampal LTP (Abraham et al., 1993; Bourtchouladze et al, 1994), which may involve pre-synaptic as well as post-synaptic changes (Bolshakov et al., 1997; Lu et al., in preparation).

(ii) Activity-dependence of the effects of NO, CO, and cGMP

Of particular interest is the finding that NO, CO, and 8-Br-cGMP require for their long-lasting effects coincident pre-synaptic activity, consistent with the idea that activity makes the pre-synaptic terminals responsive to potentiation by these messengers. According to this idea, activity-dependent long-term enhancement by NO or CO would be similar to activity-dependent pre-synaptic facilitation contributing to classical conditioning in *Aplysia* (Hawkins et al., 1983; Walters and Byrne, 1983). However, whereas the facilitating substance in *Aplysia* is released from widely projecting modulatory neurons, in the hippocampus it is presumably released from the post-synaptic cells themselves in response to NMDA receptor activation. This arrangement would allow for the use of a diffusible retrograde messenger in Hebbian potentiation at synapses onto active post-synaptic cells, and also in non-Hebbbian potentiation at synapses onto neighboring post-synaptic cells that are not active (Fig. 2). Both of these types of potentiation occur in the CA1 region of hippocampus (Wigström et al., 1986; Colino et al., 1992; Bonhoeffer et al., 1989; Schuman and Madison, 1994), where they may be used to perform computations somewhat different from the strictly Hebbian rule (Gally et al., 1990). Recent evidence indicates that activity-dependent facilitation in *Aplysia* also involves a similar combination of Hebbian and non-Hebbian components, suggesting that such hybrid mechanisms may be phylogenetically conserved (Bao et al., 1998).

At the molecular level, the results of the experiments we have described suggest that there are activity-dependent steps downstream from guanylyl cyclase (perhaps cGMP-dependent protein kinase or one of its substrate proteins), although they do not rule out the possibility that the cyclase itself might also be activity-dependent (Fig. 3). Activity-dependence of the effects of NO, CO, and cGMP might result from a dependence of one or more of the downstream enzymes on elevated Ca^{2+} levels due to Ca^{2+} influx during spike activity, as occurs for adenylyl cyclase in *Aplysia* (Kandel et al., 1983; Ocorr et al., 1985). Several of the possible targets of PKG are also regulated by Ca^{2+} and thus might contribute to the activity-dependence of potentiation by cGMP. For example, the synaptic vesicle associated protein rabphilin 3A is phosphorylated by calcium–calmodulin dependent protein kinase II (CamKII) as well as by PKG (Fyske et al., 1995). Inhibition of protein phosphatase 1 by PKG is also thought to act synergistically with activation of other kinases including CamKII to produce LTP (Blitzer et al., 1995). Similarly, Ca^{2+} can act synergistically with cyclic ADP ribose to cause release of Ca^{2+} from intracellular stores (Lee, 1993), and Ca^{2+} and 8-Br-cGMP can act synergistically to cause induction of immediate-early genes (Peunova and Enlkolopov, 1993). Alternatively, activity could lead to stimulation of autoreceptors on the pre-synaptic terminals, or it could have post-synaptic effects that have not yet been controlled for, such as activation of non-NMDA receptors.

Activity-dependence of the effects of NO, CO and cGMP in hippocampus represent new instances of the more general phenomenon of activity-dependent neuromodulation, which has now been shown to contribute to a variety of types of learning-related plasticity (Hawkins et al., 1993). In particular, cGMP and cGMP-dependent protein kinase have also been shown to produce long-lasting effects in cerebellum (long-term depression) and sensory motor cortex (increased excitability) that are enhanced by paired spike activity (Woody et al., 1978; Shibuki and Okada, 1991; Ito and Karachot, 1992; Lev-Ram et al., 1997). These effects are implicated in motor learning and classical conditioning, two types of procedural learning (Brons and Woody, 1980; Ito, 1989). Since LTP in hippocampus is thought to be involved in declarative learning (Squire, 1992), it is possible that these different cellular mechanisms and types of learning at the behavioral level may involve, in part, similar underlying biochemical mechanisms.

Acknowledgements

We thank Y.-F. Lu and E.R. Kandel for their comments, C. Lam for preparing the figures, and M. Pellan and H. Ayers for typing the manuscript. Preparation of this article was supported in part by grants from the National Institute of Mental Health (MH50733) and the Howard Hughes Medical Institute.

References

Abraham, W.C., Mason, S.E., Demmer, T., Williams, J.M., Richardson, C.L., Tate, W.P., Lawlor, P.A. and Dragonow, M. (1993) Correlations between immediate early gene induction and the persistence of long-term potentiation. *Neuroscience*, 56: 717–727.

Andersen, P., Sundberg, S.H., Sveen, O. and Wigström, H. (1977) Specific long-lasting potentiation of synaptic transmission in hippocampal slices. *Nature*, 266: 736–737.

Arancio, O., Kandel, E.R. and Hawkins, R.D. (1995) Activity-dependent long-term enhancement of transmitter release by pre-synaptic 3′,5′-cyclic cGMP in cultured hippocampal neurons. *Nature*, 376: 74–80.

Arancio, O., Kiebler, M., Lee, C.J., Lev-Ram, V., Tsien, R.Y., Kandel, E.R. and Hawkins, R.D. (1996) Nitric oxide acts directly in the pre-synaptic neuron to produce long-term potentiation in cultured hippocampal neurons. *Cell*, 87: 1025–1035.

Arancio, O., Wood, J., Lawrence, D. and Hawkins, R.D. (1997) Presynaptic cGMP-dependent protein kinase is involved in LTP in cultured hippocampal neurons. *Soc. Neurosci. Abstr.*, 23: 1393.

Bannerman, D.M., Chapman, P.F., Kelly, P.A.T. and Morris, R.G.M. (1994) Inhibition of nitric oxide synthase does not prevent induction of long-term potentiation in vivo. *J. Neuroscience*, 14: 7415–7425.

Bao, J.-X., Kandel, E.R. and Hawkins, R.D. (1998) Involvement of pre- and post-synaptic mechanisms in an analog of classical conditioning at *Aplysia* sensory-motor neuron synapses in isolated cell culture. *J. Neuroscience*, 18: 458–466.

Barati, C.M. and Kopf, S.R. (1996) A nitric oxide synthase inhibitor impairs memory storage in mice. *Neurobiol. Learning and Memory*, 65: 197–201.

Bekkers, J.M. and Stevens, C.F. (1990) Presynaptic mechanisms for long-term potentiation in the hippocampus. *Nature*, 346: 724–729.

Bliss, T.V.P. and Collingridge, G.L. (1993) A synaptic model of memory: Long-term potentiation in the hippocampus. *Nature*, 361: 31–39.

Bliss, T.V.P. and Lomo, T. (1973) Long-lasting potentiation of synaptic transmission in the dentate area of the anesthetized rabbit following stimulation of the perforant path. *J. Physiol. (Lond.)*, 232: 331–356.

Blitzer, R.D., Wong, T., Nouranifar, R., Iyengar, R. and Landau, E.M. (1995) Postsynaptic cAMP pathway gates early LTP in hippocampal CA1 region. *Neuron*, 15: 1403–1414.

Bohme, G.A., Bon, C., Stutzmann, J.M., Doble, A. and Blanchard, J.C. (1991) Possible involvement of nitric oxide in long-term potentiation. *Eur. J. Pharmacol.*, 199: 379–381.

Bonhoeffer, T., Staiger, V. and Aertsen, A. (1989) Synaptic plasticity in rat hippocampal slice cultures: Local "Hebbian" conjunction of pre- and post-synaptic stimulation leads to distributed synaptic enhancement. *Proc. Natl. Acad. Sci. USA*, 86: 8113–8117.

Bolshakov, V.Y., Golan, H., Kandel, E.R. and Siegelbaum, S.A. (1997) Recruitment of new sites of synaptic transmission during the cAMP-dependent late phase of LTP at CA3-CA1 synapses in the hippocampus. *Neuron*, 1: 635–651.

Boulton, C.L., Irving, A.J., Southam, E., Potier, B., Garthwaite, J. and Collingridge, G.L. (1994) The nitric oxide-cyclic GMP pathway and synaptic depression in rat hippocampal slices. *Eur. J. Pharmacol.*, 6: 1528–1535.

Boulton, C.L., Southam, E. and Garthwaite, J. (1995) Nitric-oxide dependent long-term potentiation is blocked by a specific inhibitor of soluble guanylyl cyclase. *Neuroscience*, 69, 699–703.

Bourtchouladze, R., Frenguelli, B., Blendy, J., Cioffi, D., Schutz, G. and Silva, A.J. (1994) Deficient long-term memory in mice with a targeted mutation of the cAMP-responsive element-binding protein. *Cell*, 79: 59–68.

Bradley, T., Zhang, Y., Bakin, R., Lester, H.A., Ronnett, G.V. and Zin, K. (1997) Functional expression of the heteromeric "olfactory" cyclic nucleotide-gated channel in the hippocampus: A potential effector of synaptic plasticity in brain neurons. *J. Neurosci.*, 17: 1993–2005.

Bredt, D.S., Glatt, C.E., Huang, P.M., Fotuhi, M., Dawson, T.M. and Snyder, S.H. (1991) Nitric oxide synthase protein and mRNA are discretely localized in neuronal populations of the mammalian CNS together with NADPH diaphorase. *Neuron*, 7: 615–624.

Brenman, J.E., Chao, D.S., Gee, S.H., McGee, A.W., Craven, S.E., Santillano, D.R., Wu, Z., Huang, F., Xia, H., Peters, M.F. et al. (1996) Interaction of nitric oxide synthase with the post-synaptic density protein PSD-95 and a-1-syntrophin mediated by PDZ domains. *Cell*, 84: 757–767.

Brons, J.F. and Woody, C.D. (1980) Long-term changes in excitability of cortical neurons after Pavlovian conditioning and extinction. *J. Neurophysiol.*, 44: 605–615.

Castillo, P.E., Janz, R., Sudhof, T.C., Tzounopoulos, T., Malenka, R.C. and Nicoll, R.A. (1997) Rab3A is essential for mossy fibre long-term potentiation in the hippocampus. *Nature*, 388: 590–593.

Chetkovich, D.M., Klann, E. and Sweatt, J.D. (1993) Nitric oxide synthase-independent long-term potentiation in area CA1 of hippocampus. *NeuroReport*, 4: 919–922.

Clark, G.D., Happel, L.T., Zorumski, C.F. and Bazan, N.G. (1992) Enhancement of hippocampal excitatory synaptic transmission by platelet-activating factor. *Neuron*, 9: 1211–1216.

Colino, A., Huang, Y.Y. and Malenka, R.C. (1992) Characterization of the integration time for the stabilization of long-term potentiation in area CA1 of the hippocampus. *J. Neurosci.*, 12: 180–197.

Collingridge, G.L., Kehl, S.J. and McLennan, H. (1983) Excitatory amino acids in synaptic transmission in the Schaffer collateral-commissural pathway of the rat hippocampus. *J. Physiol. (Lond.)*, 334: 33–46.

Cummings, J.A., Nicola, S.M. and Malenka, R.C. (1994) Induction in the rat hippocampus of long-term potentiation (LTP) and long-term depression (LTD) in the presence of a nitric oxide synthase inhibitor. *Neurosci. Lett.*, 176: 110–114.

Davis, S., Butcher, S.P. and Morris, R.G.M. (1992) The NMDA receptor antagonist D-2-amino-5-phosphonopentanoate (D-AP5) impairs spatial learning and LTP in vivo at intracerebral concentrations comparable to those that block LTP in vitro. *J. Neurosci.*, 12: 21–34.

Del Castillo, J. and Katz, B. (1954) Quantal components of the end-plate potential. *J. Physiol. (Lond.)*, 124: 560–573.

Del Cerro, S., Arai, A. and Lynch, G. (1990) Inhibition of long-term potentiation by an antagonist of platelet-activating factor receptors. *Behav. Neural. Biol.*, 54: 213–217.

DeVente, J., Bol, J.G.J.M., Hudson, L., Schipper, J. and Steinbusch, H.W.M. (1988) Atrial natriuretic factor-responding and cyclic guanosine monophosphate (cGMP)-producing cells in the rat hippocampus: A combined micropharmacological and immunocytochemical approach. *Brain Res.*, 446: 387–395.

Dinerman, J.L., Dawson, T.M., Schell, M.J., Snowman, A. and Snyder, S.H. (1994) Endothelial nitric oxide synthase localized to hippocampal pyramidal cells: Implications for synaptic plasticity. *Proc. Nat. Acad. Sci. USA*, 91: 4214–4218.

Doerner, D. and Alger, B.E. (1988) Cyclic GMP depresses hippocampal Ca^{2+} current through a mechanism independent of cGMP-dependent protein kinase. *Neuron*, 1: 693–699.

Doyle, C., Holscher, C., Rowan, M.J. and Anwyl, R. (1996) The selective neuronal NO synthase inhibitor 7-nitro-indazole blocks both long-term potentiation and depotentiation of field EPSPs in rat hippocampal CA1 in vivo. *J. Neuroscience*, 16: 418–424.

Duman, R.S., Terwillinger, R.Z. and Nestler, E.J. (1993) Alterations in nitric oxide-stimulated endogenous ADP-ribosylation associated with long-term potentiation in rat hippocampus. *J. Neurochem.*, 61: 1542–1545.

East, S.J. and Garthwaite, J. (1991) NMDA receptor activation in rat hippocampus induces cyclic GMP formation through the L-arginine-nitric oxide pathway. *Neurosci. Lett.*, 123: 17–19.

El-Husseini, A.E.-D., Bladen, C. and Vincent, S.R. (1995) Molecular characterization of type II cyclic GMP-dependent protein kinase expressed in the rat brain. *J. Neurochem.*, 64: 2814–2817.

Ewing, J.F. and Maines, M.D. (1992) In situ hybridization and immunohistochemical localization of heme oxygenase-2 mRNA and protein in normal rat brain: Differential distribution of isozyme 1 and 2. *Mol. Cell. Neurosci.*, 3: 559–570.

Furuichi, T., Furutama, D., Hakamata, Y., Nakai, T., Takeshima, H. and Mikoshiba, K. (1994) Multiple types of ryanodine receptor/Ca^{2+} release channels are differentially expressed in rabbit brain. *J. Neurosci.*, 14: 4794–4805.

Fyske, E.M., Li, C. and Sudhof, T.C. (1995) Phosphorylation of rabphilin-3A by Ca^{2+}/calmodulin- and cAMP-dependent protein kinases in vitro. *J. Neurosci.*, 15: 2385–2395.

Gage, A.T., Reyes, M. and Stanton, P.K. (1997) Nitric-oxide-guanylyl-cyclase-dependent and independent components of multiple forms of long-term synaptic depression. *Hippocampus*, 7: 286–295.

Galione, A., White, A., Willmott, N., Turner, M., Potter, B.V.L. and Watson, S.P. (1993) cGMP mobilizes intracellular Ca^{2+} in sea urchin eggs by stimulating cyclic ADP-ribose synthesis. *Nature*, 365: 456–459.

Gally, J.A., Montague, P.R., Reeke, G.N., Jr. and Edelman, G.M. (1990) The NO hypothesis: Possible effects of a short-lived, rapidly diffusible signal in the development and function of the nervous system. *Proc. Natl. Acad. Sci. USA*, 87: 3547–3551.

Garthwaite, J., Charles, S.L. and Chess-Williams, R. (1988) Endothelium-derived relaxing factor release on activation of NMDA receptors suggests role as intercellular messsenger in the brain. *Nature*, 336: 385–388.

Geppert, M., Bolshakov, V.Y., Siegelbaum, S.A., Takei, K., DeCamilli, P., Hammer, R.E. and Sudhof, T.C. (1994) The role of Rab3A in neurotransmitter release. *Nature*, 369: 493–497.

Giannini, G., Conti, A., Mammarella, S., Scrobogna, M. and Sorrentino, V. (1995) The ryanodine receptor/calcium channel genes are widely and differentially expressed in murine brain and peripheral tissues. *J. Cell Biol.*, 128: 893–904.

Gribkoff, V.K. and Lum-Ragan, J.T. (1992) Evidence for nitric oxide synthase inhibitor-sensitive and insensitive hippocampal synaptic potentiation. *J. Neurophysiol.*, 68: 639–642.

Gudi, T., Huvar, I., Meinecke, M., Lohmann, S.M., Boss, G.R. and Pilz, R.B. (1996) Regulation of gene expression by cGMP-dependent protein kinase. *J. Biol. Chem.*, 271: 4597–4600.

Haby, C., Lisovoski, F., Aunis, D. and Zwiller, J. (1994) Stimulation of the cyclic GMP pathway by NO induces expression of the immediate early genes c-fos and junB in PC12 cells. *J. Neurochem.*, 62: 496–501.

Haley, J.E., Malen, P.L. and Chapman, P.F. (1993) Nitric oxide synthase inhibitors block long-term potentiation induced by weak but not strong tetanic stimulation at physiological brain temperatures in rat hippocampal slices. *Neurosci. Lett.*, 160: 85–88.

Haley, J., Schaible, E., Paulidis, P., Murdock, A. and Madison, D.V. (1996) Basal and apical synapses of CA1 pyramidal cells employ different LTP induction mechanisms. *Learning and Memory*, 3: 289–295.

Haley, J.E., Wilcox, G.L. and Chapman, P.F. (1992) The role of nitric oxide in hippocampal long-term potentiation. *Neuron*, 8: 211–216.

Hawkins, R.D. (1996) NO Honey, I don't remember. *Neuron*, 16: 465–467.

Hawkins, R.D., Abrams, T.W., Carew, T.J. and Kandel, E.R. (1983) A cellular mechanism of classical conditioning in *Aplysia*: Activity-dependent amplification of pre-synaptic facilitation. *Science*, 219: 400–405.

Hawkins, R.D., Kandel, E.R. and Siegelbaum, S.A. (1993) Learning to modulate transmitter release: Themes and variations in synaptic plasticity. *Annu. Rev. Neurosci.*, 16: 625–665.

Hebb, D.O. (1949) *The organizatoin of behavior: A neuropsychological theory*. Wiley, New York.

Hirling, H. and Scheller, R.H. (1996) Phosphorylation of synaptic vesicle proteins: Modulation of the αSNAP interaction with the core complex. *Proc. Natl. Acad. Sci. USA*, 93: 11945–11949.

Holscher, C., McGlinchey, L, Anwyl, R. and Rowan, M.J. (1995) 7-nitro indazole, a selective neuronal niric oxide synthase inhibitor in vivo, impairs spatial learning in the rat. *Learning and Memory*, 2: 267–278.

Ignarro, L.J., Ballot, B. and Wood, K.S. (1984) Regulation of soluble guanylate cyclase activity by porphyrins and metalloporphyrins. *J. Biol. Chem.*, 259: 6201–6207.

Ikegaya, Y., Saito, H. and Matsuki, N. (1994) Involvement of carbon monoxide in long-term potentiation in the dentate gyrus of anesthetized rats. *Jpn. J. Pharmacol.*, 64: 225–227.

Ito, M. (1989) Long-term depression. *Annu. Rev. Neurosci.*, 12: 85–102.

Ito, M. and Karachot, L. (1992) Protein kinases and phosphatase inhibitors mediating long-term desensitization of glutamate receptors in cerebellar Purkinje cells. *Neuroscience Research*, 14: 27–38.

Kandel, E.R., Abrams, T., Bernier, L., Carew, T.J., Hawkins, R.D. and Shwartz, J.H. (1993) Classical conditioning and sensitization share aspects of the same molecular cascade in *Aplysia*. Cold Spring Harbor Symp. *Quant. Biol.*, 48: 821–830.

Kang, H. and Schuman, E.M. (1995) Long-lasting neurotrophin-induced enhancement of synaptic transmission in the adult hippocampus. *Science*, 267: 1658–1662.

Kantor, D.B., Lanzrein, J., Stary, S.J., Sandoval, G.M., Smith, W.B., Sullivan, B.M., Davidson, N. and Schuman, E.M. (1996) A role for endothelial NO synthase in LTP revealed by adenovirus-mediated inhibition and rescue. *Science*, 274: 1744–1748.

Kato, K. and Zorumski, C.F. (1993) Nitric oxide inhibitors facilitate the induction of hippocampal long-term potentiation by modulating NMDA response. *J. Neurophysiol.*, 70: 1260–1263.

Kelso, S.R., Ganong, A.H. and Brown, T.H. (1986) Hebbian synapses in hippocampus. *Proc. Natl. Acad. Sci. USA*, 83: 5326–5330.

Kendrick, K.M., Guevara-Guzman, R., Zorilla, T., Hinton, M.R. Broad, K.D., Mimmack, M. and Ohkura, S. (1997) Formation of olfactory memories mediated by nitric oxide. *Nature*, 388: 670–674.

Kingston, P.A., Zufall, F. and Barnstable, C.T. (1996) Rat hippocampal neurons express genes for both rod retinal and olfactory cyclic nucleotide-gated channels: Novel targets for cAMP/cGMP function. *Proc. Natl. Acad. Sci. USA*, 93: 10440–10445.

Korte, M., Griesbeck, O., Grand, C., Carroll, P., Staiger, V., Thoenen, H. and Bonhoffer, T. (1996) Virus-mediated gene transfer into hippocampal CA1 region restores long-term potentiation in brain-derived neurotrophic factor mutant mice. *Proc. Natl. Acad. Sci. USA*, 93: 12547–12552.

Kullman, D.M. and Siegelbaum, S.A. (1995) The site of expression of NMDA receptor-dependent LTP: New fuel for an old fire. *Neuron*, 15: 997–1002.

Lee, H.C. (1993) Potentiation of calcium- and caffeine-induced calcium release by cyclic ADP-ribose. *J. Biol. Chem.*, 268: 293–299.

Leinders-Zufall, T., Rosenboom, H., Barnstable, C.J., Shepherd, G.M. and Zufall, F. (1995) A calcium-permeable cGMP-activated cation conductance in hippocampal neurons. *Neuroreport*, 6: 1761–1765.

Lev-Ram, V., Jiang, T., Wood, J., Lawrence, D.S. and Tsien, R.Y. (1997) Synergies and coincidence requirements between NO, cGMP and Ca^{2+} in the induction of cerebellar long-term depression. *Neuron*, 18: 1025–1038.

Lynch, G., Larson, J., Kelso, S., Barrionuevo, G. and Schottler, F. (1983) Intracellular injection of EGTA blocks induction of hippocampal long-term potentiation. *Nature*, 305: 719–721.

Lynch, M.A., Errington, M.L. and Bliss, T.V.P. (1989) Nordihydroguaiaretic acid blocks the synaptic component of long-term potentiation and the associated increase in release of glutamate and arachidonic acid: An in vivo study in the dentate gyrus of the rat. *Neuroscience*, 30: 693–701.

Malen, P. and Chapman, P.F. (1997) Nitric oxide facilitates long-term potentiation, but not long-term depression. *J. Neurosci.*, 17: 2645–2651.

Malenka, R.C., Kauer, J.A., Zucker, R.S. and Nicoll, R.A. (1988) Postsynaptic calcium is sufficient for potentiation of hippocampal synaptic transmission. *Science*, 242: 81–84.

Malenka, R.C., Lancaster, B. and Zucker, R.S. (1992) Temporal limits on the rise in post-synaptic calcium required for the induction of long-term potentiation. *Neuron*, 9: 121–128.

Malinow, R. and Miller, J.P. (1986) Postsynaptic hyperpolarization during conditioning reversibly blocks induction of long-term potentiation. *Nature*, 320: 529–530.

Marks, G.S., Brien, J.F., Nakatsu, K. and McLaughlin, B.E. (1991) Does carbon monoxide have a physiological function? *TIPS*, 12: 185–188.

Matsuoka, J., Guili, G., Poyard, M., Stengel, D., Parma, J., Guellaen, G. and Hanoune, J. (1992) Localization of adenylyl and guanylyl cyclase in rat brain by *in situ* hybridization: Comparison with calmodulin mRNA distribution. *J. Neurosci.*, 12: 3350–3360.

Meffert, M.K., Calakos, N.C., Scheller, R.H. and Schulman, H. (1996) Nitric oxide modulates synaptic vesicle docking/fusion reactions. *Neuron*, 16: 1229–1236.

Meffert, M.K., Haley, J.E., Schuman, E.M., Schulman, H. and Madison, D.V. (1994a) Inhibition of hippocampal heme oxygenase, nitric oxide synthase and long-term potentiation by metalloporphyrins. *Neuron*, 13: 1225–1233.

Meffert, M., Premack, B.A. and Schulman, H. (1994b) Nitric oxide stimulates Ca^{2+}-independent synaptic vesicle release. *Neuron*, 12: 1235–1244.

Mizutani, A., Saito, H. and Abe, K. (1993) Involvement of nitric oxide in long-term potentiation in the dentate gyrus in vivo. *Brain Res.*, 605: 309–311.

Morris, R.G.M., Anderson, E., Lynch, G.S. and Baudry, M. (1986) Selective impairment of learning and blockade of long-term potentiation by *N*-methyl-D-aspartate antagonist, AP-5. *Nature*, 319: 774–776.

Murphy, K.P.S.J., Williams, J.H., Bettache, N. and Bliss, T.V.P. (1994) Photolytic release of nitric oxide modulates NMDA receptor-mediated transmission but does not induce long-term potentiation at hippocampal synapses. *Neuropharmacology*, 33: 1375–1385.

Myslivecek, J., Hassmannova, J., Barcal, J., Safanda, J. and Zalud, V. (1996) Inhibitory learning and memory in newborn rats influenced by nitric oxide. *Neuroscience*, 71: 299–312.

Nayak, A.S., Moore, C.I. and Browning, M.D. (1996) Ca^{2+}/calmodulin-dependent protein kinase II phosphorylation of the pre-synaptic protein synapsin I is persistently increased during long-term potentiation. *Proc. Natl. Acad. Sci. USA*, 93: 15451–15456.

Ocorr, K.A., Walters, E.T. and Byrne, J.H. (1985) Associative conditioning analog selectively increases cAMP levels of tail sensory neurons in *Aplysia*. *Proc. Natl. Acad. Sci. USA*, 82: 2548–2552.

O'Dell, T.J., Hawkins, R.D., Kandel, E.R. and Arancio, O. (1991) Tests of the roles of two diffusible substances in long-term potentiation: Evidence for nitric oxide as a possible early retrograde messenger. *Proc. Natl. Acad. Sci USA*, 88: 11285–11289.

O'Dell, T.J., Huang, P.L., Dawson, T.M., Dinerman, J.L., Snyder, S.H., Kandel E.R. and Fishman, M.C. (1994) Endothelial NOS and blockade of LTP by NOS inhibitors in mice lacking neuronal NOS. *Science*, 265: 542–546.

Okere, C.O., Kaba, H. and Higuchi, T. (1996) Formation of an olfactory recognition memory in mice: Reassessment of the role of nitric oxide. *Neuroscience*, 71: 349–354.

Peunova, N. and Enlkolopov, G. (1993) Amplication of calcium-induced gene transcription by nitric oxide in neuronal cells. *Nature*, 364: 450–453.

Poss, K.D., Thomas, M., Ebralidze, A.K., O'Dell, T.J. and Tonegawa, S. (1995) Hippocampal long-term potentiation is normal in heme oxygenase-2 mutant mice. *Neuron*, 15: 867–873.

Repasko, D.R., Corbin, J.G., Conti, M. and Goy, M.F. (1993) A cyclic GMP-stimulated cyclic nucleotide phosphodiesterase gene is highly expressed in the limbic system of the rat brain. *Neuroscience*, 56: 673–686.

Sastry, B.R., Goh, J.W. and Auyeung, A. (1986) Associative induction of post-tetanic and long-term potentiation in CA1 of rat hippocampus. *Science*, 232: 988–990.

Schuman, E.M. and Madison, D.V. (1991) A requirement for the intracellular messsenger nitric oxide in long-term potentiation. *Science*, 254: 1503–1506.

Schuman, E.M. and Madison, D.V. (1994) Locally distributed synaptic potentiation in the hippocampus. *Science*, 263: 532–536.

Schuman, E.M. Meffert, M.K., Schulman, H. and Madison, D.V. (1994) An ADP-ribosyl transferase as a potential target for nitric oxide action in hippocampal long-term potentiation. *Proc. Natl. Acad. Sci. USA*, 91: 11958–11962.

Scoville, W.B. and Milner, B. (1957) Loss of recent memory after bilateral hippocampal lesions. *J. Neurol. Neurosurg. Psych.*, 20: 11–21.

Selig, D.K., Segal, M.R., Liao, D., Malenka, R.C., Malinow, R., Nicoll, R.A. and Lisman, J.E. (1996) Examination of the role of cGMP in long-term potentiation in the CA1 region of the hippocampus. *Learning and Memory*, 3: 42–48.

Shibuki, K. and Okada, D. (1991) Endogenous nitric oxide release required for long-term synaptic depression in the cerebellum. *Nature*, 349: 326–328.

Snider, R.M., McKinney, M., Forray, C. and Richelson, E. (1984) Neurotransmitter receptors mediate cyclic GMP formation by involvement of arachidonic acid and lipoxygenase. *Proc. Natl. Acad. Sci. USA*, 81: 3905–3909.

Son, H., Hawkins, R.D., Martin, K., Kiebler, M., Huang, P.L., Fishman, M.C. and Kandel, E.R. (1996) Long-term potentiation is reduced in mice that are doubly mutant in endothelial and neuronal nitric oxide synthase. *Cell*, 87: 1015–1023.

Son, H., Lu, Y.-F., Zhuo, M., Arancio, O., Kandel, E.R. and Hawkins, R.D. (1998) The specific role of cGMP in hippocampal LTP. *Learning and Memory*, 5: 231–245.

Sporns, O. and Jenkinson, S. (1997) Potassium ion- and nitric oxide-induced exocytosis from populations of hippocampal synapses during synaptic maturation in vitro. *Neuroscience*, 80: 1057–1073.

Squire, L.R. (1992) Memory and the hippocampus: A synthesis from findings with rats, monkeys and humans. *Psychol. Rev.*, 99: 195–231.

Stevens, C.F. and Wang, Y. (1993) Reversal of long-term potentiation by inhibitors of heme oxygenase. *Nature*, 364: 147–149.

Sullivan, B.M., Wong, S. and Schuman, E.M. (1997) Modification of hippocampal synaptic proteins by nitric-oxide stimulated ADP ribosylation. *Learning and Memory*, 3: 414–424.

Thoenen, H. (1995) Neurotrophins and neuronal plasticity. *Science*, 270: 593–598.

Tokui, T., Brozovich, F. Ando, S. and Ikebe, M. (1996) Enhancement of smooth muscle contraction with protein phosphatase inhibitor 1: Activation of inhibitor 1 by cGMP-dependent protein kinase. *Biochem. and Biophys. Res. Commun.*, 220: 777–783.

Tsien, J. Z., Huerta, P.T. and Tonegawa, S. (1996) The essential role of hippocampal CA1 NMDA receptor-dependent synaptic plasticity in spatial memory. *Cell*, 87: 1327–1338.

Tsou, K.T., Snyder, G.L. and Greengard, P. (1993) Nitric oxide/cGMP pathway stimulates phosphorylation of DARPP-32, a dopamine- and cAMP-regulated phosphoprotein, in the substantia nigra. *Proc. Natl. Acad. Sci USA*, 90: 3462–3465.

Verma, A., Hirsch, D.J., Glatt, C.E., Ronnett, G.V. and Snyder, S.H. (1993) Carbon monoxide: A putative neural messenger. *Science*, 259: 381–384.

Walters, E.T. and Byrne, J.H. (1983) Associative conditioning of single sensory neurons suggests a cellular mechanism for learning. *Science*, 219: 405–408.

Wang, X. and Robinson, P.J. (1995) Cyclic GMP-dependent protein kinases in rat brain. *J. Neurochem.*, 65: 595–604.

Wang, Y., Wu, J., Rowan, M.J. and Anwyl, R. (1996) Ryanodine produces a low frequency stimulation-induced NMDA receptor-independent long-term potentiation in the rat dentate gyrus in vitro. *J. Physiol.*, 495: 755–767.

Wendland, B., Schweizer, F.E., Ryan, T.A., Nakane, M., Murad, F., Scheller, R.H. and Tsein, R.W. (1994) Existence of nitric oxide synthase in rat hippocampal pyramidal cells. *Proc. Natl. Acad. Sci. USA*, 91: 2151–2155.

Wieraszko, A., Li, G., Kornecki, E., Hogan, M.V. and Ehrlich, Y.H. (1993) Long-term potentiation in the hippocampus by platelet-activating factor. *Neuron*, 10: 553–557.

Wigström, H., Gustafsson, B., Huang, Y.Y. and Abraham, W.C. (1986) Hippocampal long-lasting potentiation is induced by pairing single afferent volley with intracellularly injected depolarizing current pulses. *Acta Physiol. Scand.*, 126: 317–319.

Williams, J.H. and Bliss, T.V.P. (1989) An in vitro study of the effect of lipoxygenase and cyclo-oxygenase inhibitors of arachidonic acid on the induction and maintenance of long-term potentiation in the hippocampus. *Neurosci. Lett.*, 107: 301–306.

Williams, J.H., Errington, M.L., Lynch, M.A. and Bliss, T.V.P. (1989) Arachidonic acid induces a long-term activity-dependent enhancement of synaptic transmission in the hippocampus. *Nature*, 341: 739–742.

Williams, J.H., Li, Y.G., Nayak, A., Errington, M.L., Murphy, K.P.S.J. and Bliss, T.V.P. (1993) The suppression of long-term potentiation in rat hippocampus by inhibitors of nitric oxide synthase is temperature and age dependent. *Neuron*, 11: 877–884.

Woody, C.D., Swartz, B.E. and Gruen, E. (1978) Effects of acetylcholine and cyclic GMP on input resistance of cortical neurons in awake cats. *Brain Res.*, 158: 373–395.

Wu, J., Wang, Y., Rowan, M.J. and Anwyl, R. (1997) Evidence for involvement of the neuronal isoform of nitric oxide synthase during induction of long-term potentiation and long-term depression in the rat dentate gyrus in vitro. *Neuroscience*, 78: 393–398.

Yamada, K. et al. (1996) Role of nitric oxide and cyclic GMP in the dizocilpine-induced impairment of spantaneous alternation behavior in mice. *Neuroscience*, 74: 365–374.

Yanagihara, D. and Kondo, I. (1996) Nitric oxide plays a key role in adaptive control of locomotion in cat. *Proc. Natl. Acad. Sci. USA*, 93: 13292–13297.

Zakhary, R., Gaine, S.P., Dinerman, J.L., Ruat, M., Flavahan, N.A. and Snyder, S.H. (1996) Heme oxygenase 2: Endothelial and neuronal localization and role in endothelium-dependent relaxation. *Proc. Natl. Acad. Sci. USA*, 93: 795–798.

Zalutsky, R.A. and Nicoll, R.A. (1990) Comparison of two forms of long-term potentiation in single hippocampal neurons. *Science*, 248: 1619–1624.

Zhuo, M., Hu, Y., Schultz, C., Kandel, E.R. and Hawkins, R.D. (1994a) Role of guanylyl cyclase and cGMP-dependent protein kinase in long-term potentiation. *Nature*, 368: 635–639.

Zhuo, M., Kandel, E.R. and Hawkins, R.D. (1994b) Nitric oxide and cGMP can produce either synaptic depression or potentiation depending on the frequency of pre-synaptic stimulation in hippocampus. *NeuroReport*, 5: 1033–1036.

Zhuo, M., Li, X.-C. and Hawkins, R.D. (1994c) Role of heme oxygenase and guanylyl cyclase in long-term potentiation by activation of metabotropic glutamate receptors in hippocampus. *Soc. Neurosci. Abstr.*, 20: 445.

Zhuo, M., Small, S.A., Kandel, ER. and Hawkins, R.D. (1993) Nitric oxide and carbon monoxide produce activity-dependent long-term synaptics enhancement in hippocampus. *Science*, 260: 1946–1950.

Zola-Morgan, S.M. and Squire, L.R. (1990) The primate hippocampal formation: Evidence for a time-limited role in memory storage. *Science*, 250: 288–290.

R.R. Mize, T.M. Dawson, V.L. Dawson and M.J. Friedlander (Eds.)
Progress in Brain Research, Vol 118

CHAPTER 12

Modulation of LTP induction by NMDA receptor activation and nitric oxide release

Charles F. Zorumski* and Yukitoshi Izumi

Departments of Psychiatry and Neurobiology, Washington University School of Medicine, 4940 Children's Place, St. Louis, MO 63110, USA

Abstract

In the CA1 hippocampal region, the induction of long-term potentiation (LTP) requires activation of *N*-methyl-D-aspartate receptors (NMDARs). However, untimely NMDAR activation either immediately prior to or following tetanic stimulation inhibits LTP generation. This NMDAR-mediated LTP inhibition is overcome by inhibitors of nitric oxide synthase (NOS) and hemoglobin, suggesting the involvement of NO. Additionally, NO inhibitors can promote the ability of weak tetanic stimuli to produce LTP under basal conditions in hippocampal slices. Recent experiments indicate that untimely NMDAR activation contributes to the failure of LTP induction during periods of low glucose exposure and hypoxia. Following hypoxia there is also a delayed form of LTP inhibition that is reversed by NMDAR antagonists and NO inhibitors. These results suggest that there are physiological and pathological conditions during which NMDAR activation and NO release modulate the induction of synaptic plasticity.

*Corresponding author. Tel.: +1314 747 2680; fax: 314 747 2682; e-mail: zorumskc@psychiatry.wustl.edu

Introduction

Long-term potentiation (LTP) is a persistent form of synaptic plasticity that results when certain synapses are used at high frequency or when lower frequency synaptic activation is coupled with strong depolarization of postsynaptic neurons. LTP satisfies a number of criteria that theorists have postulated are important for learning and memory (Bliss and Collingridge, 1993) and is presently a leading candidate to be a cellular correlate responsible for Hebbian or associative learning (Chen and Tonegawa, 1997).

Although many aspects of LTP remain controversial, a number of features seem clear. In the CA1 region of the hippocampus, LTP induction typically requires activation of *N*-methyl-D-aspartate receptors (NMDARs) and calcium influx into postsynaptic neurons (Bliss and Collingridge, 1993; Malenka, 1994). Calcium activates a host of enzymes in postsynaptic cells including calcium–calmodulin-dependent protein kinase II (CamKII), protein kinase C (PKC), and perhaps phospholipase A2 and nitric oxide synthase (NOS). Whether LTP is ultimately expressed by changes in presynaptic or postsynaptic function is less certain and reasonably strong evidence exists in favor of either possibility (Chen and Tonegawa, 1997) (Fig. 1). If LTP results from changes in the

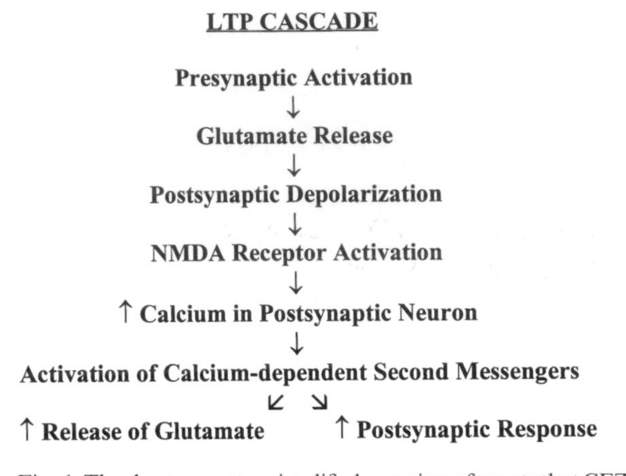

Fig. 1. The chart presents a simplified overview of events that CFZ contribute to LTP induction in the CA1 hippocampal region.

release of transmitter (glutamate) from presynaptic terminals, but requires specific postsynaptic events (e.g. NMDAR activation, membrane depolarization and calcium influx among others), a messenger is needed to communicate changes occurring in the postsynaptic cell to the presynaptic terminal (Zorumski and Izumi, 1993). Because presynaptic terminals and postsynaptic neurons are not contiguous, the "retrograde" messenger must diffuse from the postsynaptic cell and influence the presynaptic machinery involved in transmitter release. Presently, nitric oxide (NO) and arachidonic acid are the leading, although controversial, candidates to be a retrograde messenger (Zorumski and Izumi, 1993; Garthwaite and Boulton, 1995; Chen and Tonegawa, 1997). A corollary of the retrograde messenger hypothesis is that because the messenger is a diffusible substance, it is likely that expression of LTP at a single synapse will influence other synapses in a local region. Indeed, several experiments in hippocampal slices are consistent with this possibility. These experiments have shown that there is spread of enhancement to synapses that are very close to those activated during LTP induction (Bonhoeffer et al., 1989; Schuman and Madison, 1994; Engert and Bonhoeffer, 1997).

In the early 1990s, Dudek and Bear (1992) observed that prolonged low frequency activation of synapses in the CA1 hippocampal region results in homosynaptic long-term depression (LTD) of responses. Other investigators demonstrated that repeated low frequency stimulation also reverses previously established LTP, a process referred to as "depotentiation" (Fujii et al., 1991). Importantly, LTD and depotentiation, like LTP, require NMDAR activation and calcium influx into postsynaptic neurons (Mulkey and Malenka, 1992). Key factors in determining whether LTD or LTP is produced include the degree of postsynaptic membrane depolarization and perhaps the amount of calcium influx and the nature of the second messengers that are activated (Cummings et al., 1996). The activation of specific protein kinases appears to favor LTP induction while activation of protein phosphatases promotes LTD and depotentiation (Mulkey et al., 1993, 1994; O'Dell and Kandel, 1994).

To make matters even more complicated, several groups also observed that untimely activation of NMDARs prior to delivery of a high frequency tetanus that typically produces LTP, results in a failure of LTP generation. The untimely activation of NMDARs can occur during perfusion of hippocampal slices with low magnesium solutions that relieve the voltage-dependent block of NMDARs by magnesium (Coan et al., 1989), weak tetanic stimuli (Huang et al., 1992) or administration of low concentrations of agonists at NMDARs (Izumi et al., 1992a). The conditions that inhibit LTP induction produce no lasting changes in baseline synaptic transmission mediated by α-amino-3-hydroxy-5-methyl-4-isoxazole propionic acid receptors (AMPARs) and are thus distinct from the conditions that induce LTP or LTD. In all of these cases of LTP inhibition, the LTP block is reversed by low concentrations of NMDAR antagonists, strongly supporting the role of untimely NMDAR activation as a critical event. The inhibition of LTP by untimely NMDAR activation appears to reflect a shift in the threshold for LTP induction because, in some cases, higher intensity or repeated tetanic stimuli produce LTP (Huang et al., 1992). These observations indicate that not only are NMDARs important mediators of synaptic plasticity, but they are also important modulators of synaptic plasticity, depending on the timing and perhaps degree of receptor activation. In a broader context, the modulation of LTP induction by NMDARs strongly suggests that synaptic plasticity is a dynamic process subject to its own regulation. Abraham and colleagues have recently referred to this modulation of synaptic plasticity as "metaplasticity", reflecting the potential importance of the process to overall synaptic function and integration (Abraham and Bear, 1996; Abraham and Tate, 1997). In this review, we will discuss studies dealing with the modulation of LTP induction by untimely NMDAR activation produced by treatment of hippocampal slices with low concentrations of NMDAR agonists.

Factors involved in NMDAR-mediated LTP inhibition

At concentrations of about 1 μM, NMDA inhibits LTP induction when administered for 5 min either immediately before or following tetanic stimulation (Izumi et al., 1992a) (Fig. 2). The effects of NMDA are mimicked by low concentrations of aspartate and glutamate, but not by quisqualate or an agonist at metabotropic glutamate receptors (mGluRs). Although higher concentrations of NMDA depress synaptic transmission presynaptically (Manzoni et al., 1994), the concentrations required for LTP inhibition have no effect on baseline synaptic transmission. Similarly, NMDARs exhibit calcium-dependent desensitization (Clark et al., 1990; Legendre et al., 1993) and low concentrations of glutamate desensitize both the NMDARs and AMPARs that participate in fast excitatory synaptic transmission (Zorumski et al., 1996). However, depression of synaptic NMDARs does not appear to contribute to the LTP inhibition produced by 1 μM NMDA. This is based on two observations. First, administration of 1 μM NMDA for 5 min immediately following tetanic stimulation blocks LTP (Izumi et al., 1992a) and NMDAR antagonists are ineffective when administered in the post-tetanic period (Izumi et al., 1991) (Fig. 2). Second, at 1 μM, NMDA has no clear effect on NMDAR-mediated synaptic potentials or currents (Izumi et al., 1992a). In this regard the inhibition of LTP by NMDA may differ from the block of LTP produced by weak tetanic stimuli because weak tetani promote LTD of synaptic responses mediated by NMDARs (Selig et al., 1995).

The LTP inhibition produced by 1 μM NMDA appears to require calcium influx via NMDARs because perfusion of solutions with low calcium during NMDA application allows a subsequent tetanus to produce LTP (Izumi et al., 1992a). This result suggests that calcium-dependent second messengers may contribute to the LTP inhibition.

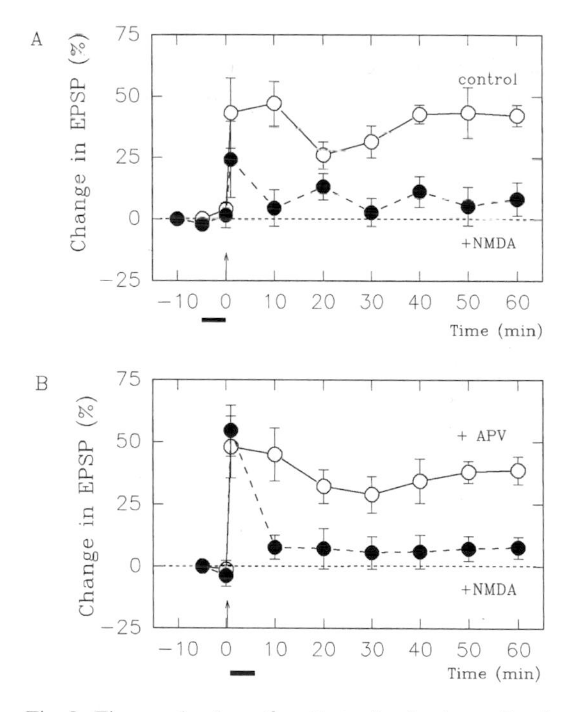

Fig. 2. The graphs show the effect of a 5 min application of 1 μM NMDA on LTP induction when NMDA was administered immediately before (A) or after (B) delivery of a 100 Hz × 1 s tetanus. NMDA was administered for the period denoted by the solid bar and the tetanus was delivered at the arrow. In panel B, the failure of 100 μM 2-amino-5-phosphoprionate (APV) to block LTP when administered after the tetanus is shown (open circles). Data points represent the mean ± SEM for 5–13 slices. In all of these experiments hippocampal slices were prepared from 30-day old albino rats.

Indeed, subsequent studies found that the LTP inhibition produced by 1 μM NMDA is reversed by inhibitors of NOS and hemoglobin, an agent that binds NO (and other chemical species) extracellularly (Izumi et al., 1992c). The LTP inhibition produced by perfusion of low magnesium solutions is also reversed by NOS inhibitors. The NO inhibitors overcome NMDA-mediated LTP inhibition regardless of whether NMDA is administered pre- or post-tetanically. Additionally, sodium nitroprusside (SNP), an agent that releases NO, inhibits LTP at concentrations that do not alter baseline synaptic transmission. The effects of SNP are inhibited by hemoglobin and are not mimicked by light-inactivated SNP, suggesting that NO release is important in the effect.

Basal NO release alters LTP threshold

Given that neurons live in sufficient extracellular glutamate to activate NMDARs tonically (Sah et al., 1989), it is possible that NMDAR-mediated LTP inhibition could shift the threshold for LTP generation under basal conditions via NO release. To examine this, we studied the effects of NO inhibitors on the ability of weak tetanic stimuli to generate LTP. Under baseline conditions, a "weak" tetanus (100 Hz × 0.3 s) produced only a transient increase in synaptic efficacy that faded to baseline over 10–20 min. When the same weak tetanus was administered in the presence of a NOS inhibitor or hemoglobin, LTP was reliably induced (Kato and Zorumski, 1993). Interestingly, the NOS inhibitors augmented synaptic responses mediated by NMDARs suggesting that NO altered LTP induction by diminishing NMDAR responses. How or whether these effects participate in the block of LTP by 1 μM NMDA remains uncertain because the effective concentrations of NMDA do not alter basal NMDAR-mediated EPSPs and NMDA blocks LTP when administered post-tetanically. At the minimum, however, the effects of the NO inhibitors on weak tetanic stimulation suggest that NO may be an important modulator of LTP threshold under certain conditions.

Age dependence of NMDAR-mediated LTP inhibition

The induction of both LTP and homosynaptic LTD in the CA1 hippocampal region shows developmental regulation (Harris and Teyler 1983; Dudek and Bear 1993; Izumi and Zorumski, 1995). When a single tetanic stimulus at an intensity that evokes a half maximal synaptic response is used for induction, the greatest degree of synaptic LTP occurs in hippocampal slices prepared from 15 to 30 day old albino rats. As animals reach adulthood (≥ 120 days of age), the single tetanus produces an early increase in synaptic efficacy that typically fades over the next 60–90 min. Similarly, several studies have found that homosynaptic LTD is easiest to induce in slices prepared from young (< 60 day old) rodents (Dudek and Bear, 1993). Consistent with these observations, we found that NMDAR-mediated LTP inhibition produced by administration of 1 μM NMDA either immediately before or following the tetanus is most prominent in slices prepared from 15 to 30 day old albino rats and that the effect is markedly diminished in slices from 60-day old animals (Fig. 3).

Interestingly, a similar developmental time course is seen in the inhibition of LTP by antagonists of mGluRs that are linked to phosphoinositide (PI) metabolism (Izumi and

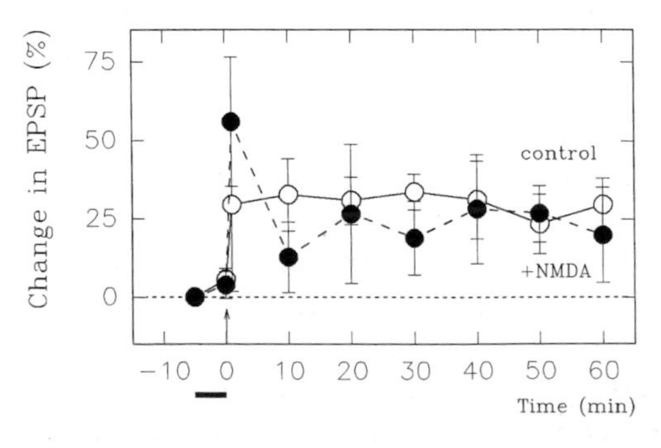

Fig. 3. The graph shows the ability of a single 100 Hz × 1 s tetanus to induce LTP in hippocampal slices prepared from 60-day old albino rats in the absence (n = 14) and presence (n = 5) of 1 μM NMDA (solid bar). At this age, NMDA had no clear effect on LTP induction.

Zorumski, 1994). NMDAR activation has previously been shown to diminish PI turnover produced by mGluRs and muscarinic acetylcholine receptors, but not that linked to α1-adrenergic receptor activation (Palmer et al., 1988; 1990). Furthermore, unlike NMDAR antagonists but like NMDAR activation, mGluR antagonists inhibit LTP induction when administered immediately following the tetanus. Neither the mGluR antagonists nor NMDA alter previously established LTP. These observations suggest that the effects of NMDA on mGluR may contribute to the alteration in LTP induction. The diminishing effect of mGluR antagonists on LTP induction during development is consistent with studies showing that the ability of mGluR agonists to promote PI turnover is greatest in young animals and fades as animals reach adulthood (Nicoletti et al., 1985).

Conditions favoring NMDAR-mediated LTP inhibition

The block of LTP by low concentrations of NMDAR agonists suggests that there are likely to be physiological or pathological conditions during which untimely activation of NMDARs by endogenous extracellular excitatory amino acids alter the ability to produce LTP. Indeed, the ability of NO inhibitors to alter LTP induction by a weak tetanus suggests one possibility (Kato and Zorumski, 1993). However, in certain pathological conditions there are increases in extracellular excitatory amino acids that participate in a cascade of events leading to neuronal death. This is believed to be the case in hypoglycemic and hypoxic neurodegeneration as well as in certain chronic neurodegenerative disorders. Early in the course or perhaps during milder insults that are not associated with significant neuronal loss, it seems possible that increases in excitatory amino acid levels could alter synaptic plasticity either in the absence of or preceding neuronal injury.

To test this, we examined the effects of relatively mild noxious conditions on the ability of a single 100 Hz × 1 s tetanus to induce LTP in the CA1 hippocampal region. The conditions chosen included exposure to low concentrations of glucose or relatively brief bouts of anoxia. Under the conditions chosen, there was no persistent change in basal synaptic transmission or evidence of histological damage to CA1 neurons. Hippocampal slice experiments are typically done in the presence of 5–10 mM extracellular glucose and under these conditions LTP is readily induced. When the glucose concentration is reduced to 2–3.3 mM, a single tetanus fails to generate LTP (Izumi and Zorumski, 1997). Reducing the glucose concentration to 2–3.3 mM after establishing LTP has no effect, suggesting that LTP induction and LTP maintenance have different glucose requirements. LTP inhibition produced by 3.3 mM glucose exposure appears to involve untimely NMDAR activation because it is reversed by low concentrations of NMDAR antagonists. Similarly, the LTP inhibition is reversed by NOS inhibitors applied before and during the low glucose exposure, again suggesting mechanisms in common with NMDAR-mediated LTP inhibition produced by exogenously applied NMDAR agonists. When the low glucose exposure is more severe (2 mM glucose), other factors appear to be involved because the LTP inhibition is unaffected by either NMDAR antagonists or NO inhibitors. At either 2 mM or 3.3 mM glucose, basal synaptic responses mediated by AMPARs or NMDARs are unaffected.

Prior studies have shown that brief exposure to anoxia at the time of tetanic stimulation inhibits LTP induction (Arai et al., 1990; Arai and Lynch, 1992). This acute LTP inhibition is overcome by NMDAR antagonists suggesting again that NMDARs participate in the LTP block (Izumi, unpublished observations). Recently, we observed that brief periods of anoxia have complex effects on synaptic plasticity. Although the periods of anoxia chosen for study do not persistently depress synaptic transmission or produce histological damage to CA1 neurons, synaptic responses mediated by either AMPARs or NMDARs are depressed completely during the anoxic period. These responses recover fully within 10–20 min following reoxygenation and a tetanus delivered 20–30 min after anoxia induces LTP. However, a tetanus delivered 60–90 min after anoxia fails to produce LTP, suggesting a "delayed" form of post-anoxic LTP inhibition (Izumi et al., 1998).

Interestingly, synaptic responses mediated by NMDARs not only recover fully after oxygen deprivation but also show a transient facilitation during the reoxygenation recovery period. This suggests that the delayed LTP inhibition might result from untimely NMDAR activation occurring during the recovery phase. However, the delayed LTP inhibition following anoxia is reversed by NMDAR antagonists and NOS inhibitors administered during the anoxic period while administration of these agents only during the reoxygenation phase fails to reverse the delayed LTP inhibition. This suggests that the NMDAR and NOS activation occurred during the anoxic insult and not during reoxygenation. Both the acute and delayed LTP inhibition as well as the intermediate period of recovery of LTP induction can be mimicked by administration of NO releasers (e.g. SNP). It is important to emphasize that the conditions associated with delayed postanoxic LTP inhibition did not produce either persistent changes in basal synaptic responses or pathomorphological changes in hippocampal slices, indicating that delayed neuronal death is unlikely to account for the change in synaptic plasticity. It remains unknown, however, whether some cell loss might result from the anoxic insult if the period of study could be extended to longer periods after the insult.

Reversal of NMDAR-mediated LTP inhibition

Although the mechanisms involved in NMDAR-mediated LTP inhibition are not completely understood, there appear to be several ways to overcome the LTP inhibition. First, Huang et al. (1992) found that following LTP inhibition by weak tetanic stimuli, delivery of a stronger (more intense) electrical tetanus allowed LTP generation. How the strong tetanus promotes LTP remains unclear and it is possible that an NMDAR-independent form of plasticity could contribute to the resulting synaptic enhancement. It is known that calcium influx via voltage-activated calcium channels can produce LTP and that this form of LTP does not require NMDAR participation (Grover and Teyler, 1990; Kato et al., 1993).

Based on earlier studies indicating that NMDAR activation interferes with mGluR-mediated PI turnover (Palmer et al., 1988, 1990), we examined whether agents that activate PI turnover alter NMDAR-mediated LTP inhibition. Although a non-selective mGluR agonist was ineffective, norepinephrine reversed the LTP block produced by 1 μM NMDA (Izumi et al., 1992b). The effects of norepinephrine are mediated by α1-adrenergic receptors because they are mimicked by phenylephrine and methoxamine, α1-receptor agonists, and are inhibited by phentolamine, an α1-receptor antagonist. Agonists or antagonists at β- or α2-adrenoceptors are ineffective against NMDA-mediated LTP inhibition. Subsequent studies suggest that the effects of norepinephrine on CA1 synaptic plasticity are complex (Thomas et al., 1996). β-adrenergic receptor activation shifts the threshold for CA1 hippocampal LTP induction and promotes LTP at stimulation frequencies of 5–10 Hz that do not typically produce LTP (Katsuki et al., 1997). Additionally, β-receptor stimulation inhibits LTD produced by 900 pulses administered at 1 Hz. α-Adrenoceptors appear to play little role in these effects. However, norepinephrine also inhibits depotentiation of previously established LTP and this effect appears to involve both α1- and β-adrenergic receptors.

Secondary mechanisms underlying the effects of NO in NMDAR-mediated LTP inhibition

It is presently unknown whether NMDAR-mediated LTP inhibition results primarily from pre- or postsynaptic effects. Because NO inhibitors overcome the LTP inhibition, it is important to consider how NO participates in the process. One possibility is that NO inhibits NMDAR function via actions at a redox modulatory site on the NMDAR complex (Lei et al., 1992; Manzoni et al., 1992; see also Lipton et al., this volume). Consistent with this, we found that NO inhibitors enhance the ability of weak tetani to generate LTP and increase synaptic NMDAR currents (Kato and Zorumski, 1993). This effect, however, cannot entirely explain the effects of 1 μM NMDA for reasons outlined above. It is important to consider that although 1 μM NMDA fails to alter synaptic NMDAR responses activated by single stimuli, nothing is presently known about changes that might occur during the train of stimuli used to induce LTP. The likelihood of changes in NMDAR function being important during repeated stimulation is increased by the observations that single synaptic activations (Mennerick and Zorumski, 1996) and trains of stimuli promote significant desensitization of synaptic NMDARs (Rosenmund et al., 1995; Tong et al., 1995). Additionally, although 1 μM NMDA blocks LTP when administered in the post-tetanic period, it is unclear whether the mechanisms involved in the LTP block are the same pre- and post-tetanically. Of interest is the observation that mGluR antagonists inhibit LTP when introduced immediately after a tetanus and the effectiveness of NMDAR and mGluR antagonists as inhibitors of LTP shows a similar developmental time course.

A second possibility is that untimely NO release activates soluble guanylyl cyclases and that cyclic GMP mediates the LTP inhibition (Garthwaite and Boulton, 1995). There is presently little support for this hypothesis although there is evidence that cyclic GMP may participate in the LTP cascade (Arancio et al., 1995). One study has reported the involvement of NO and cyclic GMP in CA1 LTD (Izumi and Zorumski, 1993) but a subsequent study failed to observe an effect of NOS inhibitors on LTD in the CA1 region (Cummings et al., 1994). Unlike its clear ability to inhibit LTP induction, 1 μM NMDA has no effect on the induction of homosynaptic LTD produced by low frequency stimulation (Fig. 4).

Given that LTP induction appears to require higher glucose concentrations than either baseline synaptic transmission or LTP maintenance (Izumi and Zorumski, 1997), a third possibility is that NO interferes with the energy metabolism needed to generate LTP. NO activates ADP-ribosyl transferases and ADP ribosylation of glyceraldehyde-3-phosphate dehydrogenase (GAPDH) inhibits glycolysis (Zhang and Snyder, 1992). In partial support of this hypothesis, preliminary studies indicate that luminol and vitamin K1, inhibitors of ADP-ribosyltransferases, overcome the effects of 1 μM NMDA on LTP. This raises the possibility that energy sources that do not require GAPDH could overcome the effects of untimely NMDAR activation. Interestingly, the end products of glycolysis, lactate and pyruvate, sustain baseline synaptic transmission and LTP induction in the absence of exogenous glucose (Izumi et al., 1994).

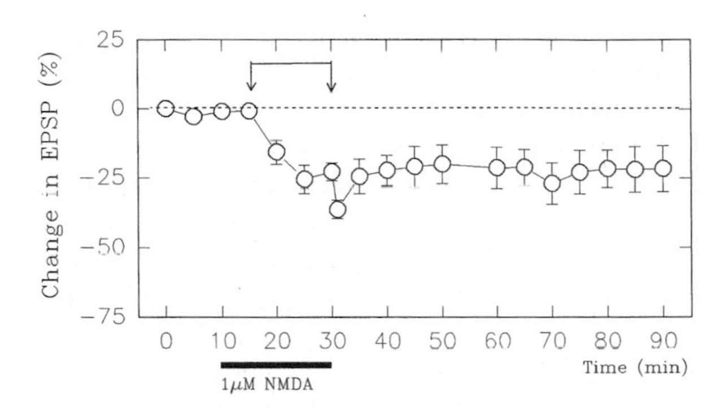

Fig. 4. The graph shows the lack of effect of 1 μM NMDA (open bar) on the induction of homosynaptic LTD by 900 pulses administered at 1 Hz (n = 5 slices). One Hz stimulation was administered during the period denoted by the solid line connecting the arrows. Compared to controls (not shown), there was no effect of NMDA on LTD induction.

Discussion

It is now clear that glutamate not only mediates fast synaptic transmission and synaptic plasticity in the CNS, but also modulates these processes. Furthermore, the modulatory role of glutamate can result from either changes in ambient extracellular levels or prior synaptic activation of glutamate-using synapses. While the details of how glutamate functions as a neurotransmitter and mediator of synaptic plasticity are becoming better understood, studies examining the mechanisms underlying this "metaplasticity" are in their infancy (Abraham and Tate, 1997).

Available data strongly suggest that the inhibition of LTP by low level tonic activation of NMDARs involves untimely NO release and perhaps activation of certain protein phosphatases. The role of NO in hippocampal LTP and LTD remains controversial (Zorumski and Izumi, 1993; see also Hawkins et al., this volume) and studies examining the role of NO in metaplasticity do not directly address this issue. Rather, the studies of metaplasticity indicate that a number of factors are involved in regulating the threshold for synaptic plasticity. This concept was originally described by Bienenstock et al. (1982) in a model for experience-dependent synaptic modification in visual cortex. Lisman (1989) subsequently postulated that specific protein phosphatases may be key molecular determinants of Hebb and anti-Hebb processes in the CNS. It remains to be determined whether NMDAR-mediated LTP inhibition and untimely NO release are critical factors in determining the dynamic range in which excitatory synapses operate and how this form of synaptic modulation participates in frequency-dependent information encoding (Abraham and Tate, 1997).

Acknowledgements

The work in the authors' lab was supported by NIMH Research Scientist Development Award MH00964, grants MH45493 and AG11355, and the Bantly Foundation.

References

Abraham, W.C. and Bear, M.F. (1996) Metaplasticity: The plasticity of synaptic plasticity. *Trends Neurosci.*, 19: 126–130.

Abraham, W.C. and Tate, W.P. (1997) Metaplasticity: A new vista across the field of synaptic plasticity. *Prog. Neurobiol.*, 52: 303–323.

Arai, A., Kessler, M. and Lynch, G. (1990) The effects of adenosine on the development of long-term potentiation. *Neurosci. Lett.*, 119: 41–44.

Arai, A. and Lynch, G. (1992) Factors regulating the magnitude of long-term potentiation induced by theta pattern stimulation. *Brain Res.*, 598: 173–184.

Arancio, O., Kandel, E.R. and Hawkins, R.D. (1995) Activity-dependent long-term enhancement of transmitter release by presynaptic 3′,5′-cyclic GMP in cultured hippocampal neurons. *Nature*, 376: 74–80.

Bienenstock, E.L., Cooper, L.N. and Munro, P.W. (1982) Theory for the development of neuron selectivity: Orientation specificity and binocular interaction in visual cortex. *J. Neurosci.*, 2: 32–48.

Bliss, T.V.P. and Collingridge, G.L. (1993) A synaptic model of memory: Long-term potentiation in the hippocampus. *Nature*, 361: 31–39.

Bonhoeffer, T., Staiger, V. and Aertsen, A. (1989) Synaptic plasticity in rat hippocampal slice cultures: Local "Hebbian" conjunction of pre- and postsynaptic stimulation leads to distributed synaptic enhancement. *Proc. Natl. Acad. Sci. USA*, 86: 8113–8117.

Chen, C. and Tonegawa, S. (1997) Molecular genetic analysis of synaptic plasticity, activity-dependent neural development, learning and memory in the mammalian brain. *Annu. Rev. Neurosci.*, 20: 157–184.

Clark, G.D., Clifford, D.B. and Zorumski C.F. (1990) The effect of agonist concentration, membrane voltage, and calcium on *N*-methyl-D-aspartate receptor desensitization. *Neuroscience*, 39: 787–797.

Coan, E.J., Irving, A.J. and Collingridge, G.L. (1989) Low frequency activation of the NMDA receptor system can prevent the induction of LTP. *Neurosci. Lett.*, 105: 205–210.

Cummings, J.A., Mulkey, R.M., Nicoll, R.A. and Malenka, R.C. (1996) Ca^{2+} signaling requirements for long-term depression in the hippocampus. *Neuron*, 16: 825–833.

Cummings, J.A., Nicola, S.M. and Malenka, R.C. (1994) Induction in the rat hippocampus of long-term potentiation (LTP) and long-term depression (LTD) in the presence of a nitric oxide inhibitor. *Neurosci. Lett.*, 176: 110–114.

Dudek, S.M. and Bear, M.F. (1992) Homosynaptic long-term depression in area CA1 of hippocampus and effects of *N*-methyl-D-aspartate receptor blockade. *Proc. Natl. Acad. Sci. USA*, 89: 4363–4367.

Dudek, S.M. and Bear, M.F. (1993) Bidirectional long-term modification of synaptic effectiveness in the adult and immature hippocampus. *J. Neurosci.*, 13: 2910–2918.

Engert, F. and Bonhoeffer, T. (1997) Synapse specificity of long-term potentiation breaks down at short distances. *Nature*, 388: 279–284.

Fujii, S., Saito, K., Miyakawa, H., Ito, K.-I. and Kato H. (1991) Reversal of long-term potentiation (depotentiation) induced by tetanus stimulation of the input to CA1 neurons of guinea pig hippocampal slices. *Brain Res.*, 555: 112–122.

Garthwaite, J. and Boulton, C.L. (1995) Nitric oxide signaling in the central nervous system. *Annu. Rev. Physiol.*, 57: 683–706.

Grover, L.M. and Teyler, T.J. (1990) Two components of long-term potentiation induced by different patterns of afferent activation. *Nature*, 347: 477–479.

Harris, K.M. and Teyler, T.J. (1983) Developmental onset of long-term potentiation in area CA1 of the rat hippocampus. *J. Physiol. (London)*, 346: 27–48.

Huang, Y.-Y., Colino, A., Selig, D.K. and Malenka, R.C. (1992) The influence of prior synaptic activity on the induction of long-term potentiation. *Science*, 255: 730–733.

Izumi, Y., Benz, A.M., Zorumski, C.F. and Olney, J.W. (1994) Effects of lactate and pyruvate on glucose deprivation in rat hippocampal slices. *NeuroReport*, 5: 617–620.

Izumi, Y., Clifford, D.B. and Zorumski, C.F. (1991) 2-amino-3-phosphonopropionate blocks the induction and maintenance of long-term potentiation in rat hippocampal slices. *Neurosci. Lett.*, 122: 187–190.

Izumi, Y., Clifford, D.B. and Zorumski, C.F. (1992a) Low concentrations of *N*-methyl-D-aspartate inhibit the induction of long-term potentiation in rat hippocampal slices. *Neurosci. Lett.*, 137: 245–248.

Izumi, Y., Clifford, D.B. and Zorumski, C.F. (1992b) Norepinephrine reverses *N*-methyl-D-aspartate-mediated inhibition of long-term potentiation in rat hippocampal slices. *Neurosci. Lett.*, 142: 163–166.

Izumi, Y., Clifford, D.B. and Zorumski, C.F. (1992c) Inhibition of long-term potentiation by NMDA-mediated nitric oxide release. *Science*, 257: 1273–1276.

Izumi, Y., Katsuki, H., Benz, A.M. and Zorumski, C.F. (1998) Oxygen deprivation produces delayed inhibition of LTP by activation of NMDA receptors and nitric oxide synthase. *J. Cereb. Bl. Flow Metab.*, 18: 97–108.

Izumi, Y. and Zorumski, C.F. (1993) Nitric oxide and long-term synaptic depression in the rat hippocampus. *NeuroReport*, 6: 1131–1134.

Izumi, Y. and Zorumski, C.F. (1994) Developmental changes in the effects of metabotropic glutamate receptor antagonists on CA1 long-term potentiation in rat hippocampal slices. *Neurosci. Lett.*, 176: 89–92.

Izumi, Y. and Zorumski, C.F. (1995) Developmental changes in long-term potentiation in CA1 of rat hippocampal slices. *Synapse*, 20: 19–23.

Izumi, Y. and Zorumski, C.F. (1997) Involvement of nitric oxide in low glucose-mediated inhibition of hippocampal long-term potentiation. *Synapse*, 25: 258–262.

Kato, K., Clifford, D.B. and Zorumski, C.F. (1993) Long-term potentiation during whole-cell recording in rat hippocampal slices. *Neuroscience*, 53: 39–47.

Kato, K. and Zorumski, C.F. (1993) Nitric oxide inhibitors facilitate the induction of hippocampal long-term potentiation by modulating NMDA responses. *J. Neurophysiol.*, 70: 1260–1263.

Katsuki, H., Izumi, Y. and Zorumski, C.F. (1997) Noradrenergic regulation of synaptic plasticity in the hippocampal CA1 region. *J. Neurophysiol.*, 77: 3013–3020.

Legendre, P., Rosenmund, C. and Westbrook, G.L. (1993) Inactivation of NMDA channels in cultured hippocampal neurons by intracellular calcium. *J. Neurosci.*, 13: 674–684.

Lei, S.Z., Pan, Z.H., Aggarwal, S.K., Chen V. H.-S., Hartman, J., Sucher, N.J. and Lipton, S.A. (1992) Effect of nitric oxide production on the redox modulatory site of the NMDA receptor-channel complex. *Neuron*, 8: 1087–1099.

Lisman, J. (1989) A mechanism for the Hebb and anti-Hebb process underlying learning and memory. *Proc. Natl. Acad. Sci. USA*, 86: 9574–9578.

Malenka, R.C. (1994) Synaptic plasticity in the hippocampus: LTP and LTD. *Cell.*, 78, 535–538.

Manzoni, O.J., Manabe, T. and Nicoll, R.A. (1994) Release of adenosine by activation of NMDA receptors in the hippocampus. *Science*, 265: 2098–2101.

Manzoni, O., Prezeau, L., Marin, P., Deshager, S., Bockaert, J. and Fagni, L. (1992) Nitric oxide-induced blockade of NMDA receptors. *Neuron*, 8: 653–662.

Mennerick, S. and Zorumski, C.F. (1996) Postsynaptic modulation of NMDA synaptic currents in rat hippocampal microcultures by paired pulse stimulation. *J. Physiol. (London)*, 490: 405–417.

Mulkey, R.M., Endo, S., Shenolikar, S. and Malenka, R.C. (1994) Involvement of a calcium-inhibitor-1 phosphatase cascade in hippocampal long-term depression. *Nature*, 369: 486–488.

Mulkey, R.M., Herron, C.E. and Malenka, R.C. (1993) An essential role for phosphatases in hippocampal long-term potentiation. *Science*, 261: 1051–1055.

Mulkey, R.M. and Malenka, R.C. (1992) Mechanisms underlying induction of homosynaptic long-term depression in area CA1 of the hippocampus. *Neuron*, 9: 967–975.

Nicoletti, F., Iadorala, M.J., Wroblewski, J.T. and Costa, E. (1985) Excitatory amino acid recognition sites coupled with phospholipid metabolism: Developmental changes and interaction with α-adrenoceptors. *Proc. Natl. Acad. Sci. USA*, 83: 1931–1935.

O'Dell, T.J. and Kandel, E.R. (1994) Low-frequency stimulation erases LTP through an NMDA receptor-mediated activation of protein phosphatases. *Learn. Memory*, 1: 129–139.

Palmer, E., Monaghan, D.T. and Cotman, C.W. (1988) Glutamate receptors and phosphoinositide metabolism: Stimulation via quisqualate receptors is inhibited by *N*-methyl-D-aspartate receptor activation. *Mol. Brain Res.*, 4: 161–165.

Palmer, E., Nangel-Taylor, K., Krause, J.P., Roxes, A. and Cotman, C.W. (1990) Changes in excitatory amino acid modulation of phoshoinositide metabolism during development. *Dev. Brain Res.*, 51: 132–134.

Rosenmund, C., Feltz, A. and Westbrook G.L. (1995) Calcium-dependent inactivation of synaptic NMDA receptors in hippocampal neurons. *J. Neurophysiol.*, 73: 427–430.

Sah, P., Hestrin, S. and Nicoll, R.A. (1989) Tonic activation of NMDA receptors by ambient glutamate enhances excitability of neurons. *Science*, 246: 815–818.

Schuman, E.M. and Madison, D.V. (1994) Locally distributed synaptic potentiation in the hippocampus. *Science*, 263: 532–536.

Selig, D.K., Hjelmstad, G.O., Herron, C., Nicoll, R.A. and Malenka, R.C. (1995) Independent mechanisms for long-term depression of AMPA and NMDA responses. *Neuron*, 15: 417–426.

Thomas, M.J., Moody, T.D., Makhinson, M. and O'Dell, T.J. (1996) Activity-dependent β-adrenergic modulation of low frequency stimulation induced LTP in the hippocampal CA1 region. *Neuron*, 17: 475–482.

Tong, G., Shepherd, D. and Jahr, C.E. (1995) Synaptic desensitization of NMDA receptors by calcineurin. *Science*, 267: 1510–1512.

Zhang, J. and Snyder, S.H. (1992) Nitric oxide stimulates auto-ADP ribosylation of glyceraldehyde-3-phosphate dehydrogenase. *Proc. Natl. Acad. Sci. USA*, 89: 9382–9385.

Zorumski, C.F. and Izumi, Y. (1993) Nitric oxide and hippocampal synaptic plasticity. *Biochem. Pharmacol.*, 46: 777–785.

Zorumski, C.F., Mennerick, S.J. and Que, J. (1996) Modulation of hippocampal excitatory synaptic transmission by low concentrations of glutamate. *J. Physiol. (London)*, 494: 465–477.

R.R. Mize, T.M. Dawson, V.L. Dawson and M.J. Friedlander (Eds.)
Progress in Brain Research, Vol 118

CHAPTER 13

Dynamic modulation of cerebral cortex synaptic function by nitric oxide

Prakash Kara and Michael J. Friedlander*

Department of Neurobiology, Department of Physiology Biophysics, University of Alabama at Birmingham, Birmingham, AL, USA

Nitric Oxide (NO) is a diffusible messenger that can effectively modify a variety of cellular functions in the central nervous system (CNS) (Snyder, 1992; Dawson et al., 1994; Garthwaite and Boulton, 1995; Jaffrey and Snyder, 1995). A number of features of NO and the constitutively expressed enzymes that control its calcium-regulated production in the central nervous system (type I and III nitric oxide synthase [NOS] – (Huang et al., 1991; Dinerman et al., 1994) confer the following functional properties on it: Constituitively expressed NOS is present in a variety of cell types in the brain, including neurons (Moncada et al., 1991; Vincent and Hope, 1992; Murphy et al., 1993); Activation of NOS and NO production in neurons is sensitive to intracellular calcium levels (see Nathan and Xie, 1994); NO readily crosses cell membranes and diffuses through extracellular space (Schuman and Madison, 1994; Montague et al., 1994); NO has a variety of molecular targets including soluble guanylate cyclase (Humbert et al., 1990; Moncada et al., 1991; Mayer, 1993), other heme-containing enzymes, e.g. cyclo-oxygenase (Salvemini and Masferre, 1996), superoxide anion (Beckman, 1991), thiol moieties and tyrosine residues on various proteins (Stamler et al., 1997; Mayer et al., 1998) and these targets are present in a variety of cells in the central nervous system including neurons, glia, cerebrovascular smooth muscle and endothelial cells (Carroll et al., 1996; Sistiaga et al., 1997). By virtue of NO's actions on these target molecules (or through effects of NO's reaction products) it can influence ion channel activity (Pape and Mager, 1992; Bolotina et al., 1994; Kurenny et al., 1994, Savtchenko et al., 1997); neurotransmitter release (Montague et al., 1994; Meffert et al., 1994, 1996; Kano et al., 1998), cerebral blood flow (Goadsby et al., 1992; Irikura et al., 1995; Wang et al., 1995; Ma et al., 1996), growth cone structure (Hess et al., 1993, Renteria and Constantine-Paton, 1996) and gene expression (Peunova and Enikolopov, 1993; Iadecola et al., 1996) in the nervous system. These actions provide potential substrates for NO to play a critical role in plasticity such as adaptive responses of the CNS to various stimuli, the achievement of developmental milestones and responses of the CNS to environmental insults. Examples include: long-term synaptic potentiation (LTP) in the hippocampus (Schuman and Madison, 1991; Malen and Chapman, 1997), long-term synaptic depression (LTD) in the cerebellum (Shibuki and Okada, 1991; Boxall and Garthwaite, 1996; Lev-Ram et al., 1997); elimination of inappropriate central projections of developing visual system afferents (Wu et al., 1994) and cellular excitotoxicity cascades (Moskowitz and Dalkara, 1996; Ayata et al., 1997).

*Corresponding author. Tel.: +1 205 934 0100; fax: +1 205 934 6571; e-mail: Mif@nrc.uab.edu

An area of particular interest and controversy regarding NO's CNS effects has been its role in the induction of synaptic plasticity in the mature nervous system, namely hippocampal LTP. Although, an initial flurry of research suggested a role for NO in this process (Bohme et al., 1991, 1993; Schuman and Madison, 1991, 1994; Zhuo et al., 1993), some studies did not support the necessity of NO production for hippocampal LTP induction (Williams et al., 1993; Chetkovich et al., 1993; Bannerman et al., 1994). However, a series of recent studies, using combined physiological, pharmacological and genetic tools has strongly established NO's role as an important component of particular forms of hippocampal LTP induction, (O'Dell et al., 1994; Son et al., 1996; Kantor et al., 1996). Its primary action in LTP induction appears to be as a retrograde messenger (Arancio et al., 1996). In cerebellar LTD, NO has been shown to be necessary for decreasing the effectiveness of parallel fiber to Purkinje cell synaptic transmission, following conjunction of parallel fiber activation or a calcium signal with climbing fiber activation (Lev-Ram et al., 1997). These electrophysiology studies of synaptic plasticity combined with behavioral studies that utilize pharmacological intervention (Fin et al., 1995) and electrophysiology studies of NOS knockout animals (Son et al., 1996), are consistent with the suggestion that NO may play a role in learning (Bohme et al., 1993; Yanagihara and Kondo, 1996).

Although NO's role in hippocampal and cerebellar plasticity has been well characterized, its action in the neocortex is less well studied. There is, however, evidence for considerable synaptic plasticity in the neocortex in sensory (Recanzone et al., 1990; Garraghty and Kaas, 1991; Gilbert and Wiesel, 1992; Recanzone et al., 1993; Elbert et al., 1995; Rauschecker, 1995), motor (Nudo et al., 1996; Huntley, 1997; Karni et al., 1995, 1998) and associational areas (Goldman-Rakic, 1982). There also is evidence to suggest that NO may be operative in neocortex including observations of: (i) a rich distribution of NOS positive neuronal processes throughout the cerebral cortex (Aoki et al., 1993, 1997; Kuchiiwa et al., 1994; Faro et al., 1995; Wallace et al., 1995; Hester et al., 1996; Bidmon et al., 1997), (ii) variations in NO production *in vivo* in cerebral cortex (Burlet and Cespuglio, 1997; Harada et al., 1997), (iii) the suggestion of a role for NO in development of cortical networks and cortical synaptic plasticity based on computer modeling (Gally et al., 1990; Montague et al., 1994); and (iv) a rich distribution in neocortex of the various molecular substrates for NO's production, e.g. NMDA receptors (Huntley et al., 1994; Petralia et al., 1994; Aoki et al., 1997; Catalano et al., 1997; Weiss et al., 1998) – NOS (see above) and for NO's action e.g. soluble guanylate cyclase (Carroll et al., 1996; Sistiaga et al., 1997) and superoxide anion sources (Lipton et al., 1993). Moreover, since neocortical circuits are likely to be under continuous dynamic modulation of their synaptic and network properties (Singer, 1995; Leopold and Logothetis, 1996), there is a requirement for molecular signaling cascades that could transiently modify synaptic weights.

In this chapter, we present a series of biochemical, *in vitro* and *in vivo* electrophysiology experiments from our laboratory on NO's potential role in cortical plasticity. The visual cortex is a particularly useful model for these experiments since its architecture and synaptic circuitry are well characterized (Peters, 1985; Martin, 1988), either electrical or visual stimulation can be used to activate synaptic pathways and the cortex is capable of various types of adaptive forms of plasticity of both short and long duration in immature (Bear and Singer, 1986; Fregnac et al., 1988; Dudek and Friedlander, 1996; Harsanyi and Friedlander, 1997b) and mature animals (Duffy et al., 1976; Fregnac et al., 1988; Gilbert and Wiesel, 1992; Harsanyi and Friedlander, 1997a). Moreover, the visual cortex is richly invested with a scaffolding of NOS positive neuronal processes (Aoki et al., 1993; Hester et al., 1996).

Forms of visual cortex plasticity in which NO is implicated

Visual cortex plasticity is most notable in mammals with overlapping binocular visual fields, as evidenced by ocular dominance plasticity within a defined critical period of sensitivity (Wiesel and Hubel, 1963; Shatz and Stryker, 1978; Wiesel, 1982; Friedlander et al., 1991; Rauschecker, 1991). However, dynamic plastic changes also occur in the mature visual cortex. These changes may underlie perceptual shifts under physiological conditions (Assad and Maunsell, 1995; DeAngelis et al., 1995; Leopold and Logothetis, 1996; Gibson and Maunsell, 1997) and functional changes in response to injury (Gilbert and Wiesel, 1992; Das and Gilbert, 1995; Storig, 1996). Cortical areas associated with other sensory modalities such as somato-sensation (Ramachandran, 1993; O'Leary et al., 1994; see references above) and audition (Ahissar et al., 1992; Recanzone et al., 1993; Cruikshank and Weinberger, 1996) exhibit robust forms of functional plasticity in adults in response to various paradigms including training, peripheral nerve injury or alternations in the behavioral salience of a stimulus. Thus, plasticity throughout life is a hallmark of cortical neural networks. Although NO has been suggested, on theoretical grounds, to be a potential source of refinement of neural circuits during development (Gally et al., 1990; Montague et al., 1991), the experimental evidence in support of its role in development is variable. For example, chronic cortical NOS inhibition does not significantly alter the elaboration of ocular dominance columns in cat visual cortex (Reid et al., 1996; Ruthazer et al., 1996) or eye-segregation of retino-geniculate afferents into eye-specific laminae in ferret thalamus (Cramer et al., 1996). NOS inhibition does however, prevent segregation of retino-geniculate afferents into appropriate on/off sublaminae in ferret thalamus (Cramer et al., 1996) and it prevents the normal retraction of ipsilateral retino-tectal projections in chick (Wu et al., 1994). In addition, NO has recently been demonstrated to play a role in the modulation of synaptic function in the thalamus (Cudeiro et al., 1994) and the primary visual cortex of mature cats (Kara and Friedlander, 1995, 1997; Cudeiro et al., 1997) and the primary visual cortex of mature guinea pigs (Harsanyi and Friedlander, 1997a). Thus, to date, NO's role in the developmental regulation of refinement of neural circuitry appears variable depending on the species and the neural structure. This may reflect real differences with other signaling pathways substituting for NO in certain cases or limitations in available experimental technology. NO's role in modulating ongoing neural signaling is becoming more clear. Our laboratory's results on NO's action in dynamically modulating visual cortical synaptic function are presented below.

In vitro model of dynamic modulation of cortical synaptic transmission

A particularly useful model for characterizing and predicting the response of visual cortical neurons to dynamic modulation of the strength and interaction of various synaptic inputs is the Bienenstock, Cooper and Munro (BCM) model (Bienenstock et al., 1982). Their model proposes a sliding threshold for modifying synaptic efficiency, it predicts both strengthening and weakening of synaptic gain based on correlated pre- and postsynaptic activity levels and it has been used to predict local and global effects of changes in activity *per se* or competitive interactions between neighboring synapses (Clothiaux et al., 1991). We investigated the applicability of this model and the cellular mechanisms that underlie this type of synaptic plasticity in the mature visual cortex *in vitro* (Friedlander et al., 1993; Fregnac et al., 1994; Harsanyi and Friedlander, 1997a) and the potential role of NO in the immature (Harsanyi and Friedlander, 1997b) and mature (Harsanyi and Friedlander, 1997a) animal's visual cortex in BCM type synaptic plasticity. Briefly, pigmented guinea pig's (> 5 weeks of age) primary visual cortex was vibro-sliced at 400 μm, slices were kept in a standard interface-type recording chamber at 35°C and were perfused with artificial cerebrospinal fluid (ACSF). Conventional sharp electrode intracellular recordings were made from primary visual cortex layer 2–3 pyramidal neurons, verified by subsequent histological examination after intracellular biocytin injection and Cresyl Violet counterstaining. Electrical stimulation was applied at low frequency (either 0.3 or 0.1 Hz) to afferents at the white matter – layer 6 border and/or in layer 3 at a distance of two mm from the recording site. After collecting a stable baseline of the postsynaptic potentials (PSPs) evoked at low frequency, the BCM model was tested by pairing each afferent stimulation-elicited PSP with a 60 ms, intracellularly applied depolarizing stimulus (usually 1.0–2.0 nA) over a 10 min period. The effects of the pairing or covariance of the PSPs with brief postsynaptic depolarizing pulses were then evaluated for subsequent PSPs elicited by afferent stimulation alone.

The amplitudes of the paired PSPs often were significantly increased for periods of 15–90 min following the pairing. Fig. 1A shows such an example. Interestingly, the potentiation of synaptic transmission was restricted to those synapses that were activated conjointly (covariance) with the postsynaptic depolarization (Fig. 1B). In this example, two discrete convergent synaptic pathways onto the same neuron were activated in an interleaved fashion with only one pathway being activated in conjunction with the postsynaptic depolarization pulses (top traces – "paired pathway") and the other being activated out of phase with the postsynaptic intracellularly applied depolarization (bottom traces – "unpaired pathway"). Note that only the "paired pathway" showed subsequent potentiation of the PSP while the "unpaired pathway" was unaffected, implying the capacity for spatial specificity of the plasticity signal or more specifically, an ability of a signal that is dependent on the conjunctive depolarization of a single neuron to recognize and target synapses (either pre-, post-synaptic sites or both) that were co-active at the time of the depolarization. Since action potential output from the depolarized postsynaptic neuron is not necessary to elicit the potentiation (Harsanyi and Friedlander, 1997a) and potentiation does not occur with the application of depolarizing pulses alone, this implies that a plasticity signal is required for induction of synaptic potentiation that is only produced during a conjunction of synaptic transmission and postsynaptic depolarization.

N-methyl-D-aspartate receptors (NMDAR) are likely candidates to serve as the coincidence detectors in this plasticity induction process. Results of experiments to evaluate their role are illustrated in Fig. 1C. The blockage of NMDARs with D-APV prevented induction of the potentiation. Moreover, restricting the rise in intracellular calcium in the postsynaptic neuron during depolarization (via intracellular application of the calcium chelator, BAPTA) also blocked the induction of synaptic potentiation by the pairing protocol (Fig. 1D). Since NMDAR activation and a postsynaptic calcium signal are necessary for the indication of this form of transient and dynamic potentiation of synaptic transmission in mature visual cortex, and since types I and III NOS are known to be calcium dependent (see Marletta, 1994 for a review) with at least type I NOS being co-localized with NMDARs (Brenman et al., 1996; Muller et al., 1996; Aoki et al., 1997; Weiss et al., 1998), these data suggest a possible role for NO production in contributing to this form of visual cortex plasticity. However, when endogenous NOS activity was blocked for 1 h of prior bath application of the NOS inhibitor, L-nitroarginine (LNA), the resulting effects were disparate. Figs. 1E, F show examples of the two predominant effects of prior LNA treatment on the induction of this form of covariance-induced synaptic potentiation in the mature visual cortex. In the example illustrated in Fig. 1E, the potentiation not only was *not* blocked (as would be expected if NO production facilitated induction of potentiation), but it increased in amplitude (+67% at 5 min post-pairing) above the normal range (+26 ± 3% increase in the PSP peak amplitude in normal ACSF vs +44% ± 7% increase in ACSF with LNA, $p < 0.01$, unpaired *t*-tests). The results suggest that for some neurons under certain conditions, a tonic level of endogenous NO production exerts an inhibitory effect on induction of synaptic potentiation. This result is consistent with observations made by several labs that NO can inhibit NMDAR calcium fluxes (Izumi et al., 1992; Lipton et al., 1993) and thus, can play a negative modulatory role on synaptic strength.

However, in other cases (Fig. 1F), pretreatment with LNA did prevent induction of potentiation, suggesting a positive modulatory role for NO, as well. This result is consistent with previous work demonstrating that an NMDAR-NO signal can enhance presynaptic release of L-glutamate in cortex (Montague et al., 1994). Thus, NO is implicated as a modulator of covariance – induced synaptic potentiation in mature visual cortex, playing at least two roles. It is not clear why under certain circumstances or for certain neurons, one effect should dominate. It may depend on variations in the redox environment in particular tissue samples or compartments (Stamler et al.,

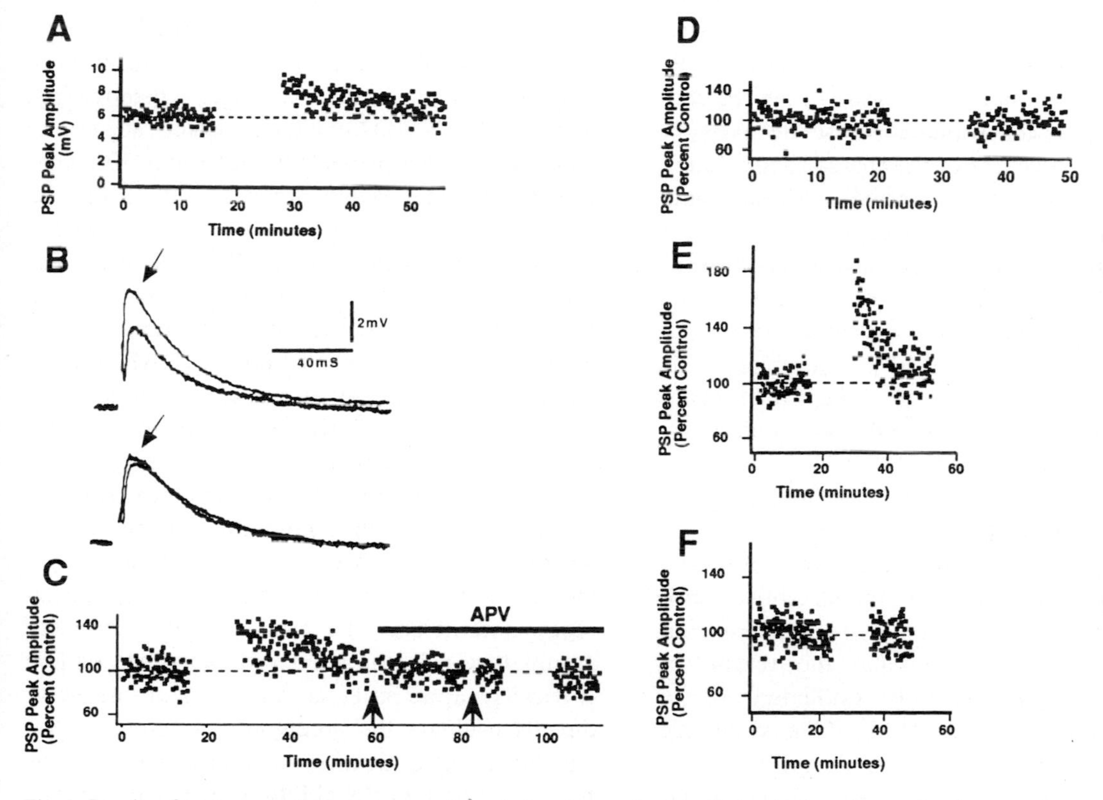

Fig. 1. Results of *in vitro* experiments on covariance-induced synaptic potentiation in the primary visual cortex of mature guinea pigs. In all experiments illustrated, intracellular recordings were made from layer 2/3 neurons and synaptic responses were evoked by electrical stimulation at the layer 6/white matter border at 0.1 Hz to elicit post-synaptic potentials (PSPs) of 3–8 mV. After a stable baseline period was established, potentiation was induced by a 10-min period of pairing 0.1 Hz afferent stimulation with an intracellularly applied, 60 ms depolarizing pulse (1.0–2.0 nA). Peak amplitudes of the PSPs (actual and normalized values) were plotted for various experiments during control and postpairing periods. (A) Typical covariance-induced potentiation of the PSP that persists for 25 min postpairing. (B) Example average (10 traces each) PSPs from a single neuron elicited prepairing (upper traces – smaller PSP) and postpairing (upper traces – larger PSP) in the paired pathway and pre- and postpairing (lower traces – similar amplitudes) for the unpaired pathway onto the same cell. Note that only the paired pathway PSP was potentiated. (C) Requirement for NMDA-receptor activation to induce potentiation. In the first 60 min NMDA receptors were not blocked and potentiation was successfully induced. At 90 min (30 min after addition of 20 μM D-APV), the subsequent attempt to induce potentiation was unsuccessful. (D) The application of the calcium chelator, BAPTA, in the recording micropipette, blocked the induction of potentiation, indicating the necessity for a rise in postsynaptic intracellular calcium in order for induction to occur. (E) and (F) The effects of inhibition of endogenous NOS in two different slices on induction of potentiation. In both cases, the NOS inhibitor, LNA (100 μM) was bath applied, beginning 1h before the records begin. In one case (E) the potentiation was successfully induced and resulted in a supernormal amplitude of potentiation (80% initially, 67% at 5 min postpairing). NOS inhibition by LNA could also block induction of potentiation, as illustrated in F.

1992; Lipton et al., 1993; Pan et al., 1996) or on differences in NO's target availability between different cells or synapses (Mothet et al., 1996). Moreover, the observation that the different isoforms of NOS may provide a NO signal that selectively modulates release of different neurotransmitters (Kano et al., 1998) may contribute to these differences.

NO's ability to contribute to potentiation of synaptic transmission in the mature cerebral cortex could be attributed to several mechanisms including a positive modulation of release of neurotransmitter. This has been evaluated, most effectively, in synaptosomal preparations prepared from a variety of brain areas that are used to evaluate the release of either exogenously applied radiolabeled neurotransmitters or endogenous neurotransmitters (Lonart et al., 1992; Sistiaga et al., 1997). For example, NO has been shown to induce or amplify the release of glutamate, aspartate, norepinephrine and dopamine (Hanbauer et al., 1992; Lonart et al., 1992) and to either facilitate (Ohkuma et al., 1995) or inhibit (Semba et al., 1995) the release of GABA. Examples of NO's role in facilitating the release of cortical (including visual cortex) L-glutamate are shown in Fig. 2. The synaptosomal preparation is illustrated in the electron micrograph in Fig. 2A. Synaptosomes are prepared from mature guinea pig cortex and the release of endogenous L-glutamate is evaluated by a luminometry assay (Fosse et al., 1986; Montague et al., 1994) in response to pulsatile application of various stimulants. The released L-glutamate is collected in sequential fractions and expressed as a percentage of the available L-glutamate in the synaptosomal preparation (obtained by adding the L-glutamate in each sample to the residual synaptosomal L-glutamate as measured by lysis of the synaptosomes with water at the end of each run). Note that the synaptosome preparation contains presynaptic elements enriched with vesicles and postsynaptic processes with associated postsynaptic densities (Fig. 2A). The synaptosomes release L-glutamate in response to stimulants (Fig. 2B), including direct depolarization induced by 50 mM KCl, or application of 100 μM of either of the glutamate receptor agonists, NMDA or AMPA, with NMDA being the more effective stimulus. These results suggest that the synaptosomal preparation has viable NMDA receptors (presumably located at the postsynaptic densities, although the existence of presynaptic NMDA receptors cannot be ruled out) and that these receptors can instigate a signaling cascade that ultimately feeds back in a positive fashion on presynaptic terminals that contain L-glutamate. Our previous results (Montague et al., 1994) demonstrated that application of the extracellular NO scavenger, hemoglobin, blocked the NMDA-mediated effect, suggesting that NO production and diffusion may be necessary components of this signaling cascade. Fig. 2C illustrates the necessity of endogenous NO production in the cascade from NMDAR activation to enhanced L-glutamate release. Pre-incubation with 1 μM of the competitive NOS inhibitor, L-nitro-arginine, completely blocks the NMDAR-mediated L-glutamate release. However, L-nitro-arginine does not interfere with KCl depolarization-induced release of L-glutamate (data not shown).

Thus, our synaptosome data are consistent with our *in vitro* electrophysiology experimental results where NO can positively modulate synaptic potentiation. However, since our slice electrophysiology data (Figs. 1E) also support a negative modulatory role for NO in NMDAR-dependent cortical synaptic potentiation, it was of interest to directly evaluate the potential interaction of NO and NMDAR activation on L-glutamate release. Fig. 2D illustrates the ability of pure NO (applied as 1.8 μM dissolved NO gas) to directly elicit L-glutamate release and NO's capacity to limit or inhibit the synaptosomes' response to direct NMDAR activation (NO + NMDA). In the synaptosome preparation, pure NO is sufficient to elicit enhanced L-glutamate release and its effect is self-limiting by virtue of NO's inhibition of subsequent NMDAR induced glutamate release. These results suggest that NO itself can act to enhance neurotransmitter release, probably by targeting some component of the presynaptic exocytosis cascade and NO can limit this feed-forward amplification process, probably by inhibiting post-synaptic NMDARs' responsiveness to subsequent activation by amplified release of ligand.

Although these results are suggestive of NO's role in cortical NMDAR-mediated amplification of glutamate release, it is necessary to directly demonstrate NO's production in cortex. This evidence is provided in Fig. 2E, utilizing a bioassay to detect NO's production in cortex in response to NMDAR activation with subsequent relaxation of vascular smooth muscle by the secreted NO signal.

The ability of NO to enhance synaptic transmission in a volume of cerebral cortex raises problems with respect to the specificity of its synaptic actions. In order to address this, we evaluated the role of calcium in NMDAR-NO-glutamate cascade. Although it was expected that NMDAR-mediated NO production and L-glutamate release would require calcium (due to the calcium flux through the NMDA-channel and the requirement for elevated intracellular calcium to activate NOS), surprisingly NO's action subsequent to NMDAR activation also required calcium. Application of pure NO elicited enhance L-glutamate release only when calcium was available (Fig. 2F). Thus, a second coincidence detector, downstream from the NMDAR is implicated. This detector senses the NO signal and responds with enhanced L-glutamate release only when elevated calcium is available. An attractive hypothesis to explain this result is that a presynaptic molecular target in the exocytosis cascade serves as an AND gate, requiring NO and elevated calcium, presumably that enters the terminal via voltage-gated calcium channels. This would confer specificity on the NO signal by targeting only the active presynaptic terminals, thus effectively linking temporally co-active synapses in a volume of cerebral cortex. This is consistent with other examples of the ability of NO to interact with calcium (Arancio et al., 1996; Lev-Ram et al., 1997).

Role, sources and targets of NO in signal processing by the visual cortex, *in vivo*

A considerable database has been established to implicate NO in a variety of cellular processes in the CNS including the type of synaptic plasticity described above. However, much of this information is derived from various types of "reduced" experimental preparations such as isolated neurons in cell culture, isolated organelles or proteins and brain slices. There is little information available from intact brain, on the physiological roles of NO (but see Cudeiro et al., 1997). Our laboratory has begun experiments to extend our *in vitro* electrophysiological studies of synaptic plasticity in the visual cortex and our biochemical/pharmacological studies on modulation of L-glutamate release in isolated cortical synapses to the intact brain. Anesthetized and paralyzed cats were used to evaluate the role of local NO effects in visual cortex in dynamic modulation of cellular responses to visual stimuli. The advantages of this experimental preparation are that (i) more natural activation of the brain's microcircuitry is used (moving visual stimuli processed through retinal circuits vs temporally punctate electrical stimulation applied to cut axons as in brain slice experiments), (ii) individual cortical neuron's responses can be tracked with *in vivo* extracellular electrophysiological recording often over periods of several hours allowing evaluation of pre- and post- NO application or NOS inhibition effects, and (iii) the extracellular milieu is in a more normal physiological state than in *in vitro* preparations, allowing for the panoply of NO's targets and compensatory mechanisms to integrate the NO signal in a more realistic fashion.

In order to evaluate the role of NO *in vivo*, a tungsten recording microelectrode is attached to a five barrel glass micropipette such that the tip of the recording electrode protrudes beyond the tips of the multi-barrel drug delivery assembly by 30–50 μm. Drugs to either facilitate endogenous NO production, inhibit endogenous NO production, apply exogenous NO and/or inhibit NMDA receptors are applied under micro-iontophoretic control while recording the extracellular single unit spike activity of individual neurons in cat area 17. Neuronal responses are monitored to sequentially presented computer-generated moving visual stimuli (drifting light or dark bars) presented monoc-

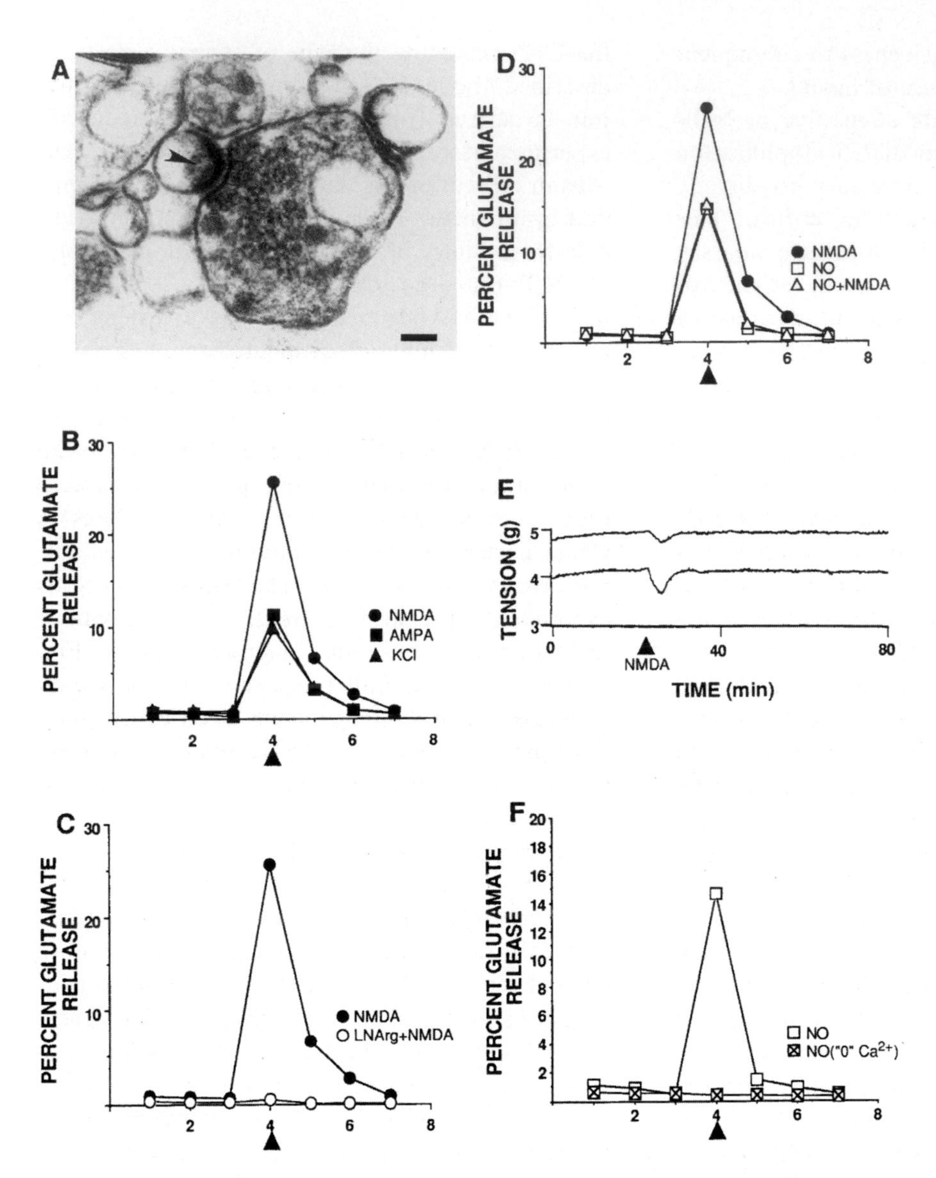

Fig. 2. Synaptosome L-glutamate release experimental results. Synaptosomes were prepared as described in Montague et al. (1994). (A) Electron micrograph of typical synaptosome preparation from adult pigmented guinea pig cerebral cortex. Scale bar = 0.1 μm. Synaptic vesicles in presynaptic terminal and a post-synaptic density (arrowhead) are apparent. (B) Normalized release of endogenous L-glutamate in sequential fractions in response to pulsatile application of 50 mM KCl, 100 μM NMDA and 100 μM AMPA (at arrowhead). Percent L-glutamate release is expressed as a fraction of the total available glutamate pool in that synaptosome preparation. (C) Necessity of endogenous NO production for NMDA mediated glutamate release. Preincubation with 1 μM of the NOS inhibitor, L-nitro arginine (LNA), completely prevented the NMDA-mediated glutamate release. (D) Sufficiency of NO to cause glutamate release. Pure NO gas dissolved in buffer (1–8 μM NO final concentration) elicited glutamate release (but somewhat less than that evoked by 100 μM NMDA). Concomitant stimulation with 100 μM NMDA in the presence of 1.8 μM NO resulted in a greatly diminished glutamate release. (E) NO released by cortical synaptosomes in response to 100 μM NMDA stimulation caused a relaxation of aortic smooth muscle rings (scarified clean of attached endothelium and precontracted – Montague et al., 1994) that are placed in parallel (two rings) on strain-gauge transducers downstream from the synaptosomes. (F) NO's ability to induce glutamate release requires extracellular calcium. Nominal zero extracellular calcium did not allow any detectable glutamate release in response to direct NO application (1.8 μM NO).

ularly and oriented in the optimal spatial and temporal configuration for each neuron, under control conditions and during facilitation and/or inhibition of local endogenous NO production and/or NMDAR blockade.

Fig. 3 is a series of peri-stimulus time histograms (PSTHs) obtained from four neurons under various conditions to manipulate NO levels and/or NMDAR availability while the cell is responding to visual stimulation. Each PSTH includes the neuron's response to between 25 and 40 trials of visual stimulation. But, for any particular cell, the identical number of trials is used for the control periods and pharmacological tests. The two upper rows (Figs. 3A–F) represent the ability of endogenous cortical NO production to facilitate or amplify the neurons' visual responses. The first cell's (Figs. 3A–C) response was facilitated by enhanced NO production (via iontophoresis of the natural substrate for NOS, L-arginine – Fig. 3B) and subsequently inhibited below control levels (Fig. 3C) when endogenous NO levels were reduced by inhibition of NOS with iontophoresis of L-mono-methyl arginine (LMMA). These effects were reversible. Such a facilitatory effect of cortical NO was seen for the majority of cells tested (66%, n = 65).

The source of cortical NO production *in vivo* has not been established. Our *in vitro* data (Fig. 2B) suggest that while NMDA receptor activation is the most effective route for NO production, activation of other receptors such as the AMPA type of glutamate receptor also can contribute to NO production. We evaluated the necessity of NMDA receptor activation for NO production *in vivo*, to determine if similar pathways are accessed under more physiological conditions. An example of such an experiment is illustrated in Figs. 3D–F. The cell's control visual response (Fig. 3D) was significantly reduced by iontophoresis of a saturating dose of D-APV. This assured that available NMDA receptors within the range of the iontophoresis procedure (both on the recorded cell and on other cells in the local volume) were blocked. The subsequent protocol combined iontophoretic application of the NOS inhibitor, L-MMA with the maximal NMDAR inhibition by D-APV (Fig. 3F). The NOS inhibition further reduced the cell's response, indicating that NOS activation (and NO production) contributes to amplification of visual responses independent of NMDA receptors. Thus, the calcium source to activate NOS does not necessarily have to pass through NMDA channels. This is consistent with our *in vitro* results that demonstrate both NMDA and AMPA components contribute to the production of cortical NO and amplification of glutamate release.

While the predominant effect of NO *in vivo* in visual cortex is the amplification of visual responses, a minority of cells (15%) show an opposite effect. In these cases, endogenous NO production inhibits the cells' visual responses. One such example is illustrated in Figs. 3G–I. Increasing NO production by L-arginine iontophoresis (Fig. 3H) reduced the cell's response while the inhibition of NOS by L-MMA application (Fig. 3I) enhanced the same cell's response. One possible mechanism for this inhibition is NO's ability to inhibit calcium flux through NMDA receptors (Lipton et al., 1993). Thus, the NMDA receptor is implicated in both the production and the targeting of NO. Since a subset of the neurons we studied were inhibited by NO, they provided a useful sample to evaluate the NMDA receptor's role as a NO target, *in vivo*. Results from one such experiment are shown in Figs. 3J–L. This cell's visual response was significantly attenuated by blocking NMDA receptors by D-APV application (Fig. 3K), similar to the cell illustrated in Fig. 3E. However, in this case, subsequent additional inhibition of NOS by iontophoresis of L-MMA in the presence of maximal NMDA receptor blockade (Fig. 3L) led to an enhancement of the visual response to a level significantly higher than the levels with NMDA receptor blockade alone (Fig. 3K). Thus, inhibition of NOS facilitated this cell's response suggesting a net inhibition by endogenous NO. But, the inhibitory effect of NO occurred even in the presence of maximal inhibition of NMDA receptors. These data do not support the NMDA receptor as the sole target for the inhibitory effects

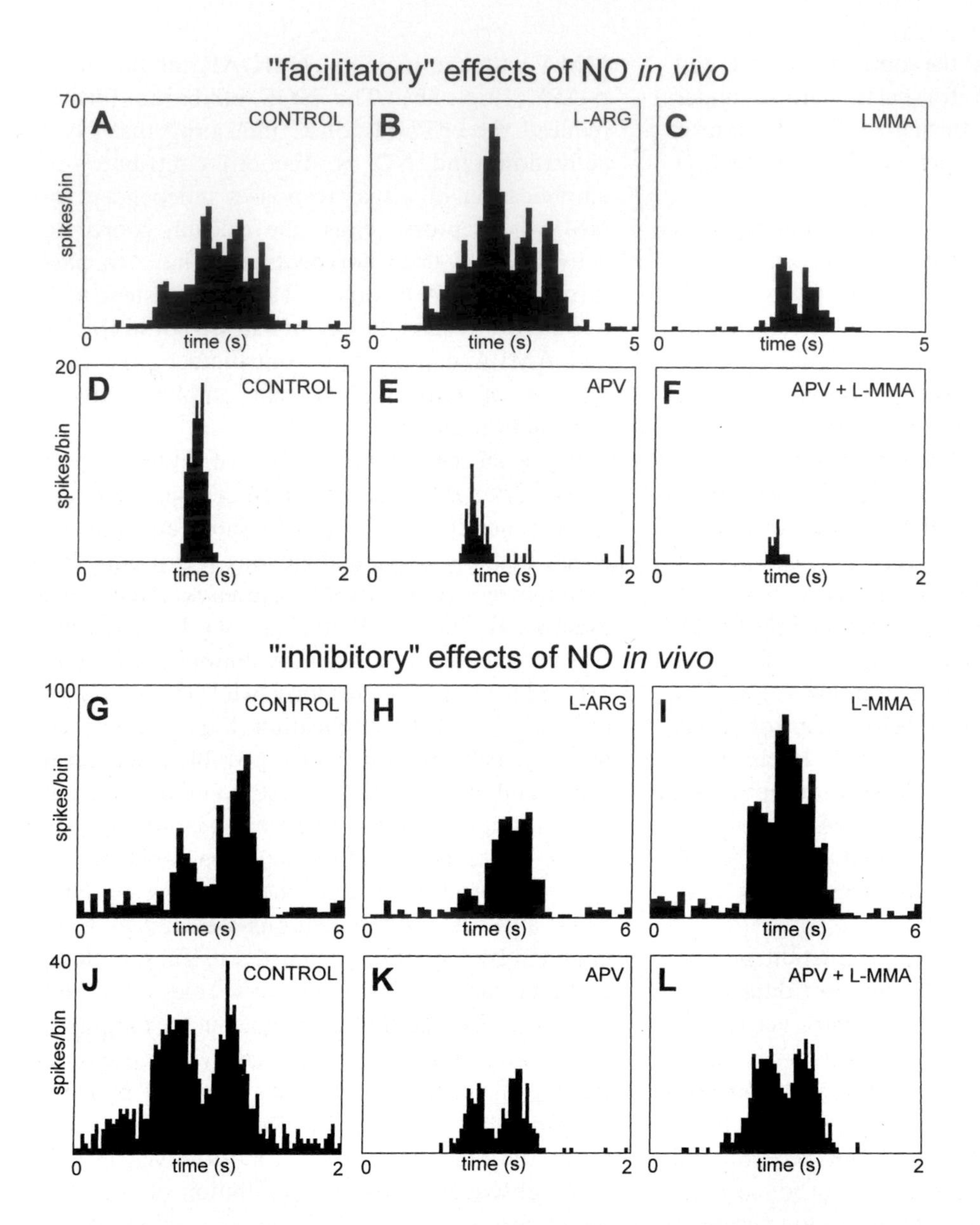

of NO, *in vivo*. Thus, we conclude that there are several targets for NO in cortex including a presynaptic positive modulatory site – perhaps an element(s) of the exocytosis cascade for excitatory glutamatergic synaptic transmission and a negative modulatory site on both NMDA receptors and some other targets, perhaps calcium channels or AMPA receptors. It is also possible that the inhibitory action of NO could be attributable to an enhancement of GABA release (Ohkuma et al., 1995; Kano et al., 1998, but see Semba et al., 1994). These *in vivo* data are consistent with our *in vitro* electrophysiology results of both a positive and negative modulatory role of NO in cortical synaptic plasticity (Figs. 1E, F) and our synaptosome data demonstrating that NO can both amplify glutamate release (Figs. 2C, D and F) and can inhibit the amplification cascade of NMDA receptor activation, NO production and glutamate release (Fig. 2D).

Fig. 3. Responses of individual visual cortical neurons in cat area 17 to repeated applications of drifting bar visual stimuli, presented monocularly, each at the preferred orientation, direction and velocity for that cell. Responses were recorded as extracellular action potentials and are shown as peristimulus time histograms (PSTH) under various control or pharmacological manipulation conditions. Cell 1: (A–C) Up-regulation of endogenous NO production via L-arginine (L-ARG) application enhanced the visual response and down-regulation of endogenous NO production via L-mono-methyl-arginine (L-MMA), attenuated the visual response. PSTH responses are shown for control conditions (A), during iontophoresis of L-ARG (B), and LMMA (C) L-ARG significantly enhanced the visual response from 9.0 spikes/trial to 14.1 spikes/trial (MW test: $Z = 3.6; p < 0.0005$). Subsequent L-MMA application reduced the visual response from 9.0 spikes/trial to 3.0 spikes/trial (MW test: $Z = 4.5; p < 0.0001$).
Cell 2: (D–F). Saturating levels of D-APV current were applied and thus, a maximal NMDA receptor blockade reduced the visual response (D–E). An additional NOS blockade in the presence of a maintained NMDA receptor inhibition further attenuated the visual response (E–F). D-APV reduced the response from 3.4 spikes/trial to 1.4 spikes/trial (MW test: $Z = 5.7; p < 0.0001$). Subsequent but simultaneous LMMA application further reduced the response from 1.4 spikes/trial to 0.5 spikes/trial (MW test: $Z = 4.3; p < 0.0001$).
Cell 3: (G–I). L-ARG inhibited and LMMA facilitated the visual response of this cortical neuron. L-ARG decreased the visual response from 16.2 spikes/trial to 11.2 spikes/trial (MW test: $Z = 3.8; p < 0.0005$). LMMA significantly increased the visual response from 16.2 spikes/trial to 23.1 spikes/trial (MW test: $Z = 3.4; p < 0.001$).
Cell 4: (J–L). Consistent with the less common effect of NO having an inhibitory effect on the visual response, a simultaneous NMDA receptor and NOS enzyme blockade enhanced (rather than reduced) the visual responses for this cell. D-APV reduced the visual response from 22.8 spikes/trial to 9.0 spikes/trial (MW test: $Z = 6.1; p < 0.0001$). An additional NOS blockade *enhanced* the visual response from 9.0 spikes/trial to 16.6 spikes/trial 9MW test: $Z = 5.5; p < 0.0001$).

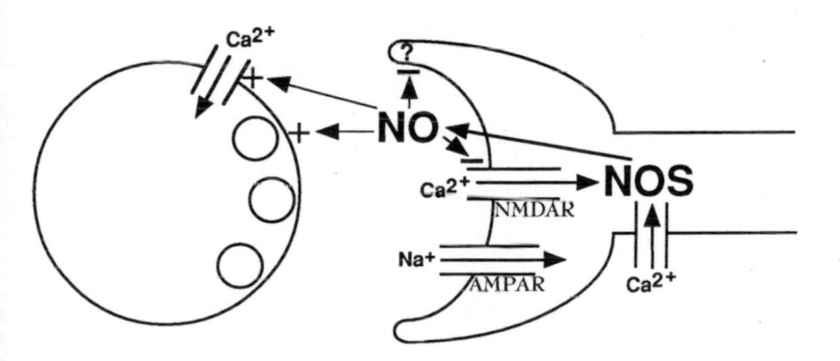

Fig. 4. Summary schematic diagram of NO's postulated actions on synaptic function in the mature visual cortex. Activation of post-synaptic NMDA receptors or AMPA receptors leads to a calcium increase (via the NMDA channel directly or voltage-gated calcium channels activated by AMPA receptor mediated depolarization); NOS is activated and NO is produced; NO (or its reaction products) diffuses retrogradely and throughout the volume of cortical tissue where they interact with various sites including pre-synaptic terminals to facilitate neurotransmitter (glutamate) release if the terminal was recently active as indicated by a calcium signal. The NO may interact with proteins of the vesicle exocytosis cascade and/or voltage-gated calcium channels at the pre-synaptic terminal (either directly, or through peroxynitrite, ONOO, or via a second messenger such as cGMP). NO also has a negative modulatory role at cortical synapses, perhaps limiting its action to plasticity and synaptic signal amplification vs inducing a run-away excitotoxicity cascade. These negative brakes include inhibition of NMDA receptors and other sites, as well.

Summary

Our experiments demonstrate that NO exerts several actions in the cerebral cortex (see Fig. 4). Its production is mediated by neuronal activity through at least two pathways, NMDA receptors and AMPA receptors. By virtue of its diffusion in extracellular space, NO can interact with synapses that are near the production site but not necessarily anatomically connected to the NO source by a conventional synaptic linkage. NO's primary action is amplification of the release of the excitatory neurotransmitter, L-glutamate, thus effectively creating a positive feed-forward gain system.

However, a number of effective brakes, presumably activated under physiological conditions, serve to limit the cascade. These include NO's ability to inhibit NMDA receptors, its negative feedback on the rate limiting enzyme, NOS (Rengasamy and Johns, 1993; Park et al., 1994; Ravichandran et al., 1995) and other inhibitory actions (Figs. 3H and L). Under conditions of extremely strong activation or curtailment of the inhibitory feedback mechanisms, as might occur with a change in the local redox milieu (see Lipton, this volume), the amplification cascade may proceed unchecked leading to neurotoxicity (see Dawson, this volume). NO's ability to modulate synaptic function is indicated by both its positive and negative modulatory role in a form of activity-dependent synaptic plasticity, covariance-induced synaptic potentiation. These opposing effects may be due to NO's ability to amplify glutamate release and inhibit NMDA receptors, respectively. The actions of endogenous NO *in vivo* are primarily facilitatory in visual cortex (Fig. 4). However, inhibitory actions also occur *in vivo*. The targets for NO *in vivo*, are potentially more diverse including the neurotransmitter release process, NMDA receptors, other receptors and ion channels and the cerebral vasculature. However, regardless of the signaling pathways, the net result of endogenous NO production in the intact visual cortex is a potent modulation of cells' responses to visual stimulation. Thus, it is likely that this signal plays an important role in ongoing information processing in the mature cerebral cortex, dynamically altering the effective strength of cortical networks.

Acknowledgement

This work was supported by NIH Grants EY-05116, (HD-32901) and the Helen Keller Eye Research Foundation. We thank Carolyn Gancayco and Felicia Hester for technical assistance and Gloria Purnell for manuscript preparation.

References

Ahissar, E., Vaadia, E., Ahissar, M., Bergman, H., Arieli, A. and Abeles, M. (1992) Dependence of cortical plasticity on correlated activity of single neurons and on behavioral context. *Science*, 257: 1412–1414.

Aoki, C., Fenstemaker, S., Lubin, M. and Go, C-G. (1993) Nitric oxide synthase in the visual cortex of monocular monkeys as revealed by light and electron microscopic immunocytochemistry. *Brain Res.*, 620: 97–113.

Aoki, C., Rhee, J., Lubin, M. and Dawson, T.M. (1997) NMDA-R1 subunit of the cerebral cortex co-localizes with neuronal nitric oxide synthase at pre- and postsynaptic sites and in spines. *Brain Res.*, 750: 25–40.

Arancio, O., Kiebler, M., Lee, C.J., Lev-Ram, V., Tsien, R.Y., Kandel, E.R. and Hawkins, R.D. (1996) Nitric oxide acts directly in the presynaptic neuron to produce long-term potentiation in cultured hippocampal neurons. *Cell*, 87: 1025–1035.

Assad, J.A. and Maunsell, J.H. (1995) Neuronal correlates of inferred motion in primate posteriorparietal cortex. *Nature*, 373: 518–521.

Ayata, C., Ayata, G., Hara, H., Matthews, R.T., Beal, M.F., Ferrante, R.J., Endres, M., Kim, A., Christie, R.H., Waeber, C., Huang, P.L., Hyman, B.T. and Moskowitz, M.A. (1997) Mechanisms of reduced striatal NMDA excitotoxicity in type I nitric oxide synthase knock-out mice. *J. Neurosci.*, 17: 6908–6917.

Bannerman, D.M., Chapman, P.F., Kelly, P.A., Butcher, S.P. and Morris, R.G. (1994) Inhibition of nitric oxide synthase does not prevent the induction of long-term potentiation *in vivo*. *J. Neurosci.*, 14: 7415–7425.

Bear, M.F. and Singer, W. (1986) Modulation of visual cortical plasticity by acetylcholine and noradrenaline. *Nature*, 320: 172–176.

Beckman, J.S. (1991) The double-edged role of nitric oxide in brain function and superoxide-mediated injury. *J. Dev. Physiol.*, 15: 53–59.

Bidmon, H.J., Wu, J., Godecke, A., Schleicher, A., Mayer, B. and Zilles, K. (1997) Nitric oxide synthase-expressing neurons are area-specifically distributed within the cerebral cortex of the rat. *Neuroscience*, 81: 321–330.

Bienenstock, E.L., Cooper, L.N. and Munro, P.W. (1982) Theory for the development of neuron selectivity: Orientation specificity and binocular interaction in visual cortex. *J. Neurosci.*, 2: 32–48.

Bohme, G.A., Bon, C., Lemaire, M., Reibaud, M., Piot, O., Stutzmann, J.M., Doble, A. and Blanchard, J.C. (1993) Altered synaptic plasticity and memory formation in nitric oxide synthase inhibitor-treated rats. *Proc. Natl. Acad. Sci. USA*, 90: 9191–9194.

Bohme, G.A., Bon, C., Stutzmann, J.M., Doble, A. and Blanchard, J.C. (1991) Possible involvement of nitric oxide in long-term potentiation. *Eur. J. Pharmacol.*, 199: 379–381.

Bolotina, V.M., Najibi, S., Palacino, J.J., Pagano, P.J. and Cohen, R.A. (1994) Nitric oxide directly activates calcium-dependent potassium channels in vascular smooth muscle. *Nature*, 368: 850–853.

Boxall, A.R. and Garthwaite, J. (1996) Long-term depression in rat cerebellum requires both NO synthase and NO-sensitive guanylyl cyclase. *Eur. J. Neurosci.*, 8: 2209–2212.

Brenman, J.E., Chao, D.S., Gee, S.H., McGee, A.W., Craven, S.E., Santillano, D.R., Wu, Z., Huang, F., Xia, H., Peters, M.F., Froehner, S.C. and Bredt, D.S. (1996) Interaction of nitric oxide synthase with the postsynaptic density protein PSD-95 and α1-syntrophin mediated by PDZ domains. *Cell*, 84: 757–767.

Burlet, S. and Cespuglio, R. (1997) Voltammetric detection of nitric oxide (NO) in the rat brain: Its variations throughout the sleep-wake cycle. *Neurosci. Lett.*, 226: 131–135.

Carroll, F.Y., Beart, P.M. and Cheung, N.S. (1996) NMDA-mediated activation of the NO/cGMP pathway: Characteristics and regulation in cultured neocortical neurons. *J. Neurosci. Res.*, 43: 623–631.

Catalano, S.M., Chang, C.K. and Shatz, C.J. (1997) Activity-dependent regulation of NMDAR1 immunoreactivity in the developing visual cortex. *J. Neurosci.*, 17: 8376–8390.

Chetkovich, D.M., Klann, E. and Sweatt, J.D. (1993) Nitric oxide synthase-independent long-term potentiation in area CA1 of hippocampus. *Neuroreport*, 4: 919–922.

Clothiaux, E.E., Bear, M.F. and Cooper, L.N. (1991) Synaptic plasticity in visual cortex: Comparison of theory with experiment. *J. Neurophysiol.*, 66: 1785–1804.

Cramer, K.S., Angelucci, A., Hahm, J.O., Bogdanov, M.B. and Sur, M. (1996) A role for nitric oxide in the development of the ferret retinogeniculate projection. *J. Neurosci.*, 16: 7995–8004.

Cruikshank, S.J. and Weinberger, N.M. (1996) Receptive-field plasticity in the adult auditory cortex induced by hebbian covariance. *J. Neurosci.* 16: 861–875.

Cudeiro, J., Rivadulla, C., Rodriguez, R., Grieve, K.L., Martinez-Conde, S. and Acuna, C. (1997) Actions of compounds manipulating the nitric oxide system in the cat primary visual cortex. *J. Physiol. (London)*, 504: 467–478.

Cudeiro, J., Rivadulla, C., Rodriguez, R., Martinez-Conde, S., Acuna, C. and Alonso, J.M. (1994) Modulatory influence of putative inhibitors of nitric oxide synthesis on visual processing in the cat lateral geniculate nucleus. *J. Neurophysiol.*, 71: 146–149.

Das, A. and Gilbert, C.D. (1995) Receptive field expansion in adult visual cortex is linked to dynamic changes in strength of cortical connections. *J. Neurophysiol.* 74: 779–792.

Dawson, T.M., Dawson, V.L. and Snyder, S.H. (1994) Molecular mechanisms of nitric oxide actions in the brain. *Ann. N. Y. Acad. Sci.*, 738: 76–85.

DeAngelis, G.C., Ohzawa, I. and Freeman, R.D. (1995) Receptive-field dynamics in the central visual pathways. *TINS.*, 18: 451–458.

Dinerman, J.L., Dawson, T.M., Schell, M.J., Snowman, A. and Snyder, S.H. (1994) Endothelial nitric oxide synthase localized to hippocampal pyramidal cells: Implications for synaptic plasticity. *Proc. Natl. Acad. Sci. USA*, 91: 4214–4218.

Dudek, S.M. and Friedlander, M.J. (1996) Developmental down-regulation of LTD in cortical layer IV and its independence of modulation by inhibition. *Neuron*, 16: 1097–1106.

Duffy, F.H., Snodgrass, S.R., Burchfiel, J.L. and Conway, J.L. (1976) Bicuculline reversal of deprivation amblyopia in the cat. *Nature*, 260: 256–257.

Elbert, T., Pantev, C., Wienbruch, C., Rockstroh, B. and Taub, E. (1995) Increased cortical representation of the fingers of the left hand in string players. *Science*, 270: 305–307.

Faro, L.R., Araujo, R., Araujo, M., Do-Nascimento, J.L., Friedlander, M.J. and Picanco-Diniz, C.W. (1995) Localization of NADPH-diaphorase activity in the human visual cortex. *Braz. J. Med. Biol. Res.*, 28: 246–251.

Fin, C., da Cunha, C., Bromberg, E., Schmitz, P.K., Bianchin, M., Medina, J.H. and Izquierdo, I. (1995) Experiments suggesting a role for nitric oxide in the hippocampus in memory processes. *Neurobiol. Learn. Mem.*, 63: 113–115.

Fosse, V.M., Kolstad, J. and Fonnum, F. (1986) A bioluminescence method for the measurement of L-glutamate: Applications to the study of changes in the release of L-glutamate from lateral geniculate nucleus and superior colliculus after visual cortex ablation in rats. *J. Neurochem.*, 47: 340–349.

Fregnac, Y., Burke, J., Smith, D.L. and Friedlander, M.J. (1994) Temporal covariance of pre- and post-synaptic activity regulates functional connectivity in the visual cortex. *J. Neurophysiol.* 71: 1403–1421.

Fregnac, Y., Shulz, D., Thorpe, S. and Bienenstock, E. (1988) A cellular analogue of visual cortical plasticity. *Nature*, 333: 367–370.

Friedlander, M.J., Fregnac, Y. and Burke, J.P. (1993) Temporal covariance of postsynaptic membrane potential and synaptic input-role in synaptic efficacy in visual cortex. *Prog. Brain. Res.*, 95: 207–223.

Friedlander, M.J., Martin, K.A.C. and Wassenhove-McCarthy, D. (1991) Effects of monocular visual deprivation on geniculocortical innervation of area 18 in cat. *J. Neurosci.*, 11: 3268–3288.

Gally, J.A., Montague, P.R., Reeke, G.N. and Edelman, G.M. (1990) The NO hypothesis: Possible effects of a short-lived rapidly diffusible signal in the development and function of the nervous system. *Proc. Natl. Acad. Sci. USA*, 87: 3547–3551.

Garraghty, P.E. and Kaas, J.H. (1991) Large-scale functional reorganization in adult monkey cortex after peripheral nerve injury. *Proc. Natl. Acad. Sci. USA*, 88: 6976–6980.

Garthwaite, J. and Boulton, C.L. (1995) Nitric oxide signaling in the central nervous system. *Annu. Rev. Physiol.*, 57: 683–706.

Gibson, J.R. and Maunsell, J.H. (1997) Sensory modality specificity of neural activity related to memory in visual cortex. *J . Neurophysiol.*, 78: 1263–1275.

Gilbert, C.D. and Wiesel, T.N. (1992) Receptive field dynamics in adult primary visual cortex. *Nature*, 356: 150–152.

Goadsby, P.J., Kaube, H. and Hoskin, K.L. (1992) Nitric oxide synthesis couples cerebral blood flow and metabolism. *Brain Res.*, 595: 167–170.

Goldman-Rakic, P.S. (1982) Neuronal development and plasticity of association cortex in primates. *Neurosci. Res. Program Bull.*, 20: 520–532.

Hanbauer, I., Wink, D., Osawa, Y., Edelman, G.M. and Gally, J.A. (1992) Role of nitric oxide in NMDA-evoked release of [3H]-dopamine from striatal slices. *Neuroreport*, 3: 409–412.

Harada, M., Fuse, A. and Tanaka, Y. (1997) Measurement of nitric oxide in the rat cerebral cortex during hypercapnoea. *Neuroreport*, 8: 999–1002.

Harsanyi, K. and Friedlander, M.J. (1997a) Transient synaptic potentiation in the visual cortex. I. Cellular mechanisms. *J. Neurophysiol.*, 77: 1269–1283.

Harsanyi, K. and Friedlander, M.J. (1997b) Transient synaptic potentiation in the visual cortex. II. Developmental regulation. *J. Neurophysiol.*, 77: 1284–1293.

Hess, D.T., Patterson, S.I., Smith, D.S. and Skene, J.H. (1993) Neuronal growth cone collapse and inhibition of protein fatty acylation by nitric oxide. *Nature*, 366: 562–565.

Hester, F., Gemmill, J., Montague, P.R. and Friedlander, M.J. (1996) Coverage of the visual cortex by nitric oxide generating profiles. *Soc. Neurosci. Abstr.*, 22: 641.

Huang, P.L., Dawson, T.M., Bredt, D.S., Snyder, S.H. and Fishman, M.C. (1993) Targeted disruption of the neuronal nitric oxide synthase gene. *Cell*, 75: 1273–1286.

Humbert, P., Niroomand, F., Fischer, G., Mayer, B., Koesling, D., Hinsch, K.D., Gausepohl, H., Frank, R., Schultz, G. and Bohme, E. (1990) Purification of soluble guanylyl cyclase from bovine lung by a new immunoaffinity chromatographic method. *Eur. J. Biochem.*, 190: 273–278.

Huntley, G.W. (1997) Correlation between patterns of horizontal connectivity and the extent of short-term representational plasticity in rat motor cortex. *Cereb. Cortex*, 7: 143–156.

Huntley, G.W., Vickers, J.C. and Morrison, J.H. (1997) Quantitative localization of NMDAR1 receptor subunit immunoreactivity in inferotemporal and prefrontal association cortices of monkey and human. *Brain Res.*, 749: 245–262.

Huntley, G.W., Vickers, J.C., Janssen, W., Brose, N., Heinemann, S.F. and Morrison, J.H. (1994) Distribution and synaptic localization of immunocytochemically identified NMDA receptor subunit proteins in sensory-motor and visual cortices of monkey and human. *J. Neurosci.*, 14: 3603–3619.

Iadecola, C., Zhang, F., Casey, R., Clark, H.B., Ross, M.E. (1996) Inducible nitric oxide synthase gene expression in vascular cells after transient focal cerebral ischemia. *Stroke*, 27: 1373–1380.

Irikura, K., Huang, P.L., Ma, J., Lee, W.S., Dalkara, T., Fishman, M.C., Dawson, T.M., Snyder, S.M. and Moskowitz, M.A. (1995) Cerebrovascular alterations in mice lacking neuronal nitric oxide synthase gene expression. *Proc. Natl. Acad. Sci. USA*, 92: 6823–6827.

Izumi, Y., Clifford, D.B. and Zorumski, C.F. (1992) Inhibition of long-term potentiation by NMDA-mediated nitric oxide release. *Science*, 257: 1273–1276.

Jaffrey, S.R. and Snyder, S.H. (1995) Nitric oxide: A neural messenger. *Ann. Rev. Cell Dev. Biol.* 11: 417–440.

Kano, T., Shimizu, F., Huang, P., Moskowitz, M., and Lo, E. (1998) Effect of nitric oxide synthase gene knockout on neurotransmitter release *in vivo*. *Neuroscience*, (In Press).

Kantor, D.B., Lanzrein, M., Stary, S.J., Sandoval, G.M., Smith, W.B., Sullivan, B.M., Davidson, N. and Schuman, E.M. (1996) A role for endothelial NO synthase in LTP revealed by adenovirus-mediated inhibition and rescue. *Science*, 274: 1744–1748.

Kara, P. and Friedlander, M.J. (1995) The role of nitric oxide in modulating the visual response of neurons in cat striate cortex. *Soc. Neurosci. Abstr.*, 21: 1753.

Kara, P. and Friedlander, M.J. (1997) NMDA receptor independent effects of nitric oxide on the visual response in cat striate cortex. *Soc. Neurosci. Abstr.*, 23: 1267.

Karni, A., Meyer, G., Jezzard, P., Adams, M.M., Turner, R. and Ungerleider, L.G. (1995) Functional MRI evidence for adult motor cortex plasticity during motor skill learning. *Nature* 14: 155–158.

Karni, A., Meyer, G., Rey-Hipolito, C., Jezzard, P., Adams, M.M., Turner, R. and Ungerleider, L.G. (1998) The acquisition of skilled motor performance: Fast and slow experience-driven changes in primary motor cortex. *Proc. Natl. Acad. Sci. USA*, 95: 861–868.

Kuchiiwa, S., Kuchiiwa, T., Mori, S. and Nakagawa, S. (1994) NADPH diaphorase neurons are evenly distributed throughout cat neocortex irrespective of functional specialization of each region. *Neuroreport*, 5: 1662–1664.

Kurenny, D.E., Moroz, L.L., Turner, R.W., Sharkey, K.A. and Barnes, S. (1994) Modulation of ion channels in rod photoreceptors by nitric oxide. *Neuron*, 13: 315–324.

Leopold, D.A. and Logothetis, N.K. (1996) Activity changes in early visual cortex reflect monkeys' percepts during binocular rivalry. *Nature*, 379: 549–553.

Lev-Ram, V., Jiang, T., Wood, J., Lawrence, D.S. and Tsien, R.Y. (1997) Synergies and coincidence requirements between NO, cGMP, and Ca^{2+} in the induction of cerebellar long-term depression. *Neuron*, 18: 1025–1038.

Lipton, S.A., Choi, Y.B., Pan, Z.H., Lei, S.Z., Chen, H.V., Sucher, N.J., Loscalzo, J., Singel, D.J. and Stalmer, J.S. (1993) A redox-based mechanism for the neuroprotective and neurodestructive effects of nitric oxide and related nitrosocompounds. *Nature*, 364: 626–632.

Lonart, G., Wang, J. and Johnson, K.M. (1992) Nitric oxide induces neurotransmitter release from hippocampal slices. *Eur. J. Pharmacol.*, 220: 271–272.

Ma, J., Ayata, C., Huang, P.L., Fishman, M.C. and Moskowitz, M.A. (1996) Regional cerebral blood flow response to vibrissal stimulation in mice lacking type I NOS gene expression. *Am. J. Physiol.*, 270: H1085–H1090.

Malen P.L., Chapman P.F. (1997) Nitric oxide facilitates long-term potentiation, but not long-term depression. *J. Neurosci.* 17: 2645–2651.

Marletta, M.A. (1994) Nitric oxide synthase: Aspects concerning structure and catalysis. *Cell*, 78: 927–930.

Martin, K.A.C. (1988) The welcome prize lecture. From single cells to simple circuits in the cerebral cortex. *Q. J. Exp. Physiol.*, 73: 637–702.

Mayer, B. (1993) Molecular characteristics and enzymology of nitric oxide synthase and soluble guanylyl cyclase in the CNS. *Seminars in the Neurosciences*, 5: 197–205.

Mayer, B., Pfeiffer, S., Schrammel, A., Koesling, D., Schmidt, K. and Brunner, F. (1998) A new pathway of nitric oxide/cyclic GMP signaling involving S-nitrosoglutathione. *J. Biol. Chem.*, 273: 3264–3270.

Meffert, M.K., Calakos, N.C., Scheller, R.H. and Schulman, H. (1996) Nitric oxide modulates synaptic vesicle docking/fusion reactions. *Neuron*, 16: 1229–1236.

Meffert, M.K., Premack, B.A. and Schulman, H. (1994) Nitric oxide stimulates Ca^{2+} - independent synaptic vesicle release. *Neuron*, 12: 1235–1244.

Moncada, S., Palmer, R.M.J. and Higgs, E.A. (1991) Nitric oxide: Physiology, pathophysiology, and pharmacology. *Pharmacol. Rev.*, 43: 109–142.

Montague, P.R., Gally, J.A. and Edelman, G.M. (1991) Spatial signaling in the development and function of neural connections. *Cereb. Cortex*, 1: 199–220.

Montague, P.R., Gancayco, C.D., Winn, M.J., Marchase, R.B. and Friedlander, M.J. (1994) Role of NO production in NMDA receptor-mediated neurotransmitter release in cerebral cortex. *Science*, 263: 973–977.

Moskowitz, M.A. and Dalkara, T. (1996) Nitric oxide and cerebral ischemia. *Adv. Neurol.*, 71: 365–367.

Mothet, J-P., Fossier, P., Tauc, L. and Baux, G. (1996) Opposite actions of nitric oxide on cholinergic synapses: Which pathways? *Proc. Natl. Acad. Sci. USA*, 93: 8721–8726.

Murphy, S., Simmons, M.L., Agullo, L., Garcia, A., Feinstein, D.L., Galea, E., Reis, D.L., Minc-Golomb and Schwartz, J.P. (1993) Synthesis of nitric oxide in CNS glial cells. *TINS* 16: 323–328.

Muller, B.M., Kistner, U., Kindler, S., Chung, W.J., Kuhlendahl, S., Fenster, S.D., Lau, L.F., Veh, R.W., Huganir, R.L., Gundelfinger, E.D. and Garner, C.C. (1996) SAP102, a novel postsynaptic protein that interacts with NMDA receptor complexes *in vivo*. *Neuron*, 17: 255–265.

Nathan, C. and Xie Q-W. (1994) Nitric oxide synthases: Rolls, tolls, and controls. *Cell*, 78: 915–918.

Nudo, R.J., Milliken, G.W., Jenkins, W.M. and Merzenich, M.M. (1996) Use-dependent alterations of movement representations in primary motor cortex of adult squirel monkeys. *J. Neurosci.*, 16: 785–807.

O'Dell, T.J., Huang, P.L., Dawson, T.M., Dinerman, J.L., Snyder, S.H., Kandel, E.R. and Fishman, M.C. (1994) Endothelial NOS and the blockade of LTP by NOS inhibitors in mice lacking neuronal NOS. *Science*, 265: 542–546.

Ohkuma, S., Narihara, H., Katsura, M., Hasegawa, T. and Kuriyama, K. (1995) Nitric oxide-induced [3H] GABA release from cerebral cortical neurons is mediated by peroxynitrite. *J. Neurochem.*, 65: 1109–1114.

O'Leary, D.D., Ruff, N.L. and Dyck, R.H. (1994) Development, critical period plasticity, and adult reorganizations of mammalian somatosensory systems. *Curr. Opin. Neurobiol.*, 4: 535–544.

Pan, Z-H., Segal, M.M. and Lipton, S.A. (1996) Nitric oxide-related species inhibit evoked neurotransmission but enhance spontaneous minature synaptic currents in central neuronal cultures. *Proc. Natl. Acad. Sci. USA*, 93: 15423–15428.

Pape, H-C. and Mager, R. (1992) Nitric oxide controls oscillatory activity in thalamocortical neurons. *Neuron*, 9: 441–448.

Park, S.K., Lin, H.L. and Murphy S. (1994) Nitric oxide limits transcriptional induction of nitric oxide synthase in CNS glial cells. *Biochem. Biophys. Res. Commun.*, 201: 762–768.

Peters, A. (1985) Neuronal composition and circuitry of rat visual cortex. In D. Rose and V.G. Dobson (Eds.). Models of the Visual Cortex, Wiley, New York, pp. 492–503.

Petralia, R.S., Yokotani, N. and Wenthold, R.J. (1994) Light and electron microscope distribution of the NMDA receptor subunit NMDAR1 in the rat nervous system using a selective anti-peptide antibody. *J. Neurosci.*, 14: 667–696.

Peunova, N. and Enikolopov, G. (1993) Amplification of calcium-induced gene transcription by nitric oxide in neuronal cells. *Nature*, 364: 450–453.

Ramachandran, V.S. (1993) Behavioral and magnetoencephalographic correlates of plasticity in the adult human brain. *Proc. Natl. Acad. Sci. USA*, 90: 10413–10420.

Rauschecker, J.P. (1991) Mechanisms of visual plasticity: Hebb Synapses, NMDA receptors, and beyond. *Physiol. Rev.*, 71: 587–615.

Rauschecker, J.P. and Kniepert, U. (1994) Auditory localization behavior in visually deprived cats. *Eur. J. Neurosci.*, 6: 149–160.

Rauschecker, J.P. (1995) Compensatory plasticity and sensory substitution in the cerebral cortex. *TINS*, 18: 36–43.

Ravichandran, L.V., Johns, R.A. and Rengasamy, A. (1995) Direct and reversible inhibition of endothelial nitric oxide synthase by nitric oxide. *Am. J. Physiol.*, 268: H2216–H2223.

Recanzone, G.H., Allard, T.T., Jenkins, W.M. and Merzenich, M.M. (1990) Receptive-field changes induced by peripheral nerve stimulation in SI of adult cats. *J. Neurophysiol.*, 63: 1213–1225.

Recanzone, G.H., Schreiner, C.E. and Merzenich, M.M. (1993) Plasticity in the frequency representation of primary auditory cortex following discrimination training in adult owl monkeys. *J. Neurosci.*, 13: 87–103.

Reid, S.N.M., Daw, N.W., Czepita, D., Flavin, H.J., and Sessa, W.C. (1996) Inhibition of nitric oxide synthase does not alter ocular dominance shifts in kitten visual cortex. *J. Physiol. (London)*, 492: 511–517.

Rengasamy, A. and Johns, R.A. (1993) Regulation of nitric oxide synthase by nitric oxide. *Mol. Pharmacol.*, 44: 124–128.

Renteria, R.C. and Constantine-Paton, M. (1996) Exogenous nitric oxide causes collapse of retinal ganglion cell axonal growth cones *in vitro*. *J. Neurobiol.*, 29: 415–428.

Ruthazer, E.S., Gillespie, D.C., Dawson, T.M., Snyder, S.H. and Stryker, M.P. (1996) Inhibition of nitric oxide synthase does not prevent ocular dominance plasticity in kitten visual cortex. *J. Physiol. (London)*, 492: 519–527.

Salvemini, D. and Masferrer, J.L. (1996) Interactions of nitric oxide with cyclooxygenase: In vitro, ex vivo, and *in vivo* studies. *Methods. Enzymol.*, 269: 12–25.

Savtchenko, A., Barnes, S. and Kramer, R.H. (1997) Cyclic-nucleotide-gated channels mediate synaptic feedback by nitric oxide. *Nature*, 390: 694–698.

Schuman, E.M. and Madison, D.V. (1991) A requirement for the intercellular messenger nitric oxide in long-term potentiation. *Science*, 254: 1503–1506.

Schuman, E.M. and Madison, D.V. (1994) Locally distributed synaptic potentiation in the hippocampus. *Science*, 263: 532–536.

Semba, J., Sakai, M., Miyoshi, R. and Kito, S. (1995) NG-monomethyl-L-arginine, an inhibitor of nitric oxide synthase, increases extracellular GABA in the striatum of the freely moving rat. *Neuroreport*, 6: 1426–1428.

Shatz, C.J. and Stryker, M.P. (1978) Ocular dominance in layer IV of the cat's visual cortex and the effects of monocular deprivation. *J. Physiol. (London)*, 281: 267–283.

Shibuki, K. and Okada, D. (1991) Endogenous nitric oxide release required for long-term synaptic depression in the cerebellum. *Nature*, 349: 326–328.

Singer, W. (1995) Development and plasticity of cortical processing architectures. *Science*, 270: 758–764.

Sistiaga, A., Miras-Portugal, M.T. and Sanchez-Prieto, J. (1997) Modulation of glutamate release by a nitric oxide/cyclic GMP-dependent pathway. *Eur. J. Pharmacol.*, 321: 247–257.

Snyder, S.H. (1992) Nitric oxide: First in a new class of neurotransmitters. *Science*, 257: 494–496.

Son, H., Hawkins, R.D., Martin, K., Kiebler, M., Huang, P.L., Fishman, M.C. and Kandel, E.R. (1996) Long-term potentiation is reduced in mice that are doubly mutant in endothelial and neuronal nitric oxide synthase. *Cell*, 87: 1015–1023.

Stamler, J.S., Singel, D.J. and Loscalzo, J. (1992) Biochemistry of nitric oxide and its redox-activated forms. *Science*, 258: 1898–1902.

Stamler, J.S., Toone, E.J., Lipton, S.A. and Sucher, N.J. (1997) (S)NO signals: Translocation, regulation, and a consensus motif. *Neuron*, 18: 691–696.

Storig, P. (1996) Varieties of vision: From blind responses to conscious recognition. *TINS*, 19: 401–406.

Vincent, S.R. and Hope, B.T. (1992) Neurons that say NO. *TINS*: 15, 108–113.

Wallace, M.N., Brown, I.E., Cox, A.T. and Harper, M.S. (1995) Pyramidal neurons in human precentral gyrus contain nitric oxide synthase. *Neuroreport*, 6: 2532–2536.

Wang, Q., Pelligrino, D.A., Baughman, V.L., Koenig, H.M. and Albrecht, R.F. (1995) The role of neuronal nitric oxide synthase in regulation of cerebral blood flow in normocapnia and hypercapnia in rats. *J. Cereb. Blood Flow Metab.*, 15: 774–778.

Weiss, S.W., Albers, D.S., Iadarola, M.J., Dawson, T.M., Dawson, V.L. and Standaert, D.G. (1998) NMDAR1 glutamate receptor subunit isoforms in neostriatal, neocortical, and hippocampal nitric oxide synthase neurons. *J. Neurosci.*, 18: 1725–1734.

Wiesel, T.N. (1982) Postnatal development of the visual cortex and the influence of the environment. *Nature*, 299: 583–591.

Wiesel, T.N. and Hubel, D.H. (1963) Single-cell responses in striate cortex of kittens deprived of vision in one eye. *J. Neurophysiol.*, 26: 1003–1017.

Williams, J.H., Li, Y-G., Nayak, A., Errington, M.L., Murphy, K.P.S.J. and Bliss, T.V.P. (1993) The suppression of long-term potentiation in rat hippocampus by inhibitors of nitric oxide synthase is temperature and age dependent. *Neuron.*, 11: 877–884.

Wu, H.H., Williams, C.V. and McLoon, S.C. (1994) Involvement of nitric oxide in the elimination of a transient retinotectal projection in development. *Science*, 265:1593–1596.

Yanagihara, D. and Kondo, I. (1996) Nitric oxide plays a key role in adaptive control of locomotion in cat. *Proc. Natl. Acad. Sci. USA*, 93: 13292–13297.

Zhuo, M., Small, S.A., Kandel, E.R. and Hawkins, R.D. (1993) Nitric oxide and carbon monoxide produce activity-dependent long-term synaptic enhancement in hippocampus. *Science*, 260: 1946–1950.

R.R. Mize, T.M. Dawson, V.L. Dawson and M.J. Friedlander (Eds.)
Progress in Brain Research, Vol 118

CHAPTER 14

Interaction of nitric oxide and external calcium fluctuations: A possible substrate for rapid information retrieval

David M. Egelman, Richard D. King and P. Read Montague*

Center for Theoretical Neuroscience, Division of Neuroscience, Baylor College of Medicine, One Baylor Plaza, Houston, TX, 77030 USA

Synapses, the circuit metaphor, and the role of spatial volume in neural processing

Traditional accounts of neurotransmission have described signals passing from one neuron to the next across a synapse, evoking images of wires, nodes, and circuit elements. From this vantage point, the individual synapse has dominated our concept of neurotransmission. In this decade, our conception of neural communication has been broadened by evidence that certain molecules may act as rapid volume signals, which diffuse freely in three dimensions to act throughout local regions of neural tissue. One such spatial signal, nitric oxide, has been implicated in learning, long-term potentiation at single synapses, blood flow control, neurotoxicity, activity-development patterning of synaptic connections, and ongoing control of synaptic transmission.

Nitric oxide (NO) is a non-polar free radical that can readily diffuse through tissue. In neural tissue, NO is known to be made on demand by at least two isoforms of a calcium-dependent enzyme, nitric oxide synthase (NOS). Its capacity to diffuse through cell membranes permits changes in NO production to be felt throughout a diffusion-defined domain. This framework departs from the circuit metaphor described above. In particular, it suggests that spatial volume may be an important parameter for the effects mediated by NO.

NO has a number of targets throughout neural tissue, and it is used in a variety of different signaling pathways. For example, it is a powerful stimulator of soluble guanylyl cyclase and ADP-ribosyltransferase. NO is also a potent vasodilator, which represents a form of negative feedback for the action of NO since (deoxy)hemoglobin is a powerful chelator of NO. NO modulates the probability of neurotransmission on short time scales, and is involved in certain forms of long-term change in synaptic function. Many of these signaling functions are discussed at length in other chapters in this volume.

Transporting coincidence detection through a tissue volume

NO is made in response to calcium fluxes through the *N*-methyl-D-aspartate (NMDA) type of glutamate receptor. NMDA receptors are both ligand- and voltage-gated, suggesting that they act as molecular coincidence detectors: presynaptic spike arrival is reported by glutamate release, and the

* Corresponding author. Tel.: 713 798 3134; fax: 713 798 3946; e-mail: read@bcm.tmc.edu

postsynaptic spike is reported by membrane voltage. The enhanced production of NO subsequent to this coincidence detection provides a means to transport this information throughout a surrounding volume of neural tissue. Furthermore, isotropic diffusion of NO establishes a fixed, radially symmetric relationship between space (r) and time (t):

$$\langle r^2 \rangle = 6Dt$$

where D is the diffusion constant, and $\langle \ \rangle$ represents the expected value operator.

In this paper, we address three domains in which a rapid volume signal could play important roles in neural tissue: (1) activity-dependent development of neural connections, (2) storage of information in volumes of neural tissue, and (3) rapid retrieval of information from volumes of neural tissue.

Role of NO in activity-dependent refinement of synaptic connections

The idea of a rapidly diffusing, membrane permeant signal was originally examined on theoretical grounds (Gally et al., 1990). In 1991, four groups demonstrated that the inhibition of nitric oxide synthase (NOS, the synthetic enzyme for NO) prevents the induction of NMDA-dependent LTP in the mammalian brain (Bohme et al., 1991; O'Dell et al., 1991; Schuman and Madison, 1991; Haley et al., 1991). Subsequently, Schuman and Madison provided an elegant demonstration that potentiation could be transported between synapses on different neurons through a diffusible messenger (Schuman and Madison, 1994). Recent work by Bonhoeffer and colleagues is consistent with this interpretation (Engert and Bonhoeffer, 1997). NO has also been shown to play an important role in long-term depression (LTD) in the cerebellum (Shibuki and Okada, 1991; Lev-Ram et al., 1995). These experiments give us clear evidence that a rapidly diffusible messenger molecule like NO influences synaptic function by direct communication through volumes of tissue. Such a signal is ideal for mediating activity-dependent patterning of synaptic connections.

Early theoretical work predicted the involvement of volume signaling in the development of neural connections (Gally et al., 1990; Montague et al., 1991). These computational studies demonstrated that a signal like NO could account for a variety of anatomical patterns of synaptic connections. These include clustered connections in the cerebral cortex, the refinement of topographic mappings in the lateral geniculate nucleus and optic tectum, the formation of ocular dominance columns, and the formation of somatotopic mappings in the cerebral cortex. An example is shown in Fig. 1. Some of these predictions are now supported by experimental data. For example, there are now direct demonstrations that NO is involved in segregation of ON/OFF layers in the ferret lateral geniculate nucleus (Cramer et al., 1996) and the refinement of topography in retino-tectal projections in chick (Wu et al., 1994). However, NO does not appear to play a role in ocular dominance shifts in kitten cortex (Reid et al., 1996), nor in the refinement of retinal ganglion cell axons to eye-specific layers in the lateral geniculate nucleus (Cramer et al., 1996). It is not yet understood why some patterns of synaptic contacts require intact NO production to develop normally and others do not. More work is needed to explore the nature of the differences in specific cases.

The NOS positive scaffold

As well as playing a role during development, NO also modulates ongoing neural processing. For example, NO enhances release of norepinephrine (NE) and glutamate (Glu) from synaptosomes in the cerebral cortex (Montague et al., 1994). The enhanced release is mediated by activation of the NMDA receptor, and is blocked by agents that inhibit nitric oxide (NO) production or remove it from the extracellular space. NO has also been shown to enhance spontaneous presynaptic transmitter release from hippocampal neurons in dissociated cell culture (O'Dell et al., 1991) and striatal slices (Hanbauer et al., 1992). Thus, NO production may link NMDA receptor activation to changes in neurotransmission in surrounding

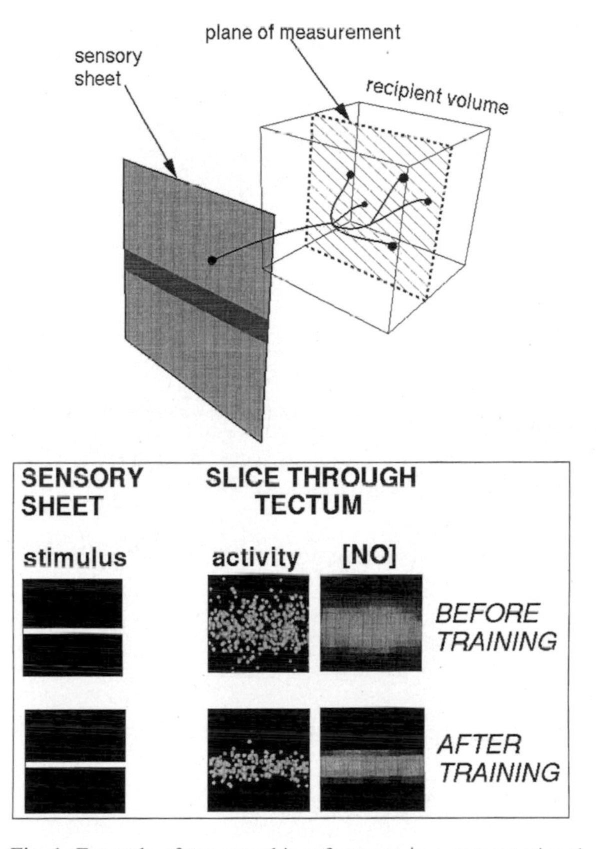

Fig. 1. Example of topographic refinement in a computational model. The development of patterns of synaptic contacts was mediated by a learning rule that stabilized synapses according to the correlation between presynaptic activity and local NO levels. A two-dimensional sensory sheet innervates a three-dimensional tectum with crude initial topography. Elongated waves of activity in the sensory sheet act to refine the topography of the projection. The figure shows a two-dimensional slice through the tectum. This figure is adapted from Montague et al., 1991.

synaptic terminals *whether or not* the synapses are made onto the same postsynaptic neuron. One major consequence of these observations is that volumes of neural tissue may come to act as integrated computational units (Montague and Sejnowski, 1994).

A volume signaling mechanism mediated by NO can play different roles depending on the spatial and temporal scales over which it acts. At small scales, the signal can function rapidly with high specificity. At larger scales, it can act more like a paracrine signal. We have begun to address these issues by examining in detail the scaffold of NOS activity expressed in the mammalian cerebral cortex. The analysis of NOS profiles has been carried out in collaboration with Michael Friedlander.

Three-dimensional distributions of NOS-positive terminals derived from guinea pig and cat cortex were generously provided by M.J. Friedlander. The assumption made here is that the range of NO action in the brain will be a function of the scaffolding of production sites and their temporal pattern of activation. One example distribution from a $93 \times 62 \times 56$ μm block of tissue is shown in Fig. 2 (top panel). As shown in Fig. 2 (bottom panel), the average first nearest neighbor distance between NOS-positive terminals is 3.9 μm. Therefore, on average, no point in the cortex is more than $\sim$2 μm away from a potential NO source. This spacing of NOS positive profiles means that every point in cortex is potentially within about 1 ms of an NO burst, i.e., the temporal width of an action potential (Fig. 3). The same kind of question can be asked in a different way. Fig. 4 shows the percentage of tissue covered by an NO signal as a function of the fraction of NOS positive profiles activated synchronously. It is important to note that we do not yet know how large a burst in NO production actually occurs in functioning tissue.

These conclusions depend on some assumptions that have not yet been tested. One major assumption is that the NOS positive profiles can act as NO sources; there are at least two ways that they could play this role: (1) action potent invading NOS positive profile (on axons or dendrites) could elicit NO production and (2) NO production could result from the nearby release of neurotransmitters. The former case suggests that our estimates of NOS sources may be a gross underestimate since we only analyzed NOS profiles larger than $\sim$0.7 μm while completely ignoring the heavily stained fibers from which the profiles originate. If these processes constitute an active source of NO, then the estimates we have made fall on the conservative side.

The latter case is intriguing because it suggests that the NOS positive profiles may literally

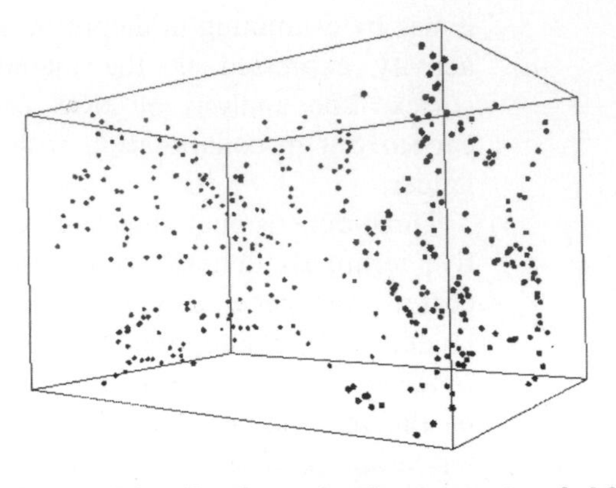

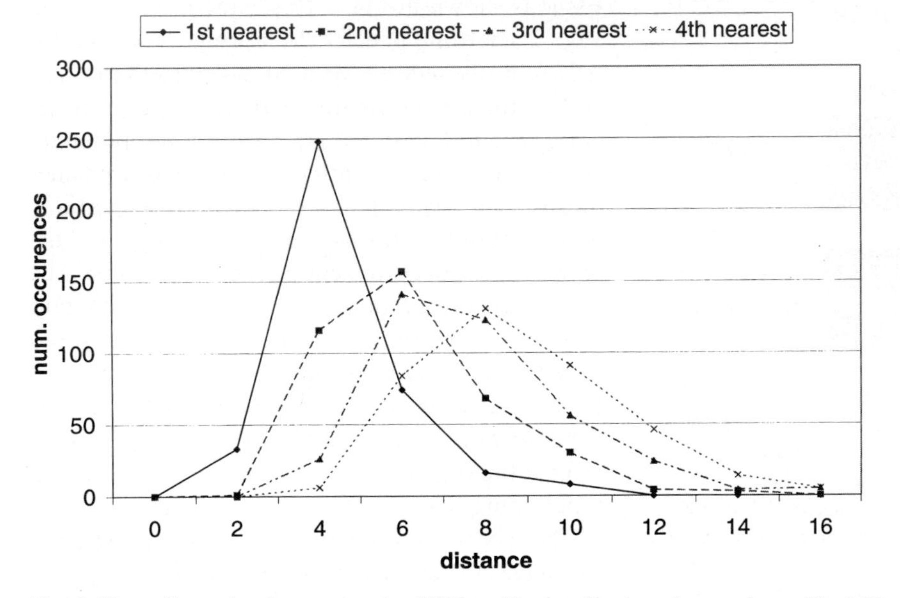

Fig. 2. Three-dimensional reconstruction NOS positive profiles in a tissue volume. (Top) 92 μm × 62 μm × 57 of cat cortical tissue showing the locations of NOS positive profiles (swellings larger than ~0.7 μm in diameter). (Bottom) Frequency distribution of nearest neighbor distances for the NOS profiles shown in top panel. 1st, 2nd, 3rd, and 4th nearest neighbor distributions (in three dimensions) are shown. The average of first nearest distances is 3.9 μm. This number is typical for the six tissue blocks analyzed to date.

provide a three-dimensional scaffold from which local signals elicit enhanced NO production. By sensing the neurotransmitters in their locale, the potential NO sources eavesdrop on ongoing activity in the vicinity, decide whether to increase production of NO, and the NO moves rapidly throughout the surrounding volume (increasing transmission probabilities at nearby synapses). Under this model, the ongoing control of NO production acts as a local gain control for transmission throughout a local volume. Experimental data suggests that the fluctuating NO signal does not distinguish one type of synapse from another, i.e., release of glutamate, dopamine, norepinephrine are all enhanced in the presence of an NO burst (O'Dell et al., 1991; Hanbauer et al., 1992; Montague et al., 1994).

The data discussed above paint a picture of positive feedback in NO production, which leads to increased transmitter release, increased calcium

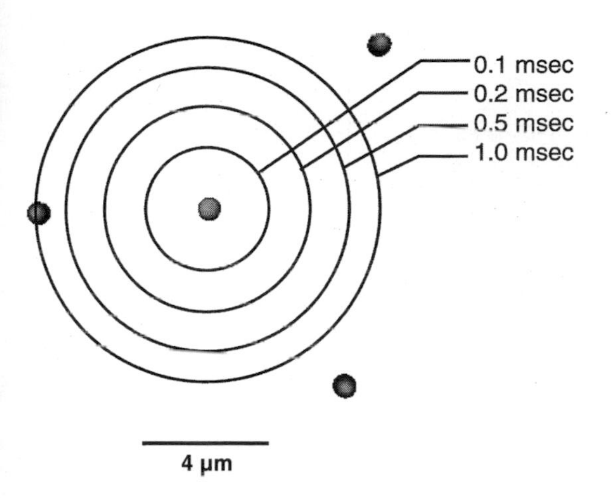

Fig. 3. The relation between root mean square distances and time for NO. The spheres represent NOS positive profiles spaced approximately 4 μm from their first nearest neighbor. The times and cantered distances are computed from the relationship:

$$r_{\mathrm{RMS}} = \sqrt{\langle r^2 \rangle} = \sqrt{6Dt}.$$

entry, and thus increased NO production. This is just one plausible positive feedback cycle. This situation strongly suggests the existence of negative feedback mechanisms on NO production. One negative pathway is the interaction of NO with changes in external calcium levels.

External calcium fluctuations

We have appealed to NO production as an information-bearing volume signal. We explore here the possibility that interaction between NO and other volume signals will carry functional significance. In particular, we examine the interaction of NO with fluctuations in external calcium. Every *influx* of calcium into intracellular compartments is mirrored by an *efflux* of calcium from the extracellular space. There is a 10 000 : 1 concentration gradient pointing from the extracellular space to intracellular compartments; therefore, any ionic channel permeable to calcium will cause an efflux of calcium from the extracellular space, *i.e.*, net calcium flux through ionic channels is unidirectional. In practice, it is almost impossible to cause calcium currents to reverse. Sodium co-transporters and ATP dependent pumps are the primary pathways for extruding calcium back into the extracellular space, and these pathways are 2–4 orders of magnitude slower than ionic channels. Thus, an open calcium conductance acts as a rapid sink for external calcium that is replenished on fast time scales by diffusion from surrounding regions of the tissue. These facts along with other considerations have led us to consider fluctuations in external calcium as an information-bearing signal important for neural function (Person et al., 1996; Montague, 1996; Egelman et al., 1998).

In Fig. 5, we show two situations in which external calcium fluctuations are significant. Using a computational model of the extracellular space (Egelman *et al*, 1998) we show the influence of an active presynaptic bouton on external calcium concentrations. The rate of calcium consumption is set to 14,000 calcium atoms per active zone per spike, as reported in the literature by Sakmann and colleagues (Helmchen et al., 1997). For isolated presynaptic terminals, the calcium recovers to baseline rapidly following the arrival of a spike (recovery in less than ~1.5 ms). Different geometrical distributions of calcium sinks will change the recovery time of external calcium. For example, back-propagating action potentials lead to calcium consumption along extended stretches of dendritic membrane (Magee and Johnston, 1997). Fig. 6 shows a small bundle of dendrite segments that are synchronously activated. In the bundle, the recovery time for external calcium is increased dramatically due to the geometry of the synchronous calcium sinks (recovery time increases about 10-fold). In the absence of some preventative mechanism, coincident back-propagating spikes are expected to occur frequently in tissue, even for neurons whose activity is uncorrelated (recall that statistically independent events are quite clumpy in time).

The interactions of external calcium and N

What is the point of focusing on external calcium fluctuations as a volume signal that interacts with NO? The short answer is that we expect external calcium fluctuations to antagonize increased NO production and its effects. The more interesting answer, discussed in the last section, is that the

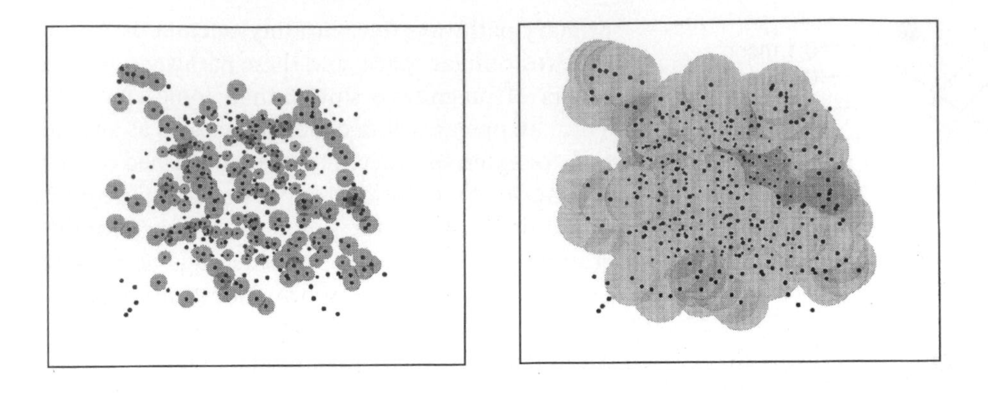

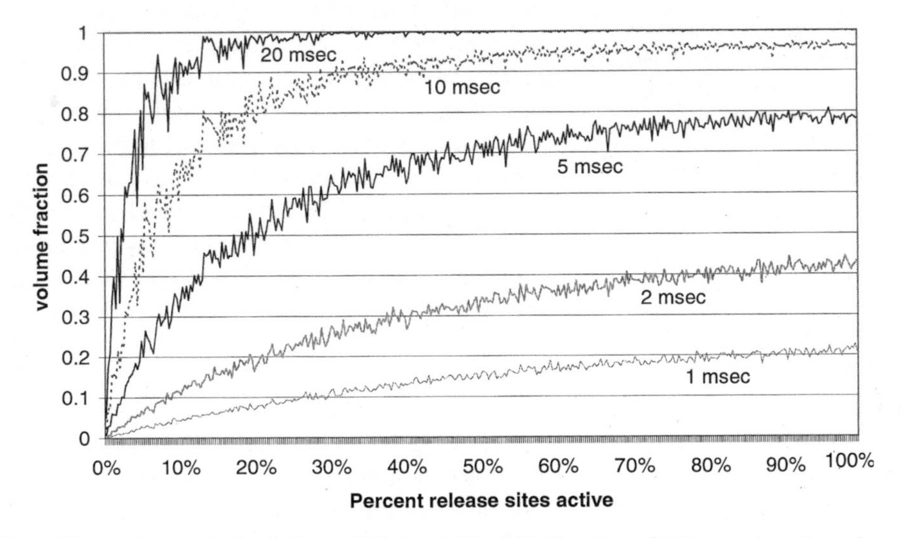

Fig. 4. Time and space tradeoffs for an NO signal. (Top) RMS sphere of NO gas released synchronously from sites defined by a 3-D reconstruction of NOS positive profiles. Left panel shows RMS spheres after 0.2 ms., and right panel shows same spheres 2 ms. (Bottom) This figure shows the fraction of a given volume that has experienced a peak in local NO levels due to synchronous activation of NOS positive profiles from real data. The fraction of the volume covered is plotted against the fraction of NOS profiles activated. Each curve represents a different time after the synchronous burst. By 10 ms, a large fraction of the volume has felt the effects of a synchronous NO burst independent of the exact fraction of NOS profile activated. The percentage volume inside the RMS spheres was calculated using Monte Carlo integration (this accounts for the jagged nature of the curves).

interaction provides a means by which the brain can index different volumes of tissue similar to the way a computer addresses the contents of its memory.

One way for changes in external calcium to antagonize the effects of NO is to decrease the capacity of a tissue volume to make NO. The sequence of events runs roughly as follows. Increased NO production causes increased neurotransmitter release throughout a diffusion-defined domain. This neurotransmitter release includes the release of L-glutamate leading to a large consumption of calcium from the extracellular space in the vicinity. This large consumption results from at least two significant sources: (1) open NMDA receptors, and (2) calcium entry into dendrites due to the effects of back-propagating dendritic spikes caused by the increased level of transmission. The calcium consumption will tend to antagonize the

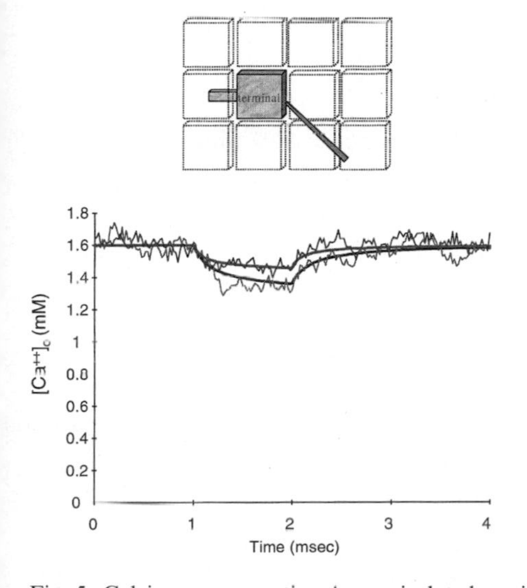

Fig. 5. Calcium consumption by an isolated, active terminal. Using two models of the extracellular space, the decrement in external calcium due to the arrival of a single action potential is shown. Cubic units were packed into a volume. Each cube measured 0.806 μm along an edge, and yielded the same volume as a 1 μm diameter sphere (about the size of a bouton or dendritic spine in the mammalian CNS). An explicit Monte Carlo model of passive calcium movement in the extracellular space was used (jagged curves) alongside a discretized model of the extracellular space. Details of the implementation are included in Egelman and Montague (1998). The two curves represent two separate diffusion constants for calcium movement in the extracellular space ($D = 600\ \mu m^2/s$, and $D = 300\ \mu m^2/s$). The lower curve is the smaller diffusion constant. The total calcium consumption during the 1 ms action potential was clamped to the value reported in the literature by Sakmann and colleagues (14 000 calcium atoms per spike per release zone). The external calcium fluctuation is brief, and relatively small (~200 μM for $D = 300$).

effect of NO on further neurotransmitter release. It is well known that small decrements in external calcium can considerably reduce the probability of NT release. The relationship between external calcium concentration and NT is approximately *quadratic*. This has been measured at the squid giant synapse (Katz and Miledi, 1970), the cerebellum (Mintz et al., 1995), and the hippocampus (Qian et al., 1997). It is therefore reasonable to suspect that long, large decrements in external calcium serve to antagonize NO production and its effects.

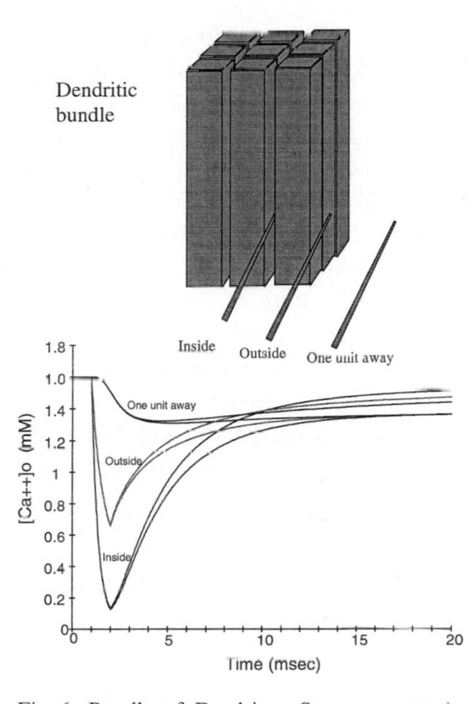

Fig. 6. Bundle of Dendrites. Same computational model as described in Fig. 5. Dendrites are defined by connecting several elementary units like the bouton in Fig. 5. The consumption of each unit area of dendritic membrane was set to 14 000 calcium atoms per spike. Each dendritic segment was 4.03 μm long. Synchronous activation of the elongated dendritic segments produces elongated calcium sinks whose effects on external calcium superimpose. Two changes result from this different arrangement of calcium sinks: (1) the amplitude of the fluctuation has increased ~6 fold, and (2) the replenishment time has increased from ~7–8 to about 7–10 ms. There are no calcium pumps or co-transporters that would significantly influence either the amplitude or the recovery time. This fact is made clear by the two curves produced for each measurement position. One curve has no calcium pumps present while the faster recovering curve results from the operation of calcium pumps with a 35-ms half-life (pumps are first order in internal calcium concentration). Since this rate is tremendously faster than pumps reported in the literature, the example serves to illustrate that pumps will not change significantly the conclusions.

Negative feedback at NMDA receptor may act as plasticity gain control

Activation of the NMDA receptor is the first step in the cascade that leads to the production of NO. It is known that NO interacts with the NMDA receptor to inhibit further influx through the channel (Hoyt et al., 1992; Manzoni et al., 1992; Manzoni and Bockaert, 1993; Tanaka et al., 1993; Akira et al., 1994). NMDA receptor populations may thus be significantly inhibited and possibly silenced under high levels of NO. Taking this possibility further, synapses may be temporarily unable to flux large amounts of calcium, making them refractory to plasticity.

From a biophysical perspective, we see a variety of homeostats in operation. Some event activates NMDA receptors, which leads to increased NO production. The increased NO production indirectly causes further NMDA receptor activation by causing increased release of glutamate from local synaptic terminals. If left unchecked, this cycle could result in two deleterious events: (1) overproduction of NO, which can be neurotoxic under appropriate oxidizing conditions, and (2) large fluxes of calcium into cells, which can lead ultimately to the induction of apoptosis. We have highlighted two negative feedback mechanisms for this positive feedback cycle: (1) NMDA receptor activation turns to a prolonged sink for external calcium, and (2) NO inhibits fluxes through the NMDA receptor thus turning off one primary stimulus for its own production. Below, we consider some computational consequences of these biophysical possibilities.

Computational consequences of the calcium–NO cycle

Push-pull character of NO production and calcium production

We have already suggested that NO and external calcium tend to move in directions that anatagonize changes in each variable. This situation suggests that the production and removal of NO and external calcium are coupled. This fact can be expressed generally:

$$\frac{\partial n}{\partial t} = f(n, c) + D_n \frac{\partial^2 n}{\partial x^2}, \tag{1}$$

$$\frac{\partial c}{\partial t} = g(n, c) + D_c \frac{\partial^2 c}{\partial x^2} \tag{2}$$

$n(x,t)$ and $c(x,t)$ are functions of time (t) and space (x) and represent respectively fluctuations of nitric oxide levels $N(x,t)$ and external calcium $C(x,t)$ about their equilibrium values N_0 and C_0. $f(n,c)$ and $g(n,c)$ are functions which may be non-linear in their arguments. The last term in the equations represents the passive diffusion of fluctuations. There are two particular facts about external calcium and nitric oxide that make these expressions more interesting; they allow a straightforward interpretation of f and g as being proportional to concentration curvatures (in the tissue space) of calcium and nitric oxide. (1) All the nitric oxide is made from enzymes located inside cells where the calcium levels are kept near zero, and (2) ionic channels that conduct calcium always conduct it *out* of the extracellular space. If external calcium goes down, internal calcium goes up. Some of the internal compartments that see this increase contain NOS, and will increase NO production. We translate these remarks into a simple linear equation by noting that a positive curvature in external calcium $\left(\partial^2 c/\partial x^2 > 0\right)$suggests that NO is getting made somewhere; giving us a specific form for $f(n,c)$:

$$\frac{\partial n}{\partial t} = K_c \frac{\partial^2 c}{\partial x^2} + D_n \frac{\partial^2 n}{\partial x^2}. \tag{3}$$

Furthermore, increases in NO production cause a negative curvature in the NO concentration, and are modeled as negatively proportional to the increase in external calcium (NO inhibits calcium fluxes out of the extracellular space thereby raising external calcium levels):

$$\frac{\partial c}{\partial t} = -K_n \frac{\partial^2 n}{\partial x^2} + D_c \frac{\partial^2 c}{\partial x^2}, \tag{4}$$

where K_c, K_n, D_c, D_n are positive constants. We are left with an interesting set of coupled equations for NO and external calcium. These equations can be re-written using a single complex variable $z = c + in$. We do not explore the biological consequences in this chapter, but this composite variable provides a novel signal that is ideally suited for the storage of important parameters (Montague, unpublished observations).

Gain control on plasticity

Our hypothesis is that calcium influx at a synapse (and thus the eligibility for plasticity) can be modulated by the ambient NO concentrations. The NMDA receptor is necessary for certain kinds of plasticity, but makes only minimal contribution to the excitatory postsynaptic potential (EPSP). It is straightforward to imagine transmission continuing unaffected at a synapse while the plasticity is shut down. We propose that ambient levels of NO in the brain may render some fraction of a synapse's NMDA receptor population silent at any given time, independent of the status of glutamate binding to the receptor. Increases or decreases of the NO concentration could modulate a synapse's eligibility for NMDA-mediated plasticity. In this framework, a sufficient rise in local NO concentration results in a greater fraction of inactivated NMDA receptors. Conversely, decreasing NO levels leads increased eligibility for plasticity.

A biological addressing scheme: Rapid indexing of small volumes of neural tissue

The possible interactions of external calcium and NO have given rise to several new suggestions about their function in neural tissue. These functions have been described only at the level of biophysical changes – when receptors may operate, when plasticity is on or off, and so forth. There is, however, a computational perspective for the meaning of these physical interactions. We propose that the interaction of external calcium fluctuations and NO provides a rapid indexing scheme used by the brain to locate specific volumes of neural tissue (presumably containing information important for some function).

Our approach here is to view the cerebral cortex as an enormous storage device that acts as a look-up table where the kinetics of the look-up process can implement a variety of functions. For example, in some cortical areas, this look-up amounts to iconic retrieval of information, *e.g.* cells in the inferior temporal cortex that respond to specific faces or views of faces. In other areas, like the visual or somatosensory cortex, the look-up process is equivalent to an information processing algorithm well fitted to some problem domain, *e.g.* the visual problem of detecting the likely presence of a contiguous border that is occluded by overlying objects. In this section, we simply detail how a specific, small volume of tissue could be indexed by a pattern of neural signals (Fig. 7).

Information arrives in a region of tissue, and the job of the biological mechanisms is to retrieve stored parameters in an appropriate amount of time. In some cases, the dynamics of the retrieval itself is the sought-after information, and in other cases, retrieval may take other forms. Here, we only address the issue of indexing the desired locations in tissue given a collection of signals impinging on a region of cortex. One of the computational problems to be solved is the natural trade-off between storage capacity and retrieval time. Greater storage capacity makes more difficult the task of keeping retrieval times short. An unlimited memory is not useful unless the information can be accessed in an appropriate amount of time, and a serial search through large memory is almost never feasible. These kinds of concerns were surely a ubiquitous evolutionary constraint: mobile organisms always had to squeeze the results of neural computations into behaviorally relevant timescales. For example, no matter how much *parallel* processing takes place in the cortex, there is an enormous *serial* processing constraint for choosing which direction to turn next when being chased by a predator. Any computational result has to be ready and available quickly if it is to be useful. This is an extreme example; however, all perceptual tasks for a mobile creature are subject to strict temporal constraints; therefore, the mechanisms employed by the cortex are also subject to these same constraints.

One way to avoid serially searching through a large memory every time a query for information is made is to use clever ways of organizing the information. The cerebral cortex solves the rapid look-up problem in part by clumping similar functions in similar volumes of tissue. This idea is interesting and suggestive, however, it falls short of specifying how specific subelements of the volumes, e.g. specific synapses or their components, could be addressed by a pattern of neural activity arriving in the form of different neurotransmitters.

The computational idea for mediating rapid look-up is outlined below, and followed by our interpretation of a possible neural substrate. Consider the cortex as a storage medium where information (or functions) are clustered in common volumes of tissues. The indexing has multiple stages, and takes advantage of this scheme for organization information (Fig. 7):

Stage I (Global query). A global query is issued through the broadcast of a vector of cues. Specific regions match this vector best. These regions are labeled as prototypes – they serve to identify the likely locations of the storage medium containing the desired information or response.

Stage II (Prototype generalization). The search is expanded rapidly around a subset of the prototypes taking advantage of the organizational scheme of clustering similar function in nearby regions of the medium.

Stage III (Refinement). Generalization about each matching prototype has identified many possible locations, and a more refined analysis is required in these regions. The refinement step executes a rapid, more detailed search only in the regions defined by the domains centered on the prototypes. This identifies even smaller volumes likely to contain the best fit to the global query.

To mark the results, a signal is generated that flags the region of the medium containing the results – the signal means something like 'I am the best fit to the query'.

Notice that Stage I imbeds an interesting assumption into the nature of the storage medium, i.e., the system cannot be sure of the exact location of the information; therefore, a broadcast to many areas must occur. This incurs a cost – the cost in time and space of constructing and distributing the broadcast. The payback for this expense is that the memory can be large, yet the system maintains recall within an acceptable time window. We suspect that in the cortex, this scheme could operate on the scale of less than 10 ms (described below).

The scheme can also be applied recursively. Once the refinement stage has located a set of smaller volumes, the cycle could repeat using these smaller volumes as new prototypes. Iterating these cycles may allow the localization of tissue volumes on the scale of just a few microns. This description has dealt very generally with locating specific volumes in a storage medium. There is a remarkable correspondence in the calcium–NO interaction that could act as a substrate for this multistage process.

The interactions of external calcium and nitric oxide give a mechanism that could solve this problem of looking-up specific volumes of tissue. The neural scheme makes use of some of the facts described earlier:

(1) The NO scaffolding places potential sources within close proximity to every point in the cortex.
(2) NO moves isotropically through tissue and increases transmission events from nearby synaptic terminals.
(3) Increasing transmission events contributes to spike production in recipient neurons, *and* directs back-propagating spikes into the region where the transmission events occurred.
(4) Back-propagating spikes that overlap in time and tissue space cause large decrement in external calcium (Fig. 6).

The neural sequence would occur as follows:

Global query to a region of tissue: A vector of transmitters is released across a fairly large region of tissue by patterns of action potentials. Some volumes match best the transmitters and their temporal variations. This locates the prototype matches.

Prototype generalization: For some prototypes, NMDA receptor activation occurs and is followed by a burst in NO production. NO moves rapidly through the surrounding tissue, increases transmission events, and turns down NMDA receptor sensitivity. The diminishment of plasticity serves to insulate the tissue from changing dramatically the contents of the storage location (we view this primarily as a look-up or 'read' mechanism).

Refinement: The increased transmission events following the NO burst serve two roles: (1) inactivate A-type potassium channels in dendrites in the region, and (2) provide input to recipient neurons that contributes to spike production. The back-propagating spikes travel into the dendrites, and are directed primarily to those regions where the transmissions occurred (Magee and Johnston, 1997). These spikes are restricted to dendrites; therefore, they represent a search through a domain vastly smaller than the one defined by the NO burst. This *'search-along-wires'* causes large peridendritic fluctuations in external calcium, and sets up physical constraints necessary for implementing the last step (flagging the results).

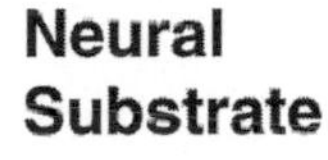

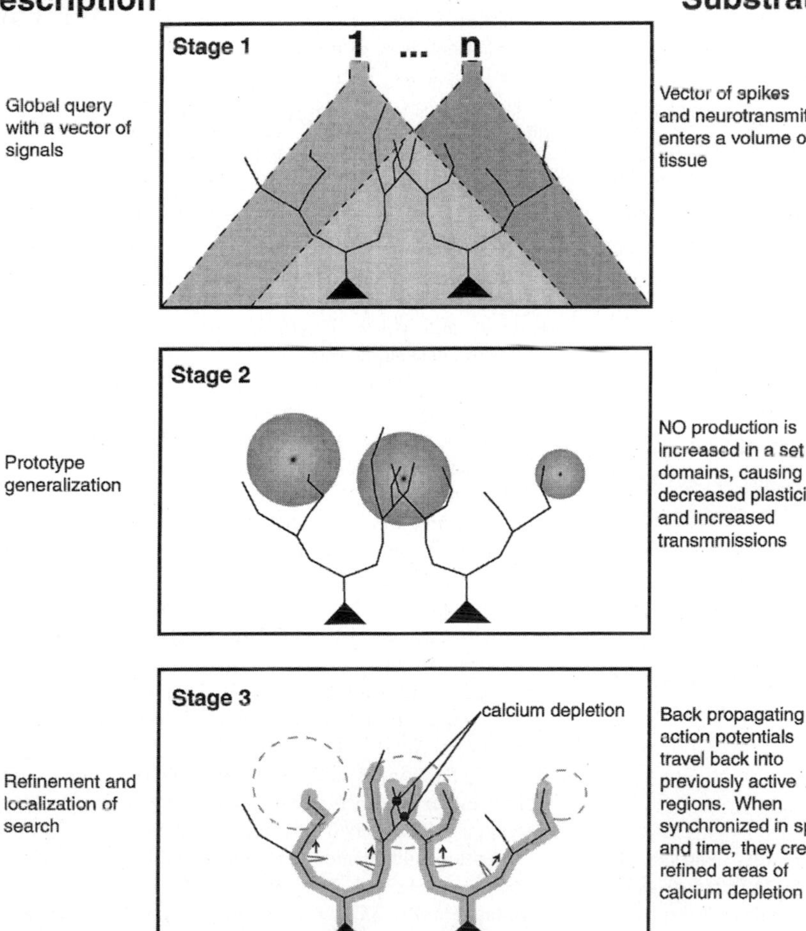

Fig. 7. NO and calcium may implement a rapid addressing scheme in neural tissue. The interaction of NO production and external calcium levels produces a plausible substrate by which patterns of neural activity are used to 'look-up' specific volumes of tissue. Details of the multistage process are given in the text.

Marking the results: If the back-propagating spikes overlap sufficiently in time and tissue space, a large and longer lasting fluctuation in external calcium 'flags' a set of much smaller volumes ('here we are'). If no such fluctuation occurs, the system has not located any sufficiently similar information matching the global query. The details of the dendritic structure and patterns of synaptic contact determine whether a given pattern of back-propagating spikes will cause a large fluctuation in external calcium. This means that information could be stored in the relative three-dimensional location of synaptic contacts, and in the relative location of dendritic branches.

Closing remarks

The recent discoveries about NO diffusion through volumes of tissue place the emphasis of neural encoding not on the single neuron, but rather on the volume of tissue in which the neurons function. Solving the problem of how to address a large parameter space may provide guidance for discovering the biological mechanisms of memory storage and retrieval. We have shown how a novel interaction of NO and external calcium levels can act together to address specific volumes of neural tissue. However, the addressing scheme proposed here is mute to the question of *what* is actually read out. Indeed, we have left this out because understanding the specifics of the read out requires some insight into how the problem domain is encoded in the first place, *i.e.*, we need a better guess at what the read-out should represent. The latter question is quite a general one for all of neuroscience, and remains open to speculation.

References

Akira, T., Henry, D. and Wasterlain, C.G. (1994) Nitric oxide mediates the sustained opening of NMDA receptor-gated ionic channels which follows transient excitotoxic exposure in hippocampal slices. *Brain Res.*, 652: 190–194.

Bohme, G.A., Bon, C., Stutzmann, J.M., Doble, A. and Blanchard, J.C. (1991) Possible involvement of nitric oxide in long-term potentiation. *Eur. J. Pharmacol.*, 199: 379–381.

Cramer, K.S., Angelucci, A., Hamn, J.O., Bogdanov, M.B. and Sur, M.B. (1996) A role for nitric oxide in the development of ferret retinogeniculate projection. *J. Neurosci.*, 16: 7995–8004.

Egelman, D.M. and Montague, P.R. (1998) Computational properties of peri-dendritic calcium fluctuation *J. Neurosci.* In press.

Engert, F. and Bonhoeffer, T. (1997) Synapse specificity of long-term potentiation breaks down at short distances. *Nature*, 388: 279–284.

Gally, J.A., Montague, P.R., Reeke, G.N. and Edelman, G.M. (1990) The NO hypothesis: Possible effects of a shortlived rapidly diffusible signal in the development and function of the nervous system. *Proc. Natl. Acad. Sci. USA*, 87: 3547–3551.

Haley, J.E., Wilcox, G.L. and Chapman, P. (1991) The role of nitric oxide in long-term potentiation. *Neuron*, 8: 211–216.

Hanbauer, I., Wink, D., Osawa, Y., Edelman, G.M. and Gally, J.A. (1992) Role of nitric oxide in NMDA-evoked release of [3H]-dopamine from striatal slices. *Neuroreport*, 3: 409–412.

Helmchen, F., Borst, G.G. and Sakmann, B. (1997) Calcium dynamics associated with a single action potential in a CNS presynaptic terminal. *Biophys. J.*, 72: 1458–1471.

Hoyt, K.R., Tang, L.H., Aizenman, E. and Reynolds, I.J. (1992). Nitric oxide modulates NMDA-induced increases in intracellular Ca^{2+} in cultured rat forebrain neurons. *Brain Res.*, 592: 310–316.

Katz, B. and Miledi, R. (1970) Further study of the role of calcium in synaptic transmission. *J. Physiol.*, 207(3) 789–801.

Lev-Ram, V., Makings, L.R., Keitz, P.F., Kao, J.P.Y. and Tsien, R.T. (1995) Long-term depression in cerebellar Purkenje neurons results from coincidence of nitric oxide and depolarization-induced Ca^{2+} transients. *Neuron*, 15: 407–415.

Magee, J. and Johnston, D. (1997) A synaptically controlled, associative signal for hebbian plasticity in hippocampal neurons. *Science*, 275: 209–213.

Manzoni, O. and Bockaert, J. (1993) Nitric oxide synthase activity endogenously modulates NMDA receptors. *J. Neurochem.*, 61: 368–370.

Manzoni, O., Prezeau, L., Marin, P., Deshager, S., Bockaert, J. and Fagni, L. (1992) NO induced blockade of NMDA receptors. *Neuron*, 8: 653–662.

Mintz, I., Sabatini, B. and Regehr, W. (1995) Calcium control of transmitter release at a central synapse. *Neuron*, 15: 675–688.

Montague, P.R. (1996) The resource consumption principle: Attention and memory in volumes of neural tissue. *Proc. Nat. Acad. Sci*, 93: 3619–3623.

Montague, P.R., Gally, J.A. and Edelman, G.M. (1991) Spatial signaling in the development and function of neural connections. *Cerebral Cortex*, 1: 199–220.

Montague, P.R., Gancayco, C.D., Winn, M.J., Marchase, R.B. and Friedlander, M.J. (1994) Role of NO production in

NMDA receptor-mediated neurotransmitter release in cerebral cortex. *Science*, 263: 973–977.

Montague, P.R. and Sejnowski, T.J. (1994) The predictive brain: Temporal coincidence and temporal order in synaptic learning mechanisms. *Learning and Memory*, 1: 1–33.

O'Dell, T.J., Hawkins, R.D., Kandel, E.R. and Arancio, O. (1991) Tests of the roles of two diffusible substances in LTP: Evidence for nitric oxide as a possible early retrograde messenger. *Proc. Natl. Acad. Sci., (USA)*, 88: 11285–11289.

Person, C., Egelman, D., King, R. and Montague, P.R. (1996) Three-dimensional synaptic distributions influence neural processing through the resource consumption principle. *J. Physiol. (Paris)*, 90: 323–325.

Qian, J., Colmers, W.F. and Saggau, P. (1997) Inhibition of synaptic transmission by Neuropeptide Y in rat hippocampal area CA1: Modulation of presynaptic calcium entry. *J. Neurosci.*, 17: 8169–8177.

Reid, S.N., Daw, N.W., Czepita, D., Flavin, H.J. and Sessa, W.C. (1996) Inhibition of nitric oxide synthase does not alter ocular dominance shifts in kitten visual cortex. *J. Physiol.*, 494: 511–517.

Schuman, E.M. and Madison, D.V. (1991) A requirement for the intercellular messenger nitric oxide in long-term potentiation. *Science*, 254: 1503–1506.

Schuman, E.M. and Madison, D.V. (1994) Locally distributed synaptic potentiation in the Hippocampus. *Science*, 263: 532–536.

Shibuki, K. and Okada, D. (1991) Endogenous nitric oxide release required for long-term synaptic depression in the cerebellum. *Nature*, 349: 326–328.

Tanaka, T., Saito, H. and Matsuki, N. (1993) Endogenous nitric oxide inhibits NMDA- and kainate- responses by a negative feedback system in rat hippocampal neurons. *Brain Res.*, 631: 72–76.

Wu, H.H., Williams, C.V. and McLoon, S.C. (1994) Involvement of nitric oxide in the elimination of a transient retinotectal projection in development. *Science*, 265: 1593–1596.

SECTION V

Nitric oxide and other signals in neurodegeneration and neuroprotection

R.R. Mize, T.M. Dawson, V.L. Dawson and M.J. Friedlander (Eds.)
Progress in Brain Research, Vol 118

CHAPTER 15

Nitric oxide in neurodegeneration

Valina L. Dawson[1,3,*] and Ted M. Dawson[1,2]

[1]*Department of Neurology,* [2]*Department of Neuroscience,* [3]*Department of Physiology,*
The Johns Hopkins University School of Medicine, 600 North Wolfe Street, Carnegie 2-214, Baltimore, MD 21287, USA

Abstract

Nitric oxide (NO) is a unique biological messenger molecule which mediates diverse physiologic roles. NO mediates blood vessel relaxation by endothelium, immune activity of macrophages and neurotransmission of central and peripheral neurons. NO is produced from three NO Synthase (NOS) isoforms: Neuronal NOS (nNOS), endothelial NOS, and inducible NOS (iNOS). In the central nervous system, NO may play important roles in neurotransmitter release, neurotransmitter reuptake, neurodevelopment, synaptic plasticity, and regulation of gene expression. However, excessive production of NO following a pathologic insult can lead to neurotoxicity. NO plays a role in mediating neurotoxicity associated with a variety of neurologic disorders, including stroke, Parkinson's Disease, and HIV dementia.

Biochemical properties of nitric oxide

In its native state, at room temperature and atmospheric pressure, NO is a gas. With the exception of the lung and paranasal sinus air where NO is present in its gaseous phase (Lundberg et al., 1995), in most biologic systems NO is a dissolved nonelectrolyte. NO is a small lipid soluble molecule with its unpaired electron in a negative π orbital which limits its potential chemical interactions with target molecules (Butler et al., 1995). Due to these physical properties, NO has the ability to diffuse approximately 40 μm/s in aqueous solutions (Lancaster, 1994; Wood and Garthwaite, 1994). The biologic activity of NO is regulated by redox-sensitive chemical reactions with target molecules. Activity is dependent on the rate of NO generation, the concentration of NO produced, the rate of reaction of NO with the target molecule and the concentration of the target molecule. As a free radical species, NO is relatively less reactive than other reactive oxygen radicals such as superoxide anion and hydroxyl radical (Beckman et al., 1994a; Butler et al., 1995). NO is a relatively stable carrier of unpaired electrons and its unique chemical properties are exploited by biologic systems allowing NO to participate in a novel form of nonclassical signaling.

Nitric oxide synthase

NO is formed by the enzymatic conversion of L-arginine by NOS in a catalytic process that consumes five electrons and results in the formation of L-citrulline in the presence of oxygen and NADPH. NOS requires co-factors including flavin mononucleotide, flavin adenine dinucleotide, heme, and tetrahydrobiopterin. In addition, NOS

*Corresponding author. Tel.: +1 410 614 3359; fax: +1 410 614 9568; e-mail: vdawson@orion.bs.jhmi.edu

isoforms require calcium and calmodulin (Dawson and Snyder, 1994; Marletta, 1993; Nathan and Xie, 1994a,b; Stuehr and Griffith, 1992). All three NOS isoforms have been cloned and sequenced and have been named by the tissue from which they were first cloned or numbered in the order in which they were cloned.

Neuronal NOS

NOS was first purified and cloned from rat brain and was termed neuronal NOS (nNOS, type I) (Bredt et al., 1991b). nNOS distribution in the brain, as revealed through immunohistochemical and *in situ* hybridization studies is unusual with only 1–2% of the total neuronal population in many brain regions expressing nNOS, with the exception of the cerebellum where most granule cells express nNOS (Bredt et al., 1991a; Bredt et al., 1990; Dawson et al., 1991a).

Inducible NOS

Inducible NOS (iNOS, Type II) was first cloned from macrophages (Lowenstein et al., 1992; Lyons et al., 1992; Xie et al., 1992) and then hepatocytes (Charles et al., 1993; Geller et al., 1993). iNOS message or protein has not been detected in healthy brain. However, following ischemic insult, trauma, viral infection, bacterial infection, or other immunologic challenges, iNOS can be expressed in the brain (Adamson et al., 1996; Bo et al., 1994; Iadecola, 1997; Koprowski et al., 1993; Nathan and Xie, 1994b; Van Dam et al., 1995). Expression of iNOS is observed predominantly in microglia and astrocytes. In culture, iNOS expression can be stimulated by exposure to lypopolysaccharide, interferon-γ, and tumor necrosis factor-α, in addition to other pathologic agents such as HIV coat proteins and fragments (Adamson et al., 1996; Chao et al., 1992; Dawson et al., 1994b; Koka et al., 1995a,b; Murphy et al., 1993). Translated iNOS monomers bind calmodulin even at very low levels of intracellular calcium (Cho et al., 1992). Dimerization and activation occurs on the binding of heme, tetrahydrobiopterin, and L-arginine, co-factors which are typically ubiquitously expressed in most cells. In the presence of NADPH and oxygen, nitric oxide is formed (Nathan and Xie, 1994b). Therefore, in the presence of all the necessary co-factors once iNOS is translated it is a functionally active enzyme and functions independently from changes in intracellular calcium. Regulation of production of NO by iNOS appears to be driven by degradation of iNOS message and iNOS protein (Park and Murphy, 1996). In some human culture systems generation of NO lags behind expression of iNOS due to insufficient biopterin synthesis (Ding and Merrill, 1997; Ding et al., 1997). However, once biopterin synthesis is restored iNOS produces NO at high levels.

Endothelial NOS

Endothelial NOS (eNOS, Type III) (Janssens et al., 1992; Lamas et al., 1992; Sessa et al., 1992) was the first NOS isoform for which NOS activity was described (Furchgott and Zawadzki, 1980; Ignarro et al., 1987; Palmer et al., 1987). eNOS is localized to both endothelial cells in the vasculature in addition to a small population of neurons in the CNS. eNOS is localized to hippocampal pyramidal cells of CA1 through CA3 region of the hippocampus and dentate gyrus granule cells (Dinerman et al., 1994). This distribution pattern is in marked contrast to the staining for nNOS which is absent from CA1 pyramidal neurons and is only expressed within GABAergic interneurons of the hippocampus (Bredt et al., 1991a; Bredt et al., 1990; Dawson et al., 1991a). eNOS is also expressed in the cerebellum and olfactory bulb. Like nNOS, eNOS is a calcium dependent NOS isoform which transiently produces NO following a rapid increase in intracellular calcium. eNOS a primary regulator of vascular hemodynamics. In the CNS, the role of eNOS expression in neurons is not entirely understood but it appears to play an essential role in the induction and maintenance of long term potentiation (Huang, 1997; Kantor et al., 1996; Son et al., 1996).

How is NO neurotoxic?

Due to its ability to modulate neurotransmitter release and reuptake, mitochondrial respiration, DNA synthesis, and energy metabolism, it is not surprising that NO is neurotoxic (Dawson and Dawson, 1996a,b). Under conditions where NO is abnormally produced, such as when iNOS expression is induced in the CNS, disregulation of normal physiologic activities of NO likely contributes to neuronal dysfunction and subsequently to neuronal death. However, acute toxicity mediated by NO appears to require production of superoxide anion (Dawson et al., 1993). NO in and of itself is a relatively nontoxic molecule which, in the absence of superoxide anion will not kill cells even at extremely high concentrations. In the presence of superoxide anion, however, NO is a potent neurotoxin. The reaction of NO with superoxide anion is the fastest biochemical rate constant currently known, resulting in the formation of the potent oxidant, peroxynitrite ($ONOO^-$) (Beckman et al., 1994a,b; Koppenol et al., 1992). It is most likely that $ONOO^-$ mediates the toxic activities of NO production. Peroxynitrite is a lipid permeable molecule with a wider range of chemical targets than NO. It can oxidize proteins, lipids, RNA, and DNA (Beckman et al., 1994a; Beckman et al., 1994b; Koppenol et al., 1992).

Neurotoxicity elicited by $ONOO^-$ formation may have a dual component (Fig. 1). Peroxynitrite can potently inhibit mitochondrial proteins (Cassina and Radi, 1996; Stuehr and Nathan, 1989). Peroxynitrite inhibits the function of manganese SOD (MnSOD) which could lead to increased superoxide anion formation and increased $ONOO^-$ formation (MacMillan-Crow et al., 1996). Additionally, $ONOO^-$ is an effective inhibitor of enzymes in the mitochondrial respiratory chain resulting in decreased ATP synthesis (Cassina and Radi, 1996; Stuehr and Nathan, 1989; Takehara et al., 1995; Yabuki et al., 1997a). Secondly, $ONOO^-$ efficiently modifies and breaks DNA strands (Derojaswalker et al., 1995; Kennedy et al., 1997) and inhibits DNA ligase which increases DNA strand breaks (Graziewicz et al.,

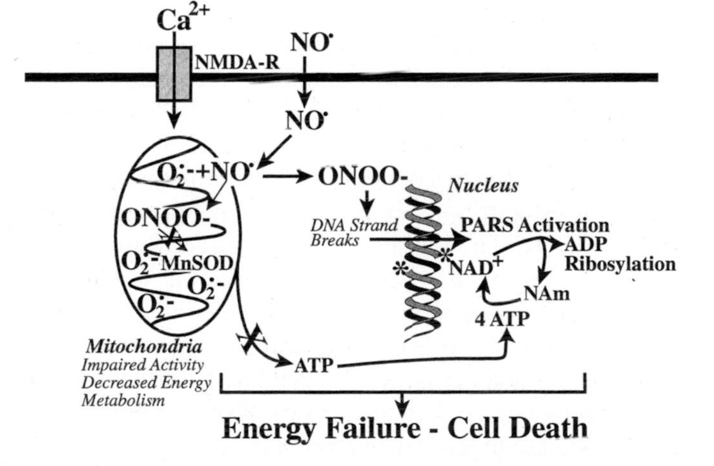

Fig. 1. PARP mediated neurotoxicity. Acting via NMDA receptors, glutamate triggers an influx of Ca^{2+} activating mitochondria and in nNOS neurons inducing the production of NO. Mitochondrial activation results in an increase in superoxide anion ($O_2^{\cdot-}$) production. NO is a diffusible molecule which combines with superoxide to form peroxynitrite ($ONOO^-$). Peroxynitrite damages mitochondrial enzymes decreasing mitochondrial respiration and the production of ATP and, peroxynitrite also damages MnSOD further increasing the production of superoxide anion. Peroxynitrite has limited diffusion across membranes and can leave the mitochondria and enter the nucleus where it damages DNA. Nicks and fragments of DNA activate PARP. Massive activation of PARP leads to ADP-ribosylation and depletion of NAD. ATP is depleted in an effort to resynthesize NAD. In the setting of impaired energy generation due to mitochondrial dysfunction this loss of NAD and ATP leads to cell death.

1996). DNA strand breaks activate DNA repair mechanisms. One of the initial proteins activated by DNA damage is the nuclear enzyme poly(ADP-ribose) polymerase (PARP) (Berger, 1985; de Murcia and Menissier de Murcia, 1994; Lautier et al., 1993). PARP catalyzes the attachment of ADP-ribose units from NAD to nuclear proteins such as histone and PARP itself. PARP can add hundreds of ADP-ribose units within seconds to minutes of being activated (Berger, 1985; de Murcia and Menissier de Murcia, 1994; Lautier et al., 1993). For every mole of ADP-ribose transferred from NAD, one mole of NAD is consumed and four free energy equivalents of ATP are required to regenerate NAD to normal cellular levels. Activation of PARP can result in a rapid drop in energy stores (Berger, 1985). If this drop is severe and sustained it can lead to impaired cellular metabolism and ultimately death. In fibroblasts and some tumor cells inhibition or genetic knockout of PARP actually accelerates the apoptotic death process by inhibiting DNA repair mechanisms (de Murcia et al., 1994; Lautier et al., 1993; Shall, 1995). In other cell lines inhibition or genetic knockout of PARP has no effect on cell death (Wang et al., 1995, 1997). However, in postmitotic cells of the CNS and heart, inhibition of PARP activity or genetic knockout of PARP potently protects neurons and cardiac myocytes against ischemic damage (Eliasson et al., 1997; Endres et al., 1997; Gilad et al., 1997; Takahashi et al., 1997; Thiemermann et al., 1997; Zhang et al., 1994). Additionally, in neurons pharmacologic inhibition of PARP or genetic knockout of PARP confers resistance to excitotoxic injury and combined oxygen-glucose deprivation (Eliasson et al., 1997). Inhibition of PARP confers a stabilization of NAD levels (Endres et al., 1997). Pancreatic β-islet cells are sensitive to NO toxicity. Inhibition of PARP or genetic knockout of PARP results in resistance of pancreatic β-islet cells to toxicity (Heller et al., 1995). However, increasing the toxic insult to pancreatic β-islet results in the recruitment of PARP-independent death pathways (Heller et al., 1995). Thus, in neurons, cardiac myocytes, and pancreatic β-islet cells, rapid drops in energy stores mediated by PARP activation is a potent mediator of cell death.

Although there is considerable evidence from multiple investigators that excess NO is neurotoxic, there are several reports which did not observe a role for NO in neurotoxicity. *In vitro* these discrepancies in the literature can largely be addressed by differences in the culture and experimental paradigms (Dawson and Dawson, 1996a, b). When neurons are co-cultured with glia taken from the same fetal animal and maintained in culture for 2–3 weeks, NOS neurons mature and express mature levels of nNOS (1–2% of the total neuronal population). Under these conditions, NO toxicity is readily observed (Dawson et al., 1993, 1991b, 1996). If fetal neurons are plated onto a mature glial feeder layer obtained from a separate animal, development and maturation of NOS neurons is inhibited (Samdani et al., 1997). Under these conditions NO neurotoxicity cannot be observed (Samdani et al., 1997). Other culture paradigms such as serum free media or a culture of neurons from postnatal animals can also effect the mature expression of nNOS neurons. If nNOS is not expressed, NO neurotoxicity cannot occur.

Recently a role for NO in apoptosis has been reported and not surprisingly some investigators report NO mediates apoptosis (Hortelano et al., 1997; Khan et al., 1997; Leist et al., 1997a; Sandau et al., 1997; von Knethen and Brune, 1997; Yabuki et al., 1997b) via caspase activation while other investigators report inhibition of apoptosis through inhibition of caspase activation (Haendeler et al., 1997; Kim et al., 1997; Li et al., 1997; Ogura et al., 1997; Tenneti et al., 1997). NO inhibits apoptosis in several cell types including endothelial cells, lymphocytes and eosinophils and hepatocytes. The anti-apoptotic concentrations of NO are in the low micromolar range (up to 1 μM). Exogenous NO donors (10 μM) inhibit TNFα-induced caspase activation in human endothelial cells (Dimmeler and Zeiher, 1997; Haendeler et al., 1997). The apparent mechanism of NO inhibition of caspase 3 activity appears to be due to a specific *S*-nitrosylation of cystine 163, an essential amino acid conserved among the caspase protein family

(Ogura et al., 1997, Dimmeler et al., 1997). Caspase 3 is thought to be the workhorse of the caspase family in mediating apoptotic cell death. It is extremely interesting that low level expression of NO has the potential to modulate caspase activity, which further increases the complexity of the general response of cells to death signals. Although low levels of NO can inhibit apoptosis simulated by other death signals, higher concentrations or progressive generation of NO can directly activate apoptosis in various cells including macrophages, β-islet cells, thymocytes, and cerebellar granule cells. In cerebellar granule cells, NO-mediated apoptosis is caspase dependent (Leist et al., 1997a,b). However, the induction of apoptosis by NO is not known. It may be the result of direct DNA damage or accumulation of the tumor suppresser protein, p53 (Sandau et al., 1997). It is however, PARP independent (Leist et al., 1997a,b). The current studies suggests that the contrasting roles for NO in apoptosis are a dose-dependent phenomena with low concentrations providing protection against apoptosis where high concentrations induce apoptosis (Dimmeler and Zeiher, 1997; Nicotera et al., 1997).

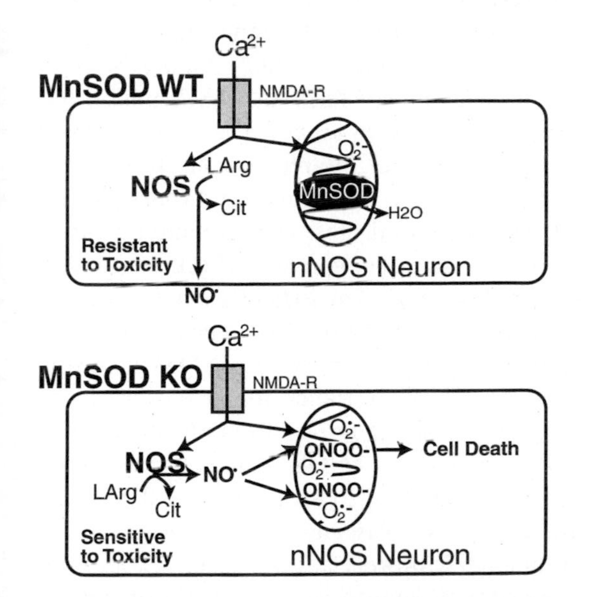

Fig. 2. MnSOD confers resistance to glutamate and nitric oxide neurotoxicity. In wild-type mice MnSOD is highly expressed in nNOS neurons in comparison to adjacent neurons. NO produced in these neurons does not have the opportunity to react with superoxide anion (O_2^-) and diffuses to adjacent neurons. In MnSOD knockout mice, superoxide anion is formed in the mitochondria following glutamate stimulation of NMDA receptors. There is no MnSOD to scavenge superoxide anion. NO produced in nNOS neurons reacts with the nearest target, O_2^-, and high concentrations of peroxynitrite ($OONO^-$) are formed in the nNOS neurons resulting in exquisite sensitivity to neurotoxic insults in the MnSOD knockout mice. Resistance to toxicity can be restored in the MnSOD knockout mice by adenoviral mediated expression of MnSOD, confirming the importance of this enzyme in the differential resistance of nNOS neurons to toxic insults.

Why are nNOS neurons resistant to toxic insults?

In 1961–1964, Thomas and Pearse described a histochemical stain, NADPH-diaphorase, which was expressed in a unique subpopulation of neurons comprising approximately 1–2% of the neuronal population (Thomas and Pearse, 1961, 1964). They termed these cells solitary neurons (Thomas and Pearse, 1964). Over the years, investigators have noted that these NADPH-diaphorase positive neurons are uniquely resistant to a variety of neurologic disorders and toxic events. They are selectively resistant to damage in Huntington's disease, Alzheimer's disease, vascular stroke, and NMDA neurotoxicity (Beal et al., 1989; Choi, 1988; Ferrante et al., 1985; Hyman et al., 1992; Koh and Choi, 1988; Koh et al., 1986; Uemura et al., 1990). With the cloning and expression of nNOS it was appreciated that this NADPH-diaphorase staining was in fact produced by nNOS (Dawson et al., 1991a; Hope et al., 1991). Thus, these nNOS neurons are selectively resistant to a variety of toxic environments and therefore must possess protective mechanisms which render them resistant. However, the molecular mechanisms which account for the selective resistance of nNOS neurons' has remained a mystery. It has been suggested that nNOS itself or production of NO confers resistance to this select neuronal population. However, pharmacologic inhibition of nNOS and genetic knockout of nNOS does not restore sensitivity to neurotoxic insults of nNOS neurons (Dawson et al., 1993, 1996). While it is always feasible that alternative splice variants of nNOS are expressed at low levels and may provide neural protection to nNOS neurons, this is unlikely. We hypothesize that nNOS neurons must have evolved unique strategies to survive the potentially toxic concentrations of NO they can produce. To identify candidate gene products which could confer neuroprotection, serial analysis of gene expression (SAGE) was performed between wild-type PC12 cells and PC12 cells which had been rendered resistant to NO toxicity (Gonzalez-Zulueta et al., 1998). The first gene we identified with the highest differential level of expression, was MnSOD. Anti-sense knockdown of MnSOD restored sensitivity to NO toxicity in the PC12 resistant cell line while adenoviral transfer of MnSOD conferred resistance to NO toxicity in wild-type PC12 cells confirming that increased expression of MnSOD was responsible for resistance in the PC12 NO resistant cell line (Gonzalez-Zulueta et al., 1998). In primary cortical cultures MnSOD is enriched in nNOS neurons. Both anti-sense knockdown or genetic knockout of MnSOD renders nNOS neurons exquisitely susceptible to NMDA or NO neurotoxicity (Fig. 2). Over-expression of MnSOD by adenoviral transfer provides dramatic protection against NMDA or NO neurotoxicity in cortical cultures. However, adenoviral over expression of MnSOD does not provide protection against kainate or AMPA neurotoxicity. Adenoviral transfer of MnSOD to MnSOD knockout cultures restores resistance to nNOS neurons to NMDA or NO neurotoxicity. Our results indicate that MnSOD is a major line of defense against excitotoxic or oxidative damage and appears to be essential for the resistance of nNOS neurons to NMDA or NO mediated neurotoxicity (Gonzalez-Zulueta et al., 1998).

NO in neurologic disease

Stroke

A role for NO in neurotoxicity was first described in an *in vitro* model of focal ischemia (Dawson et al., 1991). A five minute exposure to glutamate or NMDA in primary cortical cultures sets in motion a series of events resulting in cell death 12–18 h later following exposure to these excitatory amino acids. NOS inhibitors provide potent neuroprotection in a dose-dependent manner (Dawson et al., 1991). Furthermore, NO donors elicit neurotoxicity which develops over a similar time course of 12–18 h (Dawson et al., 1993). Both NMDA or NO neurotoxicity can be blocked by scavengers of superoxide anion or scavengers of NO (Dawson et al., 1993, 1996). These results have been replicated in *in vivo* models of focal ischemia using various pharmacologic inhibitors of NOS (Dawson et al., 1994a; Dawson and Snyder, 1994). There were, however, numerous conflicting reports in *in vivo* models of focal ischemia which were due to the use of non-selective NOS inhibitors. NOS inhibitors such as N-Arg or L-NAME inhibit not only nNOS, but eNOS as well. Decreased cerebral blood flow increases infarct volume. Concentrations of nonspecific NOS inhibitors which do not affect systemic blood pressure can inhibit cerebral blood flow by greater than 80% (Beckman, 1991). Many of the early focal ischemia studies did not control for cerebral blood flow. The development of genetic knockout mice of nNOS or eNOS allowed for elegant studies to be performed to clarify the role of NO derived from nNOS or eNOS in focal ischemia. nNOS knockout mice are dramatically resistant to focal and global ischemia (Hara et al., 1996; Panahian et al., 1996; Zaharchuk et al., 1997) however, if nNOS knockout mice are treated with nonspecific NOS inhibitors infarct volumes are equivalent to wild-type mice (Dalkara and Moskowitz, 1997). eNOS knockout mice have greatly increased infarct volumes compared to wild-type mice, however, if eNOS knockout mice are treated with NOS inhibitors infarct volumes are decreased (Huang et al., 1996). Therefore, NO derived from nNOS mediates neurotoxicity following focal ischemia, while NO derived from eNOS is critical in maintaining cerebral blood flow and has a positive effect in decreasing infarct volume following focal ischemia (Moskowitz and Dalkara, 1996). Studies using specific nNOS inhibitors have confirmed these results (Dalkara et al., 1994).

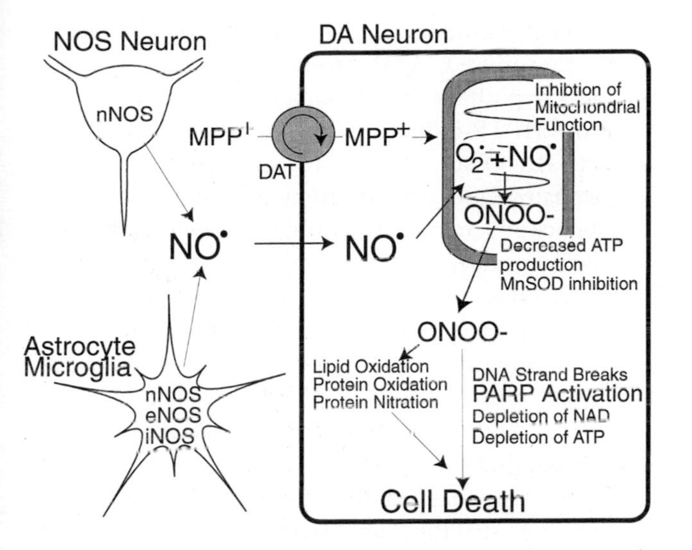

Fig. 3. MPTP-induced neurotoxicity. MPTP is converted to MPP^+ which is transported into dopamine (DA) neurons by the high affinity DA transporter (DAT). MPP^+ inhibits mitochondrial respiration resulting in formation of superoxide anion ($O_2^{\cdot-}$). NO diffuses to DA neurons from surrounding neurons or astrocytes. Astrocytes constitutively express nNOS and eNOS and during injury can express iNOS. NO reacts with $O_2^{\cdot-}$ generating peroxynitrite ($ONOO^-$), a potent oxidant which injures cells by oxidizing lipids, proteins, RNA and DNA. Deamination of DNA leads to DNA strand nicks and breaks, PARP activation, energy depletion and DA neuronal cell death.

Activation of PARP is a primary mediator of NMDA or NO neurotoxicity (Eliasson et al., 1997; Zhang et al., 1994). *In vitro* pharmacologic inhibition of PARP or genetic knockout of PARP confers dramatic protection against NMDA or NO mediated toxicity to primary cortical cultures (Eliasson et al., 1997; Zhang et al., 1994). *In vivo*, PARP knockout mice are dramatically resistant to focal ischemia (Eliasson et al., 1997; Endres et al., 1997). Infarct volumes are decreased over 80% in PARP knockout mice as compared to wild-type mice (Eliasson et al., 1997). Additionally, NAD levels are conserved in the PARP knockout mice following focal ischemia (Endres et al., 1997). Rats treated with the PARP inhibitor, DPQ, have smaller infarct volumes than saline treated rats following focal ischemia (Takahashi et al., 1997). These *in vitro* and *in vivo* studies indicate a key role for the activation of PARP in the development of infarction following focal ischemia.

Parkinson's disease

Parkinson's disease is a movement disorder characterized by the selective loss of dopamine neurons particularly in the substantia nigra which project to the striatum (Jenner et al., 1992). This human disease can be modeled in mice by injecting the toxin, 1-methyl-4-phenyl-1,2,3,6-tetrathydropyridine (MPTP) (Heikkila et al., 1989; Kopin and Markey, 1988; Langston et al., 1984). MPTP is converted to MPP^+ in the nervous system and targets dopaminergic neurons through the high affinity dopamine reuptake transporter (Kitayama et al., 1992; Pifl et al., 1993; Sanchez-Ramos et al., 1992). MPP^+ is actively transported into mitochondria where it inhibits mitochondrial respiration, in particular complex-I of the electron transport chain (Heikkila et al., 1985; Kindt et al., 1987; Ramsay et al., 1989). Inhibition of complex-I results in a decrease in ATP production and an increase in superoxide anion formation in

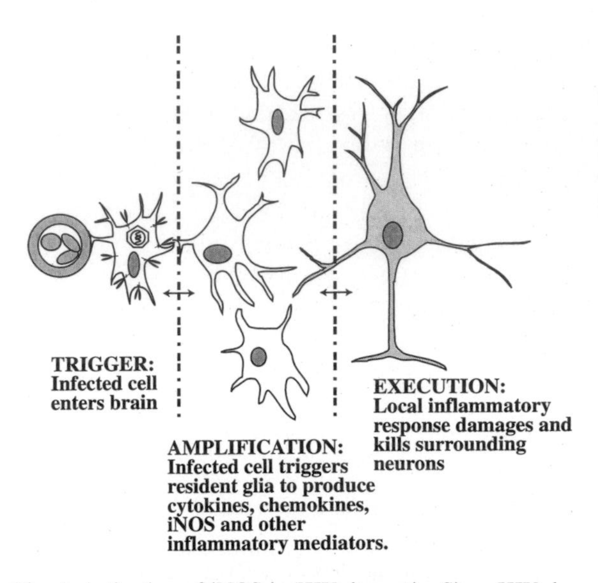

Fig. 4. Activation of iNOS in HIV-dementia. Since HIV does not directly infect and replicate in the CNS an indirect model of disease is proposed with three possible stages. The trigger phase is the entry of HIV infected into the CNS and interaction with resident glia. The infected cells express HIV proteins including HIV envelope gp41 which can stimulate adjacent cells through cell-cell contact. The amplification stage is the production of cytokines, chemokines, iNOS and other pro-inflammatory molecules by glia stimulated by the infected cell. Production of pro-inflammatory molecules would result in the recruitment of immune cells from the blood, some of them carrying additional HIV which could set up a slow feed-forward accumulation of infected cells in the CNS. Eventually, local immune responses damage neurons resulting first in synaptic pruning but as the insult increases eventually neurons are lost. iNOS is highly expressed in severe or rapidly progressing HIV dementia suggesting that the production of NO is a catalyst for neuronal death.

these cells (Ramsay and Singer, 1992). It is possible that this increase in superoxide anion production permits the formation $ONOO^-$ from endogenous NO (Fig. 3). Peroxynitrite can further inhibit mitochondrial function locally by damaging MnSOD activity (MacMillan-Crow et al., 1996) and increasing superoxide anion production as well as inhibiting enzymes in the mitochondrial respiratory chain (Henry et al., 1993; Hibbs et al., 1987; Stuehr and Nathan, 1989). Additionally, $ONOO^-$ can damage DNA activating PARP (Zhang et al., 1994, 1995). This activation of PARP would occur in a cell which already has compromised ATP production. Consistent with this hypothesis, nNOS knockout mice are resistant to MPTP-induced neurotoxicity (Przedborski et al., 1996). Levels of the neurotransmitter, dopamine, in the striatum are spared in nNOS knockout mice and neuronal degeneration visualized by silver staining is reduced in nNOS knockout mice as compared to wild-type controls treated with MPTP (Przedborski et al., 1996). Additionally, PARP knockout mice are dramatically resistant to MPTP neurotoxicity as compared to wild-type mice (Mandir, A.S., Prsedborski, S., Dawson, V.L. and Dawson, T.M., unpublished observations). Dopamine levels in the striatum are spared as are dopamine containing neurons in the substantia nigra in PARP knockout mice. Nonspecific inhibitors of PARP which partially inhibit PARP have also been shown to be protective against MPTP neurotoxicity (Cosi et al., 1996). However, these results must be interpreted with caution as nonspecific PARP inhibitors also inhibit protein synthesis and glucose metabolism (Huet and Laval, 1985; Milam and Cleaver, 1984, 1985; Redegeld et al., 1992; Shah et al., 1996). The inhibition of protein synthesis has previously been demonstrated to provide neuroprotection by either increasing in glutathione levels or preventing translation of cell death genes (Kato et al., 1997; Lobner and Choi, 1996, Ratan et al., 1994a,b). Furthermore, the bioavailability of these agents in the CNS has never been demonstrated nor has their efficacy in inhibiting PARP *in vivo* in the CNS. Although the results are consistent with those obtained in the PARP knockout mice, the neuroprotection obtained with nonspecific PARP inhibitors cannot be distinguished from inhibition of PARP or systemic affects. Recently more specific and potent PARP inhibitors have been developed which have limited systemic effects and better bioavailability which will be useful in the investigations of PARP in various models of neurologic disease (Eliasson et al., 1997; Szabo et al., 1997; Takahashi et al., 1997).

HIV Dementia

Shortly after the initial description of Acquired Immunodeficiency Syndrome (AIDS), an unusual progressive dementing illness was recognized in association with this disorder. It is estimated that 15% of Human Immunodeficiency Virus (HIV)-infected patients with AIDS are affected with varying severity of HIV dementia (Bacellar et al., 1994; Price, 1996; Wiley and Achim, 1994). Since HIV does not directly infect and kill neurons, indirect mechanisms are being sought to understand how HIV infection results in a dementing illness (Fig. 4). Pathologic studies report loss of complexity of dendritic arborizations, loss of synaptic densities and vacuolization of dendritic spines and subtle neuronal loss in various brain regions (Masliah et al., 1992a,b; Price et al., 1988). Localization of the virus in the CNS is nearly exclusively in blood-derived macrophages, microglia and multi-nucleated giant cells (Takahashi et al., 1996; Wiley et al., 1986).

Expression of iNOS is an attractive candidate for chronic neurologic disease. iNOS produces NO for sustained periods of time which could result first in neuronal dysfunction and then in neuronal death. Since NO is freely diffusible it could easily be produced by infected cells or activated astrocytes and then diffuse to affect distant neuronal populations. Abnormal production of NO could interfere with normal neuronal signaling. This interference could destabilize neuronal function and sustained NO production would lead to cell damage and ultimately neuronal cell death. Both iNOS message and iNOS protein concentrations have been examined from prospective clinically characterized postmortem tissue from HIV infected patients with or without HIV dementia. Expression of iNOS has been found to correlate with severe HIV dementia (Adamson et al., 1996). Colocalization studies have found cells expressing iNOS message are spatially localized near HIV infected cells (Nuovo and Alfieri, 1996). These data suggest that HIV can induce the expression of iNOS and CNS tissues. In human postmortem tissue the expression of the HIV coat protein, gp41, correlates with the severity of HIV dementia (Adamson et al., 1996). Recombinant gp41 exposed to primary cortical cultures induces iNOS expression over the course of 72 h which results in NO mediated neurotoxicity observed five days post exposure peaking at seven days post exposure (Adamson et al., 1996). It is likely that gp41 expressed on HIV infected cells stimulates local cells to express iNOS and NO produced from iNOS contributes significantly to neuronal damage and ultimately the severity of HIV dementia.

Conclusion

NO has revolutionized our perception of neurotransmission and neuronal signaling. It is emerging as an important regulator of a variety of physiological responses. However, under conditions of abnormal or excessive formation, NO is an important mediator of CNS damage. The full range of neurologic disease that result from derangements in NO production or action is currently not known. However, it is clear that therapies directed towards controlling the production, or biologic actions of NO will have a significant impact on the treatment of several neurologic disorders including stroke, Parkinson's Disease and HIV Dementia. Hopefully, over the course of the next few years there will be development of specific and selective inhibitors of both NOS, and PARP that are clinically tolerated and therapeutically useful.

Acknowledgements

VLD is supported by grants from USPHS NS33142, the American Heart Association, Muscular Dystrophy Association and the Amyotrophic Lateral Sclerosis Association. TMD is an Established Investigator of the American Heart Association and is supported by grants from the USPHS NS33277, and the Paul Beeson Physician Scholars in Aging Research Program. Under an agreement between the Johns Hopkins University and Guilford Pharmaceuticals, TMD and VLD are entitled to a share of sales royalty received by the University from Guilford. TMD and the University also own Guilford stock, and the University stock is subject to certain restrictions under University policy. The terms of this arrangement are being managed by the University in accordance with its conflict of interest policies.

References

Adamson, D.C., Wildemann, B., Sasaki, M., McArthur, J., Glass, J., Dawson, T.M. and Dawson, V.L. (1996) Immunologic nitric oxide synthase is elevated in HIV infected individuals with dementia. *Science*, 274: 1917–1921.

Bacellar, H., Munoz, A., Miller, E.N., Cohen, B.A., Besley, D., Selnes, O.A., Becker, J.T. and McArthur, J.C. (1994) Temporal trends in the incidence of HIV-1-related neurologic diseases: Multicenter AIDS Cohort Study, 1985–1992. *Neurology*, 44: 1892–1900.

Beal, M.F., Kowall, N.W., Swartz, K.J., Ferrante, R.J. and Martin, J.B. (1989) Differential sparing of somatostatin-neuropeptide Y and cholinergic neurons following striatal excitotoxin lesions. *Synapse*, 3: 38–47.

Beckman, J.S. (1991) The double-edged role of nitric oxide in brain function and superoxide-mediated injury. *J. Dev. Physiol.*, 15: 53–59.

Beckman, J.S., Chen, J., Crow, J.P. and Ye, Y.Z. (1994a) Reactions of nitric oxide, superoxide and peroxynitrite with superoxide dismutase in neurodegeneration. *Prog. Brain. Res.*, 103: 371–380.

Beckman, J.S., Chen, J., Ischiropoulos, H. and Crow, J.P. (1994b) Oxidative chemistry of peroxynitrite. *Methods Enzymol.*, 233: 229–240.

Berger, N.A. (1985) Poly(ADP-ribose) in the cellular response to DNA damage. *Radiat. Res.*, 101: 4–15.

Bo, L., Dawson, T.M., Wesselingh, S., Mork, S., Choi, S., Kong, P.A., Hanley, D. and Trapp, B.D. (1994) Induction of nitric oxide synthase in demyelinating regions of multiple sclerosis brains. *Ann. Neurol.*, 36: 778–786.

Bredt, D.S., Glatt, C.E., Hwang, P.M., Fotuhi, M., Dawson, T.M. and Snyder, S.H. (1991a) Nitric oxide synthase protein and mRNA are discretely localized in neuronal populations of the mammalian CNS together with NADPH diaphorase. *Neuron*, 7: 615–624.

Bredt, D.S., Hwang, P.M., Glatt, C.E., Lowenstein, C., Reed, R.R. and Snyder, S.H. (1991b) Cloned and expressed nitric oxide synthase structurally resembles cytochrome P-450 reductase. *Nature*, 351: 714–718.

Butler, A.R., Flitney, F.W. and Williams, D.L. (1995) NO, nitrosonium ions, nitroxide ions, nitrosothiols and iron-nitrosyls in biology: A chemist's perspective. *Trends Pharmacol. Sci.*, 16: 18–22.

Cassina, A. and Radi, R. (1996) Differential inhibitory action of nitric oxide and peroxynitrite on mitochondrial electron transport. *Arch. Biochem. Biophys.*, 328: 309–316.

Chao, C.C., Hu, S., Molitor, T.W., Shaskan, E.G. and Peterson, P.K. (1992) Activated microglia mediate neuronal cell injury via a nitric oxide mechanism. *J. Immunol.*, 149: 2736–2741.

Charles, I.G., Palmer, R.M., Hickery, M.S., Bayliss, M.T., Chubb, A.P., Hall, V.S., Moss, D.W. and Moncada, S. (1993) Cloning, characterization, and expression of a cDNA encoding an inducible nitric oxide synthase from the human chondrocyte. *Proc. Natl. Acad. Sci. USA*, 90: 11419–11423.

Cho, H.J., Xie, Q.W., Calaycay, J., Mumford, R.A., Swiderek, K.M., Lee, T.D. and Nathan, C. (1992) Calmodulin is a subunit of nitric oxide synthase from macrophages. *J. Exp. Med.*, 176: 599–604.

Choi, D.W. (1988) Glutamate neurotoxicity and diseases of the nervous system. *Neuron*, 1: 623–634.

Cosi, C., Colpaert, F., Koek, W., Degryse, A. and Marien, M. (1996) Poly(ADP-ribose) polymerase inhibitors protect against MPTP-induced depletions of striatal dopamine and cortical noradrenaline in C57B1/6 mice. *Brain Res*, 729: 264–269.

Dalkara, T. and Moskowitz, M.A. (1997) Neurotoxic and neuroprotective roles of nitric oxide in cerebral ischaemia. *Int. Rev. Neurobiol.*, 40: 319–336.

Dalkara, T., Yoshida, T., Irikura, K. and Moskowitz, M.A. (1994) Dual role of nitric oxide in focal cerebral ischemia. *Neuropharmacology*, 33: 1447–1452.

Dawson, T.M., Bredt, D.S., Fotuhi, M., Hwang, P.M. and Snyder, S.H. (1991a) Nitric oxide synthase and neuronal NADPH diaphorase are identical in brain and peripheral tissues. *Proc. Natl. Acad. Sci. USA*, 88: 7797–7801.

Dawson, T.M., Dawson, V.L. and Snyder, S.H. (1994a) Molecular mechanisms of nitric oxide actions in the brain. *Ann. NY. Acad. Sci.*, 738: 76–85.

Dawson, T.M. and Snyder, S.H. (1994) Gases as biological messengers: Nitric oxide and carbon monoxide in the brain. *J. Neurosci.*, 14: 5147–5159.

Dawson, V.L., Brahmbhatt, H.P., Mong, J.A. and Dawson, T.M. (1994b) Expression of inducible nitric oxide synthase causes delayed neurotoxicity in primary mixed neuronal-glial cortical cultures. *Neuropharmacology*, 33: 1425–1430.

Dawson, V.L. and Dawson, T.M. (1996a) Nitric oxide in neuronal degeneration. *Proc. Soc. Exp. Biol. Med.*, 211: 33–40.

Dawson, V.L. and Dawson, T.M. (1996b) Nitric oxide neurotoxicity. *J. Chem. Neuroanat.*, 10: 179–190.

Dawson, V.L., Dawson, T.M., Bartley, D.A., Uhl, G.R. and Snyder, S.H. (1993) Mechanisms of nitric oxide-mediated neurotoxicity in primary brain cultures. *J. Neurosci.*, 13: 2651–2661.

Dawson, V.L., Dawson, T.M., London, E.D., Bredt, D.S. and Snyder, S.H. (1991b) Nitric oxide mediates glutamate neurotoxicity in primary cortical cultures. *Proc. Natl. Acad. Sci. USA*, 88: 6368–6371.

Dawson, V.L., Kizushi, V.M., Huang, P.L., Snyder, S.H. and Dawson, T.M. (1996) Resistance to neurotoxicity in cortical cultures from neuronal nitric oxide synthase deficient mice. *J. Neurosci*, 16: 2479–2487.

de Murcia, G. and Menissier de Murcia, J. (1994) Poly(ADP-ribose) polymerase: A molecular nick-sensor. *Trends Biochem. Sci.*, 19: 172–176.

de Murcia, G., Schreiber, V., Molinete, M., Saulier, B., Poch, O., Masson, M., Niedergang, C. and Menissier de Murcia, J. (1994) Structure and function of poly(ADP-ribose) polymerase. *Mol. Cell. Biochem.*, 138: 15–24.

Derojaswalker, T., Tamir, S., Ji, H., Wishnok, J.S. and Tannenbaum, S.R. (1995) Nitric oxide induces oxidative damage in addition to deamination in macrophage DNA. *Chem. Res. Toxicol.*, 8: 473–477.

Dimmeler, S., Haendeler, J., Nehls, M. and Zeiher, A.M. (1997) Suppression of apoptosis by nitric oxide via inhibition of interleukin-1 β-converting enzyme (ICE)-like and cysteine protease protein (CPP)- 32-like proteases. *J. Exp. Med.*, 185: 601–7.

Dimmeler, S. and Zeiher, A.M. (1997) Nitric oxide and apoptosis: Another paradigm for the double-edged role of nitric oxide. *Nitric Oxide*, 1: 275–81.

Dinerman, J.L., Dawson, T.M., Schell, M.J., Snowman, A. and Snyder, S.H. (1994) Endothelial nitric oxide synthase localized to hippocampal pyramidal cells: Implications for synaptic plasticity. *Proc. Natl. Acad. Sci. USA*, 91: 4214–4218.

Ding, M. and Merrill, J.E. (1997) The kinetics and regulation of the induction of type II nitric oxide synthase and nitric oxide in human fetal glial cell cultures. *Mol. Psychiatry*, 2: 117–9.

Ding, M., St. Pierre, B.A., Parkinson, J.F., Medberry, P., Wong, J.L., Rogers, N.E., Ignarro, L.J. and Merrill, J.E. (1997) Inducible nitric-oxide synthase and nitric oxide production in human fetal astrocytes and microglia. A kinetic analysis. *J. Biol. Chem.*, 272: 11327–11335.

Eliasson, M.J.L., Sampei, K., Mandir, A.S., Hurn, P.D., Traystman, R.J., Jun Bao, J., Pieper, A., Wang, Z.-Q., Dawson, T.M., Snyder, S.H. and Dawson, V.L. (1997) Poly(ADP-Ribose) Polymerase gene disruption renders mice resistant to cerebral ischemia. *Nat. Med.*, 3: 1–8.

Endres, M., Wang, Z.Q., Namura, S., Waeber, C. and Moskowitz, M.A. (1997) Ischemic brain injury is mediated by the activation of poly(ADP-ribose)polymerase [In Process Citation]. *J. Cereb. Blood. Flow. Metab.*, 17: 1143–1151.

Ferrante, R.J., Kowall, N.W., Beal, M.F., Richardson, E.P. Jr., Bird, E.D. and Martin, J.B. (1985) Selective sparing of a class of striatal neurons in Huntington's disease. *Science*, 230: 561–563.

Furchgott, R.F. and Zawadzki, J.V. (1980) The obligatory role of endothelial cells in the relaxation of arterial smooth muscle by acetylcholine. *Nature*, 288: 373–376.

Geller, D.A., Lowenstein, C.J., Shapiro, R.A., Nussler, A.K., Di Silvio, M., Wang, S.C., Nakayama, D.K., Simmons, R.L., Snyder, S.H. and Billiar, T.R. (1993) Molecular cloning and expression of inducible nitric oxide synthase from human hepatocytes. *Proc. Natl. Acad. Sci. USA*, 90: 3491–3495.

Gilad, E., Zingarelli, B., Salzman, A.L. and Szabo, C. (1997) Protection by inhibition of poly (ADP-ribose) synthetase against oxidant injury in cardiac myoblasts in vitro. *J. Mol. Cell. Cardiol.*, 29: 2585–2597.

Gonzalez-Zulueta, M., Mukhin, G., Ensz, L., Zwacka, R.M., Engelhardt, J.F., Oberley, L.W., Dawson, V.L. and Dawson, T.M. (1998) Manganese superoxide dismutase protects nNOS neurons from NMDA and nitric oxide mediated neurotoxicity. *J. Neurosci.*, 18: 2040–2055.

Graziewicz, M., Wink, D.A. and Laval, F. (1996) Nitric oxide inhibits DNA ligase activity: potential mechanisms for NO-mediated DNA damage. *Carcinogenesis*, 17: 2501–2505.

Haendeler, J., Weiland, U., Zeiher, A.M. and Dimmeler, S. (1997) Effects of redox-related congeners of NO on apoptosis and caspase-3 activity. *Nitric Oxide*, 1: 282–293.

Hara, H., Huang, P.L., Panahian, N., Fishman, M.C. and Moskowitz, M.A. (1996) Reduced brain edema and infarction volume in mice lacking the neuronal isoform of nitric oxide synthase after transient MCA occlusion. *J. Cereb. Blood. Flow. Metab.*, 16: 605–611.

Heikkila, R.E., Nicklas, W.J., Vyas, I. and Duvoisin, R.C. (1985) Dopaminergic toxicity of rotenone and the 1-methyl-4-phenylpyridinium ion after their stereotaxic administration to rats: Implication for the mechanism of 1-methyl-4-phenyl-1,2,3,6-tetrahydropyridine toxicity. *Neurosci. Lett.*, 62: 389–394.

Heikkila, R.E., Sieber, B.A., Manzino, L. and Sonsalla, P.K. (1989) Some features of the nigrostriatal dopaminergic neurotoxin 1-methyl-4-phenyl-1,2,3,6-tetrahydropyridine (MPTP) in the mouse. *Mol. Chem. Neuropathol.*, 10: 171–183.

Heller, B., Wang, Z.Q., Wagner, E.F., Radons, J., Burkle, A., Fehsel, K., Burkart, V. and Kolb, H. (1995) Inactivation of the poly(ADP-ribose) polymerase gene affects oxygen radical and nitric oxide toxicity in islet cells. *J. Biol. Chem.*, 270: 11176–11180.

Henry, Y., Lepoivre, M., Drapier, J.C., Ducrocq, C., Boucher, J.L. and Guissani, A. (1993) EPR characterization of molecular targets for NO in mammalian cells and organelles. *FASEB J*, 7: 1124–1134.

Hibbs, J.B. Jr., Vavrin, Z. and Taintor, R.R. (1987) L-arginine is required for expression of the activated macrophage effector mechanism causing selective metabolic inhibition in target cells. *J. Immunol.*, 138: 550–565.

Hope, B.T., Michael, G.J., Knigge, K.M. and Vincent, S.R. (1991) Neuronal NADPH diaphorase is a nitric oxide synthase. *Proc. Natl. Acad. Sci. USA*, 88: 2811–2814.

Hortelano, S., Dallaporta, B., Zamzami, N., Hirsch, T., Susin, S.A., Marzo, I., Bosca, L. and Kroemer, G. (1997) Nitric oxide induces apoptosis via triggering mitochondrial permeability transition. *FEBS Lett.*, 410: 373–377.

Huang, E.P. (1997) Synaptic plasticity: A role for nitric oxide in LTP. *Curr. Biol.*, 7: R141–R143.

Huang, Z., Huang, P.L., Ma, J., Meng, W., Ayata, C., Fishman, M.C. and Moskowitz, M.A. (1996) Enlarged infarcts in endothelial nitric oxide synthase knockout mice are attenuated by nitro-L-arginine. *J. Cereb. Blood Flow Metab.*, 16: 981–987.

Huet, J. and Laval, F. (1985) Influence of poly(ADP-ribose) synthesis inhibitors on the repair of sublethal and potentially lethal damage in γ-irradiated mammalian cells. *Int J Radiat. Biol. Relat. Stud. Phys. Chem. Med.*, 47: 655–662.

Hyman, B.T., Marzloff, K., Wenniger, J.J., Dawson, T.M., Bredt, D.S. and Snyder, S.H. (1992) Relative sparing of nitric oxide synthase-containing neurons in the hippocampal formation in Alzheimer's disease. *Ann. Neurol.*, 32: 818–820.

Iadecola, C. (1997) Bright and dark sides of nitric oxide in ischemic brain injury. *Trends Neurosci.*, 20: 132–139.

Ignarro, L.J., Buga, G.M., Wood, K.S., Byrns, R.E. and Chaudhuri, G. (1987) Endothelium-derived relaxing factor produced and released from artery and vein is nitric oxide. *Proc. Natl. Acad. Sci. USA*, 84: 9265–9269.

Janssens, S.P., Simouchi, A., Quertermous, T., Bloch, D.B. and Bloch, K.D. (1992) Cloning and expression of a cDNA encoding human endothelium-derived relating factor/nitric oxide synthase. *J. Biol. Chem.*, 267: 22694.

Jenner, P., Schapira, A.H. and Marsden, C.D. (1992) New insights into the cause of Parkinson's disease. *Neurology*, 42: 2241–2250.

Kantor, D.B., Lanzrein, M., Stary, S.J., Sandoval, G.M., Smith, W.B., Sullivan, B.M., Davidson, N. and Schuman, E.M. (1996) A role for endothelial NO synthase in LTP revealed by adenovirus-mediated inhibition and rescue. *Science*, 274: 1744–1748.

Kato, H., Kanellopoulos, G.K., Matsuo, S., Wu, Y.J., Jacquin, M.F., Hsu, C.Y., Choi, D.W. and Kouchoukos, N.T. (1997) Protection of rat spinal cord from ischemia with dextrorphan and cycloheximide: Effects on necrosis and apoptosis. *J. Thorac Cardiovasc. Surg.*, 114: 609–618.

Kennedy, L.J., Moore, K., Jr., Caulfield, J.L., Tannenbaum, S.R. and Dedon, P.C. (1997) Quantitation of 8-oxoguanine and strand breaks produced by four oxidizing agents. *Chem. Res. Toxicol.*, 10: 386–392.

Khan, S., Kayahara, M., Joashi, U., Mazarakis, N.D., Sarraf, C., Edwards, A.D., Hughes, M.N. and Mehmet, H. (1997) Differential induction of apoptosis in Swiss 3T3 cells by nitric oxide and the nitrosonium cation. *J. Cell. Sci.*, 110: 2315–2322.

Kim, Y.M., Talanian, R.V. and Billiar, T.R. (1997) Nitric oxide inhibits apoptosis by preventing increases in caspase-3-like activity via two distinct mechanisms. *J. Biol. Chem.*, 272: 31138–31148.

Kindt, M.V., Heikkila, R.E. and Nicklas, W.J. (1987) Mitochondrial and metabolic toxicity of 1-methyl-4-(2′-methylphenyl)-1,2,3,6-tetrahydropyridine. *J. Pharmacol. Exp. Ther.*, 242: 858–863.

Kitayama, S., Shimada, S. and Uhl, G.R. (1992) Parkinsonism-inducing neurotoxin MPP+: Uptake and toxicity in non-neuronal COS cells expressing dopamine transporter cDNA. *Ann. Neurol.*, 32: 109–111.

Koh, J.Y. and Choi, D.W. (1988) Cultured striatal neurons containing NADPH-diaphorase or acetylcholinesterase are selectively resistant to injury by NMDA receptor agonists. *Brain Res.*, 446: 374–378.

Koh, J.Y., Peters, S. and Choi, D.W. (1986) Neurons containing NADPH-diaphorase are selectively resistant to quinolinate toxicity. *Science*, 234: 73–76.

Koka, P., He, K., Camerini, D., Tran, T., Yashar, S.S. and Merrill, J.E. (1995a) The mapping of HIV-1 gp160 epitopes required for interleukin-1 and tumor necrosis factor alpha production in glial cells. *J. Neuroimmunol.*, 57: 179–191.

Koka, P., He, K., Zack, J.A., Kitchen, S., Peacock, W., Fried, I., Tran, T., Yashar, S.S. and Merrill, J.E. (1995b) Human immunodeficiency virus 1 envelope proteins induce interleukin 1, tumor necrosis factor alpha, and nitric oxide in glial cultures derived from fetal, neonatal, and adult human brain. *J. Exp. Med.*, 182: 941–951.

Kopin, I.J. and Markey, S.P. (1988) MPTP toxicity: Implications for research in Parkinson's disease. *Annu. Rev. Neurosci.*, 11: 81–96.

Koppenol, W.H., Moreno, J.J., Pryor, W.A., Ischiropoulos, H. and Beckman, J.S. (1992) Peroxynitrite, a cloaked oxidant formed by nitric oxide and superoxide. *Chem. Res. Toxicol.*, 5: 834–842.

Koprowski, H., Zheng, Y.M., Heber-Katz, E., Fraser, N., Rorke, L., Fu, Z.F., Hanlon, C. and Dietzschold, B. (1993) In vivo expression of inducible nitric oxide synthase in experimentally induced neurologic diseases. *Proc. Natl. Acad. Sci. USA*, 90: 3024–3027.

Lamas, S., Marsden, P.A., Li, G.K., Tempst, P. and Michel, T. (1992) Endothelial nitric oxide synthase: Molecular cloning and characterization of a distinct constitutive enzyme isoform. *Proc. Natl. Acad. Sci. USA*, 89: 6348–6352.

Lancaster, J.R. Jr. (1994) Simulation of the diffusion and reaction of endogenously produced nitric oxide. *Proc. Natl. Acad. Sci. USA*, 91: 8137–8141.

Langston, J.W., Langston, E.B. and Irwin, I. (1984) MPTP-induced parkinsonism in human and non-human primates – clinical and experimental aspects. *Acta. Neurol. Scand. Suppl.*, 100: 49–54.

Lautier, D., Lagueux, J., Thibodeau, J., Menard, L. and Poirier, G.G. (1993) Molecular and biochemical features of poly (ADP-ribose) metabolism. *Mol. Cell. Biochem.*, 122: 171–193.

Leist, M., Fava, E., Montecucco, C. and Nicotera, P. (1997a) Peroxynitrite and nitric oxide donors induce neuronal apoptosis by eliciting autocrine excitotoxicity. *Eur. J. Neurosci.*, 9: 1488–1498.

Leist, M., Single, B., Kunstle, G., Volbracht, C., Hentze, H. and Nicotera, P. (1997b) Apoptosis in the absence of poly-(ADP-ribose) polymerase. *Biochem. Biophys. Res. Commun.*, 233: 518–522.

Li, J., Billiar, T.R., Talanian, R.V. and Kim, Y.M. (1997) Nitric oxide reversibly inhibits seven members of the caspase family via S-nitrosylation. *Biochem. Biophys. Res. Commun.*, 240: 419–424.

Lobner, D. and Choi, D.W. (1996) Preincubation with protein synthesis inhibitors protects cortical neurons against oxygen-glucose deprivation-induced death. *Neuroscience*, 72: 335–341.

Lowenstein, C.J., Glatt, C.S., Bredt, D.S. and Snyder, S.H. (1992) Cloned and expressed macrophage nitric oxide synthase contrasts with the brain enzyme. *Proc. Natl. Acad. Sci. USA*, 89: 6711–6715.

Lundberg, J.O., Farkas-Szallasi, T., Weitzberg, E., Rinder, J., Lidholm, J., Anggaard, A., Hokfelt, T., Lundberg, J.M. and Alving, K. (1995) High nitric oxide production in human paranasal sinuses. *Nat. Med.*, 1: 370–373.

Lyons, C.R., Orloff, G.J. and Cunningham, J.M. (1992) Molecular cloning and functional expression of an inducible nitric oxide synthase from a murine macrophage cell line. *J. Biol. Chem.*, 267: 6370–6374.

MacMillan-Crow, L.A., Crow, J.P., Kerby, J.D., Beckman, J.S. and Thompson, J.A. (1996) Nitration and inactivation of manganese superoxide dismutase in chronic rejection of human renal allografts. *Proc. Natl. Acad. Sci. USA*, 93: 11853–11858.

Marletta, M.A. (1993) Nitric oxide synthase structure and mechanism. *J. Biol. Chem.*, 268: 12231–12234.

Masliah, E., Achim, C.L., Ge, N., DeTeresa, R., Terry, R.D. and Wiley, C.A. (1992a) Spectrum of human immunodeficiency virus-associated neocortical damage. *Ann. Neurol.*, 32: 321–329.

Masliah, E., Ge, N., Morey, M., DeTeresa, R., Terry, R.D. and Wiley, C.A. (1992b) Cortical dendritic pathology in human immunodeficiency virus encephalitis. *Lab. Invest.*, 66: 285–291.

Milam, K.M. and Cleaver, J.E. (1984) Inhibitors of poly(adenosine diphosphate-ribose) synthesis: Effect on other metabolic processes. *Science*, 223: 589–591.

Milam, K.M. and Cleaver, J.E. (1985) Metabolic effects of poly (ADP-ribose) inhibitors. *Basic. Life Sci.*, 31: 25–31.

Moskowitz, M.A. and Dalkara, T. (1996) Nitric oxide and cerebral ischemia. *Adv. Neurol.*, 71: 365–367; discussion 367–369.

Murphy, S., Simmons, M.L., Agullo, L., Garcia, A., Feinstein, D.L., Galea, E., Reis, D.J., Minc-Golomb, D. and Schwartz, J.P. (1993) Synthesis of nitric oxide in CNS glial cells. *Trends. Neurosci.*, 16: 323–328.

Nathan, C. and Xie, Q.W. (1994a) Nitric oxide synthases: Roles, tolls, and controls. *Cell*, 78: 915–918.

Nathan, C. and Xie, Q.W. (1994b) Regulation of biosynthesis of nitric oxide. *J. Biol. Chem.*, 269: 13725–13728.

Nicotera, P., Brune, B. and Bagetta, G. (1997) Nitric oxide: inducer or suppressor of apoptosis? *Trends Pharmacol. Sci.*, 18: 189–190.

Nuovo, G.J. and Alfieri, M.L. (1996) AIDS dementia is associated with massive, activated HIV-1 infection and concomitant expression of several cytokines. *Mol. Med.*, 2: 358–366.

Ogura, T., Tatemichi, M. and Esumi, H. (1997) Nitric oxide inhibits CPP32-like activity under redox regulation. *Biochem. Biophys. Res. Commun.*, 236: 365–369.

Palmer, R.M., Ferrige, A.G. and Moncada, S. (1987) Nitric oxide release accounts for the biological activity of endothelium-derived relaxing factor. *Nature*, 327: 524–6.

Panahian, N., Yoshida, T., Huang, P.L., Hedley-Whyte, E.T., Dalkara, T., Fishman, M.C. and Moskowitz, M.A. (1996) Attenuated hippocampal damage after global cerebral ischemia in mice mutant in neuronal nitric oxide synthase. *Neuroscience*, 72: 343–354.

Park, S.K. and Murphy, S. (1996) Nitric oxide synthase type II mRNA stability is translation- and transcription-dependent. *J. Neurochem.*, 67: 1766–1769.

Pifl, C., Giros, B. and Caron, M.G. (1993) Dopamine transporter expression confers cytotoxicity to low doses of the parkinsonism-inducing neurotoxin 1-methyl-4-phenylpyridinium. *J. Neurosci.*, 13: 4246–4253.

Price, R.W. (1996) Neurological complications of HIV infection. *Lancet*, 348: 445–452.

Price, R.W., Brew, B., Sidtis, J., Rosenblum, M., Scheck, A.C. and Cleary, P. (1988) The brain in AIDS: Central nervous system HIV-1 infection and AIDS dementia complex. *Science*, 239: 586–592.

Przedborski, S., Jackson-Lewis, V., Yokoyama, R., Shibata, T., Dawson, V.L. and Dawson, T.M. (1996) Role of neuronal nitric oxide in 1-methyl-4-phenyl-1,2,3,6-tetrahydropyridine (MPTP)-induced dopaminergic neurotoxicity. *Proc. Natl. Acad. Sci. USA*, 93: 4565–4571.

Ramsay, R.R. and Singer, T.P. (1992) Relation of superoxide generation and lipid peroxidation to the inhibition of NADH-Q oxidoreductase by rotenone, piericidin A, and MPP+. *Biochem. Biophys. Res. Commun.*, 189: 47–52.

Ramsay, R.R., Youngster, S.K., Nicklas, W.J., McKeown, K.A., Jin, Y.Z., Heikkila, R.E. and Singer, T.P. (1989) Structural dependence of the inhibition of mitochondrial respiration and of NADH oxidase by 1-methyl-4-phenylpyridinium (MPP+) analogs and their energized accumulation by mitochondria. *Proc. Natl. Acad. Sci. USA*, 86: 9168–9172.

Ratan, R.R., Murphy, T.H. and Baraban, J.M. (1994a) Macromolecular synthesis inhibitors prevent oxidative stress-induced apoptosis in embryonic cortical neurons by shunting cysteine from protein synthesis to glutathione. *J. Neurosci.*, 14: 4385–4392.

Ratan, R.R., Murphy, T.H. and Baraban, J.M. (1994b) Oxidative stress induces apoptosis in embryonic cortical neurons. *J. Neurochem.*, 62: 376–379.

Redegeld, F.A., Chatterjee, S., Berger, N.A. and Sitkovsky, M.V. (1992) Poly-(ADP-ribose) polymerase partially contributes to target cell death triggered by cytolytic T lymphocytes. *J. Immunol.*, 149: 3509–3516.

Samdani, A.F., Newcamp, C., Resink, A., Facchinetti, F., Hoffman, B.E., Dawson, V.L. and Dawson, T.M. (1997) Differential susceptibility to neurotoxicity mediated by neurotrophins and neuronal nitric oxide synthase. *J. Neurosci.*, 17: 4633–4641.

Sanchez-Ramos, J.R., Song, S., Mash, D.C. and Weiner, W.J. (1992) 21-aminosteroids interact with the dopamine transporter to protect against 1-methyl-4-phenylpyridinium-induced neurotoxicity. *J. Neurochem*, 58: 328–334.

Sandau, K., Pfeilschifter, J. and Brune, B. (1997) Nitric oxide and superoxide induced p53 and Bax accumulation during mesangial cell apoptosis. *Kidney Int.*, 52: 378–386.

Sessa, W.C., Harrison, J.K., Barber, C.M., Zeng, D., Durieux, M.E., D'Angelo, D.D., Lynch, K.R. and Peach, M.J. (1992) Molecular cloning and expression of a cDNA encoding endothelial cell nitric oxide synthase. *J. Biol. Chem.*, 267: 15274–15276.

Shah, G.M., Poirier, D., Desnoyers, S., Saint-Martin, S., Hoflack, J.C., Rong, P., ApSimon, M., Kirkland, J.B. and Poirier, G.G. (1996) Complete inhibition of poly(ADP-ribose) polymerase activity prevents the recovery of C3H10T1/2 cells from oxidative stress. *Biochim. Biophys. Acta*, 1312: 1–7.

Shall, S. (1995) ADP-ribosylation reactions. *Biochimie*, 77: 313–318.

Son, H., Hawkins, R.D., Martin, K., Kiebler, M., Huang, P.L., Fishman, M.C. and Kandel, E.R. (1996) Long-term potentiation is reduced in mice that are doubly mutant in endothelial and neuronal nitric oxide synthase. *Cell*, 87: 1015–1023.

Stuehr, D.J. and Griffith, O.W. (1992) Mammalian nitric oxide synthases. *Adv. Enzymol. Relat. Areas. Mol. Biol.*, 65: 287–346.

Stuehr, D.J. and Nathan, C.F. (1989) Nitric oxide. A macrophage product responsible for cytostasis and respiratory inhibition in tumor target cells. *J. Exp. Med.*, 169: 1543–1555.

Szabo, C., Cuzzocrea, S., Zingarelli, B., O'Connor, M. and Salzman, A.L. (1997) Endothelial dysfunction in a rat model of endotoxic shock. Importance of the activation of poly (ADP-ribose) synthetase by peroxynitrite. *J. Clin. Invest.*, 100: 723–735.

Takahashi, K., Greenberg, J.H., Jackson, P., Maclin, K. and Zhang, J. (1997) Neuroprotective effects of inhibiting poly(ADP-ribose) synthetase on focal cerebral ischemia in rats. *J. Cereb. Blood. Flow. Metab.*, 17: 1137–1142.

Takahashi, K., Wesselingh, S.L., Griffin, D.E., McArthur, J.C., Johnson, R.T. and Glass, J.D. (1996) Localization of HIV-1 in human brain using polymerase chain reaction/in situ hybridization and immunocytochemistry. *Ann. Neurol.*, 39: 705–711.

Takehara, Y., Kanno, T., Yoshioka, T., Inoue, M. and Utsumi, K. (1995) Oxygen-dependent regulation of mitochondrial energy metabolism by nitric oxide. *Arch. Biochem. Biophys.*, 323: 27–32.

Tenneti, L., D'Emilia, D.M. and Lipton, S.A. (1997) Suppression of neuronal apoptosis by S-nitrosylation of caspases. *Neurosci. Lett.*, 236: 139–142.

Thiemermann, C., Bowes, J., Myint, F.P. and Vane, J.R. (1997) Inhibition of the activity of poly(ADP ribose) synthetase reduces ischemia-reperfusion injury in the heart and skeletal muscle. *Proc. Natl. Acad. Sci. USA*, 94: 679–683.

Thomas, E. and Pearse, A.G.E. (1961) The fine localization of dehydrogenases in the nervous system. *Histochemistry*, 2: 266–282.

Thomas, E. and Pearse, A.G.E. (1964) The solitary active cells. Histochemical demonstration of damage-resistant nerve cells with a TPN-diaphorase reaction. *Acta Neuropathology*, 3: 238–249.

Uemura, Y., Kowall, N.W. and Beal, M.F. (1990) Selective sparing of NADPH-diaphorase-somatostatin-neuropeptide Y neurons in ischemic gerbil striatum. *Ann. Neurol.*, 27: 620–625.

Van Dam, A.M., Bauer, J., Man, A.H.W.K., Marquette, C., Tilders, F.J. and Berkenbosch, F. (1995) Appearance of inducible nitric oxide synthase in the rat central nervous system after rabies virus infection and during experimental allergic encephalomyelitis but not after peripheral administration of endotoxin. *J. Neurosci. Res.*, 40: 251–260.

von Knethen, A. and Brune, B. (1997) Cyclooxygenase-2: An essential regulator of NO-mediated apoptosis. *Faseb. J*, 11: 887–95.

Wang, Z.-Q., Stingl, L., Morrison, C., Jantsch, M., Los, M., Schulze-Osthoff, K. and Wagner, E.F. (1997) PARP is important for genomic stability but dispensable in apoptosis. *Genes & Development*, 11: 2347–2358.

Wang, Z.Q., Auer, B., Stingl, L., Berghammer, H., Haidacher, D., Schweiger, M. and Wagner, E.F. (1995) Mice lacking ADPRT and poly(ADP-ribosyl)ation develop normally but are susceptible to skin disease. *Genes Dev.*, 9: 509–520.

Wiley, C.A. and Achim, C. (1994) Human immunodeficiency virus encephalitis is the pathological correlate of dementia in acquired immunodeficiency syndrome. *Ann. Neurol.*, 36: 673–676.

Wiley, C.A., Schrier, R.D., Nelson, J.A., Lampert, P.W. and Oldstone, M.B. (1986) Cellular localization of human immunodeficiency virus infection within the brains of acquired immune deficiency syndrome patients. *Proc. Natl. Acad. Sci. USA*, 83: 7089–7093.

Wood, J. and Garthwaite, J. (1994) Models of the diffusional spread of nitric oxide: Implications for neural nitric oxide signalling and its pharmacological properties. *Neuropharmacology*, 33: 1235–1244.

Xie, Q.W., Cho, H.J., Calaycay, J., Mumford, R.A., Swiderek, K.M., Lee, T.D., Ding, A., Troso, T. and Nathan, C. (1992) Cloning and characterization of inducible nitric oxide synthase from mouse macrophages. *Science*, 256: 225–228.

Yabuki, M., Inai, Y., Yoshioka, T., Hamazaki, K., Yasuda, T., Inoue, M. and Utsumi, K. (1997a) Oxygen-dependent fragmentation of cellular DNA by nitric oxide [published erratum appears in Free Radic. Res. 1997 Jun;26(6):594]. *Free Radic. Res.*, 26: 245–255.

Yabuki, M., Kariya, S., Inai, Y., Hamazaki, K., Yoshioka, T., Yasuda, T., Horton, A.A. and Utsumi, K. (1997b) Molecular mechanisms of apoptosis in HL-60 cells induced by a nitric oxide-releasing compound. *Free Radic. Res.*, 27: 325–335.

Zaharchuk, G., Hara, H., Huang, P.L., Fishman, M.C., Moskowitz, M.A., Jenkins, B.G. and Rosen, B.R. (1997) Neuronal nitric oxide synthase mutant mice show smaller infarcts and attenuated apparent diffusion coefficient changes in the peri-infarct zone during focal cerebral ischemia. *Magn. Reson. Med.*, 37: 170–175.

Zhang, J., Dawson, V.L., Dawson, T.M. and Snyder, S.H. (1994) Nitric oxide activation of poly(ADP-ribose) synthetase in neurotoxicity. *Science*, 263: 687–689.

Zhang, J., Pieper, A. and Snyder, S.H. (1995) Poly(ADP-ribose) synthetase activation: An early indicator of neurotoxic DNA damage. *J. Neurochem.*, 65: 1411–1414.

'd M.J. Friedlander (Eds.)

'd.

CHAPTER 16

anisms of NO neurotoxicity

Dalkara[†], Matthias Endres[1] and Michael A. Moskowitz[1,*]

[1,*]*Stroke and Neurovascular Regulation Laboratory, Massachusetts General Hospital, Harvard Medical School, 149 13th street, Room 6403, Charlestown, MA 02129 USA*
[†]*Department of Neurology, Hacettepe University Hospitals, Ankara, Turkey*

Introduction

Nitric oxide (NO) is synthesized from L-arginine and molecular oxygen by NO synthases (NOS). NOS has two constitutive, neuronal (Type 1) and endothelial (Type 3), and one inducible isoforms (Type 2) (for review, see Griffith and Stuehr, 1995). In the brain, neuronal NOS (nNOS) is almost exclusively expressed in neurons and perivascular nerves, whereas endothelial isoform (eNOS) is mainly detected in endothelium of cerebral vessels with expression in small populations of neurons (Iadecola et al., 1994). The constitutive isoforms are calcium/calmodulin-dependent and activated by transient elevations in intracellular calcium (Knowles et al., 1989; Knowles and Moncada, 1992). Small puffs of NO synthesized by constitutive NOS, regulates a wide variety of physiological functions such as blood pressure, vascular tone and permeability and neurotransmission. On the other hand, calmodulin is tightly bound to the inducible isoform (iNOS), and once expressed, iNOS is continuously active irrespective of intracellular calcium level and leads to a long-lasting, high output NO generation (Cho et al., 1992). iNOS can be induced in microglia, astrocytes, endothelium and vascular smooth muscle cells (Murphy et al., 1993). High output NO synthesis by the inducible isoform is cytotoxic and mediates the inflammatory actions of NO (Gross and Wolin, 1995). Like iNOS, nNOS can also generate high amounts of NO and cause cytotoxicity under pathological conditions due to an unregulated, persisting rise in intracellular calcium (Dawson et al., 1992; Dawson and Dawson, 1995). Although NOS positive neurons comprise only 1–2% of all the neurons in the cerebral cortex, corpus striatum and hippocampus, these neurons possess extensive axonal branching consistent with the idea that NOS positive cells kill neighboring neurons when NO or a reaction product is generated in excessive amounts (Bredt et al., 1991; Dawson and Dawson, 1995). Interestingly, these neurons themselves are resistant to various insults including NMDA toxicity and ischemia, perhaps related to dismutation of superoxide and attendant inhibition of peroxynitrite anion formation, a potent cytotoxic mediator (Koh and Choi, 1988; Uemura et al., 1990). NO synthesized by nNOS has recently been implicated in many pathophysiological processes including cerebral ischemia and excitotoxicity (Dawson and Dawson, 1995). Using two strains of mutant mice that do not express the gene for the neuronal or the endothelial isoforms of NOS, we have documented the detrimental role of neuronal NO in cerebral ischemia and NMDA toxicity. The evidence will be reviewed below.

*Corresponding author. Tel.: +1 617 726 8442; fax: +1 617 726 2547; e-mail: Moskowitz@helix.mgh.harvard.edu

Knockout mice

Research over the last few decades has clearly documented the importance of gene deletion or overexpression as a promising alternative to pharmaceutical agents like enzyme inhibitors or receptor antagonists in elucidating the importance of a biomolecule. Several hundred mutant strains have been developed in the last decade (Thomas, 1995; Rubin and Barsh, 1996). Knockout mutant mice are generated by targeted disruption of one (sometimes two) of estimated 50–100,000 genes in the mouse genome by homologous reconstitution (Fässler et al., 1995). The majority of the currently available mutants have "null" (or loss-of-function) alleles of the gene of interest. Null mutants have been developed to study the role of proteins during development, in adult life and disease states. The gene deletion was initially hoped to mimic specific and complete inhibition of the targeted protein. Surprisingly, many knockouts exhibited a normal phenotype due to redundancy of genes compensating for the deleted gene product (e.g. Huang et al., 1993). However, increasing evidence indicates that despite an apparently normal phenotype, knockout animals may display conspicuous abnormalities in unanticipated tissues (Huang et al., 1993) and respond differently to various manipulations than wild type animals (Irikura et al., 1995). They have been especially useful to confirm the pharmacological activity of drugs in which selectivity is in question.

nNOS mutant mice

Targeted disruption of nNOS gene was achieved by homologous recombination by substitution of a neomycin resistance gene for a critical exon (Huang et al., 1993). The mutant mouse develops normally and the endothelium of nNOS knockouts expresses eNOS immunoreactivity (Irikura et al., 1995). In nNOS mutants, neuronal NOS expression and NADPH-diaphorase staining were markedly deficient, and in vitro NOS activity was significantly reduced to less than 5% of normal (Huang et al., 1993). Of interest, alternative splice variants of NOS have been reported which generate NO in vitro, but lack the PDZ-containing domain and potential coupling to the NMDA receptor (Brenman et al., 1996; Eliasson et al., 1997a). However, the expressed splice variants do not compensate sufficiently as NOS activity in $nNOS^{-/-}$ brains is severely reduced. Consistent with this, low basal levels of $[^3H]$L-N^G-nitroarginine binding (Hara et al., 1997a) and brain cGMP levels were found in mutants and no cGMP enhancement was observed during ischemia or hypercapnia, contrary to the robust increases in wild type mice (Huang et al., 1994; Irikura et al., 1995). Consistent results were also obtained by NMR spectroscopy as NO adducts were undetectable during cerebral ischemia as measured in nNOS null mice (Mullins et al., unpublished observations). Regional cerebral blood flow responses to hypercapnia (Irikura et al., 1995) and whisker stimulation (Ma et al., 1996), and pial arterial dilation to topical ACh superfusion (Meng et al., 1996) were not significantly different from wild type mice. However, their responses to NOS inhibitors were markedly reduced.

NO and the NMDA receptor

NO was proposed as the neurotoxic agent mediating NMDA toxicity. Studies in dissociated cell cultures showed that NOS inhibitors effectively blocked NMDA-induced cell death whereas L-arginine depletion attenuated NMDA receptor-mediated toxicity (Dawson et al., 1991). NMDA toxicity was also blocked by reducing NOS catalytic activity by calmodulin antagonists, flavoprotein inhibitors, inhibitors of NOS dephosphorylation, or by scavenging NO with hemoglobin or by increasing SOD activity to thereby reduce peroxynitrite anion formation (Dawson et al., 1993). Based on these studies, it was hypothesized that an increase in intracellular calcium by ischemia-induced NMDA receptor overactivation increases NO. NO also reportedly inhibits glutamate uptake and mediates the sustained opening of NMDA receptor ion channels (Pogun et al., 1994; Akira et al., 1994).

It has recently been demonstrated that knocking out the neuronal NOS gene confers neuroprotection against NMDA receptor-mediated toxicity (Dawson et al., 1996; Ayata et al., 1997). Excitotoxic lesions produced by intrastriatal stereotactic NMDA microinjections (10–20 nmol) were 45% smaller in nNOS knockout compared to wild-type littermates. The density and distribution of striatal NMDA binding sites did not differ between strains, suggesting that the resistance to NMDA toxicity was caused by marked attenuation of NO synthesis after receptor activation and not a change in the density of NMDA receptor. In line with this view, pharmacological inhibition of nNOS by 7-nitroindazole (50 mg/kg, i.p.) also decreased NMDA lesion size by 32% in wild type mice.

NO likely mediates neurotoxicity by several mechanisms including disruption of cellular metabolism (inhibition of aconitase, complex I and II of mitochondrial electron transport and glyceraldehyde-3-phosphate dehydrogenase) and of DNA synthesis (see Dawson et al., 1992 and Gross and Wolin, 1995). NO also damages DNA structure (Liu and Hotchkiss, 1995). DNA damage may activate reparative enzymes such as poly-ADP ribose synthase (PARS), which can deplete the cell of ATP and nicotinamide dinucleotide, hence further compromising cellular energy metabolism (Zhang et al., 1994b). One of the most attractive mechanisms involves peroxynitrite formation which is initiated via NMDA receptor activation, intracellular Ca^{2+} increase, and augmented NO synthesis (Garthwaite et al., 1988). Peroxynitrite ($ONOO^-$) is formed by reaction of $NO^{\bullet}$ with $O_2^{\bullet -}$ (Lafon-Cazal et al., 1993; Dykens, 1994), and this complex rapidly decomposes into $NO_2^{\bullet}$ and hydroxyl radical ($OH^{\bullet}$), or a reactive intermediate with $OH^{\bullet}$-like activity (Crow et al., 1994). Hydroxyl radical is a highly reactive species which leads to oxidation of sulfhydryl groups, lipids, DNA, and proteins (Beckman et al., 1994, 1996). Peroxynitrite can directly inhibit glutamate transporters (Trotti et al., 1996), or produce nitronium ions, causing irreversible nitration of tyrosine residues in proteins (Beckman et al., 1996). Protein tyrosine nitration may contribute to $NO^{\bullet}$ toxicity by reducing phosphorylation by tyrosine kinases (Beckman et al., 1996) or targeting nitrated proteins for degradation (Gow et al., 1996).

We studied brain levels of free 3-nitrotyrosine as an indirect measure of protein tyrosine nitration and as a footprint of peroxynitrite formation during NMDA receptor activation. Compared to robust increases in wild type mice, brain levels of 3-nitrotyrosine and $OH^{\bullet}$ radical did not rise in $nNOS^{-/-}$ brains suggesting that the reaction of $NO^{\bullet}$ with $O_2^{\bullet -}$ is a major pathway which generates peroxynitrite and $OH^{\bullet}$ radical-like activity in brain following NMDA receptor activation. Despite greatly reduced $NO^{\bullet}$ and $OH^{\bullet}$ radical production, the protection was only about 50% in $nNOS^{-/-}$ mice, suggesting a role for additional mechanisms such as acute neuronal swelling and lysis, and $O_2^{\bullet -}$ toxicity (Chan, 1996; Kamii et al., 1996). These findings strongly suggest a detrimental role for neuronal NO synthesis in NMDA excitotoxicity which relates to oxygen radical-mediated mechanisms.

NO and cerebral ischemia

Cortical NO levels increase strikingly from approximately 10 nM to 2 μM within 3–24 min after middle cerebral artery (MCA) occlusion (Malinski et al., 1993). Brain nitrite (stable NO metabolite) and cGMP (a product of NO-mediated guanylate cyclase activation) levels also rise within the first half hour of ischemia (Kader et al., 1993). These increases are effectively blocked by prior L-N^G-nitroarginine administration, indicating an enhanced NOS activity. It is likely that constitutive NOS activity increases during ischemia due to a rise in intracellular Ca^{2+}. However, constitutive NOS activity decreases shortly after its activation at the onset of ischemia possibly due to inactivation of the enzyme by unfavorable conditions in ischemic tissue.

nNOS mutant mice and cerebral ischemia

The deficiency in neuronal NO production was associated with increased resistance to cerebral ischemia. The nNOS knockouts developed infarcts 38% smaller than the wild type mice when subjected to 24 h permanent MCA occlusion (Huang et al., 1994). Infarct size was also reduced in mutants 3–4 days after permanent MCA occlusion. Neurological deficits were less in nNOS knockout mice. Since reductions in regional cerebral blood flow (rCBF) were similar within homologous ischemic regions after MCA occlusion in both groups, the observed group differences were attributed to the consequence of neuronal NOS deletion in brain tissue and not due to hemodynamic differences between strains.

The neuroprotective action obtained after deletion of neuronal NOS has also been demonstrated in mice subjected to 3 h of ischemia and 24 h of reperfusion. Infarcts were 69% smaller in nNOS knockouts than wild type mice (Hara et al., 1996). Brain protection was greater than after permanent occlusion possibly due to greater superoxide anion production during reversible occlusion (Chan et al., 1993; Yang et al., 1994). Interestingly, quantitative [^{3}H]L-N^G-nitroarginine autoradiography demonstrated a significant increase (50–250%) in the density of [^{3}H]L-N^G-nitroarginine binding sites (B_{max}) but not the dissociation constant (K_d), during transient focal ischemia and first 3 h of reperfusion (Hara et al., 1997a). nNOS mRNA was also increased as detected by reverse transcription polymerase chain reaction. Similar results were obtained after NMDA injection, with a more extended time course (12–48 h). As noted above, [^{3}H]L-N^G-nitroarginine binding to nNOS protein was very low in nNOS mutant mice, and only a very small increase was observed after ischemia or NMDA excitotoxicity.

nNOS mutants were also protected against global ischemia (Panahian et al., 1996). Fewer dead hippocampal neurons were counted 3 days after transient global ischemia induced by bilateral common carotid plus basilar artery occlusion for 5 or 10 min. Not only were more hippocampal cells viable, but overall morbidity, weight loss, and neurological outcome were better in the mutant strain.

In addition to neurons, perivascular nerves and cerebrovascular endothelium may form NO during cerebral ischemia. A late but sustained increase in NO levels may also occur due to expression of inducible NOS within microglia and invading inflammatory cells 24–72 h after the induction of ischemia (Iadecola et al., 1995). During the immediate period following ischemia, increased NO production in vascular endothelium or perivascular nerves may improve blood flow and be neuroprotective. Indeed, infusion of L-arginine, which increases NO production and dilates pial vessels via an NO-dependent mechanism, and increases rCBF in normal as well as in ischemic brain, reduces infarct size (Morikawa et al., 1994). CBF augmentation can lead to electrocorticogram recovery if blood flow enhancement exceeds the functional flow threshold of approximately 30% of pre-ischemic level (Dalkara et al., 1994). Blood flow may also be increased within ischemic tissue by intracarotid administration of NO donors and decreases infarct size in models of focal ischemia (Zhang et al., 1994a). In addition, endothelial NO production may also improve microcirculation by reducing platelet aggregation and leukocyte adhesion (Niu et al., 1994).

eNOS mutant mice and cerebral ischemia

Consistent with the above evidence larger infarcts developed in eNOS knockout mice after 24 h permanent MCA occlusion (Huang et al., 1996). Deletion of eNOS rendered these mutants hypertensive (Huang et al., 1995). However, hypertension per se did not account for the increased susceptibility of eNOS mutants because infarct size did not decrease, but rather increases after blood pressure was reduced by hydralazine (Huang et al., 1996). eNOS mutants developed more pronounced rCBF reductions in corresponding brain regions after MCA occlusion and exhibited proportionally lower rCBFs at reduced perfusion pressures during controlled hemorrhagic hypotension (Huang et al. 1996). Dynamic CT scanning demonstrated that areas of hemodynamic penumbra were significantly smaller and core relatively larger in eNOS mutants (Lo et al., 1996). Hence, the susceptibility of eNOS mutants to ischemic injury may be due to its diminished capacity to adapt to reduced perfusion pressure (i.e., dilate) at the margins of an ischemic lesion. This coupled to enhanced platelet and neutrophil adhesion, render eNOS mutants more susceptible to injury. Consistent with this notion, L-N^G-nitroarginine administration increased infarct size in the nNOS knockout mouse, presumably due to inhibition of the constitutively expressed eNOS isoform (Huang et al., 1994).

Increases as well as reductions in the extent of tissue injury have been reported after L-NAME or L-N^G-nitroarginine administration in models of ischemia with MCA occlusion possibly because nonselective inhibition of NO synthesis within vessels and platelets may obscure neuroprotective effects of NOS inhibition in neurons during focal cerebral ischemia (for review, see Dalkara and Moskowitz, 1994). In fact, recent studies using selective nNOS inhibitors, 7-nitroindazole (Yoshida et al., 1994) or ARL17477 (Zhang et al., 1996) showed neuroprotection.

PARS activation in cerebral ischemia

As noted above, one candidate pathway for NO mediated neuronal injury is DNA damage leading to the obligatory activation of the nuclear enzyme PARS. After activation by DNA nicks, PARS catalyzes the covalent attachment of ADP-ribose subunits to nuclear proteins including PARS itself. This process depletes the intracellular substrate NAD^+ and slows the rate of glycolysis, electron transport and, ultimately, ATP formation, because 4 moles of ATP are needed to re-synthesize 1 mole of NAD^+. Hence, PARS activation may lead to cell death by energy depletion, as demonstrated in neurons, macrophages and smooth muscle cells among other cell types (Zhang et al., 1994b, Szabo et al., 1996, 1997).

PARS is activated in ischemic brain within minutes after reperfusion following MCA occlusion as evidenced by poly(ADP-ribose) immunohistochemistry. Poly(ADP-ribose) positive cells show signs of early ischemic damage. NAD^+ levels as shown by in situ histochemistry become depleted in the MCA territory. Inhibition of PARS activation or disruption of the PARS gene (PARS knockouts) confers protection after brain ischemia (Endres et al., 1997 and 1998a; Eliasson et al., 1997b) and also after myocardial ischemia (Thiemermann et al., 1997) and systemic inflammation (Szabo et al., 1997). Consistent with this protection, poly (ADP)ribose formation is inhibited and NAD levels are significantly higher in these animals, demonstrating an energy-preserving neuroprotective mechanism. Total DNA damage, upstream to PARS activation in the cascade, is not affected by PARS inhibition or gene deletion (Endres et al., 1997).

We further established the importance of NO to the PARS activation pathway in vivo by evaluating poly(ADP)ribose formation in nNOS knockout animals. Following 2 h MCA occlusion and reperfusion, nNOS knockout animals exhibited strikingly less poly(ADP)ribose positive cells compared to their littermate controls (Endres et al., 1998b). Moreover, the density of TUNEL (terminal deoxynucleotidyl transferase-mediated dUTP-biotin nick end-labeling) positive cells is decreased after ischemia in nNOS knockout mice, indicating less DNA damage (see also next paragraph). Thus, decreased DNA damage and subsequent PARS activation may confer resistance to ischemic brain damage in nNOS knockout animals.

NO-induced apoptotic cell death

At least two mechanistically distinct forms of neuronal death have been identified. Severely injured neurons that do not immediately die by swelling and lysis may ultimately undergo apoptosis. Apoptotic neuronal death contributes to infarct formation in cerebral ischemia (Li et al., 1995a,b,c, Charriaut-Marlangue et al., 1996). NMDA, or concurrent generation of nitric oxide and superoxide can cause both necrosis and apoptosis, depending on the severity of the insult and resulting mitochondrial dysfunction (Bonfoco et al., 1995; Ankarcrona et al., 1995). Both NO and $ONOO^-$ have been linked to apoptosis (Albina et al., 1993; Ankarcrona et al., 1994; Estevez et al.,

1995; Lin et al., 1995). Since blockade of neuronal NO synthesis proves to be neuroprotective in focal and global cerebral ischemia and NMDA toxicity, in vivo, it is important to understand the extent to which NO-induced apoptosis contribute to neuronal death under these pathological conditions.

We performed TUNEL staining in nNOS knockout mice 48 h following intrastriatal NMDA microinjection (Ayata et al., 1997). Fewer apoptotic neurons were found in nNOS mutants compared to wild type littermates, supporting a relationship between NO and/or its reaction products, and apoptosis. In the wild type, TUNEL (+) cells appeared at 12–24 h, reaching a peak several days thereafter, and were located predominantly at the periphery of the lesion. Accompanying these changes, DNA laddering was observed initially at approximately 12 h and was prominent at 72 hours after NMDA injection.

In another set of experiments we compared the number of apoptotic cells in wild type and nNOS mutant mice 6, 24 and 72 h after permanent MCA occlusion by filament. TUNEL positive neurons were detected at 6 h and continued to increase by 72 h. These cells were detected in both penumbra and core regions but most of them were located on the inner boundary zone of the infarct. Numbers of apoptotic cells as well as their density were significantly lower in nNOS mutants, suggesting a selective decrease in the number of TUNEL positive cells. Such findings support the notion that NO and its reaction products promote apoptosis as a mechanism of cytotoxicity as suggested by studies reporting apoptotic cell death after application of NO$^{\bullet}$ or peroxynitrite (Ratan et al., 1994; Albina et al., 1993; Estevez et al., 1995; Lin et al., 1995; Bonfoco et al., 1995; Palluy and Rigaud, 1996; for review see Nicotera et al., 1995). Precisely how the development of apoptosis in ischemia relates to caspase activation (Hara et al., 1996) and cleavage of PARS remains for further study.

Acknowledgements

Some of the studies described herein were supported by NIH Grant NS10828, the MGH Interdepartmental Stroke Program Project, NS 26361, and an unrestricted research award in neuroscience from Bristol-Myers Squibb.

References

Akira, T., Henry, B. and Wasterlain, C.G. (1994) Nitric oxide mediates the sustained opening of NMDA receptor-gated ionic channels which follows transient excitotoxic exposure in hippocampal slices. *Brain Res.*, 652: 190–194.

Albina, J.E., Cui, S., Mateo, R.B. and Reichner, J.S. (1993) Nitric oxide-mediated apoptosis in murine peritoneal macrophages. *J. Immunol.*, 150: 5080–5085.

Ankarcrona, M., Dypbukt, J.M., Bonfoco, E., Zhivotovsky, B., Orrenius, S., Lipton, S.A. and Nicotera, P. (1995) Glutamate-induced neuronal death: A succession of necrosis or apoptosis depending on mitochondrial function. *Neuron.*, 15: 961–973.

Ankarcrona, M., Dypbukt, J.M., Brune, B. and Nicotera, P. (1994) Interleukin-1β-induced nitric oxide production activates apoptosis in pancreatic RINm5F cells. *Exp. Cell. Res.*, 213, 172–177.

Ayata, C., Ayata, G., Hara, H., Matthews, R.T., Beal, M.F., Ferrante, R.J., Endres, M., Kim, A., Christie, R.H., Waeber, C., Huang, P.L., Hyman, B.T. and Moskowitz, M.A. (1997) Mechanisms of reduced striatal NMDA excitotoxicity in type I nitric oxide synthase knock-out mice. *J. Neurosci.*, 17: 6908–6917.

Beckman, J.S. (1994) Peroxynitrite versus hydroxyl radical: The role of nitric oxide in superoxide-dependent cerebral injury. *Ann. NY. Acad. Sci.*, 738: 69–75.

Beckman, J.S., Ye, Y.Z., Chen, J. and Conger, K.A. (1996) The interaction of nitric oxide with oxygen radicals and scavengers in cerebral ischemic injury. *Adv. Neurol.*, 71: 339–350.

Bonfoco, E., Krainc, D., Ankarcrona, M., Nicotera, P. and Lipton, S.A. (1995) Apoptosis and necrosis: Two distinct events induced, respectively, by mild and intense insults with *N*-methy-D-aspartate or nitric oxide/superoxide in cortical cell cultures. *Proc. Natl. Acad. Sci., U.S.A*, 92: 7162–7166.

Bredt, D.S., Glatt, E.C., Hwang, P.M., Fotuhi, M., Dawson, T.M. and Snyder, S.H. (1991) Nitric oxide synthase protein and mRNA discretely localized in neuronal populations of the mammalian CNS together with NADPH diaphorase. *Neuron*, 7: 615–624.

Brenman, J.E., Chao, D.S., Gee, S.H., McGee, A.W., Craven, S.E., Santillano, D.R., Wu, Z., Huang, F., Xia, H., Peters, M.F., Froehner, S.C. and Bredt., D.S. (1996) Interaction of nitric oxide synthase with the postsynaptic density protein PSD-95 and a1-syntrophin mediated by PDZ domains. *Cell*, 84: 757–767.

Chan, P.H. (1996) Role of oxidants in ischemic brain damage. *Stroke*, 24: 1124–1129.

Chan, P.H., Kamii, H., Yang, G., Gafni, J., Epstein, C.J., Carlson, E. and Reola, L. (1993) Brain infarction is not reduced in SOD-1 transgenic mice after a permanent focal cerebral ischemia. *NeuroReport*, 5: 293–296.

Charriaut-Marlangue, C., Margaill, I., Represa, M.A., Popovici, T., Plotkine, M. and Ben-Ari, Y. (1996) Apoptosis and necrosis after reversible focal ischemia: an in situ DNA fragmentation analysis. *J. Cereb. Blood. Flow. Metab.*, 16: 186–195.

Cho, H.J., Xie, Q.W., Calaycay, J., Mumford, R.A., Swiderek, K.M., Lee, T.D. and Nathan, C. (1992) Calmodulin as a tightly bound subunit of calcium-calmodulin independent nitric oxide synthase. *J. Exp. Med.*, 176: 599–604.

Crow, J.P., Spruell, C., Chen, J., Gunn, C., Ishciropoulos, H., Tsai, M., Smith, C.D., Radi, R., Koppenol, W.H. and Beckman, J.S. (1994) On the pH dependent yield of hydroxyl radical products from peroxynitrite. *Free Rad. Biol. Med.*, 16: 331–338.

Dalkara, T. and Moskowitz, M.A. (1994) Complex role of nitric oxide in cerebral ischemia. *Brain Pathol.*, 4: 49–57.

Dalkara, T., Morikawa, E., Moskowitz, M.A. and Panahian, N. (1994) Blood flow-dependent functional recovery in a rat model of cerebral ischemia. *Am. J. Physiol.*, 267: H837–H838.

Dawson, T.M., Dawson, V.L. and Snyder, S.H. (1992) A novel neuronal messenger in brain: The free radical, nitric oxide. *Ann. Neurol.*, 32: 297–311.

Dawson, T.M. and Dawson, V.L. (1995) Nitric oxide: Actions and pathological roles. *The Neuroscientist*, 1: 7–18.

Dawson, V.L., Dawson, T.M., London, E.D., Bredt, D.S. and Snyder, S.H. (1991) Nitric oxide mediate glutamate neurotoxicity in primary cortical culture. *Proc. Natl. Acad. Sci., USA*, 88: 6368–6371.

Dawson, V.L., Dawson, T.M., Bartley, D.A., Uhl, G.R. and Snyder, S.H. (1993) Mechanisms of nitric oxide mediated neurotoxicity in primary brain cultures. *J. Neurosci.*, 13: 2651–2661.

Dawson, V.L., Kizushi, V.M., Huang, P.L., Snyder, S.H. and Dawson, T.M. (1996) Resistance to neurotoxicity in cortical cultures from neuronal nitric oxide synthase-deficient mice. *J. Neurosci.*, 16: 2479–2487.

Dykens, J.A. (1994) Isolated cerebral and cerebellar mitochondria produce free radicals when exposed to elevated Ca^{2+} and Na^{+}: Implications for neurodegeneration. *J. Neurochem.*, 63: 584–591.

Eliasson, M.J.L., Blackshaw, S., Schell, M.J. and Snyder, S.H. (1997a) Neuronal nitric oxide synthase alternatively spliced forms: prominent functional localizations in the brain. *Proc. Natl. Acad. Sci., USA*, 94: 3396–3401.

Eliasson, M.J.L., Sampei, K., Mandir, A.S., Hurn, P.D., Traystman, R.J., Bao, J., Pieper, A., Wang, Z-Q, Dawson, T.M., Snyder, S.H. and Dawson, V.L. (1997b) Poly(ADP-ribose)polymerase gene disruption renderes mice resistant to cerbral ischemia. *Nature Med.*, 3: 1089–1095.

Endres, M., Scott, G.S., Salzman, A.L., Kun, E., Moskowitz, M.A. and Szabó, C. (1998a) Protective effects of 5-iodo-6-amino-1,2-benzopyrone, an inhibitor of poly (ADP-ribose) synthetase against peroxynitrite-induced glial damage and stroke development. Special Report, *Eur. J. Pharmacol., in press.*

Endres, M., Scott, G.S., Namura, S., Salzman, A.L., Hung, P.L., Moskowitz, M.A. and Szabó, C. (1998b) Role of peroxynitrite and neuronal nitric oxide synthatase in the activation of poly (ADP-ribose) synthetase in a murine model of cerebral ischemia-reperfusion. *Neurosci. Lettrs.*, 248: 29–32.

Endres, M., Wang, Z-Q, Namura, S., Waeber, C. and Moskowitz, M.A. (1997) Ischemic brain injury is mediated by the activation of poly(ADP-ribose) polymerase. *J. Cereb. Blood. Flow. Metab.*, 17: 1143–1151.

Estevez, A.G., Radi, R., Barbeito, L, Shin, J.T., Thompson, J.A. and Beckman, J.S. (1995) Peroxynitrite-induced cytotoxicity in PC12 cells: Evidence for an apoptotic mechanism differentially modulated by neurotrophic factors. *J. Neurochem.*, 65, 1543–1550.

Fässler, R., Martin, K., Forsberg, E., Litzenburger, T. and Iglesias, A. (1995) Knockout mice: How to make them and why. The immunological approach. *Int. Arch. Allergy Immunol.*, 106: 323–334.

Garthwaite, J., Charles, S.L. and Chess-Williams, R. (1988) Endothelium-derived relaxing factor release on activation of NMDA receptors suggests role as intercellular messenger in the brain. *Nature*, 336: 385–388.

Gow, A.J., Duran, D., Malcolm, S. and Ischiropoulos, H. (1996) Effects of peroxynitrite-induced protein modifications on tyrosine phosphorylation and degradation. FEBS Lettrs, 385: 63–66.

Griffith, O.W. and Stuehr, D.J. (1995) Nitric oxide synthases: Properties and catalytic mechanism. *Ann. Rev. Physiol.*, 57: 707–736.

Gross, S.S. and Wolin, M.S. (1995) Nitric oxide: Pathophysiological mechanisms. *Ann. Rev. Physiol.*, 57: 737–769.

Hara, H., Huang, P.L., Panahian, N., Fishman, M.C. and Moskowitz, M.A. (1996) Reduced brain edema and infarction volume in mice lacking the neuronal isoform of nitric oxide synthase after reversible MCA occlusion. *J. Cereb. Blood. Flow. Metab.*, 16: 605–611.

Hara, H., Ayata, C., Huang, P.L., Waeber, C., Ayata, G., Fujii, M. and Moskowitz, M.A. (1997a) [^{3}H]L-N^{G}-nitroarginine binding after transient focal ischemia and NMDA-induced excitotoxicity in type I and type III nitric oxide synthase null mice. *J. Cereb. Blood. Flow. Metab.*, 17: 515–526.

Hara, H., Friedlander, R.M., Gagliardini, V., Ayata, C., Fink, K., Huang, Z., Shimizu-Sasamata, M., Yuan, J. and Moskowitz, M.A. (1997b) Inhibition of interleukin 1 β-converting enzyme family proteases reduces ischemic and excitotoxic neuronal damage. *Proc. Natl. Acad. Sci., USA*, 94: 2007–2012.

Huang, P.L., Dawson, T.M., Bredt, D.S., Snyder, S.H. and Fishman, M.C. (1993) Targeted disruption of neuronal nitric oxide synthase gene. *Cell*, 75: 1773–1286.

Huang, P.L., Huang, Z., Mashimo, M., Bloch, K.D., Moskowitz, M,A., Bevan, J.S. and Fishman, M.C. (1995) Hypertension in mice lacking the gene for nitric oxide synthase. *Nature*, 377: 239–242.

Huang, Z., Huang, P.L., Panahian, N., Dalkara, T., Fishman, M.C. and Moskowitz, M.A. (1994) Effects of cerebral ischemia in mice deficient in neuronal nitric oxide synthase. *Science*, 265: 1883–1885.

Huang, Z., Huang, P.L., Ma, J., Meng, W., Ayata, C., Fishman, M.C. and Moskowitz, M.A. (1996) Enlarged infarcts in endothelial nitric oxide synthase knockout mice are attenuated by nitro-L-arginine. *J. Cereb. Blood. Flow. Metab.*, 16: 981–987.

Iadecola, C., Pelligrino, D.A., Moskowitz, M.A. and Lassen, N.A. (1994) Nitric oxide synthase inhibition and cerebrovascular regulation. *J. Cereb. Blood. Flow. Metab.*, 14: 175–192.

Iadecola, C., Xu, X., Zhang, F., el-Fakahany, E.E. and Ross, M.E. (1995) Marked induction of calcium-independent nitric oxide synthase activity after focal cerebral ischemia. *J. Cereb. Blood. Flow. Metab.*, 15: 52–59.

Irikura, K., Huang, P.L., Ma, J., Lee, W.S., Dalkara, T., Fishman, M.C., Dawson, T.M., Snyder, S.H. and Moskowitz, M.A. (1995) Cerebrovascular alterations in mice lacking neuronal nitric oxide synthase gene expression. *Proc. Natl. Acad. Sci., USA*, 92: 6823–6827.

Kader, A., Frazzini, V.I., Solomon, R.A. and Trifiletti, R.R. (1993) Nitric oxide production during focal cerebral ischemia. *Stroke*, 24: 1709–1716.

Kamii, H., Mikawa, S., Murakami, K., Kinouchi, H., Yoshimoto, T., Reola, L., Carlson, E., Epstein, C.J. and Chan, P.H. (1996) Effects of nitric oxide synthase inhibition on brain infarction in SOD-1-transgenic mice following transient focal cerebral ischemia. *J. Cereb. Blood. Flow. Metab.*, 16:1153–1157.

Knowles, R.G., Palacios, M., Palmer, R.M.J. and Moncada, S. (1989) Formation of nitric oxide from L-arginine in the central nervous system: A transduction mechanism for stimulation of the soluble guanylate cyclase. *Proc. Natl. Acad. Sci., USA*, 86: 5159–5162.

Knowles, R.G. and Moncada, S. (1992) Nitric oxide as a signal in blood vessels. *Trends. Biochem. Sci.*, 17: 399–402.

Koh, J. and Choi, D.W. (1988) Vulnerability of cultured cortical neurons to damage by excitotoxins: Differential susceptibility of neurons containing NADPH-diaphorase. *J. Neurosci.*, 8: 2153–2163.

Lafon-Cazal, M., Pietri, S., Culcasi, M. and Bockaert, J. (1993) NMDA-dependent superoxide production and neurotoxicity. *Nature*, 364: 535–537.

Li, Y., Chopp, M., Jiang, N., Yao, F. and Zaloga, C. (1995a) Temporal profile of in situ DNA fragmentation after middle cerebral artery occlusion in the rat. *J. Cereb. Blood. Flow. Metab.*, 15: 389–397.

Li, Y., Chopp, M., Jiang, N., Zhang, Z.G. and Zaloga, C. (1995b) Induction of DNA fragmentation after 10 to 120 minutes of focal cerebral ischemia in the rat. *Stroke*, 26: 1252–1258.

Li, Y., Sharov, V.G., Jiang, N., Zaloga, C., Sabbah, H.N. and Chopp, M. (1995c) Ultrastructural and light microscopic evidence of apoptosis after middle cerebral artery occlusion in the rat. *Am. J. Pathol.*, 146: 1045–1051.

Lin, K.T., Xue, J.Y., Nomen, M., Spur, B. and Wong, P.Y.K. (1995) Peroxynitrite-induced apoptosis in HL-60 cells. *J. Biol. Chem.*, 270: 16487–16490.

Liu, R.H. and Hotchkiss, J.H. (1995) Potential genotoxicity of chronically elevated nitric oxide: a review. *Mutation. Res.*, 339: 73–89.

Lo, E.H., Hara, H., Rogowska, J., Trocha, M., Pierce, A.R., Huang, P.L., Fishman, M.C., Wolf, G.L. and Moskowitz, M.A. (1996) Temporal correlation mapping analysis of the hemodynamic penumbra in mutant mice deficient in endothelial nitric oxide synthase gene. *Stroke*, 27: 1381–1385.

Ma, J., Ayata, C., Huang, P., Fishman, M.C. and Moskowitz, M.A. (1996) Regional cerebral blood flow response to vibrissal stimulation in mice lacking type I NOS gene expression. *Am. J. Physiol.*, 270: H1085–H1090.

Malinski, T., Bailey, F., Zhang, Z.G. and Chopp, M. (1993) Nitric oxide measured by a porphyrinic microsensor in rat brain after transient middle cerebral artery occlusion. *J. Cereb. Blood. Flow. Metab.*, 13: 355–358.

Meng, W., Ma, J., Ayata, C., Hara, H., Huang, P.L., Fishman, M.C. and Moskowitz, M.A. (1996) ACh dilates pial arterioles in endothelial and neuronal NOS knockout mice by NO-dependent mechanisms. *Am. J. Physiol.*, 271: H1145–H1150.

Morikawa, E., Moskowitz, M.A., Huang, Z., Yoshida, T., Irikura, K. and Dalkara, T. (1994) L-arginine infusion promotes nitric oxide-dependent vasodilatation, increases regional cerebral blood flow, and reduces infarction volume in the rat. *Stroke*, 25: 429–435.

Murphy, S., Simmons, M.L., Agullo, L., Garcia, A., Feinstein, D.L., Galea, E., Reis, D.J., Minc-Golomb, D. and Schwartz, J.P. (1993) Synthesis of nitric oxide in CNS glial cells. *Trends Neurosci.*, 16: 323–328.

Nicotera, P., Bonfoco, E. and Brune, B. (1995) Mechanisms for nitric oxide-induced cell death – involvement of apoptosis. *Adv. Neuroimmunol.*, 5: 411–420.

Niu, X.F., Smith, C.W. and Kubes, P. (1994) Intracellular oxidative stress induced by nitric oxide synthesis inhibition increases endothelial cell adhesion to neutrophils. *Circ. Res.*, 74: 1133–1140.

Palluy, O. and Rigaud, M. (1996) Nitric oxide induces cultured cortical neuron apoptosis. *Neurosci. Lett.*, 208: 1–4.

Panahian, N., Yoshida, T., Huang, P.L., Hedley-Whyte, E.T., Dalkara, T., Fishman, M.C. and Moskowitz, M.A. (1996) Attenuated hippocampal damage after global cerebral ischemia in mice mutant in neuronal nitric oxide synthase. *Neurosci.*, 72: 343–354.

Pogun, S., Dawson, V. and Kuhar, M.J. (1994) Nitric oxide inhibits ^{3}H-glutamate transport in synaptosomes. *Synapse*, 18: 21–26.

Ratan, R.R., Murphy, T.H. and Baraban, J.M. (1994) Oxidative stress induces apoptosis in embryonic cortical neurons. *J. Neurochem.*, 62: 376–379.

Rubin, E.M. and Barsh, G.S. (1996) Biological insights through genomics: Mouse to man. *J. Clin. Invest.*, 97: 275–280.

Szabó, C., Lim, L.H., Cuzzocrea, S., Getting, S.J., Zingarelli, B., Flower, R.J., Salzman, A.L. and Peretti, M. (1997) Inhibition of poly (ADP-ribose) synthetase exerts anti-inflammatory effects and inhibits neutrophil recruitment. *J. Exp. Med.*, 186: 1041–1049.

Szabó, C., Zingarelli, B., O'Connnor, M. and Salzman, A.L. (1996) DNA strand breakage, activation of poly-ADP ribosyl synthetase, and cellular energy depletion are involved in the cytotoxicity in macrophages and smooth muscle cells exposed to peroxynitrite. *Proc. Natl. Acad. Sci. USA*, 93: 1753–1758.

Thiemermann, C., Bowles, J., Myint, F.P. and Vane, J.R. (1997) Inhibition of the activity of poly (ADP-ribose) synthetase reduces ischemia-reperfusion injury in the heart and skeletal muscle. *Proc. Natl. Acad. Sci. USA*, 94: 679–683.

Thomas, K.R. (1995) The knockout mouse: Six years old and growing stronger. *Am. J. Respir. Cell. Mol. Biol.*, 12: 461–463.

Trotti, D., Rossi, D., Gjesdal, O., Levy, L.M., Racagni, G., Danbolt, N.C. and Volterra, A. (1996) Peroxynitrite inhibits glutamate transporter subtypes. *J. Biol. Chem.*, 271: 5976–5979.

Uemura, Y., Kowall, N.W. and Beal, M.F. (1990) Selective sparing of NADPH-diaphorase-somatostatin-neuropeptide Y neurons in ischemic gerbil striatum. *Ann. Neurol.*, 27: 620–625.

Yang, G., Chan, P., Chen, J., Carlson, E., Chen, S.F., Weinstein, P., Epstein, C.J. and Kamii, H. (1994) Human copper-zinc superoxide dismutase transgenic mice are highly resistant to reperfusion injury after focal cerebral ischemia. *Stroke*, 25: 165–170.

Yoshida, T., Limmroth, V., Irikura, K. and Moskowitz, M. (1994) The NOS inhibitor 7-nitroindazole decreases focal infarct volume but not the response to topical acetylcholine in pial vessels. *J. Cereb. Blood. Flow. Metab.*, 14: 924–929.

Zhang, F., White, J.G. and Iadecola, C. (1994a) Nitric oxide donors increase blood flow and reduce brain damage in focal cerebral ischemia: evidence that nitric oxide is beneficial in the early stages of cerebral ischemia. *J. Cereb. Blood. Flow. Metab.*, 14: 217–26.

Zhang, J., Dawson, V.L., Dawson, T.M. and Snyder, S.H. (1994b) Nitric oxide activation of poly (ADP-ribose) synthase in neurotoxicity. *Science*, 263: 686–689.

Zhang, Z.G., Reif, D., Macdonald, J., Tang, W.X., Kamp, D.K., Gentile, R.J., Shakespeare, W.C., Murray, R.J. and Chopp, M. (1996) ARL 17477, a potent and selective neuronal NOS inhibitor decreases infarct volume after transient middle cerebral artery occlusion in rats. *J. Cereb. Blood. Flow. Metab.*, 16: 599–604.

R.R. Mize, T.M. Dawson, V.L. Dawson and M.J. Friedlander (Eds.)
Progress in Brain Research, Vol 118

CHAPTER 17

Glial glutamate transport as target for nitric oxide: consequences for neurotoxicity

Zu-Cheng Ye and Harald Sontheimer*

Department of Neurobiology, The University of Alabama at Birmingham, AL 35294, USA

Abstract

Overwhelming evidence suggest that accumulations of extracellular glutamate are toxic to neurons. It has also been proposed that astrocytes protect neurons from glutamate toxicity by removal of glutamate from extracellular space. By using co-cultures of hippocampal neurons and astrocytes, we studied the influence of astrocytes on neuronal excitotoxicity. Moreover, we evaluated the role of nitric oxide and pro-inflammatory cytokines on astrocytic glutamate transport.

Introduction

Glutamate is the major excitatory neurotransmitter in the central nervous system. However, at the same time, elevated levels of glutamate have been demonstrated to result in wide spread neurotoxicity (for a review, see Choi, 1988; Thomas, 1995). Accumulating evidences suggest that glutamate toxicity may be a common final pathway in many CNS diseases (for a review, see Lipton and Rosenberg, 1994). Fortunately, glutamate is tightly controlled in the healthy brain and extracellular glutamate concentrations ($[Glu]_o$) are maintained close to 1 μM through the combined re-uptake of glutamate into neurons and glial cells (Nicholls and Attwell, 1990). Five types of Na^+-dependent transporters have been identified so far (Storck et al., 1992; Pines et al., 1992; Kanai and Hediger, 1992; Fairman et al., 1995; Arriza et al., 1997). They are believed to operate in parallel to sequester glutamate released during synaptic transmission and prevent excessive accumulations of glutamate in the extracellular space (Rothstein et al., 1994). Overwhelming evidence suggests that astrocytes are primarily responsible for uptake of neuronally released glutamate (Schousboe and Westergaard, 1995) and, GLT-1, primarily expressed in astrocytes (Rothstein et al., 1994), has been suggested to play a particularly important role in the control of extracellular glutamate levels. GLT-1 knockout results in elevations of $[Glu]_o$ and severe neuronal deficits (Rothstein et al., 1996; Tanaka et al., 1997).

The physiology of glial glutamate uptake has been well studied and suggests that glutamate uptake is electrogenic, most likely involving the co-transport of one glutamate with two or three Na^+ and 1 H^+ in exchange for 1 K^+ (Bouvier et al., 1992; Zerangue and Kavanaugh, 1996). This stoichiometry makes glutamate uptake sensitive to changes in membrane potential, transmembrane Na^+, K^+ and pH gradients. However, additional modulatory pathways may exist (for a review, see Gegelashvili and Schousboe, 1997). Theoretical considerations suggest that under conditions of energy failure, glial glutamate uptake can reverse,

* Corresponding author. Tel.: +1 205 934 4454; fax: +1 205 934 6571.

leading to the non-vesicular release of glutamate. Such release is believed to be contributing to excitotoxic neuronal injury (Szatkowski et al., 1990; Longuemare and Swanson, 1995) and may be a complicating factor in ischemia or trauma.

In the CNS, glial cells are important participants in the immune response. They are both target of and source for many cytokines (Benveniste, 1992). Glial-released cytokines have been implicated in the neuropathogenesis of AIDS dementia (Koka et al., 1995) and in the evolution of different acute or chronic neuropathological processes such as ischemia, brain trauma, multiple sclerosis and Alzheimer's disease (Mrak et al., 1995). Neuronal loss associated with these pathological processes displays many features that are compatible with glutamate neurotoxicity. This has led to speculations that altered glial glutamate uptake may contribute to the neuronal loss under such disease conditions. In light of the fact that cytokines can induce synthesis and release of nitric oxide (NO), which itself is believed to be a key player in excitotoxic cell injury, we were interested to learn whether astrocytic glutamate uptake may be modulated by pro-inflammatory cytokines or NO, both of which are implicated in CNS injury.

Astrocytic glutamate uptake prevents excitotoxicity

Astrocytes are highly effective glutamate sponges

We recently demonstrated that sera used as supplement in the cell culture medium of neurons and glial cells contains between 300 μM and 1.2 mM glutamate (Ye and Sontheimer, 1998). At the typical 1:10 dilution, glutamate contents from serum would be expected to be neurotoxic. It has previously been demonstrated that astrocytes can effectively transport glutamate through Na^+-dependent transporters (for reviews, see Kanai et al., 1993; Schousboe and Westergaard, 1995). We thus set out to utilize astrocytic glutamate uptake as a means to obtained glutamate-free serum-containing culture medium. Therefore, the rate of glutamate depletion from the medium by astrocytes was determined in 75 mm^2 flasks of confluent astrocytes after application of 30 ml fresh medium containing 10% fetal bovine serum (FBS). Samples were taken and glutamate content was measured using the bioluminescence method as described by Fosse et al. (1986). The resulting glutamate concentrations were plotted as a function of time after addition to the flask (Fig. 1A). Glutamate concentrations in the medium showed a rapid initial decline and, within 30 min, glutamate concentrations dropped from 85 to 20 μM. After 2–3 h, glutamate concentrations reached values below 1 μM. Repeated application of exogenous glutamate from stock solutions showed identical results (Fig. 1B). Routinely, 30–40 ml glutamate-depleted media (GDM) could be collected every 4–6 h. These data demonstrate that astrocytes are capable of sequestering glutamate in a very efficient manner and imply that in situations where neurons and astrocytes are in close proximity, astrocytes may maintain glutamate at low micromolar levels, consistent with the notion that astrocytic glutamate transport is neuroprotective.

Hippocampal neurons are highly susceptible to glutamate toxicity

Time-lapse video microscopy with images captured every second was used to assess effects of serum glutamate or glutamate added from stock solution on neuronal morphology and survival. For these experiments, mixed neuronal/glial cultures were cultured in GDM containing less than 1 μM glutamate. Representative hippocampal neurons grown for 14 days in GDM are shown in Fig. 2A. This culture was maintained in a time lapse system for 24 h without any apparent neuronal damage. Subsequently, medium was replaced with fresh GDM for another 24 h (Fig. 2B), still without any signs of neurotoxicity. Finally, the media was switched to regular 10% FBS medium which contained ~90 μM glutamate. Pronounced morphological changes including neuronal swelling and retraction of processes were observed within 5–60 min following the medium change (Fig. 2C). These signs of neuronal disinte-

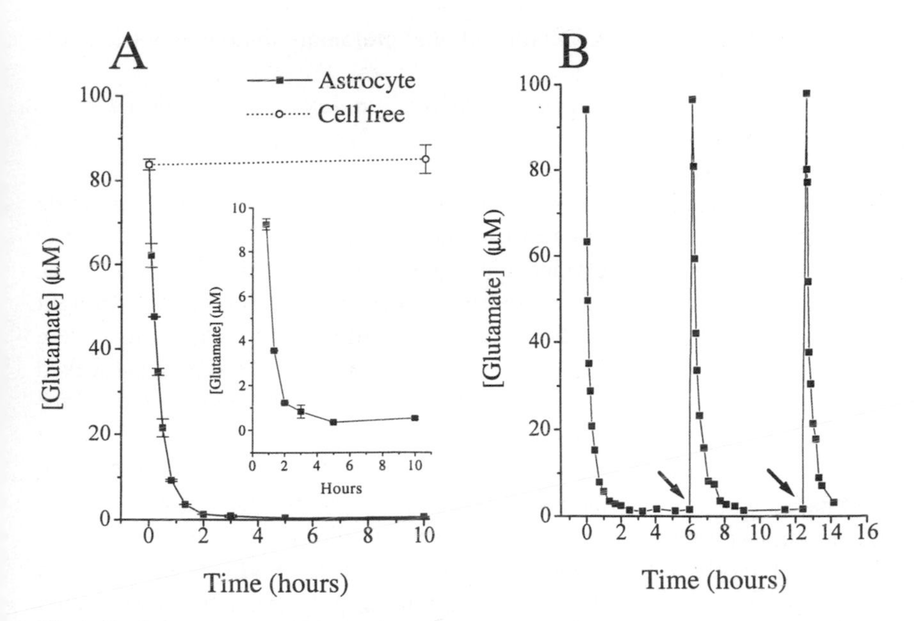

Fig. 1. Depletion of serum glutamate by astrocytes. (**A**) Culture media (30 ml) containing 10% fetal bovine serum was added into two 75 cm^2 flasks containing confluent hippocampal astrocytes from P2 rats. Samples of media (0.1 ml) were taken at the time points indicated and the glutamate content determined. [Glutamate] was close to 90 μM initially and dropped below 1 μM within 2–3 h. The inset of Fig. 1 expands the part of the curve (> 1 h) that corresponds to the most physiologically relevant glutamate concentrations (Mean ± SD, representative of seven experiments). No [Glutamate] changes were observed in cell free control flasks. (B) Glutamate depletion is reproducible and independent of glutamate source. 10% FBS media (30 ml) was applied as in (A), subsequently, two additional applications were made (arrows) from a stock glutamate solution (0.27 ml × 10 mM) at 6 h and 12.5 h, respectively. Astrocytes depleted exogenous glutamate in the same manner as serum glutamate.

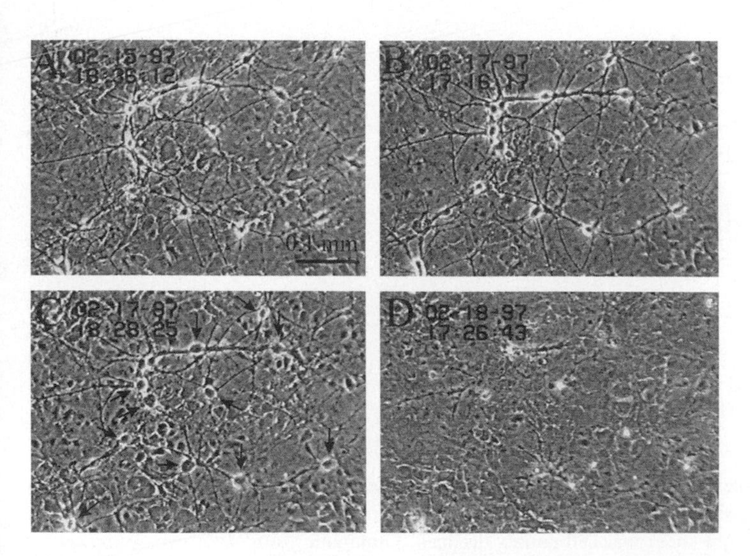

Fig. 2. Time lapse microscopy recordings of neurons before and after exposure to GDM and regular 10% FBS media. (A–B) A 14 DIV culture was maintained under time-lapse observation in GDM for 48 h, all neurons were healthy. (C–D) culture exposed to 10% FBS media, panel C (65 min after): visible morphological changes in most neurons in the field (arrows); panel D (24 h after): eventually all neurons died (Scale bar = 100 μm for all panels).

gration continued with longer exposure time and most neurons vanished after 24 h (Fig. 2D).

To assess glutamate toxicity in a more quantitative way, we used trypan blue exclusion assays to determine the relative toxicity of defined glutamate exposures by determining the ratio of live to dead neurons. The resulting dose-response curves showed a dose-dependent increase in the percentage of dead cells with a toxicity ED_{50} of ~ 10 μM (Fig. 3A). Neurotoxicity was identical whether glutamate was supplied as serum contaminant or by addition from a stock solution (Fig. 3A). Interestingly, glutamate toxicity changed during prolonged cell culturing. Initially, most neurons were relatively resistant to glutamate, whereas after >14 days in vitro (DIV), most neurons were highly susceptible to glutamate with ED_{50} values of 6–15 μM (Fig. 3B). Glutamate neurotoxicity was primarily mediated by NMDA receptors, as it could be entirely prevented if cells were treated with the NMDA antagonist MK801 (5 μM, data not shown).

Regulation of glial glutamate uptake by nitric oxide

Astrocytic glutamate uptake is compromised by inflammatory cytokines

Above data suggest that astrocytic glutamate transport effectively removes glutamate from the extracellular milieu. Modulation of astrocytic glutamate transport is poorly understood. In light of the evidences that injuries to the nervous system cause release of inflammatory cytokines, we seeked to establish whether cytokines can influence astrocytic glutamate uptake.

Primary astrocytic cultures were incubated for 24 h with three cytokines that have been implicated in the induction of NO production, namely tumor necrosis factor α (TNF-α), interferon γ (IFN-γ) and interleukin 1β (IL-1β). Subsequently total astrocytic glutamate uptake was determined and compared to untreated sister cultures. Each of the three cytokines attenuated glutamate uptake in a dose-dependent manner with effects ranging

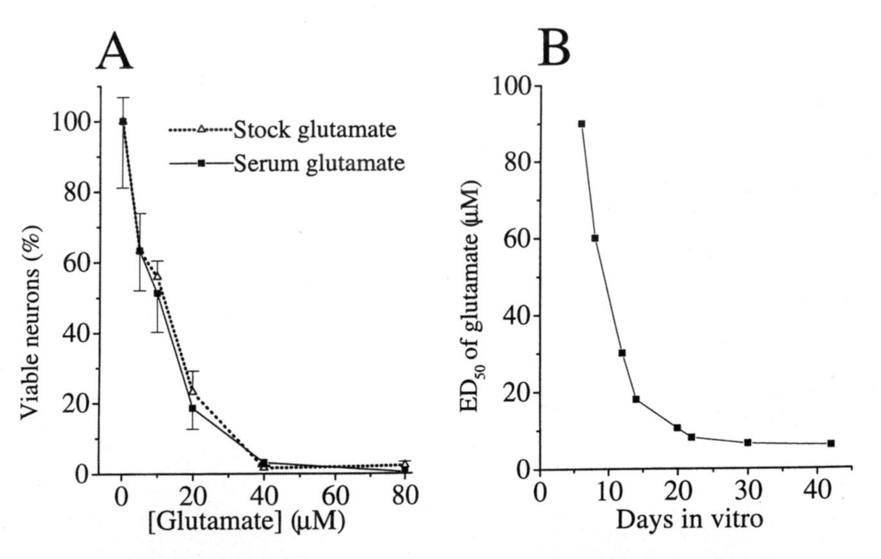

Fig. 3. (A) Serum glutamate toxicity is indistinguishable from toxicity by exogenous glutamate. Sister cultures of neurons (20 DIV) received glutamate either from serum or from glutamate stock solution. Neurotoxicity was determined 24 h after receiving glutamate at the concentrations indicated. With increasing [Glu], viable neuronal cell counts declined. Glutamate from these two sources were equally toxic. ($ED_{50} = \sim 10$ μM). (Mean ± SE, field = 14–16). (B) Vulnerability of neurons to glutamate increases with in vitro age. Relative toxicity of glutamate was assessed in sister cultures of neurons from the same preparation maintained in vitro for the times indicated. The ED_{50} for glutamate toxicity was determined for each age group and plotted as a function of in vitro age. Vulnerability to glutamate developed in the first 20 DIV.

from 26.9 ± 6.1% to 59.8 ± 4.7% for TNF-α (n = 6 for each concentration: 0.1–100 ng/ml), 15.0 ± 6.0% to 35.6 ± 2.6% for IL-1β (n = 6 for each concentration: 0.1–100 ng/ml) and 11.7 ± 5.8% to 22.6 ± 4.9% for IFN-γ (n = 6 in each concentration group: 20–500 U/ml:). Experiments were repeated in at least 3 cell preparations. Fig. 4A shows a representative experiment in which each of the cytokines was applied at its most effective concentration, which we determined to be 2 ng/ml, 1ng/ml, 100 U/ml for TNF-α, IL-1β, and IFN-γ, respectively. TNF-α exhibited by far the most potent effect and attenuated glutamate uptake by ~60%. We started to detect significant effects after 4 h of incubation with these cytokines (Fig. 4C), while shorter incubations were ineffective (data not shown). Some variability existed in the degree of effectiveness between different batches of cytokines and cell preparations. In preparation where a single cytokine application showed only moderate effect, application of two cytokines together was synergistic (Fig. 4B).

Inhibition of glutamate uptake by nitric oxide

TNF-α, IL-1β, and IFN-γ have been reported to promote the production of NO in astrocytes by stimulating Ca^{2+} independent, inducible NOS (iNOS) (Chao et al., 1996; Nomura and Kitamura, 1993). We conducted two types of experiments to determine whether the observed cytokine effects on astrocytic glutamate uptake could have been mediated by NO. In one series of experiments, astrocytes were maintained for 4 or 24 h in the presence of 1 mM N^{ω}-nitro-L-arginine (LNA), an inhibitory substrate for NOS, while cells were stimulated with cytokines as described above. LNA completely eliminated cytokine effects on glutamate uptake under these conditions (Fig. 4C), LNA restored normalized uptake from 0.769 ±

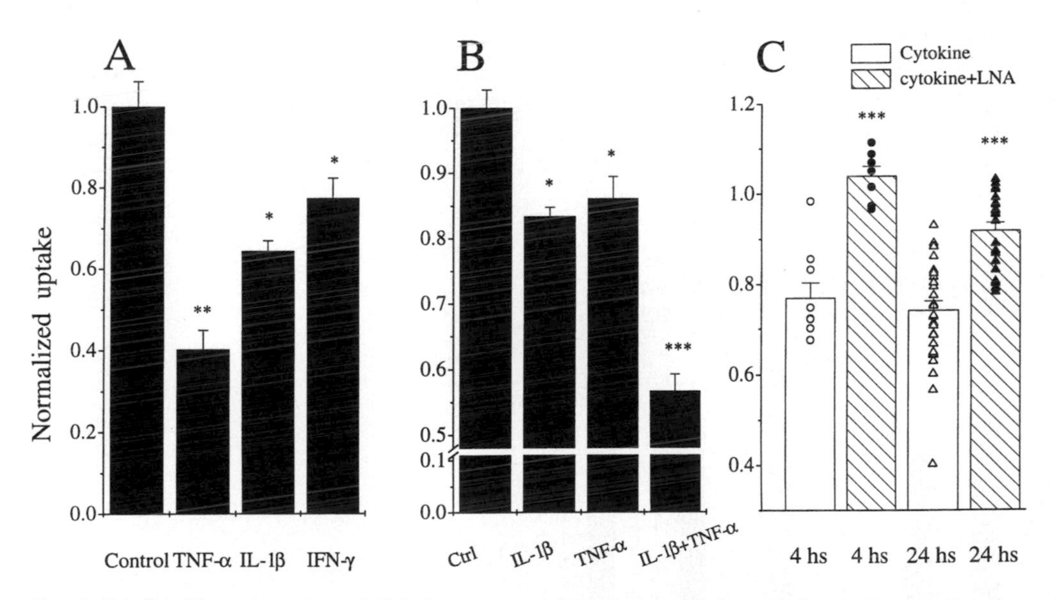

Fig. 4. (A) Cytokine attenuation of glial glutamate uptake. A representative experiment showing the effects of 24 h incubation of hippocampal astrocytes with 2 ng/ml TNF-α, 1ng/ml IL-1β or 100 U/ml IFN-γ (each n = 6) on glutamate uptake. Uptake rates were determined after 10 minutes and were normalized to untreated control sister cultures (n = 12). (*$P<0.05$; **$P<0.001$ Student's two tailed t-test) . (B) Synergistic effect between cytokines on reducting glutamate uptake. 24 h incubation with 250 ng/ml TNF-α + 20 ng/ml IL-1β exert larger attenuation than individual cytokine (n = 6, ANOVA, *$P<0.05$, **$P<0.001$). (C) Blockage of cytokine effect by coculture with the NOS inhibitor Nω-L-nitro-arginine. Data pooled from nine experiments utilizing TNF-α (0.2–250 ng/ml), IL-1β (0.1–20 ng/ml), IFN-γ (20–500 U/ml). All results had been normalized to their corresponding controls and most points in cytokine + LNA group had corresponding cytokine alone point been plotted. (***$P<0.0001$ Student's two tailed t-test).

0.034 to 1.04 ± 0.021 (average of mean values from nine and seven groups respectively, each group $n = 6$) in 4 h incubation groups. Similarly, 1mM LNA restored uptake in astrocytes incubated with cytokines for 24 h from 0.741 ± 0.021 to 0.919 ± 0.017 (pooled from the mean values of 31 and 22 groups respectively, each group with $n = 4$–6). Interestingly, if LNA was applied just 1 h prior to uptake, it could block cytokine effects as if they were applied concomitant with cytokines. However, LNA was ineffective if it was withdrawn prior to glutamate uptake. This suggests that the presence of NOS inhibitor during uptake rather than their presence during cytokine incubation were essential to be effective.

A second series of experiments was designed to further substantiate the role of NO in glutamate uptake. Since the above cytokine studies using LNA suggested that the production of NO during rather than prior to uptake is required for cytokine function on uptake, we studied glutamate uptake in the presence of two NO donors, namely 3-morpholinosydnonimine (SIN-1) and *S*-nitroso-*N*-acetylpenicillamine (SNAP). In the presence of 100 U/ml superoxide dismutase (SOD) and catalase (CAT), SIN-1 and SNAP each markedly and in a dose-dependent manner reduced glutamate uptake (Fig. 5A,B). These effects were of comparable magnitude to those observed with TNF-α, IL-1β or IFN-γ (Fig. 4A). Interestingly, SIN-1 was only effective in the simultaneous presence of 100 U/ml SOD/CAT (Fig. 5B) or 5000 U/ml SOD alone, while SNAP effects were observed with and without SOD/CAT (Fig. 5A). This is an important observation since SIN-1 is believed to simultaneously release NO and OO^- and resulting in the generation of peroxynitrite ($ONOO^-$) (Pryor and Squadrito, 1995). However, SOD inhibits the production of peroxynitrite by SIN-1 (Lipton et al., 1993). The failure of SIN-1 to affect glutamate uptake in the absence of SOD/CAT indicated that Sin-1 effects on glutamate uptake were mediated by NO rather than $ONOO^-$. Consistent with this notion, scavenging NO by exogenous hemoglobin (2 mg/ml) totally eliminated the effects of SIN-1 (Fig 5B, in the presence of SOD/CAT), again supporting the notion that the inhibitory effects on glutamate uptake were mediated by NO.

Since the effects of cytokines were presumed to have been mediated by NO, we tried to directly apply NO gas and study glutamate uptake. As

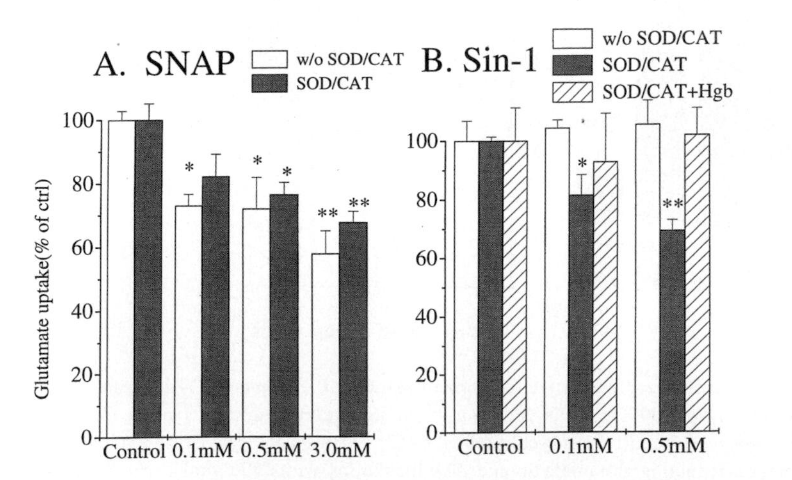

Fig. 5. NO donors mimic the inhibitory effects of cytokines on astrocytic glutamate uptake. (**A**) *S*-nitroso-*N*-acetylpenicillamine (SNAP) with or without SOD/CAT. (**B**) 3-morpholinosydnonimine (SIN-1) with or without 100 U/ml superoxide dismutase/catalase (SOD/CAT) and with hemoglobin + SOD/CAT. All data normalized to corresponding control level (all $n = 6$, $*P < 0.05$, $**P < 0.01$, student *t*-test).

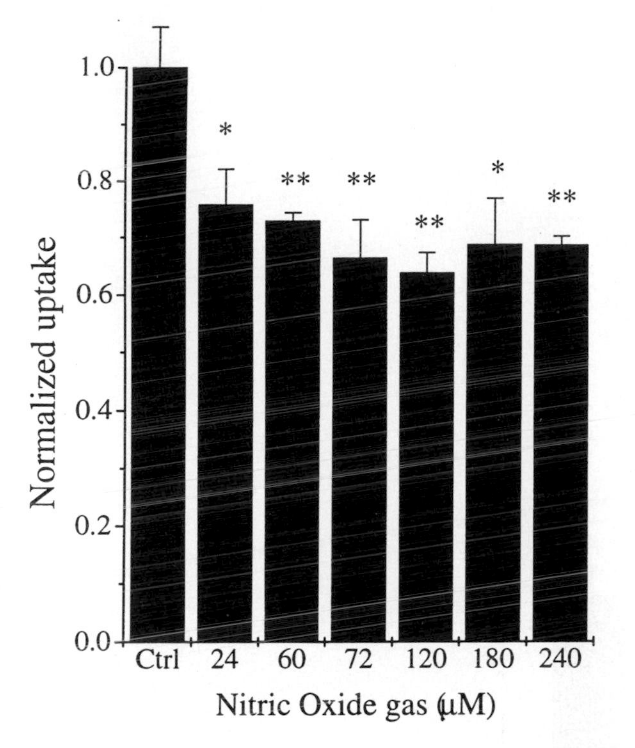

Fig. 6. NO gas attenuates glutamate uptake into astrocytes. NO gas saturated distilled water (compensated with equal amount of 2X uptake solution) was applied to HEPES buffered uptake solution at room temperature immediately followed with addition of ^{3}H-D-Asp tracer. Data were pooled from three experiments (control n = 18, other n = 6–12, $*P<0.05$, $**P<0.01$ by student's t-test).

shown in Fig. 6, NO concentrations between 24 and 240 μM significantly reduced uptake rates, with effects ranging from 24.3–36.2%.

Nitric oxide effects require activation of guanylyl cyclase

It has been well recognized that most NO effects are mediated by cGMP. We thus tested the effect of 1H-(1,2,4)oxadiazolo(4,3-a)quinoxalin-1-one (ODQ), which is a specific guanylyl cyclase inhibitor (Garthwaite et al. 1995). Preincubation with 10 μM ODQ per se reduced glutamate uptake, but after ODQ pretreatment, neither NO gas nor SNAP could further reduce glutamate uptake (Fig. 7A, B). These data suggest that the modulatory effect of NO on glutamate transport is mediated by a pathway that involves activation of guanylyl cyclase.

Furthermore, prolonged incubation ($\geqslant$90 min) with sodium nitroprusside (SNP), another commonly used NO donor, was able to induce the release of preloaded ^{3}H-D-Asp in a dose dependent manner, with the largest effects observed at 100 μM (Fig. 7C). This effect could be abolished by pretreatment of ODQ (Fig. 7D) or methylene blue (10 μM, 30 min, data not shown). In addition, potassium ferricyanide, an agent chemically similar to SNP that does not release NO had no effect.

Compared to the NO involvement in the reduction of glutamate uptake, the effects of SNP on glutamate release suggest that NO not only reduces uptake but also dramatically enhances release of intracellular glutamate (which can reach mM level in astrocytes), in a NO/guanylyl cyclase dependent manner. Similarly, induction of iNOS by lipopolysaccharides (LPS) in astrocytes was also found to increase glutamate release in a guanylyl cyclase dependent fashion (unpublished observation). Taken together, NO can induce a reduction in total net uptake of glutamate into astrocytes.

Summary and discussion

Most neurons are highly susceptible to glutamate and can die following even short exposures to glutamate. To prevent such excitotoxic neuronal injury, $[Glu]_o$ is maintained at low micromolar levels in the normal brain. Astrocytes express high affinity glutamate transporters in vitro and maintain $[Glu]_o$ at <1 μM in the culture medium, a concentration similar to that reported for glutamate in cerebral spinal fluid (CSF). If hippocampal neurons are challenged with exogeneous glutamate or serum-borne glutamate, neurons die with apparent EC_{50} values for glutamate of 6–15 μM. Healthy growth of neurons in GDM demonstrated that depletion of glutamate contents in serum-containing media is the major benefit of astrocytes to neuronal survival. Indeed, astrocytes can deplete repeatedly added glutamate to levels

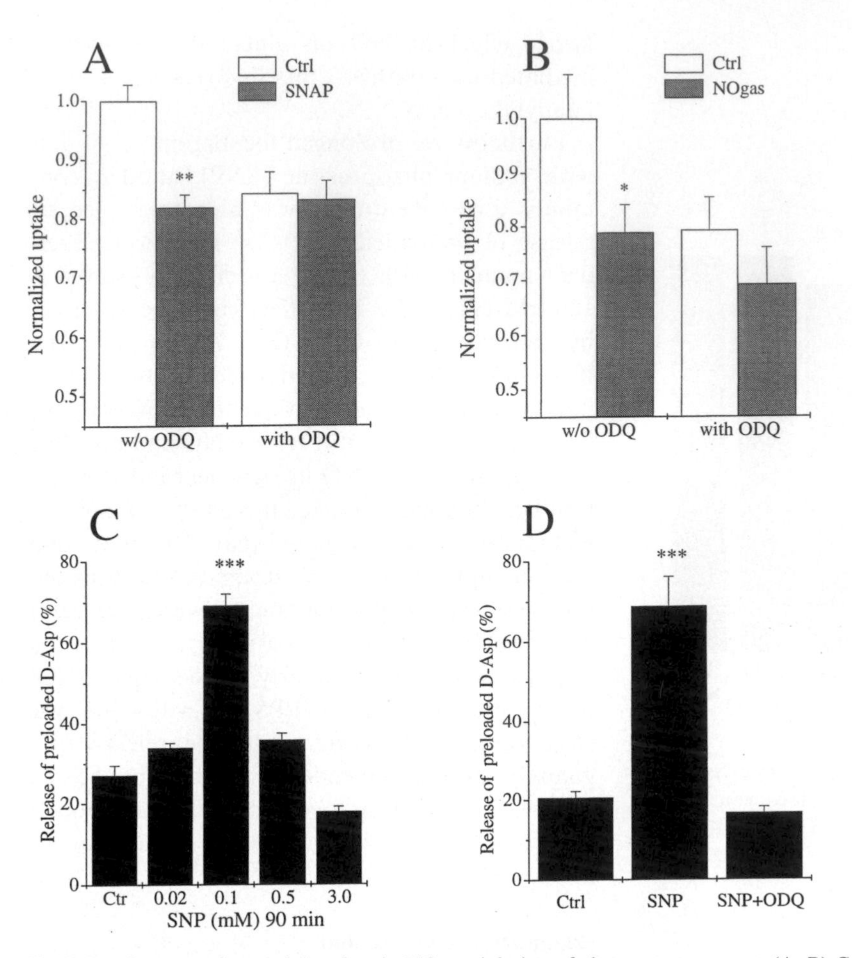

Fig. 7. Involvement of guanylyl cyclase in NO modulation of glutamate transport. (A, B) Cells were pretreated for 30 min with 10 μM 1H-(1, 2, 4)oxadiazolo(4, 3-a)quinoxalin-1-one (ODQ) before assessing uptake. Neither SNAP (1mM) nor NO gas (72 μM) can further downregulate uptake in ODQ treated cells. (C) Prolonged incubation ($\geqslant$ 90 min) with sodium nitroprusside (SNP) stimulated release of preloaded ^{3}H-D-Asp from astrocytes in a dose-dependent manner. (D) 30 min pretreatment with 10 μM ODQ blocked SNP induced release of ^{3}H-D-Asp (*$P<0.05$, **$P<0.001$ A, B by t-test, C, D by ANOVA).

$<1\mu$m in vitro, a value in close agreement with glutamate concentration in the CSF or extracellular fluid in vivo.

Pro-inflammatory cytokines, including tumor necrosis factor α (TNF-α), interleukin 1β (IL-1β) and interferon γ (IFN-γ) each reduce astrocytic glutamate transport in a dose-dependent manner. Cytokine effects could be inhibited by N^{ω}-nitro-L-arginine or by reducing the L-arginine content in the media. NO gas and several NO donors including 3-morpholinosydnonimine (SIN-1) and *S*-nitroso-*N*-acetylpenicillamine (SNAP) were found to attenuate glutamate uptake even if the generation of peroxynitrite was inhibited by application of superoxide dismutase/catalase, suggesting that NO rather than $ONOO^{-}$ was the mediator of these effects, in addition, the specific guanylyl cyclase inhibitor ODQ could block NO gas or SNAP induced attenuation of glutamate uptake. Furthermore, prolonged application of SNP (>90 min, $\sim$100 μM) was found to induce the release of preloaded ^{3}H-D-Asp from astrocytes, which were abolished by pretreatment with the guanylyl cyclase inhibitors, methylene blue or ODQ, again

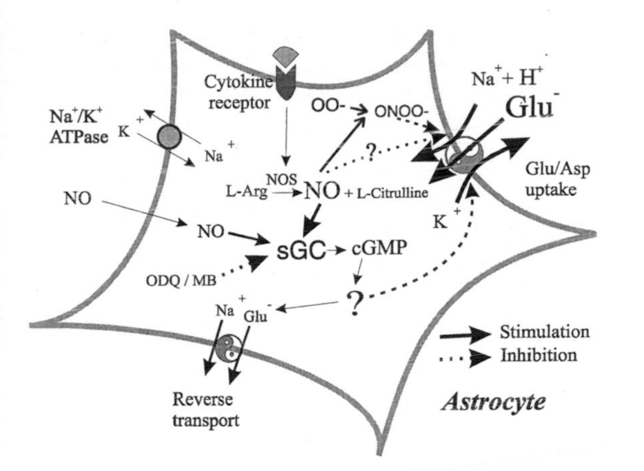

Fig. 8. Proposed pathway invloved in cytokine/nitric oxide mediated attenuation of glutamate uptake. SGC: soluble guanylyl cyclase; ODQ: 1H-(1, 2, 4)oxadiazolo(4, 3-a) quinoxalin-1-one (ODQ); MB: methylene blue; NO: nitric oxide; NOS: NO synthase.

indicating the participation of NO/cGMP in these events. Taken together, these data suggest that cytokines/NO inhibit astrocytic glutamate uptake and the underlying pathway involves activation of guanylyl cyclase/cGMP (Fig. 8).

In line with our data, previous reports showed that TNF-α and IL-1β inhibit glutamate uptake in fetal human brain cultures and contribute to neurotoxicity (Chao et al., 1995, Fine et al., 1996). Cytokine effects on astrocytic glutamate uptake required exposure of >2 h, while NO donors exerted their effect on glutamate uptake immediately. Our data suggests that cytokine effects were mediated by the release of NO through stimulation of NOS. Astrocytes express both the constitutive (cNOS) and inducible (iNOS) form of NOS. Usually, astrocytic cNOS is present at only very low levels. Cytokines, however, induce the expression of iNOS (Nomura and Kitamura, 1993). It is thus not surprising that cytokine effects develop over a period of time as their effectiveness involves induction of iNOS expression preceding the release of NO. We observed that cytokine induced attenuation of glutamate uptake decreased with in vitro age, whereas the effects of NO donors persisted, up to eight weeks in vitro. It is also worth noting that the application of hemoglobin failed to block cytokine effects while it blocked NO donors effects (data not shown), suggesting that NO produced inside astrocytes is sufficient to exert functional modulation.

It has been shown that $ONOO^-$ inhibits glutamate uptake into liposomes, with 50% inhibition after 20 s (Trotti et al., 1996). In our cultured hippocampal astrocytes, mM concentration of Sin-1, which mainly forms $ONOO^-$ in the absence of SOD, did not inhibit glutamate uptake even after 10 min., suggesting that glutamate transports in astrocytes is relatively insensitive to $ONOO^-$. However, damages in the plasma membrane and impaired glutamate uptake were also observed at prolonged incubation with Sin-1 in the absence of SOD (data not shown). Indeed, accumulating evidences suggest that protein peroxynitration may play important roles under pathological conditions (refer to Chapter 19).

Glutamate appears to be elevated in the extracellular space in animal models of stroke/ischemia and in patients with epileptic seizures (Petroff et al., 1995). These conditions, together with acute CNS injury and inflammation, are known to cause the release of cytokines from astrocytes and microglial cells (Benavides and Toulmond, 1993). Our data imply that elevated $[Glu]_o$ can be at least partially due to the cytokine-mediated down-regulation of astrocytic glutamate uptake, which could be further exacerbated by the neuronal release of NO. Since our studies were conducted on relatively pure astrocyte cultures and thus NO was mainly released from astrocytes, we can only speculate the possible in vivo effects of cytokines/NO. Neuronal injury has been associated with elevations in NO levels and several studies showed that NOS inhibitors and NO scavengers are effective in reducing neurotoxic insults (Buisson et al., 1992; Dawson et al., 1993). Under those conditions, NO may be released from both neurons and glia, which would potentiate the effects seen here. It is hard to determine the actual concentration of NO necessary to modulate astrocytic glutamate uptake. Concentrations of NO donors used were comparable to those used in the literature. We believe that NO concentrations must have been within the physiological range, since cytokine stimulation in the absence of exogenous NO resulted in similar effects on astrocytic glutamate uptake.

Astrocytic glutamate uptake has recently been implicated in the modulation of synaptic transmission (Keyser and Pellmar, 1994). Our observation that NO can attenuate astrocytic glutamate uptake suggest that NO may exert some of its documented modulatory effects on synaptic transmission (Bredt and Snyder, 1992; Montague et al., 1994) through affecting astrocytic glutamate uptake, combined with inhibition of the uptake of glutamate into synaptosomes (Pogun et al., 1994).

Concluding remarks

In summary, our data support the notion that glial glutamate uptake is critical for neuronal survival, and suggest that the high affinity astrocytic glutamate transport is a target for NO modulation. Such modulation may play important roles following CNS injury or diseases that are associated with cytokine release and NO production. It remains to be shown that whether NO/cGMP directly modulates the glutamate transporter or indirectly affects uptake by changing membrane potential or ion gradients that would secondly influence glutamate transport.

Acknowledgement

The authors thank Dr. Michael. J. Friedlander for help with bioluminescence glutamate measurements and Dr. Joseph S. Beckman for providing pure NO gas. Supported by NIH R0131234 and P50HD32301.

References

Arriza, J.L., Eliasof, S., Kavanaugh, M.P. and Amara, S.G. (1997) Excitatory amino acid transporter 5, a retinal glutamate transporter coupled to a chloride conductance. *Proc. Natl. Acad. Sci. USA*, 94: 4155–4160.

Benavides, J. and Toulmond, S. (1993) The role of cytokines in the central nervous system. *Therapie*, 48: 575–584.

Benveniste, E.N. (1992) Inflammatory cytokines within the central nervous system: Sources, function, and mechanism of action. (Review). *Am. J. Physiol.*, 263: C1–C16.

Bouvier, M., Szatkowski, M., Amato, A. and Attwell, D. (1992) The glial cell glutamate uptake carrier counter transports pH-changing anions. *Nature*, 360: 471–474.

Bredt, D.S. and Snyder, S.H. (1992) Nitric oxide, a novel neuronal messenger. (Review). *Neuron*, 8: 3–11.

Buisson, A., Plotkine, M. and Boulu, R.G. (1992) The neuroprotective effect of a nitric oxide inhibitor in a rat model of focal cerebral ischaemia. *Brit. J. Pharmacol.*, 106: 766–767.

Chao, C.C., Hu, S., Ehrlich, L. and Peterson, P.K. (1995) Interleukin-1 and tumor necrosis factor-α synergistically mediate neurotoxicity: involvement of nitric oxide and of *N*-methyl-D-aspartate receptors. *Brain, Behavior and Immunity*, 9: 355–365.

Chao, C.C., Hu, S.X., Sheng, W.S., Bu, D.F., Bukrinsky, M.I. and Peterson, P.K. (1996) Cytokine-stimulated astrocytes damage human neurons via a nitric oxide mechanism. *Glia*, 16: 276–284.

Choi, D.W. (1988) Glutamate neurotoxicity and diseases of the nervous system. (Review). *Neuron*, 1: 623–634.

Dawson, V.L., Dawson, T.M., Uhl, G.R. and Snyder, S.H. (1993) Human immunodeficiency virus type 1 coat protein neurotoxicity mediated by nitric oxide in primary cortical cultures. *Proc. Natl. Acad. Sci. USA*, 90: 3256–3259.

Fairman, W.A., Vandenberg, R.J., Arriza, J.L., Kavanaugh, M.P. and Amara, S.G. (1995) An excitatory amino-acid transporter with properties of a ligand-gated chloride channel. *Nature*, 375: 599–603.

Fine, S.M., Angel, R.A., Perry, S.W., Epstein, L.G., Rothstein, J.D., Dewhurst, S. and Gelbard, H.A. (1996) Tumor necrosis factor α inhibits glutamate uptake by primary human astrocytes – Implications for pathogenesis of HIV-1 dementia. *J. Biol. Chem.*, 271: 15303–15306.

Fosse, V.M., Kolstad, J. and Fonnum, F. (1986) A bioluminescence method for the measurement of l-glutamate: Applications to the study of changes in the release of l-glutamate from lateral geniculate nucleus and superior colliculus after visual cortex ablation in rats. *J. Neurochem.*, 47: 340–349.

Garthwaite, J., Southam, E., Boulton, C.L., Nielsen, E.B., Schmidt, K. and Mayer, B. (1995) Potent and selective inhibition of nitric oxide-sensitive guanylyl cyclase by 1H-[1, 2, 4]oxadiazolo[4, 3-a]quinoxalin-1-one. *Mol. Pharmacol.*, 48: 184–188.

Gegelashvili, G. and Schousboe, A. (1997) High affinity glutamate transporters: Regulation of expression and activity. *Mol. Pharmacol.*, 52: 6–15.

Kanai, Y. and Hediger, M.A. (1992) Primary structure and functional characterization of a high-affinity glutamate transporter. *Nature*, 360: 467–471.

Kanai, Y., Smith, C.P. and Hediger, M.A. (1993) The elusive transporters with a high affinity for glutamate. (Review). *Trends Neurosci.*, 16: 365–370.

Keyser, D.O. and Pellmar, T.C. (1994) Synaptic transmission in the hippocampus: Critical role for glial cells. *Glia*, 10: 237–243.

Kimelberg, H.K., Pang, S. and Treble, D.H. (1989) Excitatory amino acid-stimulated uptake of 22Na+ in primary astrocyte cultures. *J. Neurosci.*, 9: 1141–1149.

Koka, P., He, K., Zack, J.A., Kitchen, S., Peacock, W., Fried, I., Tran, T., Yashar, S.S. and Merrill, J.E. (1995) Human immunodeficiency virus 1 envelope proteins induce interleukin 1, tumor necrosis factor α, and nitric oxide in glial cultures derived from fetal, neonatal, and adult human brain. *J. Exp. Med.*, 182: 941–951.

Lipton, S.A., Choi, Y.B., Pan, Z.H., Lei, S.Z., Chen, H.S., Sucher, N.J., Loscalzo, J., Singel, D.J. and Stamler, J.S. (1993) A redox-based mechanism for the neuroprotective and neurodestructive effects of nitric oxide and related nitroso-compounds (see comments). *Nature*, 364: 626–632.

Lipton, S.A. and Rosenberg, P.A. (1994) Excitatory amino acids as a final common pathway for neurological disorders. *N. Eng. J. Med.*, 330, 613–622.

Longuemare, M.C. and Swanson, R.A. (1995) Excitatory amino acid release from astrocytes during energy failure by reversal of sodium-dependent uptake. *J. Neurosci. Res.*, 40: 379–386.

Montague, P.R., Gancayco, C.D., Winn, M.J., Marchase, R.B. and Friedlander, M.J. (1994) Role of NO production in NMDA receptor-mediated neurotransmitter release in cerebral cortex. *Science*, 263: 973–977.

Mrak, R.E., Sheng, J.G. and Griffin, W.S. (1995) Glial cytokines in Alzheimer's disease: Review and pathogenic implications. *Hum. Pathol.*, 26: 816–823.

Nicholls, D. and Attwell, D. (1990) The release and uptake of excitatory amino acids (see comments). (Review). *Trends Pharmacol. Sci.*, 11: 462–468.

Nomura, Y. and Kitamura, Y. (1993) Inducible nitric oxide synthase in glial cells. *Neurosci. Res.*, 18: 103–107.

Petroff, O.A.C., Pleban, L.A. and Spencer, D.D. (1995) Symbiosis between in vivo and in vitro NMR spectroscopy: The creatine, *N*-acetylaspartate, glutamate, and GABA content of the epileptic human brain. *Magn. Reson. Imaging*, 13: 1197–1211.

Pines, G., Danbolt, N.C., Bjoras, M., Zhang, Y., Bendahan, A., Eide, L., Koepsell, H., Storm-Mathisen, J., Seeberg, E. and Kanner, B.I. (1992) Cloning and expression of a rat brain l-glutamate transporter. *Nature*, 360: 464–467.

Pogun, S., Dawson, V. and Kuhar, M.J. (1994) Nitric oxide inhibits 3h-glutamate transport in synaptosomes. *Synapse*, 18: 21–26.

Pryor, W.A. and Squadrito, G.L. (1995) The chemistry of peroxynitrite: A product from the reaction of nitric oxide with superoxide (Review). *Am. J. Physiol.*, 268: Pt 1 : L699–722.

Rothstein, J.D., Dykes-Hoberg, M., Pardo, C.A., Bristol, L.A., Jin, L., Kuncl, R.W., Kanai, Y., Hediger, M.A., Wang, Y.F., Schielke, J.P. and Welty, D.F. (1996) Knockout of glutamate transporters reveals a major role for astroglial transport in excitotoxicity and clearance of glutamate. *Neuron*, 16: 675–686.

Rothstein, J.D., Martin, L., Levey, A.I., Dykes-Hoberg, M., Jin, L., Wu, D., Nash, N. and Kuncl, R.W. (1994) Localization of neuronal and glial glutamate transporters. *Neuron*, 13: 713–725.

Schousboe, A. and Westergaard, N. (1995) Transport of neuroactive amino acids in astrocytes. In: H. Kettenmann and B.R. Ransom (Eds.), *Neuroglia*, Oxford University Press, New York, pp. 246–258.

Storck, T., Schulte, S., Hofmann, K. and Stoffel, W. (1992) Structure, expression, and functional analysis of a Na(+)-dependent glutamate/aspartate transporter from rat brain. *Proc. Natl. Acad. Sci. USA*, 89: 10955–10959.

Szatkowski, M., Barbour, B. and Attwell, D. (1990) Non-vesicular release of glutamate from glial cells by reversed electrogenic glutamate uptake. *Nature*, 348: 443–446.

Tanaka, K., Watase, K., Manabe, T., Yamada, K., Watanabe, M., Takahashi, K., Iwama, H., Nishikawa, T., Ichihara, N., Kikuchi, T., Okuyama, S., Kawashima, N., Hori, S., Takimoto, M. and Wada, K. (1997) Epilepsy and exacerbation of brain injury in mice lacking the glutamate transporter glt-1. *Science*, 276: 1699–1702.

Thomas, R.J. (1995) Excitatory amino acids in health and disease. (Review). *J. Am. Geriatr. Soc*, 43: 1279–1289.

Trotti, D., Rossi, D., Gjesdal, O., Levy, L.M., Racagni, G., Danbolt, N.C. and Volterra, A. (1996) Peroxynitrite inhibits glutamate transporter subtypes. *J. Biol. Chem.*, 271: 5976–5979.

Ye, Z.C. and Sontheimer, H. (1996) Cytokine modulation of glial glutamate uptake: A possible involvement of nitric oxide. *Neuroreport*, 7: 2181–2185.

Ye, Z.C. and Sontheimer, H. (1998) Astrocytes protect neurons from neurotoxic injury by serum glutamate. *Glia*, 22: 237–248.

Zerangue, N. and Kavanaugh, M.P. (1996) Flux coupling in a neuronal glutamate transporter. *Nature*, 383: 634–637.

R.R. Mize, T.M. Dawson, V.L. Dawson and M.J. Friedlander (Eds.)
Progress in Brain Research, Vol 118

CHAPTER 18

Expression of nitric oxide synthase-2 in glia associated with CNS pathology

Angela K. Loihl and Sean Murphy*

Department of Pharmacology and Neuroscience Program, University of Iowa College of Medicine, Iowa City, IA 52242, USA

Introduction

While the production of nitric oxide (NO) in the nervous system is catalyzed by three, highly homologous isoforms of NO synthase (NOS) (Forstermann et al., 1995), in this chapter we focus exclusively on NOS-2. This dimeric, heme-containing, soluble protein whose activity is independent of a rise in intracellular calcium, is variously termed 'inducible', 'immunologic', and 'macNOS' (Griffith and Stuehr, 1995). These terms are somewhat misleading because NOS-2 has been reported in a wide variety of normal and neoplastic cell types, in some cases constitutively (Xie and Nathan 1994). All CNS parenchymal cells, together with the smooth muscle and endothelium of the vascular wall, can be transcriptionally activated by a variety of agents to express NOS-2, as well as cells (neutrophils and monocytes) infiltrating the parenchyma following changes in blood–brain barrier properties (Grzybicki et al., 1997, 1998). In addition, there are reports of developmentally related NOS-2 expression in particular brain regions (Galea et al., 1995; Arnhold et al., 1997).

Initially from studies with cultured cells (Murphy et al., 1993), but now exemplified by a host of in vivo conditions (see later), identified astrocytes have been shown to express NOS-2 upon activation. Induction can be triggered in vitro (see Table 1) by endotoxin and/or combinations of such cytokines as interleukin (IL)-1β, tumor necrosis factor (TNF)-α and interferon (IFN)-γ; HIV envelope proteins also activate the astrocyte gene (Koka et al., 1995). The promoter region of the glial NOS-2 gene displays a wide variety of potential response elements for transcription factors, such as AP-1, Hif, NFκB, NF-IL6, TNF, IRF, and GAS (Spitsin et al., 1997). In cultured astrocytes the transcript can be detected within a few hours of exposure to cytokines, is maximally transcribed by 4 h and then declines rapidly, such that it is not detectable at 12 h (Park and Murphy, 1994). In contrast, expression of NOS-2 in the CNS in vivo is an unexpectedly delayed event, which may, as we discuss later, have functional implications.

The observation that HIV envelope proteins can induce NOS-2 is important as it demonstrates that priming cells with IFN-γ (a cytokine not produced by resident CNS cells) is not obligatory for gene activation. In addition, cytokine-activated astrocytes, macrophages and cerebrovascular cells release TNFα which, in turn, induces NOS-2 in naive cells (Borgerding and Murphy, 1995; Shafer and Murphy, 1997). This implies that induction of NOS-2 in CNS parenchymal cells could occur

*Corresponding author. Tel.: 319 335 7958; fax: 319 335 8930; e-mail: sean-murphy@uiowa.edu

TABLE 1

Regulators of NOS-2 expression in vitro.

Inducers	Blockers
Proinflammatory cytokines (IL-1β, TNFα, IFNγ)	Anti-inflammatory cytokines (TGFβ1, IFNβ, IL-4, IL-10)
Endotoxin	Dexamethasone
HIV env. proteins	Chemokines (MCP-1, IL-8)
Hypoxia	Neurotransmitters (norepinephrine, glutamate, ATP, angiotensin II)
β-amyloid	Heat shock protein 70 nitric oxide (NO)

in vivo following events on the luminal side of an intact blood-brain barrier. Transcriptional activation of the cerebral vasculature by endotoxin or cytokines from adhering hematogenous cells may be sufficient to induce NOS-2 in astrocytes and microglia on the parenchymal side.

The timing and extent of NOS-2 expression associated with various neuropathologies may also be related to the prior history of the cells involved. For example, heat shock protein 70 and the anti-inflammatory cytokines transforming growth factor (TGF)-β1 and IFNβ block NOS-2 induction at the transcriptional level (Park et al., 1994; Feinstein et al., 1996; Stewart et al., 1997). Activation of glutamate, ATP, angiotensin-II and β-adrenergic receptors on astrocytes blunt the NOS-2 response to endotoxin and pro-inflammatory cytokines (Kopnisky et al., 1997; Lin and Murphy, 1997; Feinstein 1998). It is clear that some of these modulatory effects result from interference with the subsequent activation of a key transcription factor, NFκB, and/or with its binding to the promoter.

Nitric oxide inhibits not only NOS-2 activity but also regulates the level of NOS-2 mRNA expression through a mechanism involving NF-κB. Some reports suggest a blockade of NF-κB activation (Colasanti et al., 1995; Peng et al., 1995), whereas others propose inhibition of binding to its response element on the NOS-2 promoter via nitrosylation of a key cysteine residue on one of the NF-κB subunits (DelaTorre et al., 1997; Park et al., 1997). This regulation might represent a means to limit or focus NO production in discrete groups of cells in the CNS (Togashi et al., 1997). Such a feedback effect of NO has been confirmed in vivo (Luss et al., 1994), pointing to potential rebound effects with the therapeutic use of NOS-2 inhibitors.

The NOS-2 transcript is rapidly degraded, implying there may be factors co-induced in cells which are responsible for causing this instability (Park and Murphy, 1996). The 3′-untranslated region of NOS-2 has AUUUA repeats (Galea et al., 1994) which are suggested to be determinants of mRNA stability and this could represent yet another mechanism by which the amount and duration of NO production is regulated.

With all of these potential regulatory mechanisms for NOS-2 expression in mind, what is the evidence for expression in vivo? what are the roles that NO from this source play in CNS function? There are many potential targets for NO and its oxidation products (such as peroxynitrite). These include heme and non-heme iron, thiols and DNA. Such reactions account for the reported biological effects of NOS activation, such as mitochondrial dysfunction (Knowles, 1997), neuromodulation (Christopherson and Bredt, 1997), and cell and organism killing (Nathan, 1997). In addition, it is evident that NO has subtle effects on the regulation of a number of cytokine and chemokine genes (Merrill and Murphy, 1997). However, it is important to distinguish between effects due to NO and those resulting from peroxynitrite exposure. The latter is a strong oxidant with effects that, in general, are biologically deleterious; the production of NO in the absence of superoxide anion produces different effects, many of which may be beneficial (Butler et al., 1995; Nathan, 1997).

Expression of NOS-2 in vivo

Since last we reviewed this literature (Murphy and Grzybicki, 1996) there have been considerable efforts to identify specific neuropathological conditions that result in the induction of NOS-2. The earlier, suggestive reports of NOS-2 mRNA and/or protein expression associated with certain types of CNS injury have now been confirmed by in situ hybridization and immunohistochemistry and provide evidence for the cellular source. There is specific evidence for glial expression of NOS-2 associated with neuronal injury and infection of the CNS, and in neurodegenerative and demyelinating diseases (see Table 2). We refer to these most recent reports for completeness, and then focus on ischemia and the evidence for glial NOS-2 expression. Finally, we speculate on the possible roles for NO in the development of pathology, and in recovery from injury.

Trauma

Direct injury in the CNS results in a reactive gliosis, characterized by induction of the glial fibrillary acidic protein gene and changes in astrocyte morphology (Eddleston and Mucke, 1993). In a freeze-lesion model, we found that NOS-2 was induced within a few hours in cells infiltrating the lesion (Grzybicki et al., 1998). Only after 24 h could we find NOS-2 expression in astrocytes delimiting the lesion edge. In a compression injury to the spinal cord, Hamada et al. (1996) report NOS-2 mRNA, which peaks at 24 h and persists for three days. The expression was associated with motor dysfunction and application of a NOS inhibitor reduced motor disturbance. This delayed expression of NOS-2 in parenchymal cells is of interest and we speculate on its meaning later in the chapter.

There is also suggestive evidence for NOS-2 expression in glia after indirect neuronal injury, which occurs with axotomy or chemical toxicity, and results in a marked microglial, but lesser astroglial cell response. Following a local injection of kainic acid into the cortex, Lei et al. (1996) describe neuronal degeneration, followed 2–3 days later by expression of NOS-2 in reactive astrocytes and in endothelial cells at the edge of the lesion.

TABLE 2

Role of NOS-2 in neuropathologies.

Pathology	After NOS-2 block
Direct cortical injury	Reduced
Indirect injury	?
Hepatitis, Theiler's, HIV	?
LCM virus	Worsened
Endotoxin	Reduced
ALS	?
Retinal dystrophy	?
Alzheimer	?
EAE/MS	Reduced or worsened
Global ischemia	Reduced
Stroke	Reduced

Infection

How viruses cause lesions, or dysfunction without direct damage to the CNS, is unclear. Diffusible cytokines, together with reactive oxygen and nitrogen species produced as a component of the immune response presumably contribute to the pathology. In many instances, NOS-2 expression is associated with viral infection of the CNS and, in some, the resulting NO appears to be beneficial.

Intranasal inoculation of mice with the neurotropic hepatitis virus leads to expression of NOS-2 in the brain (Grzybicki et al., 1997). The cells responsible for its expression appear to be entirely infiltrating, with no evidence for parenchymal cell expression. However, this infection proves fatal within seven days. Lane et al. (1997) used a neuroattenuated strain of this virus which is cleared and the animals thus survive. They report NOS-2 in astrocytes and macrophages in the olfactory bulb. We have looked also at expression of NOS-2 in persistently infected animals, generated by raising inoculated pups with immunized dams. In these animals, which display marked demyelinating lesions and associated hindlimb paralysis, NOS-2 is expressed very prominently in astrocytes surrounding such lesions (Sun et al., 1995; Grzybicki et al., 1997). A similar expression was reported by Oleszak et al. (1997) in astrocytes surrounding necrotizing lesions in mice infected with Theiler's virus. There is also suggestive evidence that NOS-2 expression in the CNS is associated with HIV-1 (Adamson et al., 1996) and SIV infection (Lane et al., 1996).

The study by Lane et al. (1997) suggests that, while NO from NOS-2 can block viral replication in vitro, this is not the determining factor for clearing hepatitis virus in the CNS. Campbell (1996), employing intracerebral infection with lymphocytic choriomeningitis virus, found NOS-2 expression only in macrophages. While the viral burden was unaffected by a NOS inhibitor, such treatment exacerbated clinical severity and decreased survival time, suggesting a protective role for NO.

In wild-type mice early in the course of systemic inflammation, resulting from an injection of endotoxin, there is a marked expression of NOS-2 in vascular, glial and neuronal structures adjacent to areas devoid of a blood–brain barrier (Wong et al., 1996; Htain et al., 1997). Boje (1996) found that intracisternal administration of endotoxin resulted in NO production from the meninges and alteration in blood–brain barrier properties, which could be reversed by a NOS inhibitor. It is of interest to note that mice with deletions in the NOS-2 gene are protected from death after high-doses of endotoxin (MacMicking et al., 1995).

Neurodegenerative diseases

Characteristic of such diseases is the progressive loss of a subset of neurons and there is evidence to implicate NO as causal or, at least, contributory.

In a proportion (20%) of cases of familial amyotrophic lateral sclerosis (ALS) there are multiple missense mutations in cytosolic Cu/Zn superoxide dismutase (SOD) and it is proposed that these result in either a gain of function, or increase an injurious reaction catalyzed by SOD. For example, Crow and Beckman (1995) suggest that there is enhanced nitration by NO of neurofilament protein. Evidence for nitration being causal in ALS comes from the observation that nitration blocks phosphorylation of tyrosine residues and that nitrotyrosine resembles the strongly antigenic dinitrophenol, perhaps provoking an autoimmune reaction.

Inherited retinal dystrophy results from accumulation of cellular debris in the subretinal space and visual cell degeneration. Cotinet et al. (1997) looked at the retinal Muller (astrocyte-like) cell from dystrophic rats and found that they produce abnormal amounts of TNFα and NOS-2 derived NO. As NO can decrease phagocytosis of photoreceptor segments, which underlies this dystrophy, they suggest that NO may disrupt transmission in retinal neurons.

Alzheimer's disease (AD) occurs in 10–15% of individuals over 65 and is characterized by the loss of pyramidal neurons in the hippocampus and neocortex, and of cholinergic neurons in the median forebrain. The hallmark of AD is the deposition of β-amyloid aggregates which can be toxic to neurons and also leads to NO production from glial cells (Rossi and Bianchini, 1996; Barger and Harmon, 1997) and neurons (Vodovotz et al., 1996). Weldon et al. (1998) injected soluble and fibrillar β-amyloid into rat striatum. They noted that microglia and astrocytes nearby showed a marked increase in NOS-2 expression and speculate that the resultant NO may be involved in neuronal killing.

Demyelinating diseases

Demyelinating diseases are characterized by preferential damage to myelin, with relative preservation of axons. Multiple sclerosis (MS) and animal models such as experimental autoimmune encephalomyelitis (EAE) are inflammatory demyelinating diseases characterized by early damage to the blood–brain barrier. Lymphocyte and monocyte infiltration occurs, as well as activation of indigenous glia. What produces the resulting myelin damage is unknown and both indigenous as well as infiltrating cell types may contribute to myelin destruction, either directly or indirectly by secretion of proinflammatory cytokines or NO (Merrill and Benveniste, 1996).

A number of recent reports confirm a suggested role for NOS-2 in demyelinating disease. Using a spin trap method in vivo Hooper et al. (1995) could detect NO in the CNS of EAE animals, which correlated with the onset of neurological symptoms. Increased NO production occurs in MS as illustrated by elevated levels of nitrite and nitrate levels in CSF (Johnson et al., 1995). Using immunocytochemistry, NOS-2 and nitrotyrosine have been found in diseased brain (Okuda et al., 1995; Hooper et al., 1995; Bagasra et al., 1995; DeGroot et al., 1997; Merrill et al., 1998). Both the protein and the transcript for NOS-2 colocalize with markers for macrophages/microglia (Bagasra et al., 1995; Hooper et al., 1997; van der Veen, 1997) and/or astrocytes (Okuda et al., 1997; Tran et al., 1997; Merrill et al., 1998). Cross et al. (1996) showed that NOS-2 transcript and enzyme activity correlate with active EAE. It is of interest that MS patients have elevated circulating antibodies to S-nitrosocysteine, suggesting nitrosylation of proteins in vivo (Boullerne et al., 1995). Akin to the human situation with MS, Ding et al. (1997) find gender-relatedness in NOS-2 expression and NO production in mice with EAE.

Based on the evidence for NO production in MS and EAE, several studies have assessed roles for NO using NOS inhibitors. The results have been mixed. In one study of acute EAE inhibitors of NOS provided no benefit (Zielasek et al., 1995). Other EAE studies showed a small rise in clinical score (Ruuls et al., 1996) or a decrease in disease incidence, severity and duration (Cross et al., 1994; Zhao et al., 1996). Most recently, Hooper et al. (1997) report significant therapeutic effects on EAE with NOS inhibitors, and scavengers of NO and peroxynitrite.

The current therapeutic available for the treatment of MS is IFNβ, demonstrated to be a blocker of NOS-2 expression (Stewart et al., 1997). Hall et al. (1997) suggest that it is beneficial through its antagonism of endogenously produced IFNγ, hence impairing the activation of macrophages and microglia. Treatment of EAE animals with anti-inflammatory cytokines such as IL10, another NOS-2 modulator, inhibits disease (Willenborg et al., 1995). However, using the EAE model in NOS-2 gene deficient mice, Fenyk-Melody et al. (1998) report that disease is more severe and prolonged. This might suggest either that NO is immunosuppressive at the time of immunization or that, without NO, there is a loss of the normal regulation of infiltration of inflammatory cells into the CNS (De Caterina et al., 1995; Hickey et al., 1997).

Cerebral ischemia

Elucidating the sequence of physiological and biochemical events following cerebral ischemia, in an attempt to identify means to reduce the amount of damage that develops, has been the focus of extensive investigation. The early events that occur following blood flow reduction, leading to cell death, have been well characterized in animal models of cerebral ischemia (Siesjo, 1992). Briefly, ischemia induces mitochondrial injury leading to a rapid depletion of ATP levels and a disruption of normal ion homeostasis. Large amounts of glutamate are released and intracellular calcium levels rise because of NMDA-receptor activation. The massive increase in free intracellular calcium following ischemia is a precursor to a variety of events capable of promoting neuronal and glial cell death such as inflammation, gene activation, and free radical production (for a review see Iadecola, 1997). While the pathogenesis of ischemic brain injury is probably the result of the activation of multiple biochemical pathways, it is evident, given the chain of events that occur, that NO has the potential to participate in the mechanisms of cerebral ischemia.

There is sufficient evidence that NO is produced following both global and focal cerebral ischemia. Using techniques such as in vivo spin trapping (Sato et al., 1994), porphyrinic microsensor measurement (Malinski et al., 1993), and measurements of brain nitrite and cGMP levels as a reflection of increased NO production (Kader et al., 1993; Shibata et al., 1996), increased generation of NO in the brain following cerebral ischemia has been demonstrated.

NOS-2 and global ischemia

Transient global ischemia is seen clinically associated with disorders of systemic circulation, such as heart failure or other cardiac dysfunction. Blood flow to the brain may be reduced below levels of autoregulation or may be completely interrupted for a period of time. Global ischemia induces a gliotic reaction, and delayed neuronal necrosis occurs in selectively vulnerable cells such as pyramidal neurons in the CA1 subfield of the hippocampus. NOS-2 protein expression has been shown to occur in a rat model of transient global ischemia (Endoh et al., 1994). Three days after a 10 min ischemic episode, NOS-2 protein was detected in the CA1 region of the hippocampus by immunohistochemistry, with immunoreactivity progressively increasing to day 30. Double-labeling experiments identified these NOS-2-expressing cells as reactive astrocytes, and ruled out microglial expression.

In our own studies with spontaneously hypertensive rats, we can detect NOS-2 protein by 72 h of reperfusion following 30 min of global ischemia. These cells morphologically resemble astrocytes in the corpus callosum, basal ganglia, and the hippocampus (Fig. 1). In response to global ischemia, NOS-2 induction parallels, spatially and temporally, the gliotic reaction. The induction of NOS-2 appears to be restricted to astrocytes in regions of neuronal damage, as there is no evidence for NOS-2 expression in other cell types or other regions. The mechanism of delayed NOS-2 induction, and the role of NO generated by this isoform following global ischemia remains to be determined. To date, no studies have been performed investigating global ischemic damage in mice deficient in NOS-2 expression. This may be due to the difficulties associated with mice not surviving the ischemic episode (Yang et al., 1997), or it may be due to the complications of strain variability in global ischemic outcome (Fujii et al., 1997).

NOS-2 and focal ischemia

Focal cerebral ischemia, or stroke, occurs as a regional injury and results from the prolonged interruption of blood flow due to the occlusion of a single CNS artery. A focal vascular occlusion produces a localized, destructive lesion or infarct, with necrosis of all cellular elements in the territory normally perfused by the occluded vessel. Ischemic stroke occurs most often in humans as the result of a thromboembolic vascular occlusion, secondary to atherosclerosis, in which there is a focal deposition of plasma-derived lipids that inhibits normal flow. In rodent models, focal ischemia can be induced by either permanently or transiently occluding the middle cerebral artery (MCA).

Fig. 1. *Expression of NOS-2 in astrocytes following global ischemia.* In tissue sections stained with hematoxylin and eosin, neuronal cell death can be seen in the CA1 region of the rat hippocampus three days following 30 min of transient global ischemia (A, B). Astrogliosis, indicated by an increase in GFAP immunoreactivity, is also evident three days after the insult (C, D). Immunohistochemistry reveals NOS-2 protein expressed after three days in cells morphologically resembling astrocytes (E, F). Monoclonal GFAP antibody (Sigma, 1:200) and polyclonal NOS-2 antibody (Transduction Labs., 1:100) labelling was detected using standard immunoperoxidase techniques with 3, 3′diaminobezidine as substrate. All sections were counterstained with hematoxylin and eosin. Magnfication: A, 40x; B, 200x; C, 400x; D–F, 600x.

Several studies have provided biochemical, molecular, and immuno-histochemical evidence of NOS-2 induction following focal cerebral ischemia. Two separate reports demonstrated the induction of NOS-2 activity following permanent MCA occlusion in rats (Yoshida et al., 1995; Iadecola 1995a). Both studies found an increase in NOS-2 activity at 2–3 days post-occlusion, with a return to baseline by seven days. The time course of NOS-2 induction suggests that NO generated from this isoform does not contribute to the early stages of ischemic injury. Iadecola et al. (1995c; 1996) have also investigated NOS-2 mRNA and protein expression in response to focal ischemia. Following permanent MCA occlusion in rats, NOS-2 mRNA was detected by 12 h, peaked at 48 h and returned to baseline by seven days post-occlusion. Immunoreactivity for NOS-2, observed at 48–96 hours after ischemia, was detected only in neutrophils invading the infarct and was not seen in resident cells (Iadecola et al., 1996). Transient

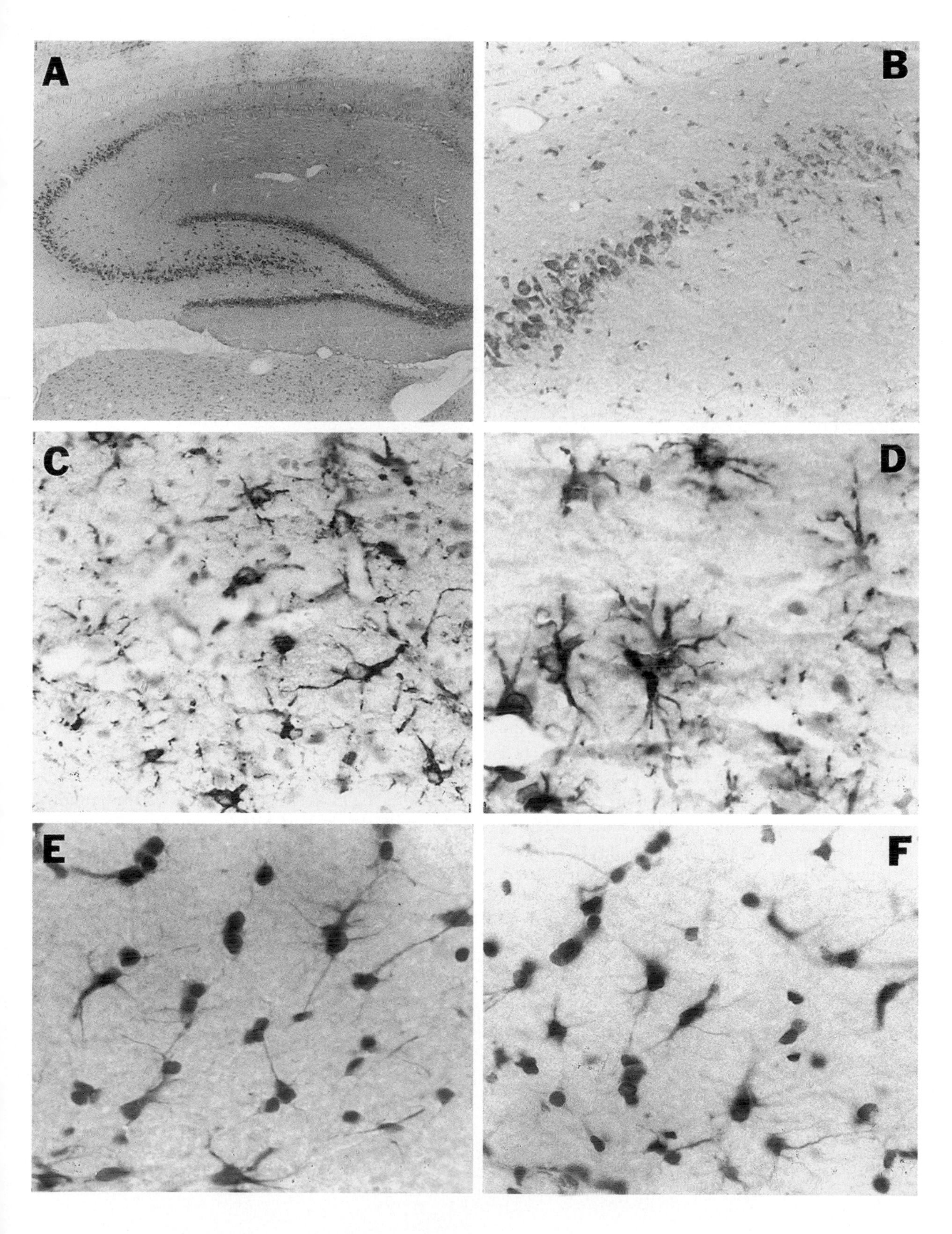
A
B
C
D
E
F

focal ischemia in rats also leads to NOS-2 induction, though with different spatial and temporal patterns of expression than seen following permanent occlusion. Two hours of MCA occlusion, followed by reperfusion, resulted in maximal expression of NOS2 mRNA at an earlier time point, and detection of protein predominantly in vascular cells, rather than in neutrophils (Iadecola et al., 1996).

In mice, permanent MCA occlusion using the filament model of ZeaLonga et al. (1989) leads to the induction of NOS-2 mRNA and protein expression. Using in situ hybridization, we were able to detect NOS-2 message at 24, 48 and 72 h (latest point studied), but not at 6 h post-occlusion (Fig. 2). At 24 hours after ischemia, NOS-2 immunoreactivity was observed primarily in what appeared, morphologically, to be inflammatory cells within the infarct (Figure 2). However, at 48 and 72 h NOS-2 immunoreactivity was also observed in cells morphologically resembling astrocytes, as well as by GFAP staining on adjacent sections (Fig. 2). These NOS-2 positive cells were located primarily in regions surrounding the infarct. In contrast to previously published reports, these data indicate that resident glial cells, in addition to infiltrating cells, express NOS-2 in the ischemic brain. The differences in our findings compared to previous reports may reflect the differences in total infarct volumes produced as a result of using different surgical procedures (cauterization versus filament occlusion). The infarct volume generated with the filament model of MCA occlusion was typically greater than 80 mm^3 (Loihl and Murphy, 1997) while infarcts produced by cauterization were generally smaller ($< 40\ mm^3$), and predominantly cortical (Iadecola et al., 1997).

Role of NOS-2 in cerebral ischemia

While it is clear that NO participates in the mechanisms of ischemia, much of the early work investigating the precise role of NO yielded contradictory results. This controversy stemmed largely from the fact that much of the research being done involved the use of pharmacological inhibitors that, while somewhat selective, are not actually specific for a particular isoform of NOS. Experiments in vivo have shown that inhibition of NO synthesis ameliorates (Nowicki et al., 1991; Buisson et al., 1992; Nagafugi et al., 1992; Ashwal et al., 1993; Iadecola et al., 1995b, 1996; Zhang et al., 1996) and worsens (Yamamoto et al., 1992; Dawson et al., 1992; Kuluz et al., 1993) tissue damage in animal models of cerebral ischemia. Given the scope of potential biological actions of NO, and the different spatial and temporal patterns of expression of the NOS isoforms, it is likely that NO generated by the different isoforms is subserving different roles at different times following ischemic insult. Therefore, non-specifically blocking NOS activity overshadows the potential for multiple roles of NO in ischemia. To circumvent the limitations of the currently available pharmacological inhibitors, and to elucidate the contributions of the different isoforms of NOS, ischemic brain damage has been investigated in genetically altered mice, which are deficient in a particular isoform of NOS.

Evidence provided in the last few years using gene targeting strategies has resolved the debate over the roles of the constitutive isoforms of NOS in the pathophysiology of focal cerebral ischemia (Iadecola, 1997). Infarct volume, resulting from MCA occlusion, in mice deficient for NOS-1 is reduced compared to wild type mice. Infarct

Fig. 2. *Expression of NOS-2 revealed by in situ hybridization and immunohistochemistry at 48 and 72 h following middle cerebral artery occlusion in mice.* NOS-2 mRNAS, detected with a 35S-labelled cRNA probe (Grzybicki et al., 1997), was observed at 48 and 72 h following permanent MCA occlusion (A, B). NOS-2 immunoreactivity was evident at 48 h post-occlusion, in what appear to be infiltrating inflammatory cells within the infarct (C). NOS-2 immunoreactivity at 72 h post-occlusion was also seen in cells morphologically resembling astrocytes located at the border of the infarct (D). GFAP immunoreactivity reveals astrocytes located at the infarct border at 48 h post-occlusion (E). NOS-2 immunoreactivity at 48 h post-occlusion, in an adjeacent section to that shown in E, in cells resembling astrocytes (F). See legend to Fig. 1 for details of methods. Magnification: A, B, 200x; C–F, 400x.

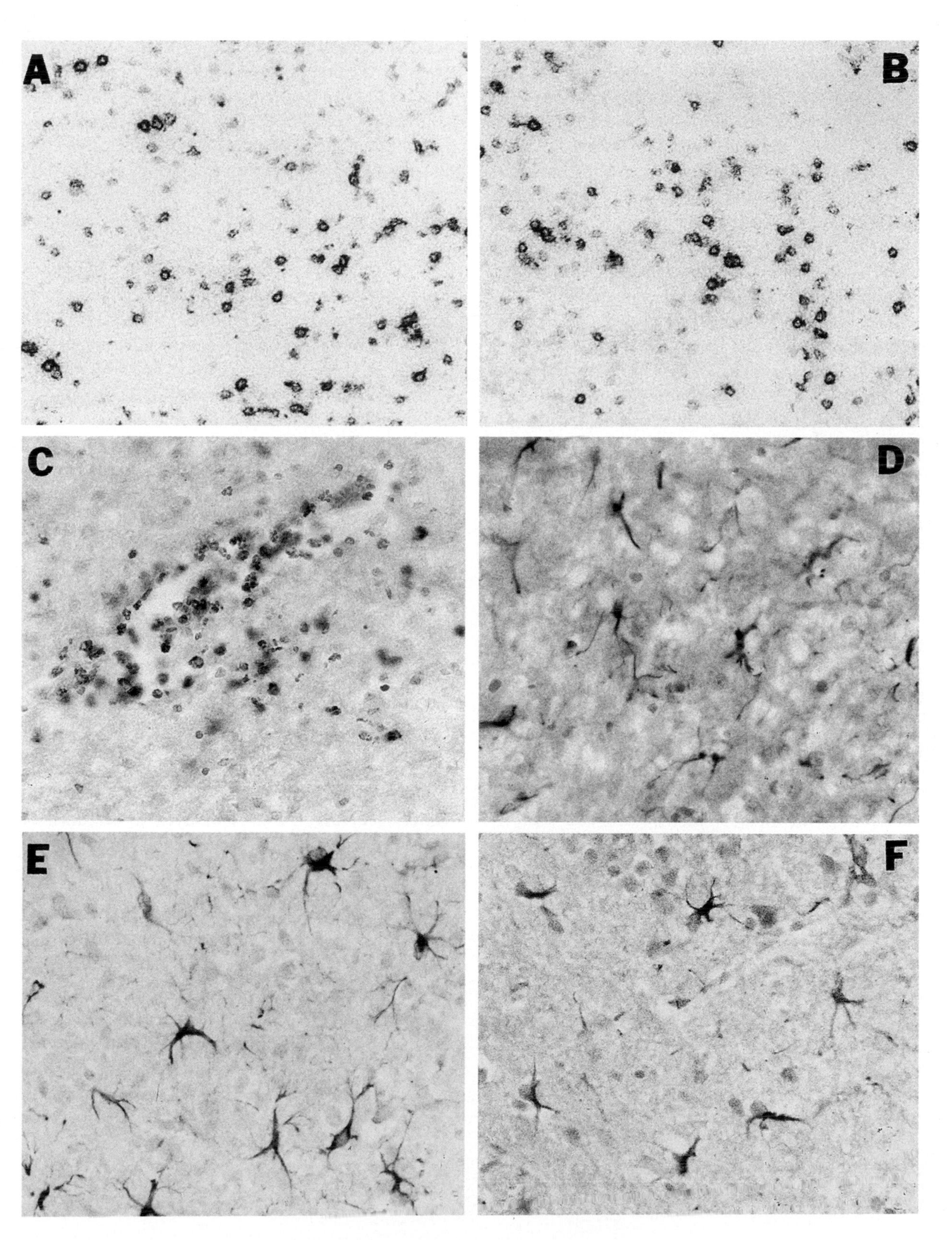
A
B
C
D
E
F

volume, however, in these mutant mice increased following administration of nitro-L-arginine, a NOS inhibitor, presumably due to inhibition of NOS-3 (Huang et al., 1994). In mice deficient for NOS-3, infarct volumes were larger than in wild type littermates. These results indicate that NO derived from NOS-1 acts early following cerebral ischemia and contributes to the development of tissue damage, while NO generated from NOS-3 serves a protective function, most likely by regulating blood flow to the ischemic penumbra.

More recently, evidence has been provided that begins to elucidate the role of the NOS-2 isoform in focal cerebral ischemia. A report by Iadecola et al. (1997) demonstrated that 24 h following MCA occlusion, infarct volumes in mice deficient for NOS-2 did not vary from infarct volumes in wildtype animals. We have found similar results in our own studies (Loihl and Murphy, 1997). This is consistent with the idea, suggested by enzyme activity studies, that NOS-2 does not contribute to the early stages of ischemic damage, and is not particularly surprising given the time course of NOS-2 induction. Iadecola et al. (1997) did find, however, that 96 h after MCA occlusion infarct volumes in NOS-2 deficient mice were 28% smaller than in wild-type controls. This smaller infarct at 96 h post-occlusion was accompanied by less severe motor deficits in mutant mice, indicating that the lack of NOS-2 correlates with improved functional outcome.

Regulation and role of astroglial NOS-2 in neuropathologies

Perhaps the only conclusions that can be drawn from the neuropathological evidence presented above are that NOS-2 is sometimes expressed in astroglial cells and that, when there is expression, its appearance is delayed (> 1 day) beyond what we might predict from in vitro observations (a few hours). This tardy appearance of NOS-2 might simply reflect development of the appropriate conditions for gene activation, such as infiltration of inflammatory cells either producing the relevant cytokines or initiating their production in resident cells. Indeed, in some pathologies other stimuli may predominate, such as hypoxia. On the other hand, we could interpret the delay in NOS-2 expression as a reflection of the presence of factors that suppress gene activation. In this case, the timing of NOS-2 expression and the production of NO from this source may represent an adaptive response, and thus hint at a beneficial function.

If NOS-2 gene expression is initially suppressed following damage, then the in vitro evidence points to a number of likely candidates. While the in vivo data are not definitive, they are at least consistent. Levels of transforming growth factor $\beta 1$ rise following ischemia (Wang et al., 1995; Krupinski et al., 1996). Extracellular glutamate and ATP are elevated and persist for a few hours following CNS injury and the activation of selective chemokine genes is an early event. Specific chemokines are elevated following trauma (Ghirnikar et al., 1996; Grzybicki et al., 1998), exposure to β-amyloid (Meda et al., 1996), with bacterial infection (Spanaus et al., 1997), in EAE and MS (Hulkower et al., 1993; Karpus et al., 1995; Miyagishi et al., 1995; Ransohoff et al., 1996) and in ischemia (Kim et al., 1995; Yamasaki et al., 1995, 1997; Takami et al., 1997; Yoshimoto et al., 1997). Not only will chemokines such as IL-8 and MCP-1 block NOS-2 induction (McCall et al., 1992; Rojas et al., 1993) but NO can also inhibit the expression of IL-8 and MCP-1 (Andrew et al., 1995; Zeiher et al, 1995; De Caterina et al., 1995). A delayed appearance of NOS-2 might therefore ensure an environment in which chemokine gene expression can proceed and the entry of hematogenous cells is unhindered by NO effects on endothelial adhesion molecule expression (Khan et al., 1996; Tsao et al., 1996; Hickey et al., 1997). That the later production of NO might then terminate infiltration of inflammatory cells into the parenchyma, or downregulate expression of pro-inflammatory cytokines, are intriguing possibilities. Together with its general role as a vasodilator, thus improving perfusion in compromised tissue and leading to remodelling, NO may therefore prove beneficial by limiting the inflammatory response.

From functional and genetic studies, it seems that the effects of NO from NOS-2 can either worsen or improve recovery from damage. Evidence from the various enzyme inhibition studies is mixed, fraught with problems regarding specificity. Evidence from studies with NOS-2 gene-deficient mice is emerging to suggest that the presence of the enzyme can be beneficial, deleterious or without consequence (Nathan 1997). In cerebral ischemia, while NOS-2 clearly does not contribute to pathology within the first 24 h (Iadecola et al., 1997; Loihl and Murphy, 1997), gene-deficient mice are reported to have smaller infarcts at 4 days (Iadecola et al., 1997). Such NOS-2 deficient animals have a greater leukocyte response in endotoxemia (Hickey et al., 1997) which might explain why the condition of such mice with EAE is made worse (Fenyk-Melody et al., 1998).

The interpretation of results from gene-deficient mice is itself not without problems because of the possibility of developmental compensation and the real potential for expression of a truncated protein, perhaps with novel or unregulated functions. However, and in the absence of truly selective enzyme inhibitors, observing neuropathological outcome in experimental models employing such NOS-2 deficient mice is a powerful indicator of the roles of NOS-2. Verification could then be sought by virally re-introducing the gene in deficient animals. It would also be informative to observe pathological outcomes in transgenic mice in which NOS-2 can be overexpressed. However, there are no reports of the generation of such animals to date.

Acknowledgements

We recognize the contributions of Sherry Kardos and Hsin Lee Lin to this work, which is supported by the NIH (NS29226) and American Heart Association (Grant-in-Aid).

References

Adamson, D.C., Wildemann, B., Sasaki, M., Glass, J.D., McCarthur, J.C., Christov, V.I., Dawson, T.M. and Dawson, V.L. (1996) Immunologic NO synthase: Elevation in severe AIDS dementia and induction by HIV-1 gp41. *Science*, 274: 1917–1921.

Andrew, P.J., Harant, H. and Lindley, I.J.D. (1995) Nitric oxide regulates IL-8 expression in melanoma cells at the transcriptional level. *Biochem. Biophys. Res. Commun.*, 214: 949–956.

Arnhold, S., Andressen, C., Bloch, W., Mai, J.K. and Addicks, K. (1997) NO synthase II is transiently expressed in embryonic mouse olfactroy receptor neurions. *Neurosci. Lett.*, 229: 165–168.

Ashwal, S., Cole, D.J., Osborne, T.N. and Pearce, W.J. (1993) Low dose L-NAME reduces infarct volume in the rat MCA/O reperfusion model. *J. Neurosurg. Anesthesiol.*, 5: 241–259.

Bagasra, O., Michaels, F.H., Zheng, Y.M., Bobroski, L.E., Spitsin, S.V., Fu, Z.F., Tawadros, R. and Koprowski, H. (1995) Activation of the inducible form of nitric oxide synthase in the brains of patients with multiple sclerosis. *Proc. Natl. Acad. Sci. USA*, 92: 12041–12045.

Barger, S.W. and Harmon, A.D. (1997) Microglial activation by Alzheimer amyloid precursor protein and modulation by apolipoprotein E. *Nature*, 388: 878–881.

Boje, K.M.K. (1996) Inhibition of nitric oxide synthase attenuates blood–brain barrier disruption during experimental meningitis. *Brain Res.*, 720: 75–83.

Borgerding, R. and Murphy, S. (1995) Expression of NO synthase in cerebral endothelial cells is regulated by cytokine-activated astrocytes. *J. Neurochem.*, 65: 1342–1347.

Boullerne, A.I., Petry, K.G., Meynard, M. and Geffard, M. (1995) Indirect evidence for nitric oxide involvement in multiple sclerosis by characterization of circulating antibodies directed against conjugated S-nitrosocysteine. *J. Neuroimmunol.*, 60: 117–124.

Buisson. A., Plotkine, M. and Boulu, R.G. (1992) The neuroprotective effect of a nitric oxide inhibitor in a rat model of focal cerebral ischemia. *Br. J. Pharmacol.*, 106: 766–767.

Butler, A.R., Flitney, F.W. and Williams, D.L.H. (1995) NO, nitrosonium ions, nitroxide ions, nitrosothiols and irronitrosyls in biology. *Trends Pharmacol. Sci.*, 16: 18–22.

Campbell, I.L. (1996) Exacerbation of lymphocytic choriomeningitis in mice treated with the inducible nitric oxide synthase inhibitor aminoguanidine. *J. Neuroimmunol.*, 71: 31–36.

Christopherson, K.S. and Bredt, D.S. (1997) Nitric oxide in excitable tissues. *J. Clin. Invest.*, 100: 2424–2429.

Colasanti, M., Persichini, T., Menegazzi, M., Mariotto, S., Giodano, E., Caldarera, C.M., Sogos, V., Lauro, G.M. and Suzuki, H. (1995) Induction of nitric oxide synthase mRNA expression. *J. Biol. Chem.*, 270: 26731–26733.

Cotinet, A., Goureau, O., Hicks, D., Thillaye-Goldenberg, B. and de Kozak, Y. (1997) TNF and NO production by retinal Muller glial cells from rats exhibiting inherited retinal dystrophy. *Glia*, 20: 59–69.

Cross, A.H., Misko, T.P., Lin, R.F., Hickey, W.F., Trotter, J.L. and Tilton, R.G. (1994) Aminoguanidine, an inhibitor of inducible nitric oxide synthase, ameliorates experimental autoimmune encephalomyelitis in SJL mice. *J. Clin. Invest.*, 93: 2684–2690.

Cross, A.H., Keeling, R.M., Goorha, S., San, M., Rodi, C., Wyatt, P.S., Manning, P.T. and Misko, T.P. (1996) Inducible nitric oxide synthase gene expression and enzyme activity correlate with disease activity in murine EAE. *J. Neuroimmunol.*, 71: 145–153.

Crow, J.P. and Beckman, J.S. (1995) Reactions between nitric oxide, superoxide and peroxynitrite. In: L. Ignarro and F. Murad (Eds.), *Nitric Oxide.*, Academic Press, San Diego, pp. 17–43.

Dawson, D.A., Kusumoto, K., Graham, D.I., McCulloch, J. and Macrae, I.M. (1992) Inhibition of nitric oxide synthesis does not reduce infarct volume in a rat model of focal cerebral ischemia. *Neurosci. Lett.*, 142: 151–154.

De Caterina, R., Libby, P., Peng, H-B., Thannickal, V.J., Rajavshisth, T.B., Gimbrone, M.A. Jr., Shin, W.S. and Liao, J.K. (1995) Nitric oxide decreases cytokine-induced endothelial activation. *J. Clin. Invest.*, 96: 60–68.

De Groot, C.J.A., Ruuls, S.R., Theeuwes, J.W.M., Dijkstra, C.D. and Van der Walk, P. (1997) Immunocytochemical characterization of the expression of inducible and constitutive isoforms of nitric oxide synthase in demyelinating multiple sclerosis lesions. *J. Neuropathol. Exp. Neurol.*, 56: 10–20.

DelaTorre, A., Schroeder, R.A. and Kuo, P.C. (1997) Alteration of NF-κB p50 DNA binding kinetics by S-nitrosylation. *Biochem. Biophys. Res. Commun.*, 238: 703–706.

Ding, M., Wong, J.L., Rogers, N.E., Ignarro, L.J. and Voslhul, R.R. (1997) Gender differences of inducible nitric oxide production in SJL/J mice with EAE. *J. Neuroimmunol.*, 77: 99–106.

Eddleston, M. and Mucke, L. (1993) Molecular profile of reactive astrocytes-implications for their role in neurologic disease. *Neuroscience*, 54: 15–36.

Endoh, M., Maiese, K. and Wagner, J.A. (1994) Expression of the inducible form of nitric oxide synthase by reactive astrocytes after transient global ischemia. *Brain Res.*, 651: 92–100.

Feinstein, D.L. (1998) Suppression of astroglial nitric oxide synthase expression by norepinepohrine results from decreased NOS-2 promoter activity. *J. Neurochem.* 70: 1484–1496.

Feinstein, D.L., Galea, E., Aquino, D.A., Li, G.C., Xu, H. and Reis, D.J. (1996) Heat shock protein 70 suppresses astroglial inducible nitric oxide synthase expression by decreasing NFκB activation. *J. Biol. Chem.*, 271: 17724–17732.

Fenyk-Melody, J., Garrison, A., Brunnert, S., Weidner, J., Shen, F., Shelton, B. and Mudgett, J.S. (1998) Experimental autoimmune encephalomyelitis is exacerbated in mice lacking the NOS-2 gene. *J. Immunol.* 160: 2940–2949.

Forstermann, U., Gath, I., Schwarz, P., Closs, E.I. and Kleinert, H. (1995) Isoforms of nitric oxide synthase. *Biochem. Pharmacol.*, 50: 1321–1332.

Fujii, M., Hara, H., Meng, W., Vonsattel, J.P., Huang, Z. and Moskowitz, M.A. (1997) Strain-related differences in susceptibility to transient forebrain ischemia in SV-129 and C57.BL6 mice. *Stroke*, 28: 1805–1810.

Galea, E., Reis, D.J. and Feinstein, D.L. (1994) Cloning and expression of inducible nitric oxide synthase from rat astrocytes. *J. Neurosci. Res.*, 37: 406–414.

Galea, E., Reis, D.J., Xu, H. and Feinstein, D.L. (1995) Transient expression of calcium-independent nitric oxide synthase activity in brain blood vessels during development. *FASEB J.*, 9: 1632–1637.

Ghirnikar, R.S., Lee, Y.L., He, T.R. and Eng, L.F. (1996) Chemokine expression in rat stab wound brain injury. *J. Neurosic. Res.*, 46: 727–733.

Griffith, O.W. and Stuehr, D.J. (1995) Nitric oxide synthases. *Annu. Rev. Physiol.*, 57: 707–736.

Grzybicki, D., Kwack, K.B., Perlman, S. and Murphy, S. (1997) Nitric oxide synthase type II expression by different cell types in MHV-JHM encephalitis suggests distinct roles for nitric oxide in acute vs. persistent virus infection. *J. Neuroimmunol.*, 73: 15–27.

Grzybicki, D., Moore, S.A., Schelper, R., Glabinski, A., Ransohoff, R.M. and Murphy, S. (1998) Expression of monocyte chemoattractant protein (MCP-1) and nitric oxide synthase-2 following cerebral trauma. *Acta Neuropathol.*, 95: 98–103.

Hall, G.L., Compston, A. and Scolding, N.J. (1997) β-interferon and multiple sclerosis. *Trends Neurosci.*, 20: 63–67.

Hamada, Y., Ikata, T., Katoh, S., Tsuchiya, K., Niwa, M., Tsutsumishata, Y. and Fukuzawa, K. (1996) Roles of nitric oxide in compression injury of rat spinal cord. *Free Rad. Biol. Med.*, 20: 1–9.

Hickey, M.J., Sharkey, K.A., Sihota, E.G., Reinhardt, P.H., MacMicking, J.D., Nathan, C. and Kubes, P. (1997) Inducible nitric oxide synthase-deficient mice have enhanced leukocyte-endothelium interactions in endotoxemia. *FASEB J.*, 11: 955–964.

Hooper, D.C., Ohnishi, S.T., Kean, R., Numagami, Y., Dietzschold, B. and Koprowski, H. (1995). Local nitric oxide production in viral and autoimmune diseases of the central nervous system. *Proc. Natl. Acad. Sci. USA*, 92: 5312–5316.

Hooper, D.C., Bagasra, O., Marini, J.C., Zborek, A., Ohnishi, S.T., Kean, R., Champoin, J.M., Sarler, A.B., Bobroski, L., Farber, J.L., Akaike, T., Maeda, H. and Koprowski, H. (1997) Prevention of EAE by targeting nitric oxide and peroxynitrie. *Proc. Natl. Acd. Sci. USA*, 94: 2528–2533.

Htain, W.W., Leong, S.K. and Ling, E.A. (1997) In vivo expression of inducible nitric oxide synthase in supraventicular amoeboid microglial cells in neonatal BALB/c and athymic mice. *Neurosci. Lett.*, 223: 53–56.

Huang, Z., Huang, P.L., Panahian, N., Dalkara, T., Fishman, M.C. and Moskowitz, M.A. (1994) Effects of cerbral ischemia in mice deficient in neuronal nitric oxide synthase. *Science*, 265: 1883–1885.

Hulkower, K., Brosnan, C.F., Aquino, D.A., Cammer, W., Kulshrestha, S., Guida, M.P., Rapoport D.A. and Berman, J.W. (1993) Expression of CSF-1, c-fms, and MCP-1 in the central nervous system of rats with experimental allergic encephalomyelitis. *J. Immunol.*, 150: 2525–2533.

Iadecola, C. (1997) Bright and dark sides of nitric oxide in ischemic brain injury. *Trends Neurosci.*, 20: 132–139.

Iadecola, C., Zhang, F., Xu, S., Casey, R. and Ross, M.E. (1995a) Inducible nitric oxide synthase expression in brain following cerebral ischemia. *J. Cerebral Blood Flow Metab.*, 15: 378–384.

Iadecola, C., Zhang, F. and Xu, X. (1995b) Inhibition of inducible nitric oxide synthase ameliorates cerebral ischemic damage. *Am. J. Physiol.*, 268: R286–R292.

Iadecola, C., Xu, X., Zhang, F., El-Fakahany, E.E. and Ross, M.E. (1995c) Marked induction of calcium-independent nitric oxide synthase activity after focal cerebral ischemia. *J. Cereb. Blood Flow Metab.*, 15: 52–59.

Iadecola, C., Zhang, F., Casey, R., Clark, H.B. and Ross M.E. (1996) Inducible nitric oxide synthase gene expression in vascular cells after transient focal cerebral ischemia. *Stroke*, 27: 1373–1380.

Iadecola, C. Zhang, F., Casey, R., Nagayama, M. and Ross, M.E. (1997) Delayed reduction of ischemic brain injury and neurological deficits in mice lacking the inducible nitric oxide synthase gene. *J. Neurosci.*, 17: 9157–9164.

Johnson, A.W., Land, J.M., Thompson, E.J., Bolanos, J.P., Clark, J.B. and Heales, S.J. (1995) Evidence for increased nitric oxide production in multiple sclerosis. *J. Neurol. Neurosur. Psych.*, 58: 107–116.

Kader, A., Frazzini, V.I., Solomon, R.A. and Trifiletti, R.R. (1993) Nitric oxide production during focal cerebral ischemia in rats. *Stroke*, 24: 1709–1716.

Karpus, W.J., Lukacs, N.W., McRae, B.L., Strieter, R.M., Kunkel, S.L. and Miller, S.D. (1995) An important role for the chemokine macrophage inflammatory protein-1a in the pathogenesis of the T cell-mediated autoimmune disease, experimental autoimmune encephalomyelitis. *J. Immunol.*, 155: 5003–5010.

Khan, B.V., Harrison, D.G., Olbrych, M.T., Alexander, R.W. and Medford, R.M. (1996) Nitric oxide regulates vascular cell adhesion molecule 1 gene expression and redox-sensitive transcriptional events in human vascular cells. *Proc. Natl. Acad. Sci. USA*, 93: 9114–9119.

Kim, J.S., Gautam, S.C., Chopp, M., Zaloga, C., Jones, M.L., Ward, P.A. and Welch, K.M.A. (1995) Expression of monocyte chemoattractant protein-1 and macrophage inflammatory protein-1 after focal cerebral ischemia in the rat. *J. Neuroimmunol.*, 56: 127–134.

Knowles, R.G. (1997) Nitric oxide, mitochondria and metabolism. *Trans. Biochem. Soc.*, 25: 895–901.

Koka, P., He, K., Zack, J.A., Kitchen, S., Peacock, W., Fried, I., Tran, T., Yashar, S.S. and Merrill, J.E. (1995) Human immunodeficiency virus 1 envelope proteins induce interleukin 1, tumor necrosis factor α and nitric oxide in glial culture from fetal, neonatal, and adult human brain. *J. Exp. Med.*, 182: 941–952.

Kopnisky, K.L., Sumners, C. and Chandler, L.J. (1997) Cytokine- and endotoxin-induced nitric oxide synthase in rat astroglial cultures. *J. Neurochem.*, 68: 935–944.

Kossman, T., Stahel, P.F., Lenzlinger, P.M., Redl, H., Dubs, R.W., Trentz, O., Schalg, G. and Morganti-Kossman, M.C. (1997) Interleukin 8 released into the CSF after brain injury is associated with blood-brain barrier dysfunction and NGF production. *J. Cereb. Blood Flow Metab.*, 17: 280–289.

Krupinski, J., Kumar, P., Kumar, S. and Kaluza, J. (1996) Increased expression of TGF-b1 in brain tissue after ischemic stroke in humans. *Stroke*, 27: 852–857.

Kuluz, J.W., Prado, R.J., Dietrich, W.D., Schleine, C.L. and Watson, B.D. (1993) The effect of nitric oxide synthase inhibition on infarct volume after reversible focal cerebral ischemia in conscious rats. *Stroke*, 24: 2023–2029.

Lane, T.E., Buchmeier, M.J., Watry, D.D. and Fox, H.S. (1996) Expression of inflammatory cytokines and inducible nitric oxide synthase in brains of SIV-infected Rhesus monkeys: Applications to HIV-induced central nervous system disease. *Molec. Med.*, 2: 27–37.

Lane, T.E., Paoletti, A.D. and Buchmeier, M.J. (1997) Disassociation between the invitro and in vivo effects of nitric oxide on a neurotropic murine coronavirus. *J. Virol.*, 71: 2202–2210.

Lei, D.L., Yang, D.L. and Liu, H.M. (1996) Local injection of kainic acid causes widespread degeneration of NADPH-d neurons and induction of NADPH-d in neurons, endothelial cells and reactive astrocytes. *Brain Res.*, 730: 199–206.

Lin, H.L. and Murphy, S. (1997) Regulation of astrocyte nitric oxide synthase type II expression by ATP and glutamate involves loss of transcription factor binding to DNA. *J. Neurochem.*, 69: 612–616.

Loihl, A.K. and Murphy, S. (1997) Expression of nitric oxide synthase-2 in various cell types following focal cerebral ischemia in mice. *Abst. Soc. Neurosci.*, 23: 2437.

Luss, H., DiDilvio, M., Litton, A.L., Molin y Vedia, L., Nussler, A.K. and Billiar, T.R. (1994) Inhibition of nitric oxide synthesis enhances the expression of inducible nitric oxide synthase mRNA and protein in a model of chronic liver inflammation. *Biochem. Biophys. Res. Commun.*, 204: 635–640.

MacMicking, J.D., Nathan, C., Hom, G., Chartrain, N., Fletcher, D.S., Trumbauer, M., Stevens, K., Xie, Q., Sokol, K., Hutchinson, N., Chen, H. and Mudgett, J.S. (1995) Altered responses to bacterial infection and endotoxic shock in mice lacking inducible nitric oxide synthase. *Cell*, 81: 641–650.

Malinski, T., Bailey, F., Zhang, Z.G. and Chopp, M. (1993) Nitric oxide measured by a porphyrinic microsensor in rat brain after transient middle cerebral artery occlusion. *J. Cereb. Blood Flow Metab.*, 13: 355–358.

McCall, T.B., Palmer, R.M.J. and Moncada, S. (1992) Interleukin-8 inhibits the induction of nitric oxide synthase in rat peritoneal neutrophils. *Biochem. Biophys. Res. Commun.*, 186: 680–685.

Meda, L., Bernasconi, S., Bonaiuto, C., Sozzani, S., Zhou, D., Otvos, L., Mantovani, A., Rossi, F. and Cassatella, M.A. (1996) β-amyloid (25–35) peptide and IFN-γ synergistically induce the production of the chemotactic cytokine MCP-1/JE in monocytes and microglial cells. *J. Immunol.*, 157: 1213–1218.

Merrill, J.E., Genain, C.P., Parkinson, J.F., Medberry, P., Halks-Miller, M., DelVecchio, V., Kardos, S. and Murphy, S. (1998) iNOS and nitrotyrosine in macrophages and glia in demyelinating lesions. Submitted

Merrill, J.E. and Murphy, S. (1997) Regulation of gene expression in the nervous system by reactive oxygen and nitrogen species. *Metab. Brain Disease*, 12: 97–112.

Merrill, J.E. and Benveniste, E.N. (1996) Cytokines in inflammatory brain lesions: Helpful and harmful. *Trends Neurosci.*, 19: 331–338.

Miyagishi, R., Kikuchi, S., Fukazawa, T. and Tashiro, K. (1995) Macrophage inflammatory protein-1a in the cerebrospinal fluid of patients with multiple sclerosis and other inflammatory neurological diseases. *J. Neurol. Sci.*, 129: 223–227.

Murphy, S. and Grzybicki, D. (1996) Glial NO: Normal and pathological roles. *Neuroscientist*, 2: 91–100.

Murphy, S., Simmons, M.S., Agullo, L., Garcia, A., Feistein, D.L., Gallea, E., Reis, D.L., Minc-Golomb, D. and Schwartz, J.P. (1993) Synthesis of nitric oxide in CNS glial cells. *Trends Neurosci.*, 16: 323–328.

Nagafugi, T., Matsui, T., Koide, T. and Asano, T. (1992) Blockade of nitric oxide formation by N^{w}-nitro-L-arginine mitigates ischemic brain edema and subsequent cerebral infarction in rats. *Neurosci. Lett.*, 147: 159–162.

Nathan, C. (1997) Inducible nitric oxide synthase. *J. Clin. Invest.*, 100: 2417–2423.

Nowicki, J.P., Duval, D., Poignet, H. and Scatton, B. (1991) Nitric oxide mediates neuronal death after focal cerebral ischemia in the mouse. *Eur. J. Pharmacol.*, 204: 339–340.

Okuda, Y., Nakatsuji, Y., Fujimura, H., Esumi, H., Ogura, T., Yanagihara, T. and Sakoda, S. (1995) Expression of the inducible isoform of nitric oxide synthase in the central nervous system of mice correlates with the severity of actively induced experimental allergic encephalomyelitis. *J. Neuroimmunol.*, 62: 103–12.

Okuda, Y., Sakoda, S., Fujimura, H. and Yanagihara, T. (1997) Nitric oxide via an inducible isoform of nitirc oxide synthase is a spossible factor to eliminate inflammatory cells from the CNS of mice with EAE. *J. Neuroimmunol.*, 73: 107–116.

Oleszak, E.L., Katsetos, C.D., Kuzmak, J. and Varadhachary, A. (1997) Inducible nitric oxide synthase in Theiler's murine encephalomyelitis virus infection. *J. Virol.*, 71: 3228–3235.

Park, S.K. and Murphy, S. (1994) Duration of expression of inducible nitric oxide synthase in glial cells. *J. Neurosci. Res.*, 39: 405–411.

Park, S.K. and Murphy, S. (1996) Nitric oxide synthase type II mRNA stability is translation- and transcription-dependent. *J. Neurochem.*, 67: 1766–1769.

Park, S.K., Lin, H.L. and Murphy, S. (1994) Nitric oxide limits transcriptional induction of nitric oxide synthase in CNS glial cells. *Biochem. Biophys. Res. Commun.*, 201: 762–768.

Park, S.K., Lin, H.L. and Murphy, S. (1997) Nitric oxide regulates nitric oxide synthase-2 gene expression by inhibiting NF-κB binding to DNA. *Biochem. J.*, 322: 609–613.

Peng, H., Libby, P. and Liao, J.K. (1995) Induction and stabilization of IκBα by nitric oxide mediates inhibition of NF-κB. *J. Biol. Chem.*, 270: 14214–14219.

Ransohoff, R.M., Glabinski, A. and Tani, M. (1996) Chemokines in immune-mediated inflammation of the central nervous system. *Cytokine and Growth Factor Rev.*, 7: 35–46.

Rojas, A., Delgado, R., Glaria, L. and Palacios, M. (1993) MCP-1 inhibits the induction of nitric oxide synthase in J774 cells. *Biochem. Biophys. Res. Commun.*, 196: 274–279.

Rossi, F. and Bianchini, E. (1996) Synergistic induction of nitric oxide by β-amyloid and cytokines in astrocytes. *Biochem. Biophys. Res. Commun.*, 225: 474–478.

Ruuls, S.R., Van Der Linden, S., Sontrop, K., Huitinga, I. and Dijkstra, C.D. (1996) Aggravation of experimental allergic encephalomyelitis by administration of nitric oxide synthase inhibitors. *Clin. Exp. Immunol.*, 103: 467–476.

Sato, S., Tominaga, T., Ohnishi, T. and Ohnishi, S.T. (1994) Role of nitric oxide in brain ischemia. *Ann. NY. Acad. Sci.*, 738: 369–373.

Shafer, R. and Murphy, S. (1997) Activated astrocytes induce nitric oxide synthase-2 in cerebral endothelium via TNFa. *GLIA*, 21: 370–379.

Shibata, M., Araki, N., Hamada, J., Sasaki, T., Shimazu, K. and Fukuuchi, Y. (1996) Brain nitrite production during global ischemia and reperfusion. *Brain Res.*, 735: 86–90.

Siesjö, B.K. (1992) Pathophysiology and treatment of focal cerebral ischemia. *J. Neurosurg.*, 77: 169–184.

Spanaus, K-S., Nadal, D., Pfister, H-W., Seebach, J., Widmer, U., Frei, K., Gloor, S. and Fontana, A. (1997) C-X-C and C-C chemoklines are expressed in the CSF in bacterial meningitis and mediate chemotactic activity on peripheral blood-derived polymorphonclear and mononuclear cells in vitro. *J. Immunol.*, 158: 1956–1964.

Spitsin, S.V., Koprowski, H. and Michaels, F.H. (1996) Characterization and functional analysis of the human inducible nitric oxide synthaser gene promoter. *Molec. Med.*, 2: 226–235.

Stewart, V.C., Giovannoni, G., Land, J.M., McDonald, W.I., Clark, J.B. and Heales, S.J.R. (1997) Pretreatment of astrocytes with interferon α/β impairs interferon γ induction of nitric oxide synthase. *J. Neurochem.*, 68: 2547–2551.

Sun, N., Grzybicki D., Castro, R.F., Murphy, S. and Perlman, S. (1995) Activation of astrocytes in the spinal cord of mice chronically infected with a neurotropic coronavirus. *Virology*, 213: 482–493.

Takami, S., Nishikawa, H., Minami, M., Nishiyori, A., Sato, M., Akaike, A. and Satoh, M. (1997) Induction of MIP-1a mRNA on glial cells after focal ischemia in the rat. *Neurosci. Lett.*, 227: 173–176.

Togashi, H., Sasaki, M., Frohman, E., Taira, E., Ratan, R.R., Dawson, T.M. and Dawson, V.L. (1997) Neuronal (type I) nitric oxide synthase regulates nuclear factor κB activity and immunologic (type II) nitric oxide synthase expression. *Proc. Natl. Acad. Sci. USA*, 94: 2676–2680.

Tran, E.H., Hardin-Pouzet, H., Verge, G. and Owens, T. (1997) Astrocytes and microglia exporess inducible nitric oxide synthase in mice with EAE. *J. Neuroimmunol.*, 74: 121–129.

Tsao, P.S., Buitrago, R., Chan, J.R. and Cooke, J.P. (1996) Fluid flow inhibits endothelial adhesiveness. *Circulation*, 94: 1682–1689.

van der Veen, R.C., Hinton, D.R., Incardonna, F. and Hofman, F.M. (1997) Extensive peroxynitrite actvivity during progressive stages of CNS inflammation. *J. Immunol.*, 77: 1–7.

Vodovotz, Y., Lucia, M.S., Flanders, K.C., Chesler, L., Xie, Q.W., Smith, T.W., Weidner, J., Mumford, R., Webber, R., Nathan, C., Roberts, A.B., Lippa, C.F. and Sporn, M.B. (1996) Inducible nitric oxide synthase in tangle-bearing neurons of patients with Alzheimer's disease. *J. Exp. Med.*, 184: 1425–1433.

Wang, X., Yue, T., White, R.F., Baroen, F.C. and Feuerstein, G.Z. (1995) Transforming growth factor β1 exhibits delayed gene expression following focal cerebral ischemia. *Brain Res. Bull.*, 36: 607–609.

Weldon, D.T., Rogers, S.D., Ghilardi, J.R., Finke, M.P., Cleary, J.P., O'Hare, E., Esler, W.P., Maggio, J.E. and Mantyh, P.W. (1998) Fibrillar β-amyloid induces microglial phagocytosis expression of inducible nitric oxide synthase and loss of a select population of neurons in the rat CNS in vivo. *J. Neurosci.*, 18: 2161–2173.

Willenborg, D.O., Fordham, S.A., Cowden, W.B. and Ramshaw, I.A. (1995) Cytokines and murine autoimmune encephalomyelitis: Inhibition or enhancement of disease with antibodies to select cytokines, or by delivery of exogenous cytokines using a recombinant vaccinia virus system. *Scand. J. Immunol.*, 41: 31–41.

Wong, M.L., Rettori, V., Al-Shekhlee, A., Bongiorno, P.B., Canteros, G., McCann, S.M., Gold, P.W. and Licinio, J. (1996) Inducible nitric oxide synthase gene expression in the brain during systemic inflammation. *Nature Med.*, 2: 581–584.

Xie, Q-W. and Nathan C. (1994) The high output nitric oxide pathway: Role and regulation. *J. Leuk. Biol.*, 56: 576–582.

Yamamoto, S., Golanov, E.V., Berger, S.B. and Reis, D.J. (1992) Inhibition of nitric oxide synthase increases focal ischemic infarction in rat. *J. Cereb. Blood Flow Metab.*, 12: 717–726.

Yamasaki, Y., Matsuo, Y., Matsuura, N., Onodera, H., Itoyama, Y. and Kogure, K. (1995) Transient increase of cytokine-induced neutrophil chemoattractant, a member of the interleukin-8 family, in ischemic brain areas after focal ischemia in rats. *Stroke*, 26: 318–323.

Yamasaki, Y., Matsuo, Y., Zagorski, J., Matsuura, N., Onodera, H., Itoyama, Y. and Kogure, K. (1997) New therapeutic possibility of blocking cytokine-induced neurtrophil chemoattractant on tranisent ischemic brain damage in rats. *Brain Res.*, 759: 103–111.

Yang, G., Kitagawa, K., Matsushita, K., Mabuchi, T., Yagita, Y., Yanagihara, T. and Matsumoto, I. (1997) C57BL/6 strain is most susceptible to cerebral ischemia following bilateral common carotid occlusion among seven mouse strains. *Brain Res.*, 752: 209–218.

Yoshida, T., Waeber, C., Huang, Z. and Moskowitz, M.A. (1995) Induction of nitric oxide synthase activity in rodent brain following middle cerebral artery occlusion. *Neurosci. Lett.*, 194: 214–218.

Yoshimoto, T., Houkin, K., Tada, M. and Abe, H. (1997) Induction of cytokines, chemokines and adhesion molecule mRNA in a rat forebrain reperfusion model. *Acta Neuropathol.*, 93: 154–158.

Zea Longa, E., Weinstein, P.R., Carlson, S. and Cummins, R. (1989) Reversible middle cerebral artery occlusion without craniectomy in rats. *Stroke*, 20: 84–91.

Zeiher, A.M., Fisslthaler, B., Schray-Utz, B. and Busse, R. (1995) Nitric oxide modulates the expression of monocyte chemoattractant protein 1 in cultured human endothelial cells. *Circ. Res.*, 76: 980–986.

Zhang, F., Casey, R., Ross, M.E. and Iadecola, C. (1996) Aminoguanidine ameliorates and L-arginine worsens brain damage from intraluminal middle cerebral artery occlusion. *Stroke*, 27: 317–323.

Zhao, W., Tilton, R.G., Corbett, J.A., McDaniel, M.L., Misko, T.P., Williamson, J.R., Cross, A.H. and Hickey, W.F. (1996) Experimental allergic encephalomyelitis in the rat is inhibited by aminoguanidine, an inhibitor of nitric oxide synthase. *J. Neuroimmunol.*, 64: 123–133.

Zielasek, J., Jung, S., Gold, R., Liew, F.Y., Toyka, K.V. and Hartung, H.-P. (1995). Administration of nitric oxide synthase inhibitors in experimental autoimmune neuritis and experimental autoimmune encephalitis. *J. Neuroimmunol.*, 58: 81–88.

R.R. Mize, T.M. Dawson, V.L. Dawson and M.J. Friedlander (Eds.)
Progress in Brain Research, Vol 118

CHAPTER 19

Role of endogenous nitric oxide and peroxynitrite formation in the survival and death of motor neurons in culture

Alvaro G. Estévez[1,4,6], Nathan Spear[1,4], S. Machelle Manuel[5], Luis Barbeito[6,8], Rafael Radi[7] and Joseph S. Beckman[1,2,3,4,*]

[1]*Department of Anesthesiology,* [2]*Department of Biochemistry and Molecular Genetics,* [3]*Department of Neurobiology and* [4]*The UAB Center For Free Radical Biology, The University of Alabama at Birmingham, Birmingham, AL, USA*
[5]*Schoool of Pharmacy, Texas Tech University Health Sciences Center, Amarillo, Texas, USA*
[6]*Sección Neurociencias, Facultad de Ciencias,* [7]*Departamento de Bioquímica, Facultad de Medicina, Universidad de la República*
[8]*División Neurobiología Celular y Molecular, Instituto Clemente Estable, Montevideo, Uruguay*

The selective degeneration of motor neurons is the predominant pathological finding in amyotrophic lateral sclerosis (ALS) and a variety of spinal atrophies. In spite of recent advances implicating mutations to superoxide dismutase (SOD) with familial ALS, the mechanisms leading to motor neuron degeneration remain mysterious. The recent development of an efficient method to purify and culture motor neurons provides a powerful tool to study the requirements for the survival of motor neurons and the mechanisms responsible for their death in vitro (Camu and Henderson, 1994; Henderson et al., 1995). We have found trophic factor deprivation of cultured motor neurons to be a useful model for exploring two divergent effects of nitric oxide (NO). The production of NO is important for the survival of motor neurons through a cGMP-dependent mechanism. Paradoxically NO can also serve as the precursor for peroxynitrite, which induces cell death by apoptosis when motor neurons are deprived of trophic factors. The behavior of motor neurons in vitro can provide important clues concerning the role of nitric oxide and superoxide in their selective degeneration in ALS.

Embryonic rat spinal motor neurons can be prepared free of other neurons and glia by a combination of centrifugation with metrizamide density gradient and immunopanning with the monoclonal antibody IgG192 against the p75 low affinity neurotrophin receptor (Henderson et al., 1995). During the development of the spinal cord, the p75 receptor is selectively expressed by motor neurons (Yan and Johnson, 1988). With appropriate trophic support, motor neurons develop long branched neurites after plating on a substrate of poly-ornithine and laminin. Under these conditions, more than 94% of the cells are immunoreactive for the motor neuron markers p75 neurotrophin receptor and Islet 1/2 (Henderson et al., 1993, 1994; Pennica et al., 1996; Estévez et al., 1998a). Originally, motor neurons were cultured in L15 medium supplemented with neurotrophins or GDNF (Henderson et al., 1993, 1994), and remained viable for only 72 h, after

*Corresponding author. Tel.: (205) 934 5422; fax: (205) 934 7437; e-mail: joe.beckman@ccc.uab.edu

which ~70% of the neurons underwent apoptosis in the next 3 days (Estévez et al., 1998a). An important recent advance has resulted from changing the culture media to B27-supplemented Neurobasal medium (Gibco, Grand Island, NY), optimized originally for the culture of hippocampal neurons (Brewer et al., 1993). In Neurobasal media supplemented with cardiotrophin 1, motor neurons remain viable for more than 6 days. Even after 16 days, 43% of the motor neurons remain alive (Pennica et al., 1996). We have found that BDNF and GDNF can replace cardiotrophin 1 in Neurobasal medium supporting motor neuron survival for at least 9 days.

Survival of motor neurons in culture depends strongly upon an appropriate supply of trophic factors (Arakawa et al., 1990, Sendtner et al., 1990, 1991; Henderson et al., 1993, 1994; Hughes et al., 1993; Pennica et al., 1996; Yamamoto et al., 1997). Motor neurons cultured without trophic factors will initially attach and extend neurites for the first 18 h after isolation, but then become round with shrunken soma and withdrawn neurites by 24 h. Under these conditions, the cells showed clear evidence of nuclear condensation with DNA fragmentation visualized by the TUNEL method (Estévez et al., 1998a). Motor neuron death in culture is dependent upon protein synthesis (Milligan et al., 1994) and can be prevented by caspase inhibitors (Milligan et al., 1995; Estévez et al., 1998a), indicating that motor neurons deprived of trophic support undergo apoptosis.

Nitric oxide participates in the induction of motor neuron apoptosis

We have been interested in what role the nitric oxide-derived oxidants like peroxynitrite may play in motor neuron death (Beckman et al., 1993). Significant evidence suggests that neuronal NOS can be induced in motor neurons and contribute to motor neuron death in vivo. Proximal injury to motor neuron axons in the ventral root induces NOS expression (Wu, 1993; Wu and Li, 1993; Yu, 1994; Novikov et al., 1995, 1997) and L-NAME prevents the subsequent motor neuron degeneration (Wu and Li, 1993). Continuous infusion of BDNF in the same model blocks NOS expression and prevents motor neuron death (Novikov et al., 1995, 1997). In addition, axotomy induces NOS and apoptosis in facial motor neurons (Dubois-Dauphin et al., 1994; Ruan et al., 1995; de Bilbao and Dubois-Dauphin, 1996), suggesting that impairment of target-derived trophic factor is sufficient to induce NOS and apoptosis. Increased neuronal NOS is also present in motor neuron degeneration after trauma (Wu et al., 1994) and in ALS (Chou et al., 1996a,b; Abe, 1997).

Motor neurons cultured with BDNF do not express neuronal nitric oxide synthase (NOS), but trophic factor deprivation is sufficient to induce the enzyme (Estévez et al., 1998a). Inhibition of NOS activity largely prevented motor neuron death for up to 3 days after plating in L15 medium without trophic factors. These results suggest that the induction of neuronal NOS plays a role in the initiation of motor neuron apoptosis. Because of the extremely low density of the cultures, NO is most likely acting within the same cell where it is produced. After 3 days, the survival-promoting effects of BDNF and L-NAME decreased in parallel (Estévez et al., 1998a). The protective effects of NOS inhibition were overcome by generating a low steady state concentration (< 100 nM) of nitric oxide in the culture media. The NO was generated by the spontaneous dissociation of 20 μM DETA-NONOate, which has a half-life of 56 h at pH 7.4. DETA-NONOate is a much cleaner means to generate authenthic NO compared to nitrosothiols or sodium nitroprusside. Nitric oxide generated under these conditions was not toxic to motor neurons cultured with BDNF and can even improve the survival of motor neurons by stimulating the formation of cyclic guanosine monophosphate (cGMP).

Nitric oxide helps BDNF-support of motor neuron survival

Although inhibition of NO production is sufficient to prevent motor neuron death in culture after

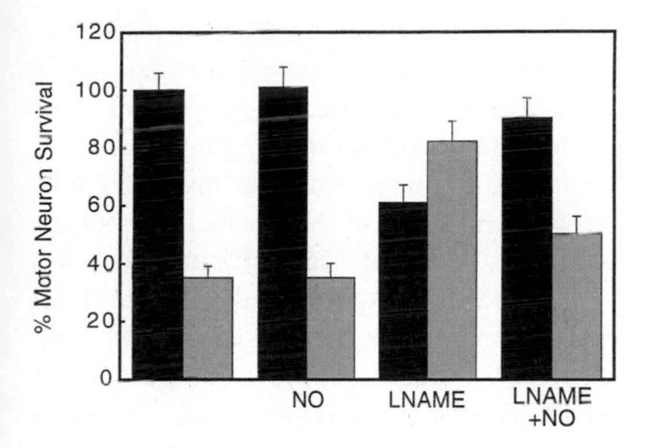

Fig. 1. Contrasting actions of NO on motor neuron survival. Motor neurons isolated from E15 rat embryonic spinal cords were plated at a density of 2000 cells/35 mm dish in L15 medium supplemented as previously described (Estévez, et al., 1998a) deprived of trophic factors (open bars) or with 100 pg/ml BDNF (filled bars). Motor neurons survival was assessed by counting all neurons with neurites longer than 4 soma diameter in a 1 cm^2 field in the center of the dish 3 days after plating. NO at a 100 nM steady state concentration was generated from 20 μM DETA-NONOate (Estévez, et al., 1998a). NO production was inhibited with 1 mM L-NAME in these experiments.

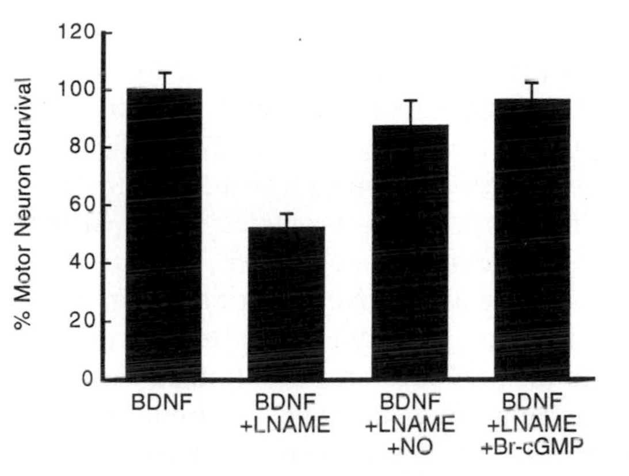

Fig. 2. cGMP prevents L-NAME-induced apoptosis of BDNF-treated motor neurons. BDNF was used at a 100 pg/ml. L-NAME and Br-cGMP were added at a final concentration of 1 mM.

trophic factor deprivation, NO itself is not directly toxic to motor neurons in culture (Estévez et al., 1998a). NO production can help to maintain motor neuron survival in the presence of trophic factors. L-NAME decreased cell viability of BDNF-treated motor neurons (Fig. 1) suggesting that trophic effects of BDNF require NO production. Nitric oxide generation by DETA-NONOate prevented L-NAME-induced death of BDNF-treated motor neurons at the same concentration that caused the death of motor neurons cultured with L-NAME but without BDNF (Fig. 1).

Nitric oxide is an important intercellular messenger in the central nervous system involved in the regulation of long term potentiation and depression. The best characterized physiological action of nitric oxide is to stimulate cGMP synthesis through the activation of the soluble guanylate cyclase (Ignarro, 1989; Mayer, 1994; Garthwaite and Boulton, 1995; Farinelli et al., 1996). Although, the protective and regulatory activities of NO have been suggested to involve nitrosylation of thiol groups (Lipton et al., 1993; Stamler et al., 1997), the survival of embryonic motor neurons depends upon the activation of the soluble guanylate cyclase (Fig. 3). Before NO was recognized to activate guanylate cyclase in vivo, cGMP was shown to reduce programmed cell death of motor neurons during development (Weill and Greene, 1984, 1990). We found that the membrane-permeable analog 8Br-cGMP protected BDNF-treated motor neurons from L-NAME-induced cell death (Fig. 2). Inhibition of the NO stimulated soluble guanylate cyclase with 1H-[1,2,4]oxadiazolo[4,3-a]quinoxalin-1-one (ODQ) also results in motor neuron death in the presence

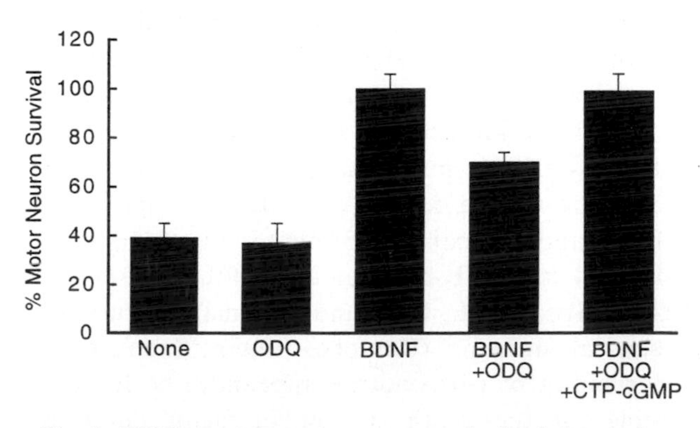

Fig. 3. BDNF-treated motor neuron death induced by inhibition of NO-stimulted cGMP synthesis. BDNF, ODQ and CTP-cGMP were added at 100 pg/ml, 2 and 100 μM respectively.

of BDNF, but did not affect the survival of trophic factor-deprived motor neurons (Fig. 3). Analogs of cGMP reversed the cell death resulting from inhibition of soluble guanylate cyclase (Fig. 3). These results strongly suggest that NO stimulation of cGMP production promotes motor neuron survival in the presence of BDNF.

Surprisingly, the source of nitric oxide in freshly isolated motor neurons is due to the type III isoform of NOS most commonly found in endothelium rather than the type I neuronal isoform. Both the mRNA and immunoreactivity for endothelial NOS are present in motor neurons cultured with BDNF (Estévez et al., 1998b). In contrast, the neuronal NOS mRNA and protein can only be detected in motor neurons cultured without trophic factors (Estévez et al., 1998a). Abe et al. (1997) has recently reported that adult human motor neurons are immunoreactive for endothelial NOS. In the presence of trophic factors, the activity of the constitutively expressed endothelial NOS may produce sufficient endogenous NO to prevent motor neuron death in culture.

Peroxynitrite mediates NO-dependent induction of apoptosis

The previous results indicate that in conditions of trophic factor deprivation the cell is producing something that makes NO toxic. Formation of peroxynitrite most likely accounts for the NO toxicity (Beckman, 1991, Beckman et al., 1993; Lipton et al., 1993; Dawson et al., 1993, Dawson and Dawson, 1996; Bonfoco et al., 1995; Estévez et al., 1995; Troy et al., 1996; Strijbos et al., 1996). Peroxynitrite is a strong oxidant formed by the diffusion-limited reaction of NO and superoxide (Huie and Padmaja, 1993), that induces apoptosis in PC12 cells (Estévez et al., 1995; Troy et al., 1996; Spear et al., 1997) and in primary cultures of cortical neurons (Bonfoco, 1995). Endogenous formation of peroxynitrite appears to be responsible for the death of trophic factor deprived motor neurons in culture (Estévez et al., 1998a) and for the excitotoxic death of cultured granule cells (Tabuchi et al., 1996).

Nitric oxide is the only known biological molecule that reacts faster with superoxide than SOD can scavenge superoxide (Beckman, 1996). Under normal physiological conditions, the molar concentration of SOD is substantially higher in cells than the concentration of NO, which minimizes the formation of peroxynitrite. Decreased scavenging of superoxide often leads to neuronal degeneration. Inhibition of SOD expression with antisense RNA induces apoptosis of PC12 cells (Troy and Shelanski, 1994; Troy et al., 1996) and motor neurons in organotypic cultures of spinal cord (Rothstein et al., 1994). Mice homozygous and heterozygous for an SOD mutation that impairs superoxide scavenging activity show increased infarct volume and neuronal apoptosis after ischemic injury (Kondo et al., 1997). Overexpression of human wild type SOD prevents apoptosis of sympathetic neurons induced by trophic factor deprivation (Jordan et al., 1995; Greenlund et al., 1995), and the apoptotic degeneration of rat hippocampal neurons in culture induced by staurosporine (Prehn et al., 1997). Although extracellular human wild type SOD was not protective for motor neurons, treatment of motor neurons with manganese TBAP, a membrane permeant SOD mimetic (Faulkner et al., 1994; Szabo et al., 1996), largely prevents motor neuron apoptosis induced by trophic factor deprivation (Estévez et al., 1998a). In addition, SOD delivered in liposomes largely prevents motor neuron apoptosis induced by trophic factor deprivation (Estévez A.G. and Beckman J.S., unpublished observations). The protection by SOD suggests that intracellular superoxide formation is necessary for the induction of motor neuron apoptosis, suggesting the formation of peroxynitrite may be the responsible agent.

Nitrotyrosine in motor neuron pathology

Peroxynitrite reacts with both free and protein-bound tyrosine to form nitrotyrosine in a reaction catalyzed by SOD (Beckman et al., 1992; Ischiropoulos et al., 1992). Nitrotyrosine immunoreactivity is a marker for peroxynitrite and possibly other NO-derived oxidants in vivo, but is not formed by NO itself (Beckman, 1996). Nitrotyrosine immunoreactivity has been found in Alzheimer's disease (Good et al., 1996; Smith et al., 1997), multiple sclerosis (Bagasra, 1995), ALS (Abe et al., 1995, 1997; Chou et al., 1996a,b; Beal et al., 1997) and in animal models of Parkinson's disease (Hantraye et al., 1996) and Huntington's disease (Schulz et al., 1996).

We found increased nitrotyrosine immunoreactivity in trophic factor-deprived motor neurons compared with BDNF-treated cultures that further supports for peroxynitrite formation. Either inhibition of NO formation or increased superoxide scavenging can prevent nitrotyrosine accumulation and motor neuron death (Estévez et al., 1998a). These results suggest that a common mechanism involving peroxynitrite formation may account for the neuronal death in neurodegenerative diseases and in cell culture.

Whether nitrotyrosine formation is involved in peroxynitrite toxicity or can be used only as a marker of oxidant formation is an important issue that will require further investigation. Although the mechanism of peroxynitrite-induced apoptosis remains unknown, peroxynitrite is a complex oxidant that can attack a variety of biological targets, including thiols (Radi et al., 1991a), thiol ethers (Moreno and Peyor, 1992), iron–sulfur centers (Castro et al., 1994; Hausladen and Fridovich, 1994) and zinc fingers (Crow, 1995). It can also initiate lipid peroxidation (Radi et al., 1991b). These other oxidative reactions may well account for the toxicity of peroxynitrite. However, the association of SOD mutations with ALS and their propensity to become zinc deficient and serve as better catalysts of tyrosine nitration (Crow et al., 1997a) suggests that tyrosine nitration might have a causal connection with motor neuron degeneration.

SOD mutations in ALS

About 20% of the cases of the familial ALS (FALS) are associated with mutations the cytoplasmic Cu/Zn SOD (Rosen et al., 1993; Deng, et al., 1993, 1995). Transgenic mice carrying out the FALS mutant SOD develop the disease (Gurney et al., 1994; Price et al., 1994; Dal Canto and Gurney, 1995; Bruijn et al., 1997; Ferrante et al., 1997), indicating that the mutant SODs lead to the death of motor neurons. Overexpression of the antiapoptotic protein Bcl-2 delays the onset of the disease and prolonged the life of transgenic mice carrying the G93A SOD mutant (Kostic et al., 1997). The expression of a dominant negative inhibitor of ICE in transgenic mice carrying the same mutation slows the progression of the disease (Friedlander et al., 1997). These results strongly suggest that the mechanism for motor neuron death in the transgenic mice for FALS SOD mutations involves apoptosis.

Substantial evidence indicates that a gain in function is responsible by the selective death of motor neurons (Beckman et al., 1993; Rabizadeh et al., 1995). Mutations in the SOD weaken the structure of the protein decreasing the affinity for zinc up to 50-fold compared to wild type with little effect on the affinity for copper (Lyons et al., 1996; Crow et al., 1997a). The affinity of neurofilament-L (NF-L) for zinc is sufficient to remove zinc from mutant SODs. The loss of zinc increases the efficiency of the SOD to catalyze nitration of proteins including NF-L (Crow et al., 1997a,b). The assembly of unmodified NF-L is inhibited by a small proportion of nitrated NF-L (Crow et al., 1997b). In addition, zinc loss decreases the superoxide dismutation activity of the SOD (Forman and Fridovich, 1973; Pantoliano et al., 1982). Increased levels of free and protein-bound nitrotyrosine in the spinal cord of the transgenic mouse with the G93A mutant SOD has been reported by Ferrante et al., (1997). Bruijn et al., (1997) found an increase in free nitrotyrosine concentrations at the beginning of the disease in the transgenic mouse with the G37R mutant SOD, but could not demonstrate greater nitration of protein in transgenic mice. In spite of the form in which nitrotyrosine is found, it is clear that an increased nitration of tyrosines occurs early at the beginning of the disease in transgenic mice and is found in human ALS patients.

A possible mechanism for motor neuron pathology

A hypothesis that may explain the mechanism of ALS involves an initial generation of NO at concentrations that compete with SOD for superoxide leading to the formation of peroxynitrite. NF-L is nitrated in a low proportion but potentially enough to disrupt the assembly of unmodified neurofilament subunits into intermediate filaments (Crow et al., 1997b). Aberrant polymerized NF-L can potentially pull zinc from SOD, which potentiates the capacity of the enzyme to catalyze tyrosine nitration (Crow et al., 1997a). In addition, the loss of zinc decreases the scavenging of superoxide by SOD (Forman and Fridovich, 1973; Pantoliano et al., 1982), which indirectly increases peroxynitrite formation. Disruption of neurofilaments in the axon may inhibit the retrograde transport (Sasaki and Iwata, 1997), leading to a functional deprivation of muscle and Schwann cell-derived trophic factors. Trophic factor deprivation is sufficient to stimulate de novo synthesis of NOS in cultured motor neurons (Estévez et al., 1998a) and a similar mechanism may be occurring in vivo (Wu, 1993; Wu and Li, 1993; Yu, 1994). This may result in an increased NO production that competes with SOD to form more peroxynitrite, leading to the death of the cell. At the same time, NO may diffuse to neighboring motor neurons at high enough concentrations to form peroxynitrite, initiating the death process in a new motor neuron. Alternatively, an injured motor neuron may release cytokines or other factors that induce nitric oxide in other motor neurons.

Induction of neuron death by trophic factor deprivation is a useful model for the study of neuronal apoptosis (for a review see Deckwerth 1997). The same trophic factors that support the survival of embryonic motor neurons in culture (Arakawa et al., 1990; Sendtner et al., 1991; Henderson et al., 1993, 1994; Hughes et al., 1993; Pennica et al., 1996) also prevent their degeneration in vivo during the period of programmed cell death (Oppenheim et al., 1991, 1992, 1995; Houenou et al., 1994). Also, the same trophic factors prevent the loss of adult motor neurons after axotomy (Sendtner et al., 1990, 1991, 1992; Yan et al., 1992; Koliatsos et al., 1993; Henderson et al., 1994; Li et al., 1995; Novikov et al., 1995, 1997; Schmalbruch and Rosenthal, 1995; Vejsada et al., 1995; Pennica et al., 1996). Apoptosis appears to be the most likely mechanism involved in motor neuron death during embryonic development, in pathology, after injury or in culture (Lo et al., 1995; Kostic et al., 1997). A growing number of similarities are emerging among the different models of motor neuron death and in pathology including induction of neuronal NOS and increased contents of free and protein-bound nitrotyrosine. The use of cultured embryonic motor neurons allows the study of the intracellular pathway and specific responses of the motor neurons that otherwise will be impossible. In motor neurons, BDNF inhibit the expression of neuronal NOS and an unknown source of super-

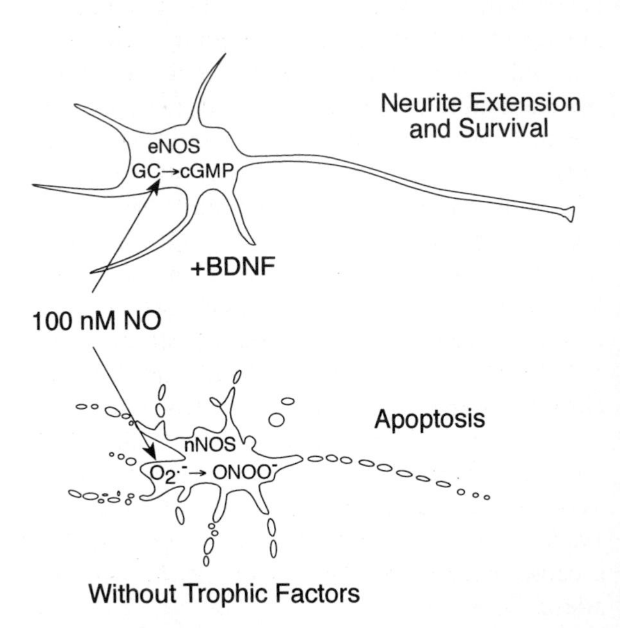

Fig. 4. NO effects in motor neuron survival. Trophic factor deprivation induces de novo synthesis of neuronal NOS and stimulates increased superoxide production, producing peroxynitrite and initiating to motor neuron apoptosis. With BDNF present, NO production by the endothelial NOS stimulates cGMP synthesis that helps to support the survival of the motor neurons.

oxide that make the NO toxic for the cells, probably by the formation of peroxynitrite, but it stimulates the fomation of NO by the endothelial NOS that activates the soluble guanylate cyclase to produce cGMP (Fig. 4). In contrast, BDNF can stimulate neuronal NOS expression in primary cultures of cortical neurons and retinal ganglion cells (Samdani et al., 1997; Klocker et al., 1998). In axotomized retinal ganglion cells, BDNF increases neuron death by a mechanism that is diminished by free radical scavengers and by NOS inhibitors (Klocker et al., 1998).

Other factors affecting motor neuron survival

Cultures of purified motor neurons have many limitations, ignoring for example the interaction of motor neurons with muscle and Schwann cells outside the spinal cord or with astrocytes, interneurons and cortical imputs in the spinal cord. These cells are clearly important for motor neuron survival in vivo (Oppenheim, 1996). The muscle is generally accepted as the major source of trophic factors for motor neurons (Henderson et al., 1981, 1983; Smith and Appel, 1983; Oppenheim, 1991; 1996). However, Schwann cells produced GDNF and CNTF (Williams et al., 1984; Manthorpe et al., 1986; Henderson et al., 1994), which can support motor neuron survival in culture (Arakawa et al., 1990; Sendtner et al., 1991; Henderson et al., 1994), after axotomy (Sendtner et al., 1990, 1991; Henderson et al., 1994; Vejsada et al., 1995) and prevent programmed cell death during development (Oppenheim et al., 1991, 1995; Sendtner et al., 1991). Astrocytic production of trophic factors for motor neurons is not well documented. Conditioned media from astrocytic monolayer increases the survival of motor neurons in culture. Riluzole, a drug used in ALS treatment, potentiates the production of trophic factors from astrocytes but it has no direct effects on motor neuron survival (Peluffo et al., 1997). Astrocytes also mediated the trophic effects of NGF, insulin and IGF-1 on motor neurons (Ang et al., 1993).

TABLE 1

Trophic factors modulate peroxynitrite-induced apoptosis of PC12 cells

Treatment	% Survival	
	Added before $ONOO^-$	Added after $ONOO^-$
Control	98 ± 1	98 ± 2
Peroxynitrite	62 ± 12	60 ± 10
NGF	98 ± 2	25 ± 8
Insulin	94 ± 3	92 ± 6
EGF	72 ± 8	63 ± 6
FGF2	36 ± 8	66 ± 10
FGF1 + heparin	27 ± 12	60 ± 13
FGF	71 ± 5	62 ± 7

PC12 cultures were treated with peroxynitrite for 5 min and cell viability was determined 24 h later using fluorescein diacetate and propidium iodide as described (Estévez et al., 1995; Spear et al., 1997). NGF was added 2 h before or immediately after peroxynitrite treatment. The values are the mean ± SEM. Concentrations used were: 1 mM peroxynitrite; 100 ng/ml NGF; 5 μg/ml insulin; 40 ng/ml EGF; 50 ng/ml FGF1 and FGF2; 50 U/ml heparin.

Summary

Motor neuron survival is highly dependent on trophic factor supply. Deprivation of trophic factors results in induction of neuronal NOS, which is also found in pathological conditions. Growing evidence suggests that motor neuron degeneration involves peroxynitrite formation. Trophic factors modulate peroxynitrite toxicity (Estévez et al., 1995; Shin et al., 1996; Spear et al., 1997). Whether a trophic factor prevents or potentiates peroxynitrite toxicity depends upon when the cells are exposed to the trophic factor (Table 1). These results strongly suggest that a trophic factor that can protect neurons under optimal conditions, but under stressful conditions can increase cell death. In this context, it is possible that trophic factors or cytokines produced as a response to damage may potentiate rather than prevent motor neuron death. A similar argument may apply to the therapeutic administration of trophic factors to treat neurodegenerative diseases. Similarly, the contrasting actions of

NO on motor neurons may have important consequences for the potential use of nitric oxide synthase inhibitors in the treatment of ALS and other related neurodegenerative diseases.

Acknowledgements

This work was supported by grants from the National Institutes of Health and the American ALS Association.

References

Abe, K., Pan, L.-H., Watanabe, M., Kato, T. and Itoyama, Y. (1995) Induction of nitrotyrosine-like immunoreactivity in the lower motor neuron of amyotrophic lateral sclerosis. *Neurosci. Lett.*, 199: 152–154.

Abe, K., Pan, L.H., Watanabe, M., Konno, H., Kato, T. and Itoyama, Y. (1997) Upregulation of protein-tyrosine nitration in the anterior horn of amyotrophic lateral sclerosis. *Neurol. Res.*, 19: 124–128.

Ang, L.C., Bhaumick, B. and Juurlink, B.H.J. (1993) Neurite promoting activity of insulin, insulin-like growth factor I and nerve growth factor on spinal motoneurons is astrocyte dependent. *Dev. Brain Res.*, 74: 83–88.

Arakawa, Y., Sendtner, M. and Thoenen, H. (1990) Survival effect of ciliary neurotrophic factor (CNTF) on chick embryonic motoneurons in culture: Comparison with other neurotrophic factor and cytokines. *J. Neurosci.*, 10: 3507–3515.

Bagasra, O., Michaels, F.H., Zheng, Y.M., Bobroski, L.E., Spitsin, S.V., Fu, Z.F., Tawadros, R. and Koprowski, H. (1995) Activation of the inducible form of nitric oxide synthase in the brains of patients with multiple sclerosis. *Proc. Natl. Acad. Sci. USA*, 92: 12041–12045.

Beal, M.F., Ferrante, R.J., Browne, S.E., Matthews, R.T., Kowall, N.W. and Brown, R.H. Jr. (1997) Increased 3-nitrotyrosine in both sporadic and familial amyotrophic lateral sclerosis. *Ann. Neurol.*, 42: 644–654.

Beckman, J.S. (1991) The double edged role of nitric oxide in brain function and superoxide-mediated pathology. *J. Dev. Physiol.*, 15: 53–59.

Beckman, J.S. (1996) Oxidative damage and tyrosine nitration from peroxynitrite. *Chem. Res. Toxicol.*, 9: 836–844.

Beckman, J.S., Carson, M., Smith, C.D. and Koppenol, W.H. (1993) ALS, SOD and peroxynitrite. *Nature*, 364: 584.

Beckman, J.S., Ischiropoulos, H., Zhu, L., van der Woerd, M., Smith, C., Chen, J., Harrison, J., Martin, J.C. and Tsai, M. (1992) Kinetics of superoxide dismutase- and iron-catalyzed nitration of phenolics by peroxynitrite. *Arch. Biochem. Biophys.*, 298: 438–445.

Bonfoco, E., Krainc, D., Ankarcrona, M., Nicotera, P. and Lipton, S.A. (1995) Apoptosis and necrosis: Two distinct events induced, respectively, by mild and intense insults with *N*-methyl-D-aspartate or nitric oxide/superoxide in cortical cell cultures. *Proc. Natl. Acad. Sci. USA*, 92: 7162–7166.

Brewer, G.J., Torricelli, J.R., Evege, E.K. and Price, P.J. (1993) Optimized survival of hippocampal neurons in B27-supplemented Neurobasal™, a new serum-free medium combination. *J. Neurosci. Res.*, 35: 567–576.

Bruijn, L.I., Beal, M.F., Becher, M.W., Schulz, J.B., Wong, P.C., Price, D.L. and Cleveland, D.W. (1997) Elevated free nitrotyrosine levels, but not protein-bound nitrotyrosine or hydroxyl radicals, throughout amyotrophic lateral sclerosis (ALS)-like disease implicate tyrosine nitration as an aberrant in vivo property of one familial ALS-linked superoxide dismutase 1 mutant. *Proc. Natl. Acad. Sci. USA*, 94: 7606–7611.

Camu, W. and Henderson, C.E. (1994) Rapid purification of embryonic rat motoneurons: An *in vitro* model for studying MND/ALS pathogenesis. *J. Neurol. Sci.*, 124(Suppl): 73–74.

Castro, L., Rodriguez, M. and Radi, R. (1994) Aconitase is readily inactivated by peroxynitrite, but not by its precursor, nitric oxide. *J. Biol. Chem.*, 269: 29409–29415.

Chou, S.M., Wang, H.S. and Taniguchi, A. (1996a) Role of SOD-1 and nitric oxide/cyclic GMP cascade on neurofilament aggregation in ALS/MND. *J. Neurol. Sci.*, 139(Suppl): 16–26.

Chou, S.M., Wang, H.S. and Komai, K. (1996b) Colocalization of NOS and SOD1 in neurofilament accumulation within motor neurons of amyotrophic lateral sclerosis: An immunohistochemical study. *J. Chem. Neuroanat.*, 10: 249–258.

Crow, J.P., Beckman, J.S. and McCord, J.M. (1995) Sensitivity of the essential zinc-thiolate moiety of yeast alcohol dehydrogenase to hypochlorite and peroxynitrite. *Biochemistry*, 34: 3544–3552.

Crow, J.P., Sampson, J.B., Zhuang, Y., Thompson, J.A. and Beckman, J.S. (1997a) Decreased zinc affinity of amyotrophic lateral sclerosis-associated superoxide dismutase mutants leads to enhanced catalysis of tyrosine nitration by peroxynitrite. *J. Neurochem.*, 69: 1936–1944.

Crow, J.P., Strong, M.J., Zhuang, Y., Ye, Y. and Beckman, J.S. (1997b) Superoxide dismutase catalyzes nitration of tyrosines by peroxynitrite in the rod and head domains of neurofilament L. *J. Neurochem.*, 69: 1945–1953.

Dal Canto, M.C. and Gurney, M.E. (1995) Neuropathological changes in two lines of mice carrying a transgene for mutant human Cu,Zn SOD, and in mice overexpressing wild type human SOD: A model of familial amyotrophic lateral sclerosis (FALS). *Brain Res.*, 676: 25–40.

Dawson, V.L., Dawson, T.M., Bartley, D.A., Uhl, G.R. and Snyder, S.H. (1993) Mechanisms of nitric oxide-mediated neurotoxicity in primary brain cultures. *J. Neurosci.*, 13: 2651–2661.

Dawson, V.L. and Dawson, T.M. (1996) Nitric oxide neurotoxicity. *J. Chem. Neuroanat.*, 10: 179–190.

de Bilbao, F. and Dubois-Dauphin, M. (1996) Time course of axotomy-induced apoptotic cell death in facial motoneurons of neonatal wild type and *bcl*-2 transgenic mice. *Neuroscience*, 71: 1111–1119.

Deckwerth, T.L. (1997) Molecular mechanisms of neuroprotection from neuronal death by triphic factor deprivation. In: M.P. Mattson (Ed.) *Neuroprotective Signal Transduction*. Humana Press Inc. Totowa, NJ. pp 61–82.

Deng, H.-X., Hentati, A., Tainer, J., Igbal, Z., Cayabyab, A., Hung, W.-Y., Getzoff, E., Hu, P., Herzfeldt, B., Roos, R., Warner, C., Deng, G., Soriano, E., Smyth, C., Parge, H., Ahmed, A., Roses, A., Hallewell, R., Pericak-Vance, M. and Siddique, T. (1993) Amyotrophic lateral sclerosis and structural defects in Cu,Zn superoxide dismutase. *Science*, 261: 1047–1051.

Deng, H.-X., Tainer, J.A., Mitsumoto, H., Ohnishi, A., He, X., Hung, W.-Y., Zhao, Y., Juncja, T., Hentati, A. and Siddique, T. (1995) Two novel SOD1 mutations in patients with familial amyotrophic lateral sclerosis. *Hum. Mol. Genet.*, 4: 1113–1116.

Doubois-Dauphin, M., Frankowski, H., Tsujimoto, Y., Huarte, J. and Martinou, J. (1994) Neonatal motoneurons overexpressing the *bcl-2* protooncogene in transgenic mice are protected from axotomy-induced cell death. *Proc. Natl. Acad. Sci. USA*, 91: 3309–3313.

Estévez, A.G., Radi, R., Barbeito, L., Shin, J.T., Thompson, J.A. and Beckman, J.S. (1995) Peroxynitrite-induced cytotoxicity in PC12 cells: Evidence for an apoptotic mechanism differentially modulated by neurotrophic factors. *J. Neurochem.*, 65: 1543–1550.

Estévez, A.G., Spear, N., Manuel, S.M., Radi, R., Henderson, C.E., Barbeito, L. and Beckman, J.S. (1998a) Nitric oxide and superoxide contribute to motor neuron apoptosis induced by trophic factor deprivation. *J. Neurosci.*, 18: 923–931.

Estévez, A.G., Spear, N., Thompson, J.A., Cornwell, T.L., Radi, R., Barbeito, L. and Beckman, J.S. (1998b) Nitric oxide dependent production of cGMP supports the survival of rat embryonic motor neurons cultured with brain derived neurotrophic factor. *J. Neurosci.*, 18: 3708–3714.

Farinelli, S.E., Park, D.S. and Greene, L.A. (1996) Nitric oxide delays the death of trophic factor-drprived PC12 cells and sympathetic neurons by a cGMP-mediated mechanism. *J. Neurosci.*, 16: 2325–2334.

Faulkner, K.M., Liochev, S.I. and Fridovich, I. (1994) Stable Mn(III) porphyrins mimic superoxide dismutase in vitro and substitute for it in vivo. *J. Biol. Chem.*, 269: 23471–23476.

Ferrante, R.J., Shinobu, L.A., Schulz, J.B., Matthews, R.T., Thomas, C.E., Kowall, N.W., Gurney, M.E. and Beal, M.F. (1997) Increased 3-nitrotyrosine and oxidative damage in mice with a human copper/zinc superoxide dismutase mutation. *Ann. Neurol.*, 42: 326–334.

Forman, H.J. and Fridovich, I. (1973) On the stability of bovine superoxide dismutase. The effects of metals. *J. Biol. Chem.*, 248: 2645–2649.

Friedlander, R.M., Brown, R.H., Gagliardini, V., Wang, J. and Yuan, J. (1997) Inhibition of ICE slows ALS in mice. *Nature*, 388: 31.

Garthwaite, J. and Boulton, C.L. (1995) Nitric oxide signaling in the central nervous system. *Annu. Rev. Physiol.*, 57: 683–706.

Good, P.F., Werner, P., Hsu, A., Olanow, C.W. and Perl, D.P. (1996) Evidence for neuronal oxidative damage in Alzheimer's disease. *Am. J. Pathol.*, 149: 21–28.

Greenlund, L.J., Deckwerth, T.L. and Johnson, E.M. (1995) Superoxide dismutase delays neuronal apoptosis: A role for reactive oxygen species in programmed neuronal death. *Neuron*, 14: 303–315.

Gurney, M.E., Pu, H., Chiu, A.Y., Dal Corto, M.C., Polchow, C.Y., Alexander, D.D., Caliendo, J., Hentati, A., Kwon, Y.W., Deng, H.-X., Chen, W., Zhai, P., Sufit, R.L. and Siddique, T. (1994) Motor neuron degeneration in mice that express a human Cu,Zn superoxide dismutase mutation. *Science*, 264: 1772–1775.

Hantraye, P., Brouillet, E., Ferrante, R., Palfi, S., Bolan, R., Matthews, R.T. and Beal, M.F. (1996) Inhibition of neuronal nitric oxide synthase prevents MPTP-induced parkinsonism in baboons. *Nature Med.*, 2: 1017–1021.

Hausladen, A. and Fridovich, I. (1994) Superoxide and peroxynitrite inactivate aconitases, nitric oxide does not. *J. Biol. Chem.*, 269: 29405–29408.

Henderson, C.E., Bloch-Gallego, E. and Camu, W. (1995) Purification and culture of embryonic motorneurons. In: J. Cohen and G. Wilkin (Eds.). *Neural Cell Culture: A Practical Approach*. IRL Press, Oxford, UK. pp. 69–81.

Henderson, C.E., Camu, W., Mettling, C., Gouin, A., Poulsen, K., Karihaloo, M., Rullamas, J., Evans, T., McMahon, S.B., Armanini, M.P., Berkemeier, L., Phillips, H.S. and Rosenthal, A. (1993) Neurotrophins promote motor neuron survival and are present in embryonic limb bud. *Nature*, 363: 266–270.

Henderson, C.E., Huchet, M. and Changeux, J.-P. (1981) Neurite outgrowth from embryonic chicken spinal neurons is promoted by media conditioned by muscle cells. *Proc. Natl. Acad. Sci. USA*, 78: 2625–2629.

Henderson, C.E., Huchet, M. and Changeux, J.-P. (1983) Denervation increases a neurite-promoting activity in extracts of skeletal muscle. *Nature*, 302: 609–611.

Henderson, C.E., Phillips, H.S., Pollock, R.A., Davies, A.M., Lemeulle, C., M.A., Simpson, L.C., Moffet, B., Vandlen, R.A., Koliatsos, V.E. and Rosenthal, A. (1994) GDNF: A potent survival factor for motoneurons present in peripheral nerve and muscle. *Science*, 266: 1062–1064.

Houenou, L.J., Li, L., Lo, A.C., Yan, Q. and Oppenheim, R.W. (1994) Naturally occurring and axotomy-induced motoneuron death and its prevention by neurotrophic agents: A comparison between chick and mouse. *Prog. Brain. Res.*, 102: 217–226.

Hughes, R.A., Sendtner, M. and Thoenen, H. (1993) Members of several gene families influence survival of rat motoneurons in vitro and in vivo. *J. Neurosci. Res.*, 36: 663–671.

Huie, R.E. and Padmaja, S. (1993) The reaction rate of nitric oxide with superoxide. *Free Rad. Res. Commun.*, 18: 195–199.

Ignarro, L.J. (1989) Heme-dependent activation of soluble guanylate cyclase by nitric oxide: Regulation of enzyme activity by porphyrins and metalloporphyrins. *Seminars in Hematology*, 26: 63–76.

Ischiropoulos, H., Zhu, L., Chen, J., Tsai, H.M., Martin, J.C., Smith, C.D. and Beckman, J.S. (1992) Peroxynitrite-mediated tyrosine nitration catalyzed by superoxide dismutase. *Arch. Biochem. Biophys.*, 298: 431–437.

Jordan, J., Ghadge, G.D., Prehn, J.H.M., Toth, P.T., Roos, R.P. and Miller, R.J. (1995) Expresion of human copper/zinc–superoxide dismutase inhibits the death of rat sympathetic neurons caused by withdrawal of nerve growth factor. *Mol. Pharmacol.*, 47: 1095–1100.

Klocker, N., Cellerino, A. and Bahr, M. (1998) Free radical scavenging and inhibition of nitric oxide synthase potentiates the neurotrophic effects of brain-derived neurotrophic factor on axotomized retinal ganglion cells *in vivo*. *J. Neurosci.*, 18: 1038–1046.

Koliatsos, V.E., Clatterbuck, R.E., Winslow, J.W., Cayouette, M.H. and Price, D.L. (1993) Evidence that brain-derived neurotrophic factor is a trophic factor for motor neurons in vivo. *Neuron*, 10: 359–367.

Kondo, T., Reaume, A.G., Huang. T.T., Carlson, E., Murakami, K., Chen, S.F., Hoffman, E.K., Scott, R.W., Epstein, C.J. and Chan, P.H. (1997) Reduction of CuZn–superoxide dismutase activity exacerbates neuronal cell injury and edema formation after transcient focal cerebral ischemia. *J. Neurosci.*, 17: 4180–4189.

Kostic, V., Jackson-Lewis, V., de Bilbao, F., Dubois-Dauphin, M. and Przedborski, S. (1997) *Bcl*-2: Prolonging life in a transgenic mouse model of familial amyotrophic lateral sclerosis. *Science*, 277: 559–562.

Li, L., Wu, W., Lin, L.-F.H., Lei, M., Oppenheim, R.W. and Houenou, L.J. (1995) Rescue of adult mouse motoneurons from injury-induced cell death by glial cell line-derived neurotrophic factor. *Proc. Natl. Acad. Sci. USA*, 92: 9771–9775.

Lipton, S.A., Choi, Y.B., Pan, Z.H., Lei, S.Z., Chen, H.S.V., Sucher, N.J., Loscalzo, J., Singel, D.J. and Stamler, J.S. (1993) A redox-based mechanism for the neuroprotective and neurodestructive effects of nitric oxide and related nitrosocompounds. *Nature*, 364: 626–631.

Lo, A.C., Houenou, L.J. and Oppenheim, R.W. (1995) Apototosis in the nervous system: Morphological features, methods, pathology, and prevention. *Arch. Hist. Cytol.*, 58: 139–149.

Lyons, T.J., Liu, H., Goto, J.J., Nersissian, A., Roe, J.A., Graden, J.A., Café, C., Ellerby, L.M., Bredesen, D.E., Gralla, E.B. and Valentine, J.S. (1996) Mutations in copper–zinc superoxide dismutase that cause amyotrophic lateral sclerosis alter the zinc binding site and the redox behavior of the protein. *Proc. Natl. Acad. Sci. USA*, 93: 12240–12244.

Mayer, B. (1994) Nitric oxide/cyclic GMP-mediated signal transduction. *Ann. NY. Acad. Sci.*, 733: 357–364.

Manthorpe, M., Skaper, S.D., Williams, L.R. and Varon, S. (1986) Purification of adult rat sciatic nerve ciliary neurotrophic factor. *Brain Res.*, 367: 282–286.

Martinou, J.C., Dubois-Douphin, M., Staple, J.K., Rodriguez, I., Frankowski, H., Missotten, M., Albertini, P., Talabot, D., Catsicas, S., Pietra, C. and Huarte, J. (1994) Overexpression of *BCL*-2 in transgenic mice protects neurons from naturally occurring cell death and experimental ischemia. *Neuron*, 13: 1017–1030.

Martinou, J.C., Le Van, A., Thai, A., Cassar, G., Roubinet, F. and Weber, M.J. (1989) Characterization of two factors enhancing choline acetyltransferase activity in cultures of purified rat motoneurons. *J. Neurosci.*, 9: 3645–3656.

Milligan, C.E., Oppenheim, R.W. and Schwartz, L.M. (1994) Motoneurons deprived of trophic support in vitro require new gene expression to undergo programmed cell death. *J. Neurobiol.*, 25: 1005–1016.

Milligan, C.E., Prevette, D., Yaginuma, H., Homma, S., Cardwell, C., Fritz, L.C., Tomaselli, K.J., Oppenheim, R.W. and Scwartz, L.M. (1995) Peptide inhibitors of the ICE protease family arrest programmed cell death of motoneurons in vivo and in vitro. *Neuron*, 15: 385–393.

Moreno, J.J. and Pryor, W.A. (1992) Inactivation of α-1-proteinase inhibitor by peroxynitrite. *Chem. Res. Toxicol.*, 5: 425–431.

Novikov, L., Novikova, L. and Kellerth, J.-O. (1995) Brain-derived neurotrophic factor promotes survival and blocks nitric oxide synthase expression in adult rat spinal motoneurons after ventral root avulsion. *Neurosci. Lett.*, 200: 45–48.

Novikov, L., Novikova, L. and Kellerth, J.-O. (1997) Brain-derived neurtrophic factor promotes axonal regenearation and long-term survival of adult rat spinal motoneurons in vivo. *Neuroscience*, 79: 765–774.

Oppenheim, R.W. (1991) Cell death during development of the nervous system. *Annu. Rev. Neurosci.*, 14: 453–501.

Oppenheim, R.W. (1996) Neurotrophic survial molecules for motor neurons: An embarrassment of riches. *Neuron*, 17: 195–197.

Oppenheim, R.W., Houenou, L.J., Johnson, J.E., Lin, L.F.H., Li, L., Lo, A.C., Newsome, A.L., Prevette, D.M. and Wang, S. (1995) Developing motor neurons rescued from programmed cell death and axotomy-induced cell death by GDNF. *Nature*, 373: 344–346.

Oppenheim, R.W., Prevette, D., Qin-Wei, Y., Collins, F. and MacDonald, J. (1991) Control of embryonic motoneuron survival in vivo by ciliary neurotrophic factor. *Science*, 251: 1616–1618.

Oppenheim, R.W., Qin-Wei, Y., Prevette, D. and Yan, Q. (1992) Brain-derived neurotrophic factor rescues developing avian motoneurons from cell death. *Nature*, 360: 755–757.

Pantoliano, M.W., Valentine, J.S., Burger, A.R. and Lippard, S.J. (1982) A pH-dependent superoxide dismutase activity for zinc-free bovine erythrocuprein. Reexamination of the role of zinc in the holoprotein. *J. Inorg. Chem.*, 17: 325–341.

Peluffo, H., Estevez, A., Barbeito, L. and Stutzmann, J.M. (1997) Riluzole promotes motoneuron survival by stimulating trophic activity produced by spinal astrocyte monolayers. *Neurosci. Lett.*, 228: 207–211.

Pennica, D., Arce, V., Swanson, T.A., Vejsada, R., Pollock, R.A., Armanini, M., Dudley, K., Phillips, H.S., Rosenthal, A., Kato, A.C. and Henderson, C.E. (1996) Cardiotrophin-1, a cytokine present in embryonic muscle, supports long-term survival of spinal motoneurons. *Neuron*, 17: 63–74.

Prehn, J.H., Jordan, J., Ghadge, G.D., Preis, E., Galindo, M.F., Roos, R.P., Krieglstein, J. and Miller, R.J. (1997) Ca^{2+} and reactive oxygen species in staurosporine-induced neuronal apoptosis. *J. Neurochem.*, 68: 1679–1685.

Price, D.L., Cleveland, D.W. and Koliatsos, V.E. (1994) Motor neurone disease and animal models. *Neurobiol. Dis.*, 1: 3–11.

Rabizadeh, S., Gralla, E.B., Borchelt, D.R., Gwinn, R., Valentine, J. S., Sisodia, S., Wong, P., Lee, M., Hahn, H. and Bredesen, D.E. (1995) Mutations associated with amyotrophic lateral sclerosis convert superoxide dismutase from an antiapoptotic gene to a proapoptotic gene: Studies in yeast and neural cells. *Proc. Natl. Acad. Sci. USA*, 92: 3024–3028.

Radi, R., Beckman, J.S., Bush, K.M. and Freeman, B.A. (1991a) Peroxynitrite-mediated sulfhydryl oxidation: The cytotoxic potential of superoxide and nitric oxide. *J. Biol. Chem.*, 266: 4244–4250.

Radi, R., Beckman, J.S., Bush, K.M. and Freeman, B.A. (1991b) Peroxynitrite-induced membrane lipid peroxidation: The cytotoxic potential of superoxide and nitric oxide. *Arch. Biochem. Biophys.*, 288: 481–487.

Rosen, D.R., Siddique, T., Patterson, D., Figlewicz, D.A., Sapp, P., Hentati, A., Donaldson, D., Goto, J., O'Regan, J.P., Deng, H.-X., Rahmani, Z., Krizus, A., McKenna-Yasek, D., Cayabyab, A., Gaston, S.M., Berger, R., Tanszi, R.E., Halperin, J.J., Herzfeldt, B., Van den Bergh, R., Hung, W.-Y., Bird, T., Deng, G., Mulder, D.W., Smyth, C., Lang, N.G., Soriana, E., Pericak-Vance, M.A., Haines, J., Rouleau, G.A., Gusella, J.S., Horvitz, H.R. and Brown, R.H., Jr (1993) Mutations in Cu/Zn superoxide dismutase gene are associated with familial amyotrophic lateral sclerosis. *Nature*, 362: 59–62.

Rothstein, J.D., Bristol, L.A., Hosler, B., Brown Jr., R.H. and Kuncl, R.W. (1994) Chronic inhibition of superoxide dismutase produces apoptotic death of spinal neurons. *Proc. Natl. Acad. Sci. USA*, 91: 4155–4159.

Ruan, R.S., Leong, S.K. and Yeoh, K.H. (1995) The role of nitric oxide in facial motoneuronal death. *Brain Res.*, 698: 163–168.

Samdani, A.F., Newcamp, C., Resink, A., Facchinetti, F., Hoffman, B.E., Dawson, V.L. and Dawson, T.M. (1997) Differential susceptibility to neurotoxicity medaited by neurotrophins and neuronal nitric oxide synthase. *J. Neurosci.* 17: 4633–4641.

Sasaki, S. and Iwata, M. (1997) Impairment of fast azonal transport in the proximal axons of anterior horn neurons in amyotrophic lateral sclerosis. *Neurology*, 47: 535–540.

Schmalbruch, H. and Rosenthal, A. (1995) Neurotrophin 4/5 postpones the death of injuried motoneurons in newborn rats. *Brain Res.*, 700: 254–260.

Schulz, J.B., Huang, P.L., Matthews, R.T., Passov, D., Fishman, M.C. and Beal, M.F. (1996) Striatal malonate lesions are attenuated in neuronal nitric oxide synthase knockout mice. *J. Neurochem.*, 67: 430–433.

Sendtner, M., Arakawa, Y., Stöckli, K.A., Kreutzberg, G.W. and Thoenen, H. (1991) Effect of ciliary neurotrophic factor (CNTF) on motoneuron survival. *J. Cell Sci.*, (Suppl.) 15: 103–109.

Sendtner, M., Holtmann, B., Kolbeck, R., Thoenen, H. and Barde, Y.A. (1992) Brain-derived neurotrophic factor prevents the death of motoneurons in newborn rats after nerve section. *Nature*, 360: 757–759.

Sendtner, M., Kreutzbert, G.W. and Thoenen, H. (1990) Ciliary neurotrophic factor prevents the degeneration of motor neurons after axotomy. *Nature*, 345: 440–441.

Shin, J.T., Barbeito, L., MacMillan-Crow, L.A., Beckman, J.S. and Thompson, J.A. (1996) Acidic fibroblast growth factor enhances peroxynitrite-induced apoptosis in primary murine fibroblasts. *Arch. Biochem. Biophys.*, 335: 32–41.

Smith, M.A., Harris, P.L., Sayre, L.M., Beckman, J.S. and Perry, G. (1997) Widespread peroxynitrite-mediated damage in Alzheimer disease. *J. Neurosci.*, 17: 2653–2657.

Smith, R.G. and Appel, S.H. (1983) Extracts of skeletal muscle increase neurite outgrowth and cholinergic activity of fetal rat spinal motor neurons. *Science*, 219: 1079–1081.

Spear, N., Estevez, A.G., Barbeito, L., Beckman, J.S. and Johnson, G.V.W. (1997) Nerve growth factor protects PC12 cells against peroxynitrite-induced apoptosis via a mechanism dependent on phosphatidylinositol-3 kinase. *J. Neurochem.*, 69: 53–59.

Stamler, J.S., Toone, E.J., Lipton, S.A. and Sucher, N.J. (1997) (S)NO signals: Translocation, regulation, and a consensus motif. *Neuron*, 18: 691–696.

Strijbos, P.J.L.M., Leach, M.J. and Garthwaite, J. (1996) Vicious cycle involving Na^+ channels, glutamate release, and NMDA receptors mediates delayed neurodegeneration through nitric oxide formation. *J. Neurosci.*, 15: 5004–5013.

Szabo, C., Day, B.J. and Salzman, A.L. (1996) Evaluation of the relative contribution of nitric oxide and peroxynitrite to the suppression of mitochondrial respiration in immunostimulated macrophages using a manganese mesoporohyrin superoxide dismutase mimetic and peroxynitrite scavenger. *FEBS Lett.*, 381: 82–86.

Tabuchi, A., Oh, E., Taoka, A., Sakurai, H., Tsuchida, T. and Tsuda, M. (1996) Rapid attenuation of AP-1 transcriptional factors associated with nitric oxide (NO)-mediated neuronal cell death. *J. Biol. Chem.*, 271: 31061–31067.

Troy, C.M. and Shelanski, M.L. (1994) Down-regulation of copper/zinc superoxide dismutase causes apoptotic death in PC12 neuronal cells. *Proc. Natl. Acad. Sci. USA*, 91: 6384–6387.

Troy, C.M., Derossi, D., Prochiantz, A., Greene, L.A. and Shelanski, M.L. (1996) Downregulation of Cu/Zn superoxide dismutase leads to cell death via the nitric oxide-peroxynitrite pathway. *J. Neurosci.*, 16: 253–261.

Vejsada, R., Sagot, Y. and Kato, A.C. (1995) Quantitative comparison of the transcient rescue effects on neurotrophic factors on axotomized motoneurons in vivo. *Eur. J. Neurosci.*, 7: 108–115.

Williams, L,R., Manthorpe, M., Barbin, G., Nieto-Sampedro, M., Cotman, C.W. and Varon, S. (1984) High ciliary neurotrophic factor specific activity in rat peripheral nerve. *Int. J. Dev. Neurosci.*, 2: 177–180.

Weill, C.L. and Greene, D.P. (1984) Prevention of natural motoneurone cell death by dibutyryl cyclic GMP. *Nature*, 308: 482–482.

Weill, C.L. and Greene, D.P. (1990) Prevention of natural motoneurone cell death by dibutyryl-cyclic GMP in the spinal cord of White Leghorn chick embryos. *Dev. Brain Res.*, 55: 143–146.

Wu, W. (1993) Expression of nitric-oxide synthase (NOS) in injured CNS neurons as shown by NADPH diaphorase histochemistry. *Exp. Neurol.*, 120: 153–159.

Wu, W. and Li, L. (1993) Inhibition of nitric oxide synthase reduces motoneuron death due to spinal root avulsion. *Neurosci. Lett.*, 153: 121–124.

Wu, W., Liuzzi, F.J., Schinco, F.P., Dpeto, A.S., Li, Y., Mong, J.A., Dawson, T.M. and Snyder, S.H. (1994) Neuronal nitric oxide synthase is induced in spinal neurons by traumatic injury. *Neuroscience*, 61: 719–726.

Yamamoto, Y, Livet J, Pollock, R.A., Garces, A., Arce, V., deLapeyrière O. and Henderson, C.E. (1997) Hepatocyte growth factor (HGF/SF) is a muscle-derived survival factor for a subpopulation of embryonic motoneruons. *Development*, 124: 2903–2913.

Yan, Q., Elliott, J. and Snider, W.D. (1992) Brain-derived neurotrohic factor rescues spinal motor neurons from axotomy-induced cell death. *Nature*, 360: 753–755.

Yan, Q. and Johnson, E.M., Jr. (1988) An immunohistochemical study of the nerve growth factor receptor in developing rats. *J. Neurosci.*, 8: 3481–3498.

Yu, W.H.A. (1994) Nitric oxide synthase in motor neurons after axotomy. *J. Histochem. Cytochem.*, 42: 451–457

R.R. Mize, T.M. Dawson, V.L. Dawson and M.J. Friedlander (Eds.)
Progress in Brain Research, Vol 118

CHAPTER 20

The neuromessenger platelet-activating factor in plasticity and neurodegeneration

Nicolas G. Bazan

Louisiana State University Medical Center, School of Medicine, Neuroscience Center of Excellence, 2020 Gravier street, Suite D, New Orleans, LA 70112, USA

Abstract

Synaptic activation leads to the formation of arachidonic acid, platelet-activating factor (PAF, 1-O-alkyl-2-acyl-sn-3-phosphocholine) and other lipid messengers. PAF is a potent bioactive phospholipid in synaptic plasticity. PAF enhances presynaptic glutamate release, is a retrograde messenger in long-term potentiation and enhances memory formation. PAF also couples synaptic events with gene expression by stimulating a FOS/JUN/AP-1 transcriptional signaling system, as well as transcription of COX-2 (inducible prostaglandin synthase). Since the COX-2 gene is also involved in synaptic plasticity, the PAF-COX-2 pathway may have physiological significance. Seizures, ischemia and other forms of brain injury promote phospholipase A_2 (PLA_2) overactivation, resulting in the accumulation of bioactive lipids at the synapse. PAF, under these pathological conditions, behaves as a neuronal injury messenger by at least two mechanisms: (a) enhancing glutamate release; and, (b) by sustained augmentation of COX-2 transcription. These events link PAF with neurodegeneration. The upstream intracellular pathways of signal transduction involved in neuronal or photoreceptor cell apoptosis are not well understood and involve stress sensitive kinases. PAF is a transcriptional activator of the COX-2 gene. BN 50730, a potent intracellular PAF antagonist, blocks COX-2 induction. COX-2 transcription and protein expression are upregulated in the hippocampus in kainic acid induced epileptogenesis. There is a selectively elevated induction of COX-2 (72-fold) by kainic acid preceding neuronal cell death. BN 50730 administered by icv injection blocks seizure-induced COX-2 induction. Overall, PAF is a dual modulator of neural function and becomes an endogenous neurotoxin when over produced.

Introduction

In neuronal and glial cell signaling, several diffusible messengers other than nitric oxide (NO) are active participants, e.g., PAF, prostaglandins and lipoxygenase products are prominent examples. These bioactive lipids share the following: (a) they are stored in the structure of membrane phospholipids; (b) synaptic activation promotes their release; (c) phospholipase A_2 (PLA_2) activation catalyzes the initiation of the pathway leading to their synthesis; and, (d) seizures and ischemia activate their release. Prostaglandins (Bazan and Rodriguez de Turco, 1980) and lipoxygenase reactions products (Moscowitz et al., 1984) are synthetized by cyclooxygenases and lipoxygenases and rapidly accumulate in brain during stimulation or in ischemia. These biologically active metabolites are released upon depolarization from neural tissue (Birkle and Bazan, 1984; Claeys et al., 1986). 12-hydroxyeicosatrienoic (12-HETE) is a prominent diffusible arachidonic acid metabolite in Aplysia Californica neurons (Piomelli et al., 1987a,b), during K^+-induced depolarization in the retina (Birkle and Bazan, 1984), as well as in synaptosomes from brains of rats undergoing bicuculline induced seizures (Birkle and Bazan, 1987). 12-HETE in Aplysia Californica has been implicated in the modulation of synaptic transmission through ion channels and in behavior (Piomelli et al., 1987a,b). PLA_2 activation is an early event in the brain response to injury, such as ischemia or at the onset of seizures (Bazan, 1970). Endogenous free arachidonic acid accumulates in synaptosomes from rats undergoing seizures (Birkle and Bazan, 1987), strongly suggesting a synaptic location for the phospholipase A_2 activated during brain stimulation.

Although the oxygenated metabolites of arachidonic acid (e.g., prostaglandins and lipoxygenase reaction products) have been found to change during stimulation and injury, only recently defined sites of action and their potential significance has been uncovered.

The cloning of prostaglandin receptors and the finding that *adenylcylcylase* activation follows prostaglandin receptor occupancy has defined a signaling route whereby the diffusable prostaglandins modulate protein kinase A and the transcription factor, CREB (cyclic response element binding protein), leading to the control of gene expression (Bazan, in press). Moreover, it has recently been found that prostaglandins modulate glutamate release from glial cells (Bonventre et al., 1997) suggesting that intercellular prostaglandins (derived, perhaps, from the activation of certain neurons) may modulate excitatory synaptic transmission.

Since the diffusible lipid mediators, prostaglandins, PAF and others are generated by PLA_2 activation, it is important to highlight the significance of this enzyme (Fig. 1). First, there are several PLA_2's: cytosolic ($cPLA_2$), secretory

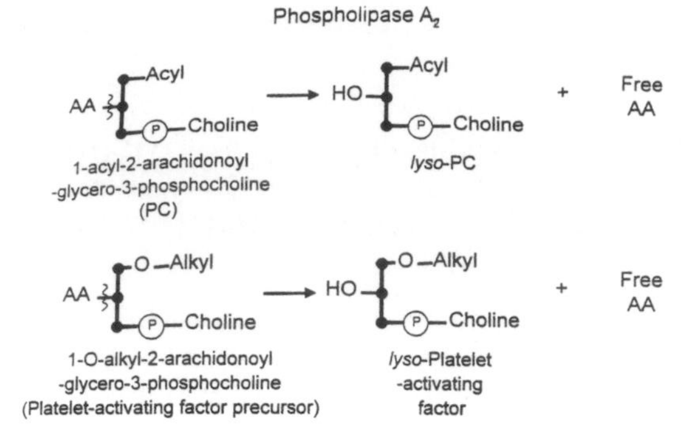

Fig. 1. Sites of phospholipid cleavage by the major classes of phospholipases, showing some of the important reaction products. The top two reactions show the phospholipase A_2-catalyzed degradation of choline phospholipids. In the first reaction, the C2 arachidonoyl chain of PC is cleaved to liberate free arachidonic acid (AA). This reaction is probably the quantitatively most important source of free AA for the synthesis of eicosanoids. The second reaction involves the quantitatively much less significant ether lipid analogue of PC, alkyl arachidonoyl GPC. Although this reaction probably does not contribute significantly to overall AA release, the lysophospholipid formed is the direct precursor of the highly potent platelet-activating factor. The third reaction shows the action of inositol phospholipid (PI)-specific phospholipase C on PIP_2. This classical bifurcating signaling reaction yields DAG, an activator of protein kinase C, and IP_3, a mobilizer of intracellular calcium stores. The bottom reaction shows the action of phospholipase D on PC. The PA released can provide a source of DAG quantitatively much more significant than that derived from the smaller PI pool, and is also a precursor of the novel bioactive lipid lyso-PA. The free choline released is thought to be an important source of precursor for the neurotransmitter acetylcholine.

($sPLA_2$) and calcium-independent ($iPLA_2$). Limitations of the present chapter do not permit an in-depth discussion of each form. However, their relevance to brain function and, particularly, in cerebrovascular diseases was recently reviewed (Bazan and Allan, 1997). Some highlights follow. $sPLA_2$ is released along with neurotransmitters (Matsuzawa et al., 1996) and strongly potentiates glutamate excitotoxicity (Koko et al., 1996). $cPLA_2$ may be critical in spreading depression through the generation of PAF (Miettinen et al., 1997). The initiating event for the synthesis of several of the diffusible lipid messengers seems to be $cPLA_2$. This enzyme is activated by the

NMDA-mediated rise in intracellular calcium and is further enhanced by ERK phosphorylation. $cPLA_2$ has a preference for arachidonoyl phospholipids. The important function of $cPLA_2$ in brain stimulation is supported by the recent finding that $cPLA_2$ knockout mice have substantially reduced infarcts and neurological deficits following transient middle cerebral artery occlusion (Bonventre et al., 1997). This strongly implicates $cPLA_2$ in postischemic brain injury, as previously suggested for a PLA_2 (Bazan, 1970, 1976).

The plenary lecture at this symposium on diffusible messengers will focus on the physiological significance and pathological role of the neuromessenger, PAF, closely related by its origins to the other lipids messengers briefly described above.

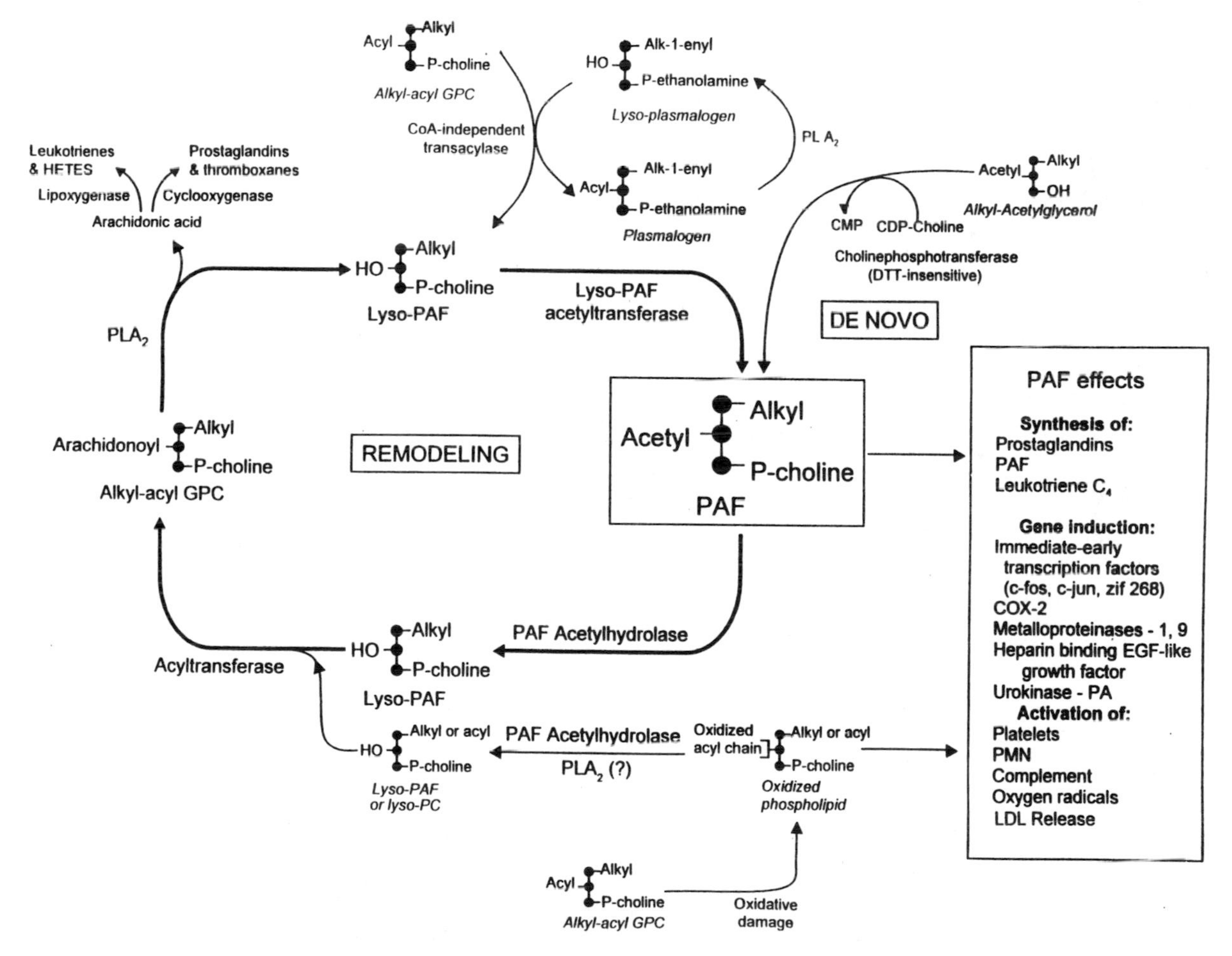

Fig. 2. Summary of the pathways of PAF synthesis and degradation. The central portion of this figure represents the "PAF cycle" between the membrane phospholipid PAF precursor alkylacyl GPC and biologically active PAF. The remodeling synthesis involves the production of lyso-PAF, generated from the PAF precursor alkylacyl GPC either by the direct action of phospholipase A_2 or by the transfer of the *sn*-2 acyl chain to a "donor" lysoplasmalogen, which is itself mobilized from membrane plasmalogen by phospholipase A_2 action. The de novo pathway involves the direct transfer of a choline moiety to alkyl-acetylglycerol. Note that PAF acetylhydrolase inactivates all PAF molecules regardless of their biosynthetic route, and additionally inactivates oxidatively damaged phospholipids that exhibit biologic activity at the PAF receptor.

PAF enhances presynaptic glutamate release and modulates plasticity

PAF is one of the most potent bioactive lipids known. Fig. 2 outlines the metabolic pathways for its synthesis and degradation. PAF is implicated in the inflammatory/injury response (Bazan, 1995), as well as in physiological processes (Prescott et al., 1990). PAF enhances glutamate release in synaptically paired rat hippocampal neurons in culture (Clark et al., 1992). The PAF analog methylcarbamyl (mc-PAF), but not biologically inactive lyso PAF, increases excitatory synaptic responses. The inhibitory neurotransmitter γ-aminobutyric acid is unaffected by mc-PAF under these conditions. The presynaptic PAF receptor antagonist BN 52021, a terpenoid extracted from the leaf of the *Ginkgo biloba* tree which binds preferentially to the synaptosomal PAF binding site (Marcheselli et al., 1990), blocks the mc-PAF enhancement of glutamatergic neurotransmitter release. PAF enhancement of glutamatergic synaptic function has been shown to endure for periods relevant to the long-term potentiation of neurotransmission thought to underlie memory formation (Kato et al., 1994). Indeed, in vivo administration of PAF has been shown to disrupt memory (Packard et al., 1996). PAF has been shown to enhance memory in rats performing an inhibitory avoidance task (Izquierdo et al., 1995) and in a water maze task (Packard et al., 1996). In these instances, specificity for the synaptic PAF receptor was shown, as was neuroanatomical and temporal selectivity. For instance, in the inhibitory avoidance task, PAF only had an effect when infused at specific times after training into defined areas of the limbic system, and, in the water maze task, PAF was only effective when infused into the striatum. Therefore,

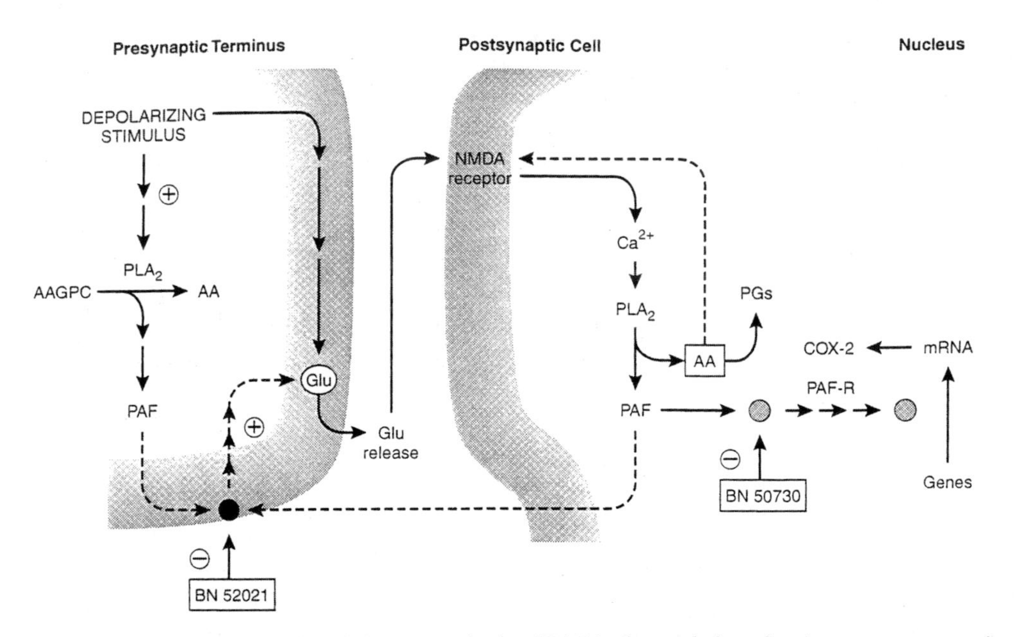

Fig. 3. Schematic representation of the proposed role of PAF in the regulation of excitatory neurotransmitter release and how, during cerebral trauma, it may play a role in the excessive release of neurotransmitter. Under normal physiological conditions, glutamate is released from a presynaptic terminal and stimulates NMDA receptors at the postsynaptic terminal. The resultant elevation of intracellular calcium activates cytoplasmic PLA_2 and initiates accumulation of PAF and 20:4. Upon sufficient stimulation there is enough PAF to diffuse back across the synapse and provide a sufficient concentration at the presynaptic terminal to stimulate the BN 52021-sensitive, presynaptic PAF receptor and, via a mechanism as yet undefined, stimulate glutamate release. At the same time, 20:4 accumulates in the postsynaptic terminal and have an positive effect on glutamate neurotransmission involving the NMDA receptor. Under pathophysiological conditions such as during cerebral ischemia, glutamate neurotransmission could be elevated by a superinduction of these pathways where an initially elevated signal at the presynaptic terminal could initiate a positive feedback loop.

as a modulator of glutamate release, PAF participates in long-term potentiation (LTP), synaptic plasticity and memory formation. Fig. 3 represents the presynaptic and postsynaptic regulatory sites involving PAF.

Activation of synaptic phospholipases A_2 during seizures

One of the earliest consequence of seizures is cleavage of phospholipids in excitable membranes. Phospholipid molecules of membranes from neurons and glial cells store a wide variety of lipid messengers. Receptor-mediated events and changes in intracellular [Ca^{2+}], as occurs during excitatory neurotransmission and in activity-dependent synaptic plasticity, activates phospholipases that catalyze the release of bioactive moieties from phospholipids. These messengers then participate in intracellular and/or intercellular signaling pathways. Bioactive lipids have significant neurobiological actions in neurotransmitter release, synaptic plasticity, and programs of neuronal gene expression. Accordingly, contemporary research into bioactive lipids has focused on their neurobiological significance and a major area to be developed is epileptic brain damage and epileptogenesis.

Seizures disrupt the tightly regulated events that control the production and accumulation of lipid messengers, such as free arachidonic acid, diacylglycerol and PAF under physiological conditions. Rapid activation of phospholipases, particularly of PLA_2, occurs at the onset of seizures (Bazan et al., 1995). There are a wide variety of PLA_2s (Dennis, 1994), and current investigations aim to define those affected by epileptic seizures. For example, in addition to the role(s) of intracellular PLA_2s in lipid messenger formation, it has recently been discovered that a low molecular weight, secretory PLA_2 similar to that released from astrocytes (Lauritzen et al., 1994) and neurons (Matsuzawa et al., 1996) has been recently shown to synergize glutamate-induced neuronal damage (Kolko et al., 1996). Whereas pathways leading to PLA_2 activation/release are part of normal neuronal function, seizures enhance these events, overproducing PLA_2-derived lipid messengers, (e.g., enzymatically produced arachidonic acid oxygenation metabolites, non-enzymatically generated lipid peroxidation products and other reactive oxygen species), involved in neuronal damage. Among the consequences of PLA_2 activation by seizures are alterations in mitochondrial function by the rapid increase in the brain free fatty acid pool size (e.g., uncoupling of oxidative phosphorylation from respiratory chain) (Bazan et al., 1971; Bazan, 1971) and the generation of lipid messengers. Arachidonic acid and its prostaglandin metabolites may also acutely distort signal transduction by altering glutamate receptor affinity (Bazan et al., 1995; Bernard et al., 1995) and sphingomyelinase enzyme activity (Jayadev et al., 1994). Prostaglandins are also known to regulate transcription of immediate early genes (Niimi et al., 1996), possibly by binding to nuclear transcription factor receptors (Kliewer et al., 1995; Kainu et al., 1994).

PAF is a very potent and short-lived lipid messenger. It is known to have a wide range of actions: as a mediator of inflammatory and immune responses, as a second messenger, and as a potent inducer of gene expression. Thus, in addition to its acute roles, PAF can potentially mediate longer-term effects on cellular physiology and brain function. In this article, the significance of PAF in synaptic function and neuronal gene expression relevant to epilepsy is discussed.

PAF, a neuronal injury messenger, contributes to excitotoxicity by enhancing glutamate release

Seizures and ischemia increase PAF content in brain (Kumar et al., 1988). Furthermore, brain is endowed with a variety of degradative enzymes that rapidly convert PAF to biologically inactive lyso PAF (Prescott et al., 1990; Hattori et al., 1993). Presynaptic membranes display PAF binding that can be displaced by BN 52021 (Kliewer et al., 1995). It is likely that this PAF binding site is the seven transmembrane PAF receptor that has been cloned by Shimizu, et al. (for reference see Bazan et al., 1995). BN 52021 inhibits both PAF-induced glutamate release (Bazan et al., 1995) and long-term potentiation (Clark et al., 1992). Moreover, this antagonist is neuroprotective in ischemia-reperfusion damage in the gerbil brain (for reference see Bazan et al., 1995). Taking these finding together, PAF, when overproduced at the synapse during seizures, will promote enhanced glutamate release that in turn, through the activation of post-synaptic receptors, will contribute to excitotoxicity.

PAF is a transcriptional activator of prostaglandin endoperoxide synthase-2

In addition to its modulatory effect on synaptic transmission and neural plasticity, PAF activates receptor-mediated immediate early gene expression (Squinto et al., 1989, 1990). Since PAF is a phospholipid and can pass through membranes, it is rapidly taken up by cells. An intracellular binding site with very high affinity, yet pharmacologically distinct from the presynaptic site, was found in brain (Marcheselli et al., 1990). The synthetic hetrazepine, BN 50730, is selective for this intracellular site and blocks PAF-induced gene expression of the inducible form of prostaglandin G synthase in transfected cells (Bazan et al., 1994).

Prostaglandin G/H synthase-2 (PGS-2, COX-2, TIS-10) catalyzes the cyclooxygenation and peroxidation of arachidonic acid into PGH_2, the precursor of biologically active prostaglandins, thromboxanes and prostacyclin. COX-1 also catalyzes the same first committed step of the arachidonic acid cascade. COX-2, however, is expressed in response to mitogenic and inflammatory stimuli and is encoded by an early-response gene. In contrast, COX-1 expression is not subject to short-term regulation. Neurons in the hippocampus, as well as in a few other brain regions, are unlike other cells in that they display basal levels of COX-2 expression (Yamagata et al., 1993). This expression is modulated by synaptic activity, LTP and involves the *N*-methyl-D-aspartate class of glutamate receptors (Yamagata et al., 1993; Kaufmann et al., 1996).

PAF is a transcriptional activator of COX-2, as PAF induces mouse COX-2 promoter-driven luciferase activity transfected in neuroblastoma cells (NG108-15 or SH-SY5Y) and in NIH 3T3 cells. The intracellular PAF antagonist, BN 50730, inhibits PAF activation of this construct (Bazan et al., 1994). Fig. 1 outlines the role of PAF as a presynaptic messenger.

Sustained transcriptional upregulation of COX-2 during kainic acid-induced seizures in hippocampus

The abundance in brain of several early-response gene transcripts shows rapid and transient increases during cerebral ischemia and after seizures. Several early-response genes encode transcription factors which in turn modulate the expression of other genes, whereas others encode inducible enzymes. The glutamate analog, kainic acid, promotes extensive neuronal damage, particularly in the hippocampus, and also induces early-response genes such as the transcription factor *zif*-268. COX-2 is also induced under these conditions, but there are striking differences in the magnitude and duration of the induction of COX-2 as compared with *zif*-268 (Figs. 4 and 5). COX-2 mRNA, 2 h after kainic acid injection, showed a 35-fold increase in hippocampus as compared with only a 5.5-fold increase in *zif*-268 (Marcheselli and Bazan, 1996). Also COX-2 peak in mRNA abundance was evident at 3 h (71-fold increase) as compared with 1 h for *zif*-268 (10-fold increase). *Zif*-268 mRNA time-course of changes in the hippocampus corresponds to the expected profile of early-response genes, i.e., a rapid decrease in abundance after the peak. COX-2 on the other hand, displayed sustained upregulation for several hours after kainic acid injection (5.2-fold increase at 12 h) (Marcheselli and Bazan, 1996) (Fig. 6).

A sustained upregulation of COX-2 mRNA also follows KCl-induced spreading depression in brain slices, a model of the depolarization-mediated neurotoxicity that contributes to penumbral brain damage following focal ischemia (Miettinen et al., 1997). As in the kainic acid model, the peak in neocortical COX-2 mRNA abundance occurred at 4 h post-treatment and returned to baseline only after 24 h. This upregulation was abolished by the

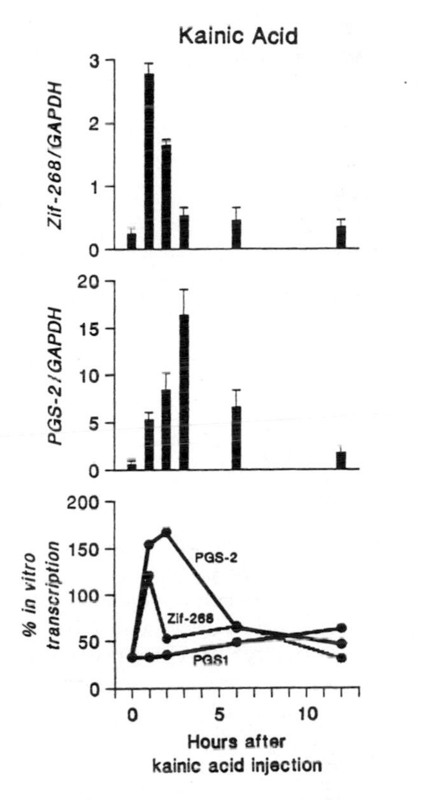

Fig. 4. Time course of changes in relative mRNA abundance and transcriptional activity in hippocampus after KA treatment or a single electroconvulsive shock (ECS). (Top and Middle) Relative abundance of COX-2 and *zif*-268 mRNAs compared to GAPDH mRNA, as assessed by northern analysis. ($n = 9$–12 for each time point from three separate experiments, error bars ± 1 s.d.) (Bottom) Transcriptional activity of COX-2, COX-1 and *zif*-268 genes assessed by nuclear run-on transcription. ($n = 3$–4 from three separate experiments). Data are normalized to transcriptional activity of GAPDH. (from Marcheselli and Bazan, Printed by permission, J. Biol. Chem)

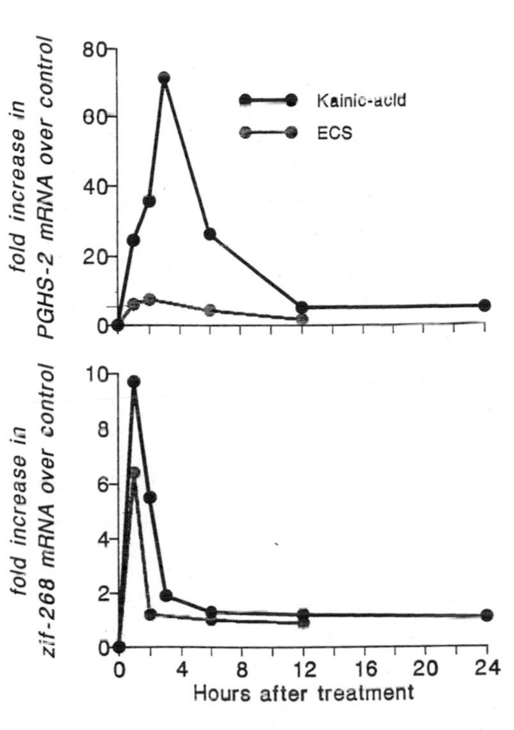

Fig. 5. Prostaglandin endoperoxide synthase-2 (PGHS-2 or COX-2) and *zif*-268 in hippocampus: Time course of changes in relative mRNA abundance in hippocampus after KA treatment. Relative abundance of COX-2 and *zif*-268 mRNAs compared to GAPDH mRNA, as assessed by northern analysis. ($n = 9$–12 for each time point from three separate experiments, error bars ± 1 s.d.). (from Marcheselli and Bazan, Printed by permission, J. Biol. Chem)

NMDA glutamate receptor antagonist MK-801. In an in vivo ischemic model of penumbral neurotoxicity, as sustained induction of COX-2 mRNA was also attenuated by MK-801, and by inhibiting PLA_2 activity (Miettinen et al., 1997).

The platelet activating factor-prostaglandin G synthase-2 intracellular signaling pathway in epileptogenesis

A PAF-stimulated signal transduction pathway is a major component of the kainic acid-induced COX-2 expression in hippocampus. This conclusion is based upon the finding that (a) PAF induces mouse COX-2 promoter-driven luciferase activity in transfected cells, and BN 50730 inhibits this effect (Marcheselli et al., 1990); and (b) BN 50730 (given intracerebroventricularly 15 min prior to kainic acid) inhibits kainic acid-induced COX-2 mRNA accumulation in hippocampus by 90% (Squinto et al., 1989) (Fig. 7). Both PAF (Prescott et al., 1990) and COX-2 (Bazan et al., 1996) are potent mediators of the injury/inflammatory response. PAF (Kolko et al., 1996) and COX-2 (Bazan et al., 1995, Kaufmann et al., 1996) are also interrelated in neuronal plasticity. The PAF transcriptional activation of COX-2 may provide clues about novel neuronal cell death pathways. The antagonist BN 50730 was much less effective against *zif*-268

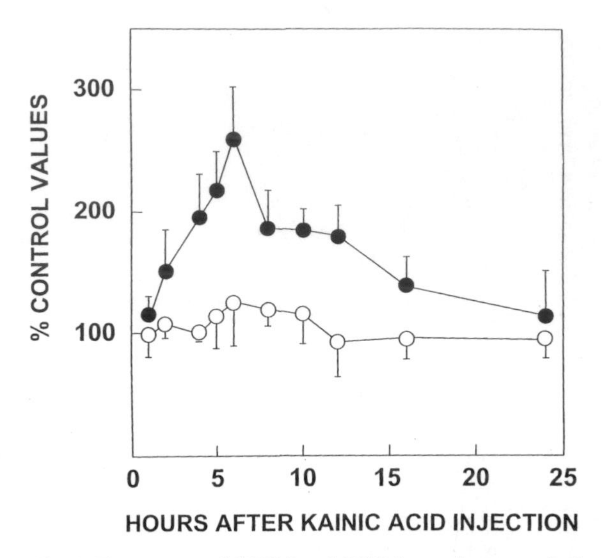

Fig. 6. Time course of COX-2 and COX-1 protein accumulation in rat hippocampus after kainic acid injection. Quantification of Western blots to assess relative induction of COX-2 (black circles) and COX-1 (open circles) protein in experimental (kainic acid treated) vs. control (saline-treated). ($n = 8$–12, from three separate experiments, error bars ± 1 s.d.)

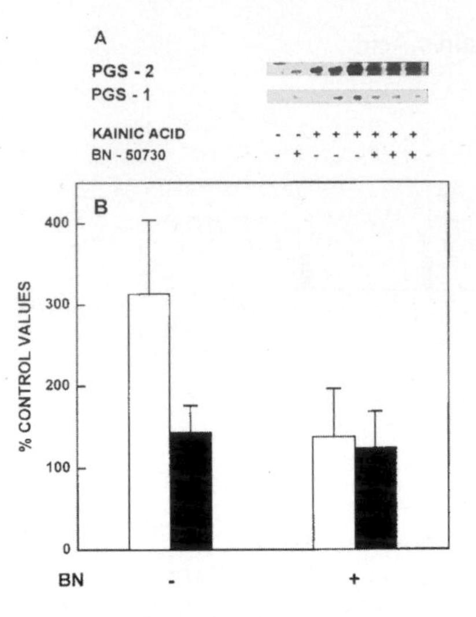

Fig. 7. Inhibition by BN 50730 pretreatment of KA-induced COX-2 protein accumulation (open bars), but not of COX-1 (closed bars) in rat hippocampus. (A) Representative Western blots. BN 50730 or vehicle treatments were as for Fig. 4. Samples were collected 6 h after KA treatment. (B) Quantification of Western blot data expressed as a percent increase over control (vehicle-pretreated) values. ($n = 10$–12 from three separate experiments, error bars ± 1 s.d.)

expression. In fact, the delayed hippocampal induction of COX-2 by kainic acid precedes selective neuronal apoptosis by this agonist in this neuroanatomical region. Furthermore, recent experiments demonstrate that the COX-2 specific antagonist NS-398, but not the COX-1 selective antagonist valeryl salicylate, attenuates the NMDA-induced neuronal death in primary cortical cell cultures (Hewett and Hewett, 1997). Fig. 8 outlines how the signaling events evolving from the synapse may modulate COX-2 expression.

For example, the significance of the PLA_2-related signaling triggered by ischemia reperfusion may be part of events finely balanced between neuroprotection and neuronal cell death (Fig. 9). The precise events that would tilt this balance toward the latter are currently being explored. It is interesting to note that PAF, being short-lived and rapidly degraded by PAF acetylhydrolase (Bazan, 1995), is a long-term signal with consequences to neurons though COX-2 sustained expression. COX-2 is localized in the nuclear envelope and perinuclear endoplasmic reticulum. The overexpression of hippocampal COX-2 during cerebral

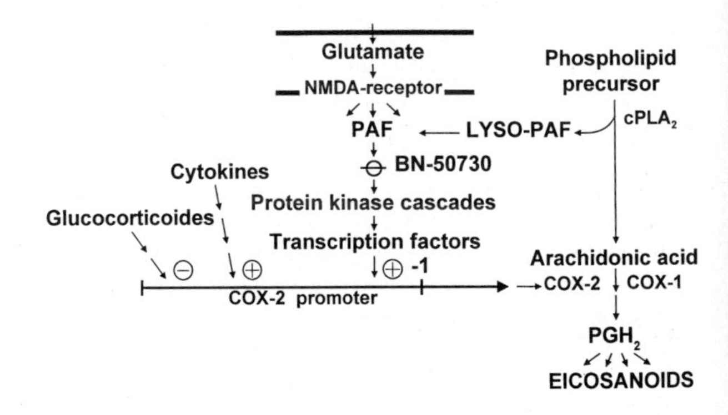

Fig. 8. Seizure-triggered signaling events linking synapse activation and COX-2 gene expression in neurons. NMDA-receptor activation by glutamate leads to phospholipase A_2 activation and the generation of PAF and of arachidonic acid. PAF is synthesized through other metabolic routes as well. PAF activates COX-2 gene expression through a BN-50730-sensitive intracellular site, protein kinase cascades and transcription factors. The COX-2 promoter is also a target for cytokines (activation) and glucocorticoids (inhibition). COX-2 protein then catalyzes the conversion of arachidonic acid into PGH_2, the precursor of eicosanoids. Constitutive COX-1 also catalyzes this metabolic step (modified from N. Bazan, *Primer Cerebrovascular Diseases*. M. Welsh, L. Chaplan, D. Reis, B. Siesjö, B. Weir (Eds.), Academic Press, 1997, with permission).

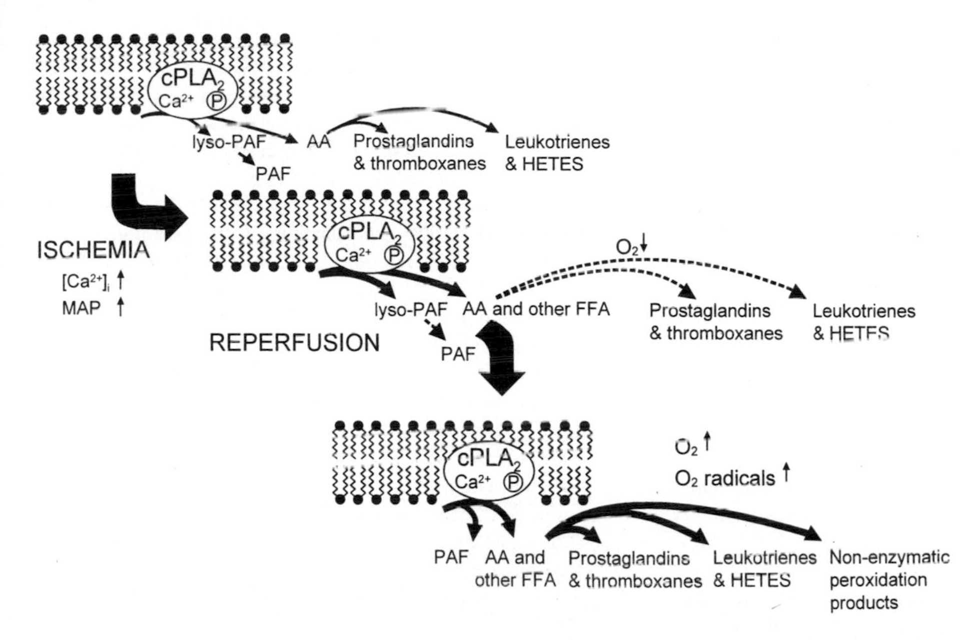

Fig. 9. $cPLA_2$ and the generation of bioactive lipids during ischemia and reperfusion. During the ischemic phase, phospholipase overactivation and the down regulation of oxidative and energy metabolism, and hence reincorporation of $cPLA_2$ metabolites, promote the accumulation of AA and lysophospholipids such as lyso PAF. The reperfusion state permits the completion of PAF and eicosanoid synthesis, but at the expense of the accumulation of pathophysiologically high levels of these mediators. Reactive oxygen radicals are generated at rates that can overload the antioxidant and free radical scavenger systems of the brain, thus allowing free radical damage to a range of molecules, including peroxidation of polyunsaturated fatty acids.

ischemia and seizures may in turn lead to the formation of neurotoxic metabolites (e.g., superoxide). Current investigations aim to determine whether or not other messengers cooperate to enhance neuronal damage (e.g., nitric oxide) and the possible involvement of astrocytes and microglial cells. Further understanding of these potentially neurotoxic events involving lipid messengers and COX-2 will permit the identification of new strategies and define therapeutic windows for the management of epilepsy.

Acknowledgement

This work was supported by NIH Grant NS23002.

References

Bazan, N.G. (1970) Effects of ischemia and electroconvulsive shock on free fatty acid pool in the brain. *Biochim. Biophys. Acta.*, 218: 1–10.

Bazan, N.G. (1971) Changes in free fatty acids of brain by drug-induced convulsions, electroshock and anesthesia. *J. Neurochem.*, 18: 1379–1385.

Bazan, N.G. (1976) Free arachidonic acid and other lipids in the nervous system during early ischemia and after electroshock. In: G. Porcelati, L. Amaducci, C. Galli (Eds) *Functional and Metabolism of Phospholipids in the Central and Peripheral Nervous System*, Vol 72, Plenum Press, New York, pp. 317–335.

Bazan, N.G. (1995) Inflammation: A signal terminator. *Nature*, 374: 501–502.

Bazan, N.G. Brain response to injury: Eicosanoids and inflammation. In: B. Siesjö and T. Wieloch (Eds.), *Basic Neurochemistry*, Vol. 72, Lippincott-Raven Publishers, Philadelphia, 1998, in press .

Bazan, N.G. and Allan, G. (1997) Platelet-activating factor and other bioactive lipids. In: *Cerebrovascular Disease, Pathophysiology, Diagnosis and Management*, Blackwell Scientific Publishers, Cambridge, MA, pps 532–555, 1998.

Bazan, N.G., Bazan, H.E.P., Kennedy, W.G. and Joel, C.D. (1971) Regional distribution and rate of production of free fatty acids in rat brain. *J. Neurochem.*, 18: 1387–1393.

Bazan, N.G., Botting, J. and Vane, J.R. (Eds.) (1996) New targets in inflammation: Inhibitors of COX-2 or adhesion molecules. William Harvey Press and Kluwer Academic Publishers, UK.

Bazan, N.G., Fletcher, B.S., Herschman, H.R. and Mukherjee, P.K. (1994) Platelet-activating factor and retinoic acid synergistically activate the inducible prostaglandin synthase gene. *Proc. Natl. Acad. Sci.*, 91: 5252–5256.

Bazan, N.G. and Rodriguez de Turco, E.B. (1980) Membrane lipids in the pathogenesis of brain edema: Phospholipids and arachidonic acid, the earliest membrane components changed at the onset of ischemia. In: J. Cervós-Navarro, R. Ferszt (Eds.), *Advances in Neurology*, Vol 28: Brain Edema. Raven Press, New York, pp. 197–205.

Bazan, N.G., Rodriguez de Turco, E.B. and Allan, G. (1995) Mediators of injury in neurotrauma: Intracellular signal transduction and gene expression. *J. Neurotrauma*, 12: 789–911.

Bernard, J., Chabot, C., Gagne, J., Baudry, M. and Massicotte, G. (1995) Melittin increases AMPA receptor affinity in rat brain synaptoneurosomes. *Brain Research*, 671: 195–200.

Birkle, D.L. and Bazan, N.G. (1984) Effects of K+ depolarization on the synthesis of prostaglandins and hydroxyeicosatetra(5,8,11,14)enoic acids (HETE) in the rat retina. Evidence for esterification of 12-HETE in lipids. *Biochim. Biophys. Acta.*, 795: 564–573.

Birkle, D.L. and Bazan, N.G. (1987) Effect of bicuculline-induced status epilepticus on prostaglandins and hydroxyeicosatetraenoic acids in rat brain subcellular fractions. *J. Neurochem.*, 48: 1768–1778.

Bonventre, J.V., Huang, Z., Taheri, M.R., O'Leary, E., Li, E., Moskowitz, M.A. and Sapirstein, A. (1997) Reduced fertility and postischaemic brain injury in mice deficient in cytosolic phospholipase A_2. *Nature.*, 390: 622–625.

Claeys, M., Bazan, H.E.P., Birkle, D.L. and Bazan, N.G. (1986) Diacylglycerols interfere in nonnal phase HPLC analysis of lipoxygenase products of docosahexaenoic or arachidonic acids. *Prostaglandins*, 32: 813–827.

Clark, G.D., Happel, L.T., Zorumski, C.F. and Bazan, N.G. (1992) Enhancement of hippocampal excitatory synaptic transmission by platelet-activating factor. *Neuron.*, 9: 1211–1216.

Dennis, E.A. (1994) Diversity of group types, regulation and function of phospholipase A_2. *J. Biol. Chem.*, 269: 13057–13060.

Hattori, M., Arai, H. and Inoue, K. (1993) Purification and characterization of bovine brain platelet-activating factor acetylhydrolase. *J. Biol. Chem.*, 268: 18748–18753.

Hewett, S.J. and Hewett, J.A. (1997) COX-2 contributes to NMDA-induced neuronal death in cortical cell cultures. *Soc. Neurosci.* (abstract) Vol. 23, p.

Izquierdo, I., Fin, C., Schmitz, P.K., Da Silva, R.C., Jerusalinsky, O., Quillfeldt, J.A., Ferreira, M.B., Medina, J.H. and Bazan, N.G. (1995) Memory enhancement by intrahippocampal, intraamygdala, or intraentorhinal infusion of platelet-activating factor measured in an inhibitory avoidance task, *Proc. Natl. Acad. Sci. USA*, 92: 5047–5051.

Jayadev, S., Linardic, C.M. and Hannun, Y.A. (1994) Identification of arachidonic acid as a mediator of sphingomyelin hydrolysis in response to tumor necrosis factor alpha. *J. Biol. Chem.*, 269: 5757–5763.

Kainu, T., Wikstrom, A.C., Gustafsson, J.A. and Pelto Huikko, M. (1994) Localization of the peroxisome proliferator-activated receptor in brain. *Neuroreport*, 5: 2481–2485.

Kaufmann, W.E., Worley, P.F., Pegg, P.J., Bremer, M. and Isakson, P. (1996) COX-2, a synaptically induced enzyme, is expressed by excitatory neurons at postsynaptic sites in rat cerebral cortex. *Proc. Natl. Acad. Sci.*, 93: 2317–2321.

Kato, K., Clark, G.D., Bazan, N.G. and Zorumski, C.F. (1994) Platelet-activating factor as a potential retrograde messenger in Ca^1 hippocampal long-term potentiation. *Nature*, 367: 175–179.

Kliewer, S.A., Lenhard, J.M., Willson, T.M., Patel, I., Morris, D.C. and Lehmann, J.M. (1995) A prostaglandin J_2 metabolite binds peroxisome proliferator-activated receptor gamma and promotes adipocyte differentiation. *Cell*, 83: 813–819.

Kolko, M., DeCoster, M.A., Rodriguez de Turco, E.B. and Bazan, N.G. (1996) Synergy by secretory phospholipase A_2 and glutamate on inducing cell death and sustained arachidonic acid metabolic changes in primary cortical neuronal cultures. *J. Biol. Chem.*, 271: 32722–32728.

Kumar, R., Harvey, S., Kester, N., Hanahan, D. and Olsen, M. (1988) Production and effects of platelet-activating factor in the rat brain. *Biochim. Biophys. Acta.*, 963: 375–383.

Lauritzen, I., Heurteaux, C. and Lazdunski, M. (1994) Expression of group II phospholipase A_2 in rat brain after severe forebrain ischemia and in endotoxic shock. *Brain Research*, 651: 353–356.

Marcheselli, V.L., and Bazan, N.G. (1996) Sustained induction of prostaglandin endoperoxide synthase-2 by seizures in hippocampus: Inhibition by a platelet-activating factor antagonist. *J. Biol. Chem.*, 271: 24794–24799.

Marcheselli, V.L., Rossowska, M., Domingo, M.T., Braquet, P. and Bazan, N.G. (1990) Distinct platelet-activating factor binding sites in synaptic endings and in intracellular membranes of rat cerebral cortex. *J. Biol. Chem.*, 265: 9140–9145.

Moskowitz, M.A., Kiwak, K.J., Hekiman, J. and Levine, L. (1984) Synthesis of compounds with the properties of leukotrienes C_4 and D_4 in gerbil brains after ischemia and reperfusion. *Science*, 224: 886–889.

Matsuzawa, A., Murakami, M., Atsumi, G., Imai, K., Prados, P., Inoue, K. and Kudo, I. (1996) Release of secretory phospholipase A_2 from rat neuronal cells and its possible function in the regulation of catecholamine secretion. *Biochem. J.*, 318: 701–709.

Miettinen, S., Fusco, F.R., Yrjanheikki, J., Keinanen, R., Hivonen, T., Roivainen, R., Narhi, M., Hokfelt, T. and Koistinaho, J. (1997) Spreading depression and focal brain ischemia induce cyclooxygenase-2 in cortical neurons through *N*-methyl-D-aspartic acid receptors and phospholipase A_2. *Proc. Natl. Acad. Sci.*, 94: 6500–6505.

Niimi, M., Sato, M., Wada, Y., Takahara, J. and Kawanishi, K. (1996) Effect of central and continuous intravenous injection of interleukin-1 β on brain c-fos expression in the rat: Involvement of prostaglandins. *Neuroimmunomodulation*, 3: 87–92.

Packard, M.G., Teather, L. and Bazan, N.G. (1996) Effects of intrastriatal injections of platelet-activating factor and the PAF antagonist BN 52021 on memory. *Neurobiol. Learn. Mem.*, 66: 177–182.

Piomelli, D., Shapiro, E., Feinmark, S.J. and Schwartz, J.H. (1987a) Metabolites of arachidonic acid in the nervous system of *Aplysia*: Possible mediators of synaptic modulation. *J. Neurosci.*, 7: 3675–3686.

Piomelli, D., Shapiro, E., Feinmark, S.J. and Schwartz, J.H. (1987b) Formation of 8-hydroxy-11,12-epoxy-eicosatetraenoic acid in *Aplysia* neurons: A potential second messenger for neurotransmission. *Proc. Natl. Acad. Sci.*, 86: 1721–1725.

Prescott, S.M., Zimmerman, G.A and McIntyre, T.M. (1990) Platelet-activating factor. *J. Biol. Chem.*, 265: 17381–17384.

Squinto, S.P., Block, A.L., Braquet, P. and Bazan, N.G. (1989) Platelet-activating factor stimulates a Fos/Jun/AP-1 transcriptional signaling system in human neuroblastoma cells. *J. Neurosci. Res.*, 24: 558–566.

Squinto, S.P., Braquet, P., Block, A.L. and Bazan, N.G. (1990) Platelet-activating factor activates HIV promoter in transfected SH-SY5Y neuroblastoma cells and MOLT-4 T lymphocytes. *J. Mol. Neurosci.*, 2: 79–84.

Yamagata, K., Andreasson, K.I., Kaufmann, W.E., Barnes, C.A. and Worley, P.F. (1993) Expression of a mitogen-inducible cyclooxygenase in brain neurons: Regulation by synaptic activity and glucocorticoids. *Neuron*, 11: 371–386.

Subject Index